T0296841

Fractional Difference, Differential Equations, and Inclusions

Fractional Difference, Differential Equations, and Inclusions
Analysis and Stability

Saïd Abbas
Department of Electronics
Tahar Moulay University of Saïda
Saïda, Algeria

Bashir Ahmad
Department of Mathematics
King Abdulaziz University
Jeddah, Saudi Arabia

Mouffak Benchohra
Department of Mathematics
Djillali Liabes University of Sidi Bel Abbes
Sidi Bel Abbes, Algeria

Abdelkrim Salim
Faculty of Technology
Hassiba Benbouali University of Chlef
Chlef, Algeria

MK
MORGAN KAUFMANN PUBLISHERS
ELSEVIER   AN IMPRINT OF ELSEVIER

Morgan Kaufmann is an imprint of Elsevier
50 Hampshire Street, 5th Floor, Cambridge, MA 02139, United States

Copyright © 2024 Elsevier Inc. All rights are reserved, including those for text and data mining, AI training, and similar technologies.

No part of this publication may be reproduced or transmitted in any form or by any means, electronic or mechanical, including photocopying, recording, or any information storage and retrieval system, without permission in writing from the publisher. Details on how to seek permission, further information about the Publisher's permissions policies and our arrangements with organizations such as the Copyright Clearance Center and the Copyright Licensing Agency, can be found at our website: www.elsevier.com/permissions.

This book and the individual contributions contained in it are protected under copyright by the Publisher (other than as may be noted herein).

Notices

Knowledge and best practice in this field are constantly changing. As new research and experience broaden our understanding, changes in research methods, professional practices, or medical treatment may become necessary.

Practitioners and researchers must always rely on their own experience and knowledge in evaluating and using any information, methods, compounds, or experiments described herein. In using such information or methods they should be mindful of their own safety and the safety of others, including parties for whom they have a professional responsibility.

To the fullest extent of the law, neither the Publisher nor the authors, contributors, or editors, assume any liability for any injury and/or damage to persons or property as a matter of products liability, negligence or otherwise, or from any use or operation of any methods, products, instructions, or ideas contained in the material herein.

ISBN: 978-0-443-23601-3

For information on all Morgan Kaufmann publications
visit our website at https://www.elsevier.com/books-and-journals

Publisher: Mara Conner
Acquisitions Editor: Chris Katsaropoulos
Editorial Project Manager: Shivangi Mishra
Production Project Manager: Neena S. Maheen
Cover Designer: Vicky Pearson Esser

Typeset by VTeX

Working together
to grow libraries in
developing countries

www.elsevier.com • www.bookaid.org

We dedicate this book to our family members. In particular, Saïd Abbas dedicates it to the memory of his father Abdelkader Abbas (1926–2008), his mother, his wife, and his children Mourad, Amina, and Ilyes; Mouffak Benchohra makes his dedication to the memory of his father Yahia Benchohra and his wife Kheira Bencherif; Abdelkrim Salim makes his dedication to his mother, his brother, and his sisters.

Contents

Biography

Saïd Abbas

Dr. Saïd Abbas is a full professor in the Department of Mathematics at Tahar Moulay University of Saida since October 2006. Abbas received the master's degree in functional analysis from Mostaganem University, Algeria, 2006, and the doctorate's degree in Differential Equations from Djillali Liabes University of Sidi Bel Abbes, Algeria, 2011. His research fields include fractional differential equations and inclusions, evolution equations and inclusions, control theory and applications, etc. Abbas has published four monographs and more than 160 papers.

Bashir Ahmad

Dr. Bashir Ahmad is a full professor in the Department of Mathematics at King Abdulaziz University, Jeddah, Saudi Arabia. He received his PhD degree from Quaid-i-Azam University, Islamabad, Pakistan in 1995. His research interest includes approximate/numerical methods for nonlinear problems involving a variety of differential equations, existence theory of fractional differential equations, stability and instability properties of dynamical systems, impulsive systems, control theory, and fluid mechanics. He was honored with "Best Researcher of King Abdulaziz University" award in 2009. He has published 5 books and more than 650 research articles in JCR journals. He is the managing editor of the Journal Bulletin of Mathematical Sciences and member of editorial boards of several journals. He has been "Highly-Cited Researcher" in the category of Mathematics from 2014 to 2019 (Clarivate Analytics database). He was the top 1% of reviewers in Mathematics on Publons (Web of Science) for the 2018 and 2019 global Peer Review Awards. He has been among the world's top 2% researchers in 2020, 2021, and 2022, according to Stanford University databases.

Mouffak Benchohra

Dr. Mouffak Benchohra is a full professor in the Department of Mathematics, at the Djillali Liabes University of Sidi Bel Abbes since October 1994. Benchohra received the master's degree in nonlinear analysis from Tlemcen University, Algeria, 1994 and PhD degree in mathematics from Djillali Liabes University, Sidi Bel Abbes, Algeria. His research fields include fractional differential equations, evolution equations and inclusions, control theory and applications, etc. Benchohra has published more than 500 papers and five monographs. He is a Highly Cited Researcher in mathematics according to Thompson Reuters (2014) and Clarivate Analytics (2017–2018). Benchohra has also occupied the position of head of department of mathematics at Djillali Liabes University, Sidi Bel Abbes. He is in the editorial board of 10 international journals. He was also among the world top 2% researchers in 2020, 2021, and 2022.

Abdelkrim Salim

Dr. Abdelkrim Salim is an associate professor at the Faculty of Technology, Hassiba Benbouali University of Chlef since 2022. Salim received the master's degree in functional analysis and differential equations from Djillali Liabes University, Algeria, 2016, and the doctorate's degree in mathematical analysis and applications from Djillali Liabes University of Sidi Bel Abbes, Algeria, 2021. His research fields include fractional differential equations and inclusions, control theory and applications, etc.

Preface

The idea of fractional calculus (FC) is more than 300 years old and it presumably stemmed from a question about a fractional-order derivative raised in communication between L'Hopital and Leibniz in the year 1695. This branch of mathematical analysis is regarded as the generalization of classical calculus, since it deals with the derivative and integral operators of fractional order. The tools of fractional calculus are have proved very useful in improving the mathematical modeling of many natural phenomena and processes in engineering, social, natural, and bio-medical sciences, etc. This led to a volcanic growth of the subject and many researchers contributed to its development by producing scores of research articles and books.

This book is devoted to the existence and stability (Ulam–Hyers–Rassias stability and asymptotic stability) of solutions for various classes of functional fractional difference equations or inclusions. Some equations present delay that may be finite, infinite, or state-dependent. Others are subject to an impulsive effect that may be fixed or non-instantaneous. Tools used include some fixed-point theorems, densifiability techniques, monotone iterative technique, as well as some notions of Ulam stability, attractivity, and the measures of noncompacteness and weak noncompacteness. Each chapter concludes with a section devoted to notes and bibliographical remarks and all abstract results are illustrated by examples.

The content of this book is completely new and complements the existing literature in fractional calculus. It is useful for researchers and graduate students for research, seminars, and advanced graduate courses in pure and applied mathematics, engineering, biology, and all other applied sciences.

We owe a great deal to N. Al Arifi, B. Alqahtani, S. S. Alzaid, Z. Baitiche, A. Cabada, C. Derbazi, J.R. Graef, J. Henderson, N. Laledj, G. M. N'Guérékata, J. J. Nieto, S. Ntouyas, A. Petrusel, B. Samet, and Y. Zhou for their collaboration in research related to the problems considered in this book.

S. Abbas
Saïda, Algeria
B. Ahmad
Jeddah, Saudi Arabia
M. Benchohra
Sidi Bel Abbes, Algeria
A. Salim
Chlef, Algeria

1

Introduction

Fractional calculus may be considered an old and yet novel topic. It is an old topic because, starting from some speculations of G.W. Leibniz (1695, 1697) and L. Euler (1730), it has been developed progressively up to now. A list of mathematicians who have provided important contributions up to the middle of the twentieth century includes P.S. Laplace (1812), S.F. Lacroix (1819), J.B.J. Fourier (1822), N.H. Abel (1823–26), I. Liouville (1832–73), B. Riemann (1847), H. Holmgren (1865–67), A.K. Grünwald (1867–72), A.V. Letnikov (1868–72), H. Laurent (1884), P.A. Nekrassov (1888), A. Krug (1890), I. Hadamard (1892), O. Heaviside (1892–1912), S. Pincherle (1902), G.H. Hardy and I.E. Littlewood (1917–28), H. Weyl (1917), P. Lévy (1923), A. Marchaud (1927), H.T. Davis (1924–36), E.L. Post (1930), A. Zygmund (1935–45), E.R. Love (1938–96), A. Erdélyi (1939–65), H. Kober (1940), D.V. Widder (1941), M. Riesz (1949), W. Feller (1952). However, it may be considered a novel topic as well. Only since the seventies it has been the object of specialized conferences and treatises. For the first conference, the merit is due to B. Ross (1975), who, shortly after his PhD dissertation on fractional calculus, organized the First Conference on Fractional Calculus and its Applications at the University of New Haven in June 1974 and edited the proceedings, see [353]. For the first monograph, the merit is ascribed to K.B. Oldham and I. Spanier, see [321] who, after a joint collaboration that began in 1968, published a book on fractional calculus in 1974. Nowadays, the series of texts devoted to fractional calculus and its applications includes over ten titles, including the books of Kilbas et al. [281], Miller and Ross [312], Oldham and Spanier [321], Podlubny [332], Rubin (1996); Samko et al. [385], West et al. [419], Guo et al. [255], Chakraverty et al. [206], Yang et al. [424], Anastassiou et al. [135], Georgiev [248] and Ray and Gupta [344]. This list is expected to grow in the forthcoming years.

Reviewing its history of three centuries, we find that for a long time fractional calculus was of interest mainly to mathematicians because it lacked an application background. However, in the recent decades more and more researchers have paid attention to fractional calculus, finding that the fractional-order integrals and derivatives are more suitable for describing real-world phenomena, such as viscoelastic systems, dielectric polarization, electromagnetic waves, heat conduction, robotics, biological systems, finance, and so on; see, for example, [266,281,332,385]. Owing to great efforts of researchers, there have been rapid developments in the theory of fractional calculus and its applications, including well-posedness, stability, bifurcation, and chaos in fractional differential equations and their control. Several useful tools for solving fractional-order equations have been discovered, of which Laplace transform is frequently applied. Furthermore, it is showed to be most helpful in analysis and applications of fractional-order systems, since some results can be derived immediately from it. For instance, in [296,297], the au-

Fractional Difference, Differential Equations, and Inclusions. https://doi.org/10.1016/B978-0-44-323601-3.00008-3
Copyright © 2024 Elsevier Inc. All rights are reserved, including those for text and data mining, AI training, and similar technologies.

thors investigated stability of fractional-order nonlinear dynamical systems using Laplace transform method and Lyapunov direct method, introducing the Mittag–Leffler stability and generalized Mittag–Leffler stability concepts. The Laplace transform was also used in [241,428].

Fractional calculus is relative to the traditional integer-order calculus, which extends the order of calculus from integer orders extended to any order of the mathematical promotion. From the theoretical point of view, the fractional differential calculus signal processing order is extended by an arbitrary number from an integer, the ways and means of information processing has been extended. Fractional differential equations and inclusions have recently been applied in various areas of engineering, mathematics, physics, bio-engineering, and other applied sciences [392]. For some fundamental results in the theory of fractional calculus and fractional differential equations, we refer the reader to the monographs of Abbas et al. [19,48,66,69,70], Ahmad et al. [97], Anastassiou et al. [135], Atangana [138], Chakraverty et al. [206], Georgiev [248], Geo et al. [255], Kilbas et al. [281], Ray et al. [344], Samko et al. [385], Tas et al. [393], Vyawahare et al. [408], Yang et al. [423–425] and Zhou [438,440], the papers by Abbas et al. [20,21,58,59,62–65,73,79,80], Alsaedi et al. [128], Agarwal et al. [86,87,90,92], Ahmad et al. [99,101,102,104–106,108,110,111], Benchohra et al. [161,171–173,175–178,180,181], Lakshmikantham et al. [290–292], Salim et al. [157,184,222,223,264,295,359,360,362,365–368,370–375,377,382], Vityuk et al. [406], Wang et al. [415], and the references therein.

Recently, considerable attention has been given to the existence of solutions of initial and boundary value problems for fractional differential and partial differential equations and inclusions with Caputo fractional derivative; see [73,161,172,173,180,181], and Hadamard fractional integral equations; see [20,21,58,116,117,162,177,258,280], and the references therein.

In [196], Butzer et al. investigated properties of the Hadamard fractional integral and derivative. In [197], they obtained the Mellin transform of the Hadamard fractional integral and differential operators, and in [333], Pooseh et al. obtained expansion formulas of the Hadamard operators in terms of integer order derivatives. Many other interesting properties of these and other operators are summarized in [385] and the references therein.

Impulsive differential equations are well known for modeling problems from many areas of science and engineering. The theory of impulsive differential equations is the subject of much research activity; see [41,69,70,90,114,175,207,261,311,409,430,434,436] and the references therein.

Implicit functional differential equations have been considered by many authors [19–26,34,69,78,114,171–173,175–178]. Our intention is to extend the results to implicit differential equations of fractional order.

The stability of functional equations was originally raised by Ulam [396] and then by Hyers [269]. Thereafter, this type of stability is called Ulam–Hyers stability. In 1978, Rassias [343] provided a remarkable generalization of Ulam–Hyers stability of mappings by considering variables. The concept of stability of a functional equation arises when we replace the functional equation by an inequality, which acts as a perturbation of the equation. Considerable attention has been given to the study of the Ulam–Hyers and Ulam–Hyers–Rassias stability of all kinds of functional equations; one can see the monographs of [69,272], and the papers of Abbas et al. [20,34,62,73,176,177], Benchohra and Lazreg [176,177], Petru et al. [329], and Rus [345,346] discussed the Ulam–Hyers stability for operatorial equations and inclusions. More details from historical point of view and recent developments of such stabilities are reported in [193,273,345].

2

Preliminary background

2.1 Notations and definitions

Let $J := [a, T]$; $0 \leq a < T$ be a compact interval of $\mathbb{R}$, and denote by $\mathcal{C}$ the Banach space of all continuous functions v from I into a Banach space E with the supremum (uniform) norm

$$\|v\|_\infty := \sup_{t \in J} \|v(t)\|_E.$$

As usual, $AC(J)$ denotes the space of absolutely continuous functions from J into E, and $L^1(J)$ denotes the space of Bochner-integrable functions $v : J \to E$ with the norm

$$\|v\|_1 = \int_a^T \|v(t)\|_E dt.$$

2.2 Elements from fractional calculus theory

In the following we first define ψ-Riemann–Liouville fractional integrals and derivatives.

Definition 2.1 ([119,281]). For $\alpha > 0$, the left-sided ψ-Riemann–Liouville fractional integral of order α for an integrable function $u : J \longrightarrow \mathbb{R}$ with respect to another function $\psi : J \longrightarrow \mathbb{R}$ that is an increasing differentiable function such that $\psi'(t) \neq 0$, for all $t \in J$ is defined as follows:

$$I_{a^+}^{\alpha; \psi} u(t) = \frac{1}{\Gamma(\alpha)} \int_a^t \psi'(s)(\psi(t) - \psi(s))^{\alpha-1} u(s) \mathrm{d}s, \tag{2.1}$$

where $\Gamma(\cdot)$ is the (Euler's) Gamma function $\Gamma(\alpha) = \int_0^{+\infty} e^{-t} t^{\alpha-1} \mathrm{d}t$, $\alpha > 0$.

Definition 2.2 ([119]). Let $n \in \mathbb{N}$ and let $\psi, u \in C^n(J, \mathbb{R})$ be two functions such that ψ is increasing and $\psi'(t) \neq 0$, for all $t \in J$. The left-sided ψ-Riemann–Liouville fractional derivative of a function u of order α is defined by

$$D_{a^+}^{\alpha; \psi} u(t) = \left(\frac{1}{\psi'(t)} \frac{d}{dt} \right)^n I_{a^+}^{n-\alpha; \psi} u(t)$$

$$= \frac{1}{\Gamma(n-\alpha)} \left(\frac{1}{\psi'(t)} \frac{d}{dt} \right)^n \int_a^t \psi'(s)(\psi(t) - \psi(s))^{n-\alpha-1} u(s) \mathrm{d}s,$$

where $n = [\alpha] + 1$.

Fractional Difference, Differential Equations, and Inclusions. https://doi.org/10.1016/B978-0-44-323601-3.00009-5
Copyright © 2024 Elsevier Inc. All rights reserved, including those for text and data mining, AI training, and similar technologies.

Definition 2.3 ([119]). Let $n \in \mathbb{N}$ and let $\psi, u \in C^n(J, \mathbb{R})$ be two functions such that ψ is increasing and $\psi'(t) \neq 0$, for all $t \in J$. The left-sided ψ-Caputo fractional derivative of u of order α is defined by

$$^c D_{a+}^{\alpha;\psi} u(t) = I_{a+}^{n-\alpha;\psi} \left(\frac{1}{\psi'(t)} \frac{d}{dt} \right)^n u(t),$$

where $n = [\alpha] + 1$ for $\alpha \notin \mathbb{N}$, $n = \alpha$ for $\alpha \in \mathbb{N}$.

To simplify notation, we will use the abbreviated symbol

$$u_\psi^{[n]} u(t) = \left(\frac{1}{\psi'(t)} \frac{d}{dt} \right)^n u(t). \tag{2.2}$$

From the definition, it is clear that

$$^c D_{a+}^{\alpha;\psi} u(t) = \begin{cases} \int_a^t \frac{\psi'(s)(\psi(t)-\psi(s))^{n-\alpha-1}}{\Gamma(n-\alpha)} u_\psi^{[n]}(s)ds, & \text{if } \alpha \notin \mathbb{N}, \\ u_\psi^{[n]}(t), & \text{if } \alpha \in \mathbb{N}. \end{cases} \tag{2.3}$$

This generalization (2.3) yields the Caputo fractional derivative operator when $\psi(t) = t$. Moreover, for $\psi(t) = \ln t$, it gives the Caputo–Hadamard fractional derivative.

We note that if $u \in C^n(J, \mathbb{R})$, the ψ–Caputo fractional derivative of order α of u is determined as

$$^c D_{a+}^{\alpha;\psi} u(t) = D_{a+}^{\alpha;\psi} \left[u(t) - \sum_{k=0}^{n-1} \frac{u_\psi^{[k]} u(a)}{k!} (\psi(t) - \psi(a))^k \right].$$

(see, for instance, [119, Theorem 3]).

Lemma 2.4 ([122]). *Let $\alpha, \beta > 0$ and $u \in L^1(J, \mathbb{R})$. Then*

$$I_{a+}^{\alpha;\psi} I_{a+}^{\beta;\psi} u(t) = I_{a+}^{\alpha+\beta;\psi} u(t), \ a.e. \ t \in J.$$

In particular, if $u \in C(J, \mathbb{R})$, then $I_{a+}^{\alpha;\psi} I_{a+}^{\beta;\psi} u(t) = I_{a+}^{\alpha+\beta;\psi} u(t), t \in J$.

Next, we recall the property describing the composition rules for fractional ψ-integrals and ψ-derivatives.

Lemma 2.5 ([122]). *Let $\alpha > 0$, then the following holds:*

- *If $u \in C(J, \mathbb{R})$, then*

$$^c D_{a+}^{\alpha;\psi} I_{a+}^{\alpha;\psi} u(t) = u(t), \ t \in J.$$

- *If $u \in C^n(J, \mathbb{R})$, $n - 1 < \alpha < n$, then*

$$I_{a+}^{\alpha;\psi} {}^c D_{a+}^{\alpha;\psi} u(t) = u(t) - \sum_{k=0}^{n-1} \frac{u_\psi^{[k]}(a)}{k!} [\psi(t) - \psi(a)]^k,$$

for all $t \in J$.

Lemma 2.6 ([122,281]). *Let $t > a$, $\alpha \geq 0$ and $\beta > 0$. Then*

- $I_{a+}^{\alpha;\psi}(\psi(t) - \psi(a))^{\beta-1} = \frac{\Gamma(\beta)}{\Gamma(\beta+\alpha)}(\psi(t) - \psi(a))^{\beta+\alpha-1}$,
- $^c D_{a+}^{\alpha;\psi}(\psi(t) - \psi(a))^{\beta-1} = \frac{\Gamma(\beta)}{\Gamma(\beta-\alpha)}(\psi(t) - \psi(a))^{\beta-\alpha-1}$,
- $^c D_{a+}^{\alpha;\psi}(\psi(t) - \psi(a))^k = 0$, *for all* $k \in \{0, \dots, n-1\}, n \in \mathbb{N}$.

Definition 2.7 ([403]). Let $\theta = (a, a)$ and $\alpha = (\alpha_1, \alpha_2)$ where $\alpha_1, \alpha_2 > 0$. Also let $\psi(\cdot)$ be an increasing positive monotone function on each of $(a, b]$ and $(a, c]$ and having continuous derivative $\psi'(\cdot)$ on each of $(a, b]$ and $(a, c]$. The ψ-Riemann–Liouville partial integral of a function $x \in L^1(\tilde{\mathrm{I}}, \mathbb{R})$, is defined as

$$\left(I_\theta^{\alpha;\psi} x\right)(\tau, \mathrm{u}) = \int_a^\tau \int_a^{\mathrm{u}} \frac{\psi'(s)\psi'(t)(\psi(\tau) - \psi(s))^{\alpha_1-1}(\psi(\mathrm{u}) - \psi(t))^{\alpha_2-1}}{\Gamma(\alpha_1)\Gamma(\alpha_2)} x(s, t)\,dt\,ds. \quad (2.4)$$

Definition 2.8 ([403]). Let $\theta = (a, a)$ and $\alpha = (\alpha_1, \alpha_2)$ where $0 < \alpha_1, \alpha_2 \leq 1$. Also let $x \in C^1(\tilde{\mathrm{I}}, \mathbb{R})$ and let ψ belong to both $C^1([a, b], \mathbb{R})$ and $C^1([a, c], \mathbb{R})$ such that ψ is increasing in both cases, and $\psi'(\tau) \neq 0$, $\psi'(\mathrm{u}) \neq 0$, for all $(\tau, \mathrm{u}) \in \tilde{\mathrm{I}}$. The ψ-Caputo fractional partial derivative of functions of two variables of order α is given by

$$\left(^c D_\theta^{\alpha;\psi} x\right)(\tau, \mathrm{u}) = \left(I_\theta^{1-\alpha;\psi} x\right)\left(\frac{1}{\psi'(\tau)\psi'(\mathrm{u})} \frac{\partial^2}{\partial\tau\partial\mathrm{u}}\right) x(\tau, \mathrm{u}).$$

Lemma 2.9 ([403]). *Let $\sigma_1, \sigma_2 \in (-1, +\infty)$ and $\alpha = (\alpha_1, \alpha_2) \in (0, \infty) \times (0, \infty)$. Then*

$$I_\theta^{\alpha;\psi}\left((\psi(\tau) - \psi(a))^{\sigma_1} \cdot (\psi(\mathrm{u}) - \psi(a))^{\sigma_2}\right)$$
$$= \frac{\Gamma(1+\sigma_1)\Gamma(1+\sigma_2)}{\Gamma(\alpha_1 + \sigma_1 + 1)\Gamma(\alpha_2 + \sigma_2 + 1)}(\psi(\tau) - \psi(a))^{\alpha_1+\sigma_1}(\psi(\mathrm{u}) - \psi(a))^{\alpha_2+\sigma_2}.$$

The next function is a direct generalization of the exponential series.

Definition 2.10 ([249]). The one-parameter Mittag–Leffler function $\mathbb{E}_\alpha(\cdot)$ is defined as

$$\mathbb{E}_\alpha(z) = \sum_{k=0}^\infty \frac{z^k}{\Gamma(\alpha k + 1)}, \quad (z \in \mathbb{R}, \ \alpha > 0).$$

For $\alpha = 1$, this function coincides with the series expansion of e^z, i.e.,

$$\mathbb{E}_1(z) = \sum_{k=0}^\infty \frac{z^k}{\Gamma(k + 1)} = \sum_{k=0}^\infty \frac{z^k}{k!} = e^z.$$

Definition 2.11 ([249]). The two-parameter Mittag–Leffler function $\mathbb{E}_{\alpha,\beta}(\cdot)$ is defined as

$$\mathbb{E}_{\alpha,\beta}(z) = \sum_{k=0}^\infty \frac{z^k}{\Gamma(\alpha k + \beta)}, \quad \alpha, \beta > 0 \text{ and } z \in \mathbb{R}. \quad (2.5)$$

2.3 Fractional q-calculus

Now, we give some definitions and properties of fractional q-calculus.

For $a \in \mathbb{R}$, we set

$$[a]_q = \frac{1 - q^a}{1 - q}.$$

The q analogue of the power $(a - b)^n$ is

$$(a - b)^{(0)} = 1, \quad (a - b)^{(n)} = \Pi_{k=0}^{n-1}(a - bq^k); \quad a, b \in \mathbb{R}, \ n \in \mathbb{N}.$$

In general,

$$(a - b)^{(\alpha)} = a^\alpha \, \Pi_{k=0}^\infty \left(\frac{a - bq^k}{a - bq^{k+\alpha}} \right); \quad a, b, \alpha \in \mathbb{R}.$$

Definition 2.12 ([274]). The q-gamma function is defined by

$$\Gamma_q(\xi) = \frac{(1 - q)^{(\xi - 1)}}{(1 - q)^{\xi - 1}}; \quad \xi \in \mathbb{R} - \{0, -1, -2, \ldots\}.$$

Definition 2.13 ([274]). The q-beta function is defined by

$$\beta_q(x, y) = \int_0^1 t^{x-1}(1 - t)^{(y-1)} d_q t$$

Notice that the q-gamma and q-beta functions satisfy

$$\Gamma_q(1 + \xi) = [\xi]_q \Gamma_q(\xi), \quad and \quad \beta_q(x, y) = \frac{\Gamma_q(x)\Gamma_q(y)}{\Gamma_q(x + y)}.$$

Definition 2.14 ([274]). The q-derivative of order $n \in \mathbb{N}$ of a function $u : I \to E$ is defined by $(D_q^0 u)(t) = u(t)$,

$$(D_q u)(t) := (D_q^1 u)(t) = \frac{u(t) - u(qt)}{(1 - q)t}; \quad t \neq 0, \quad (D_q u)(0) = \lim_{t \to 0}(D_q u)(t),$$

and

$$(D_q^n u)(t) = (D_q D_q^{n-1} u)(t); \quad t \in I, \ n \in \{1, 2, \ldots\}.$$

Set $I_t := \{tq^n : n \in \mathbb{N}\} \cup \{0\}$.

Definition 2.15 ([274]). The q-integral of a function $u : I_t \to E$ is defined by

$$(I_q u)(t) = \int_0^t u(s) d_q s = \sum_{n=0}^\infty t(1 - q)q^n f(tq^n),$$

provided that the series converges.

We note that $(D_q I_q u)(t) = u(t)$, while if u is continuous at 0, then

$$(I_q D_q u)(t) = u(t) - u(0).$$

Definition 2.16 ([85]). The Riemann–Liouville fractional q-integral of order $\alpha \in \mathbb{R}_+ := [0, \infty)$ of a function $u : I \to E$ is defined by $(I_q^0 u)(t) = u(t)$ and

$$(I_q^\alpha u)(t) = \int_0^t \frac{(t - qs)^{(\alpha-1)}}{\Gamma_q(\alpha)} u(s) d_q s; \ t \in I.$$

Lemma 2.17 ([341]). *For $\alpha \in \mathbb{R}_+ := [0, \infty)$ and $\lambda \in (-1, \infty)$, we have*

$$(I_q^\alpha (t - a)^{(\lambda)})(t) = \frac{\Gamma_q(1 + \lambda)}{\Gamma(1 + \lambda + \alpha)}(t - a)^{(\lambda+\alpha)}; \ 0 < a < t < T.$$

In particular,

$$(I_q^\alpha 1)(t) = \frac{1}{\Gamma_q(1 + \alpha)} t^{(\alpha)}.$$

Definition 2.18 ([342]). The Riemann–Liouville fractional q-derivative of order $\alpha \in \mathbb{R}_+$ of a function $u : I \to E$ is defined by $(D_q^0 u)(t) = u(t)$ and

$$(D_q^\alpha u)(t) = (D_q^{[\alpha]} I_q^{[\alpha]-\alpha} u)(t); \ t \in I,$$

where $[\alpha]$ is the integer part of α.

Definition 2.19 ([342]). The Caputo fractional q-derivative of order $\alpha \in \mathbb{R}_+$ of a function $u : I \to E$ is defined by $(^C D_q^0 u)(t) = u(t)$ and

$$(^C D_q^\alpha u)(t) = (I_q^{[\alpha]-\alpha} D_q^{[\alpha]} u)(t); \ t \in I.$$

Lemma 2.20 ([342]). *Let $\alpha \in \mathbb{R}_+$, then the following equality holds:*

$$(I_q^\alpha \, ^C D_q^\alpha u)(t) = u(t) - \sum_{k=0}^{[\alpha]-1} \frac{t^k}{\Gamma_q(1 + k)} (D_q^k u)(0).$$

In particular, if $\alpha \in (0, 1)$, then

$$(I_q^\alpha \, ^C D_q^\alpha u)(t) = u(t) - u(0).$$

2.4 Multi-valued analysis

Let $(X, \| \cdot \|)$ be a Banach space and K be a subset of X. We denote:

$$\mathcal{P}(X) = \{K \subset X : K \neq \emptyset\};$$
$$\mathcal{P}_{cl}(X) = \{K \subset \mathcal{P}(X) : K \text{ is closed}\};$$

$$\mathcal{P}_b(X) = \{K \subset \mathcal{P}(X) : K \text{ is bounded}\};$$
$$\mathcal{P}_{cv}(X) = \{K \subset \mathcal{P}(X) : K \text{ is convex}\};$$
$$\mathcal{P}_{cp}(X) = \{K \subset \mathcal{P}(X) : K \text{ is compact}\};$$
$$\mathcal{P}_{cv,cp}(X) = \mathcal{P}_{cv}(X) \cap \mathcal{P}_{cp}(X).$$

Let $A, B \in \mathcal{P}(X)$. Consider $H_d : \mathcal{P}(X) \times \mathcal{P}(X) \to \mathbb{R}_+ \cup \{\infty\}$ the Hausdorff distance between A and B given by

$$H_d(A, B) = max\{\sup_{a \in A} d(a, B), \sup_{b \in B} d(A, b)\},$$

where $d(A, b) = \inf_{a \in A} d(a, b)$ and $d(a, B) = \inf_{b \in B} d(a, b)$. As usual, $d(x, \emptyset) = +\infty$.

Then $(\mathcal{P}_{b,cl}(X), H_d)$ is a metric space and $(\mathcal{P}_{cl}(X), H_d)$ is a generalized (complete) metric space (see [282]).

Definition 2.21. A multi-valued operator $N : X \to \mathcal{P}_{cl}(X)$ is called:

(a) γ-Lipschitz if there exists $\gamma > 0$ such that

$$H_d(N(x), N(y)) \leq \gamma d(x, y), \text{ for all } x, \ y \in X;$$

(b) a contraction if it is γ-Lipschitz with $\gamma < 1$.

Definition 2.22. A multi-valued map $F : J \to \mathcal{P}_{cl}(X)$ is said to be measurable if, for each $y \in X$, the function

$$t \longmapsto d(y, F(t)) = inf\{d(y, z) : z \in F(t)\}$$

is measurable.

Definition 2.23. The selection set of a multi-valued map $G : J \to \mathcal{P}(X)$ is defined by

$$S_G = \{u \in L^1(J) : u(t) \in G(t) , \ a.e. \ t \in J\}.$$

For each $u \in C$, the set S_{Fou} known as the set of selectors from F is defined by

$$S_{Fou} = \{v \in L^1(J) : v(t) \in F(t, u(t)) , \ a.e. \ t \in J\}.$$

Definition 2.24. Let X and Y be two sets. The graph of a set-valued map $N : X \to \mathcal{P}(Y)$ is defined by

$$graph(N) = \{(x, y) : x \in X, \ y \in N(x)\}.$$

There are more details about multi-valued functions in [137,139,217,225,268].

Definition 2.25. Let $(X, \|\cdot\|)$ be a Banach space. A multi-valued map $F : X \to \mathcal{P}(X)$ is convex (closed) if $F(x)$ is convex (closed) for all $x \in X$.

The map F is bounded on bounded sets if $F(\mathcal{B}) = \cup_{x \in \mathcal{B}} F(x)$ is bounded in X for all $\mathcal{B} \in \mathcal{P}_b(X)$, i.e., $sup_{x \in \mathcal{B}}\{sup\{|y| : y \in F(x)\}\} < \infty.$

Definition 2.26. A multi-valued map F is called upper semi-continuous (u.s.c. for short) on X if for each $x_0 \in X$ the set $F(x_0)$ is a nonempty, closed subset of X, and for each open set U of X containing $F(x_0)$, there exists an open neighborhood V of x_0 such that $F(V) \subset U$. A set-valued map F is said to be upper semi-continuous if it is so at every point $x_0 \in X$. F is said to be completely continuous if $F(\mathcal{B})$ is relatively compact for every $\mathcal{B} \in \mathcal{P}_b(X)$.

If the multi-valued map F is completely continuous with nonempty compact values, then F is upper semi-continuous if and only if F has closed graph (i.e., $x_n \to x_*$, $y_n \to y_*$, $y_n \in F(x_n)$ imply $y_* \in F(x_*)$).

The map F has a fixed point if there exists $x \in X$ such that $x \in Fx$. The set of fixed points of the multi-valued operator F will be denoted by $Fix\,F$.

Definition 2.27. A measurable multi-valued function $F : J \to \mathcal{P}_{b,cl}(X)$ is said to be integrable bounded if there exists a function $g \in L^1(J, \mathbb{R}_+)$ such that $|f| \leq g(t)$ for a.e. $t \in J$ for all $f \in F(t)$.

Lemma 2.28 ([268]). *Let G be a completely continuous multi-valued map with nonempty compact values. Then G is u.s.c. if and only if G has a closed graph (i.e., $u_n \to u$, $w_n \to w$, $w_n \in G(u_n)$ imply $w \in G(u)$).*

Lemma 2.29 ([294]). *Let X be a Banach space. Let $F : J \times X \longrightarrow \mathcal{P}_{cp,cv}(X)$ be an L^1-Carathéodory multi-valued map and let Λ be a linear continuous mapping from $L^1(J, X)$ to $C(J, X)$. Then the operator*

$$
\begin{aligned}
\Lambda \circ S_{F \circ u} : C(J, X) &\longrightarrow & \mathcal{P}_{cp,cv}(C(J, X)), \\
w &\longmapsto & (\Lambda \circ S_{F \circ u})(w) := (\Lambda S_{F \circ u})(w)
\end{aligned}
$$

is a closed graph operator in $C(J, X) \times C(J, X)$.

Proposition 2.30 ([268]). *Let $F : X \to Y$ be an u.s.c map with closed values. Then $Gr(F)$ is closed.*

Definition 2.31. A multi-valued map $F : J \times \mathbb{R} \times \mathbb{R} \to \mathcal{P}(\mathbb{R})$ is said to be L^1-Carathéodory if

(i) $t \to F(t, x, y)$ is measurable for each $x, y \in \mathbb{R}$;
(ii) $(x, y) \to F(t, x, y)$ is upper semicontinuous for a.e. $t \in J$;
(iii) For each $q > 0$, there exists $\varphi_q \in L^1(J, \mathbb{R}_+)$ such that

$$
\|F(t, x, y)\|_{\mathcal{P}} = sup\{|f| : f \in F(t, x, y)\} \leq \varphi_q(t)
$$

for all $|x| \leq q$, $|y| \leq q$ and for a.e. $t \in J$.

The multi-valued map F is said to be Carathéodory if it satisfies (i) and (ii).

Lemma 2.32 ([250]). *Let X be a separable metric space. Then every measurable multi-valued map $F : X \to \mathcal{P}_{cl}(X)$ has a measurable selection.*

For more details on multi-valued maps and the proof of the known results cited in this section, we refer interested reader to the books of Aubin and Cellina [137], Deimling [218], Gorniewicz [250], and Hu and Papageorgiou [268].

2.5 Measure of noncompactness

We will define the Kuratowski (1896–1980) and Hausdorf (1868–1942) measures of noncompactness ($\mathcal{MNC}$ for short) and give their basic properties. Let us recall some fundamental facts of the notion of measure of noncompactness in a Banach space.

Let (X, d) be a complete metric space and $\mathcal{P}_{bd}(X)$ be the family of all bounded subsets of X. Analogously denote by $\mathcal{P}_{rcp}(X)$ the family of all relatively compact and nonempty subsets of X. Recall that $B \subset X$ is said to be bounded if B is contained in some ball. If $B \subset \mathcal{P}_{bd}(X)$ is not relatively compact (precompact), then there exists an $\epsilon > 0$ such that B cannot be covered by a finite number of ϵ-balls, and it is then also impossible to cover B by finitely many sets of diameter $< \epsilon$. Recall that the diameter of B is given by

$$diam(B) := \begin{cases} \sup\limits_{(x,y) \in B^2} d(x, y), & \text{if} \quad B \neq \phi, \\ 0, & \text{if} \quad B = \phi. \end{cases}$$

Definition 2.33 ([283]). Let (X, d) be a complete metric space and $\mathcal{P}_{bd}(X)$ be the family of bounded subsets of X. For every $B \in \mathcal{P}_{bd}(X)$, we define the Kuratowski measure of noncompactness $\alpha(B)$ of the set B as the infimum of the numbers d such that B admits a finite covering by sets of diameter smaller than d.

Remark 2.34. It is clear that $0 \leq \alpha(B) \leq diam(B) < +\infty$ for each nonempty bounded subset B of X and that $diam(B) = 0$ if and only if B is an empty set or consists of exactly one point.

Definition 2.35 ([153]). Let X be a Banach space and $\mathcal{P}_{bd}(X)$ be the family of bounded subsets of X. For every $B \in \mathcal{P}_{bd}(X)$, the Kuratowski measure of noncompactness is the map $\alpha : \mathcal{P}_{bd}(X) \to [0, +\infty)$ defined by

$$\alpha(B) = inf\{r > 0 : B \subseteq \cup_{i=1}^{n} B_i \text{ and } diam(B_i) < r\}.$$

Definition 2.36 ([153]). Let E be a Banach space and B a bounded subsets of E. Then the Hausdorff measure of noncompactness of B is defined by

$$\chi(B) = inf \left\{ \begin{array}{l} \varepsilon > 0 : B \text{ can be covered by} \\ \text{finitely many balls with radius } < \varepsilon \end{array} \right\}.$$

The Kuratowski measure of noncompactness satisfies the following properties:

Proposition 2.37 ([153,154,283]). *Let X be a Banach space. Then for all bounded subsets A, B of X the following assertions hold:*

- $\alpha(B) = 0$ *implies $\overline{B}$ is compact (B is relatively compact), where $\overline{B}$ denotes the closure of B.*

- $\alpha(\phi) = 0$.
- $\alpha(B) = \alpha(\overline{B}) = \alpha(conv B)$, *where $conv B$ is the convex hull of B.*
- *monotonicity: $A \subset B$ implies $\alpha(A) \leq \alpha(B)$.*
- *algebraic semi-additivity: $\alpha(A + B) \leq \alpha(A) + \alpha(B)$, where $A + B = \{x + y : x \in A; y \in B\}$.*
- *semi-homogencity: $\alpha(\lambda B) = |\lambda|\alpha(B)$, $\lambda \in \mathbb{R}$, where $\lambda(B) = \{\lambda x : x \in B\}$.*
- *semi-additivity: $\alpha(A \cup B) = max\{\alpha(A), \alpha(B)\}$.*
- *semi-additivity: $\alpha(A \cap B) = min\{\alpha(A), \alpha(B)\}$.*
- *invariance under translations: $\alpha(B + x_0) = \alpha(B)$ for any $x_0 \in X$.*

The following definition of measure of noncompactness appeared in Banas and Goebel [153].

Definition 2.38. A function $\mu : \mathcal{P}_{bd}(X) \longrightarrow [0, \infty)$ is called a measure of noncompactness if it satisfies the following conditions:

- $Ker \mu(A) = \{A \in \mathcal{P}_{bd}(X) : \mu(A) = 0\}$ is nonempty and $ker \mu(A) \subset \mathcal{P}_{rcp}(X)$.
- $A \subset B$ implies $\mu(A) \leq \mu(B)$.
- $\mu(\overline{A}) = \mu(A)$.
- $\mu(conv A) = \mu(A)$.
- $\mu(\lambda A + (1 - \lambda)B) \leq \lambda\mu(A) + (1 - \lambda)\mu(B)$ pour $\lambda \in [0, 1]$.
- If $(A_n)_{n \in \geq 1}$ be a sequence of closed sets in $\mathcal{P}_{bd}(X)$ such that

$$X_{n+1} \subset A_n \ (n = 1, 2, \ldots.)$$

and

$$\lim_{n \to +\infty} \mu(A_n) = 0,$$

then the intersection set $A_\infty = \bigcap_{n=1}^{\infty} A_n$ is nonempty.

Remark 2.39. The family $Ker \mu$ described in Definition 2.38 is said to be the kernel of the measure of noncompactness μ. Observe that the intersection set A_∞ in Definition 2.38 is a member of the family $Ker \mu$. Since $\mu(A_\infty) \leq \mu(A_n)$ for any n, we infer that $\mu(A_\infty) = 0$. This yields that $\mu(A_\infty) \in Ker \mu$. This simple observation will be essential for our further investigations.

Moreover, introduce the notion of a measure of noncompactness in $L^1(J)$, we let $\mathcal{P}_{bd}(J)$ be the family of all bounded subsets of $L^1(J)$. Analogously denote by $\mathcal{P}_{rcp}(J)$ the family of all relatively compact and nonempty subsets of $L^1(J)$. In particular, the measure of non-compactness in $L^1(J)$ is defined as follows. Let X be a fixed nonempty and bounded subset of $L^1(J)$. Set

$$\mu(X) = \lim_{\delta \to 0} \left\{ \sup \left\{ \sup \left(\int_0^T |x(t + h) - x(t)|dt \right), |h| \leq \delta \right\}, x \in X \right\}. \tag{2.6}$$

It can be easily shown that μ is measure of noncompactness in $L^1(J)$ (see [153]).

For more details on measure of noncompactness and the proof of the known results cited in this section, we refer the reader to Akhmerov et al. [113] and Banas et al. [153,154].

Lemma 2.40 ([189]). *If Y is a bounded subset of Banach space X, then for each $\epsilon > 0$, there is a sequence $\{y_k\}_{k=1}^{\infty} \subset Y$ such that*

$$\mu(Y) \le 2\mu(\{y_k\}_{k=1}^{\infty}) + \epsilon.$$

Lemma 2.41 ([313]). *If $\{u_k\}_{k=1}^{\infty} \subset L^1(I)$ is uniformly integrable, then we have that $\mu(\{u_k\}_{k=1}^{\infty})$ is measurable and for each $t \in I$, and*

$$\mu\left(\left\{\int_0^t u_k(s)ds\right\}_{k=1}^{\infty}\right) \le 2\int_0^t \mu(\{u_k(s)\}_{k=1}^{\infty})ds.$$

Lemma 2.42 ([256]). *If $V \subset C(I, E)$ is a bounded and equicontinuous set, then*

(i) *the function $t \to \alpha(V(t))$ is continuous on I, and*

$$\alpha_c(V) = \sup_{t \in I} \alpha(V(t)).$$

(ii) $\alpha\left(\int_0^T x(s)ds : x \in V\right) \le \int_0^T \alpha(V(s))ds$, *where*

$$V(s) = \{x(s) : x \in V\}, \ s \in I.$$

2.6 Measure of weak noncompactness

The measure of weak noncompactness was introduced by DeBlasi [217]. The strong measure of noncompactness was developed first by Banaš and Goebel [153] and subsequently developed and used in many papers; see, for example, Alvàrez [133], Benchohra et al. [170], Guo et al. [256], and the references therein. In [170,324] the authors considered some existence results by applying the techniques of the measure of noncompactness. Recently, several researchers obtained other results by application of the technique of measure of weak noncompactness; see [70,164,168] and the references therein.

Let $(E, w) = (E, \sigma(E, E^*))$ be the Banach space E with its weak topology.

Definition 2.43. A Banach space X is called weakly compactly generated (WCG for short) if it contains a weakly compact set whose linear span is dense in X.

Definition 2.44. A function $h : E \to E$ is said to be weakly sequentially continuous if h takes each weakly convergent sequence in E to a weakly convergent sequence in E (i.e., for any (u_n) in E with $u_n \to u$ in (E, w) then $h(u_n) \to h(u)$ in (E, w)).

Definition 2.45 ([328]). The function $u : I \to E$ is said to be Pettis integrable on I if and only if there is an element $u_J \in E$, corresponding to each $J \subset I$ such that $\phi(u_J) = \int_J \phi(u(s))ds$

for all $\phi \in E^*$, where the integral on the right-hand side is assumed to exist in the sense of Lebesgue (by definition, $u_J = \int_J u(s)ds$).

Let $P(I, E)$ be the space of all E-valued Pettis integrable functions on I and $L^1(I, E)$ be the Banach space of Bochner measurable functions $u : I \to E$. Define the class $P_1(I, E)$ by

$$P_1(I, E) = \{u \in P(I, E) : \varphi(u) \in L^1(I, \mathbb{R}); \ for \ every \ \varphi \in E^*\}.$$

The space $P_1(I, E)$ is normed by

$$\|u\|_{P_1} = \sup_{\varphi \in E^*, \ \|\varphi\| \le 1} \int_1^T |\varphi(u(x))| d\lambda x,$$

where λ stands for a Bochner measure on I.

The following result is due to Pettis (see [328], Theorem 3.4 and Corollary 3.41).

Proposition 2.46 ([328]). *If $u \in P_1(I, E)$ and h is a measurable and essentially bounded real-valued function, then $uh \in P_1(J, E)$.*

For all what follows, the sign "$\int$" denotes the Pettis integral.

Remark 2.47. Let $g \in P_1([1, T], E)$. For every $\varphi \in E^*$, we have

$$\varphi(^H I_1^q g)(x) = (^H I_1^q \varphi g)(x); \ for \ a.e. \ x \in I.$$

Definition 2.48 ([217]). Let E be a Banach space, Ω_E the bounded subsets of E, and B_1 the unit ball of E. The De Blasi measure of weak noncompactness is the map $\beta : \Omega_E \to [0, \infty)$ defined by

$$\beta(X) = \inf\{\epsilon > 0 : \text{there exists a weakly compact subset } \Omega \text{ of } E : X \subset \epsilon B_1 + \Omega\}.$$

The De Blasi measure of weak noncompactness satisfies the following properties:

(a) $A \subset B \Rightarrow \beta(A) \le \beta(B)$,
(b) $\beta(A) = 0 \Leftrightarrow A$ is relatively weakly compact,
(c) $\beta(A \cup B) = \max\{\beta(A), \beta(B)\}$,
(d) $\beta(\overline{A}^\omega) = \beta(A)$, ($\overline{A}^\omega$ denotes the weak closure of A),
(e) $\beta(A + B) \le \beta(A) + \beta(B)$,
(f) $\beta(\lambda A) = |\lambda|\beta(A)$,
(g) $\beta(conv(A)) = \beta(A)$,
(h) $\beta(\cup_{|\lambda| \le h} \lambda A) = h\beta(A)$.

The next result follows directly from the Hahn–Banach theorem.

Proposition 2.49. *Let E be a normed space and $x_0 \in E$ with $x_0 \ne 0$. Then there exists $\varphi \in E^*$ with $\|\varphi\| = 1$ and $|\varphi(x_0)| = \|x_0\|$.*

For a given set V of functions $v : I \to E$, let us denote by

$$V(t) = \{v(t) : v \in V\}; \ t \in I,$$

and

$$V(I) = \{v(t) : v \in V, \ t \in I\}.$$

Lemma 2.50 ([256]). *Let $H \subset C$ be a bounded and equicontinuous. Then the function $t \to \beta(H(t)$ is continuous on I, and*

$$\beta_C(H) = \max_{t \in I} \beta(H(t)),$$

and

$$\beta\left(\int_I u(s)ds\right) \leq \int_I \beta(H(s))ds,$$

where $H(s) = \{u(s) : u \in H, \ s \in I\}$, and β_C is the De Blasi measure of weak noncompactness defined on the bounded sets of C.

2.7 Degree of nondensifiability

Assume that $(E, \|\cdot\|)$ is a Banach space and $\mathfrak{M}_E$ be the class of nonempty and bounded subsets of E. Moreover, we denote by $(C(J, E), \|\cdot\|_\infty)$ and $(L^1(J, E), \|\cdot\|_1)$ the Banach spaces of continuous and Bochner-integrable mappings u from J into E, respectively.

We begin with the following definitions that are adapted from [314,316]:

Definition 2.51. Assume that $\epsilon \geq 0$ and $\mathbb{P} \in \mathfrak{M}_E$, a continuous mapping $\sigma : \Delta := [0, 1] \to E$ is said to be an ϵ-dense curve in $\mathbb{P}$ if the following conditions hold:

- $\sigma(\Delta) \subset \mathbb{P}$.
- For any $z_1 \in \mathbb{P}$, there is $z_2 \in \sigma(\Delta)$ such that $\|z_1 - z_2\| \leq \epsilon$.

Moreover, if for every $\epsilon > 0$ there is an ϵ-dense curve in $\mathbb{P}$, then $\mathbb{P}$ is said to be densifiable.

From the concept of ϵ-dense curve, we can define the degree of nondensifiability (DND for short).

Definition 2.52 ([247,315]). Let $\epsilon \geq 0$ and denote by $\Gamma_{\epsilon,\mathbb{P}}$ the class of all ϵ-dense curves in $\mathbb{P} \in \mathfrak{M}_E$. The degree of nondensifiability is a mapping $\omega : \mathfrak{M}_E \to \mathbb{R}_+$ defined as follows

$$\omega(\mathbb{P}) = \inf\left\{\epsilon \geq 0 : \Gamma_{\epsilon,\mathbb{P}} \neq \emptyset\right\},$$

for each $\mathbb{P} \in \mathfrak{M}_E$.

Remark 2.53. It is worth noting that in [247], a careful study about the DND was made and, in particular, it was proved that the DND is not a measure of noncompactness (MNC) (see

[247] for its definition and properties). However, it has characteristics that are extremely similar to it (see [247, Proposition 2.6]).

Certain characteristics of the DND proven in [244,247] are presented in the following lemma.

Lemma 2.54. *Let* $\mathbb{P}_1, \mathbb{P}_2 \in \mathfrak{M}_E$. *Then the DND has the following properties:*

(1) $\omega(\mathbb{P}) = 0 \Longleftrightarrow \mathbb{P}$ *is a precompact set, for each nonempty, bounded and arc-connected subset* $\mathbb{P}$ *of* E,
(2) $\omega(\overline{\mathbb{P}_1}) = \omega(\mathbb{P}_1)$, *where* $\overline{\mathbb{P}_1}$ *denotes the closure of* $\mathbb{P}_1$,
(3) $\omega(\upsilon\mathbb{P}_1) = |\upsilon|\omega(\mathbb{P}_1)$ *for* $\upsilon \in \mathbb{R}$,
(4) $\omega(x + \mathbb{P}_1) = \omega(\mathbb{P}_1)$, *for all* $x \in E$,
(5) $\omega(Conv(\mathbb{P}_1)) \leq \omega(\mathbb{P}_1)$, *and*

$$\omega(Conv(\mathbb{P}_1 \cup \mathbb{P}_2)) \leq \max\{\omega(Conv(\mathbb{P}_1)), \omega(Conv(\mathbb{P}_2))\},$$

where $Conv(\mathbb{P}_1)$ *represent the convex hull of* $\mathbb{P}_1$,
(6) $\omega(\mathbb{P}_1 + \mathbb{P}_2) \leq \omega(\mathbb{P}_1) + \omega(\mathbb{P}_2)$.

On the other hand, we introduce the following class of functions given by

$$\mathcal{A} = \left\{ \begin{matrix} \beta \colon \mathbb{R}_+ \to \mathbb{R}_+ : \beta \text{ is a monotone increasing} \\ \text{and } \lim_{n\to\infty} \beta^n(t) = 0 \text{ for any } t \in \mathbb{R}_+ \end{matrix} \right\},$$

where $\beta^n(t)$ denotes the n-th composition of β with itself.

Lemma 2.55 ([243]). *Let* $\mathbb{P} \subset C(J, E)$ *be nonempty and bounded. Then*

$$\sup_{t \in J} \omega(\mathbb{P}(t)) \leq \omega(\mathbb{P}).$$

2.8 Some attractivity concepts

In this section, we present some results concerning the attractivity concepts of fixed-point equations.

Denote by $BC := BC(\mathbb{R}_+)$ is the Banach space of all bounded and continuous functions from $\mathbb{R}_+$ into $\mathbb{R}$. Let $\emptyset \neq \Omega \subset BC$ and let $N : \Omega \to \Omega$, and consider the solutions of equation

$$(Nu)(t) = u(t). \tag{2.7}$$

We introduce the following concept of attractivity of solutions for Eq. (2.7).

Definition 2.56. A solution of Eq. (2.7) is locally attractive if there exists a ball $B(u_0, \eta)$ in the space BC such that, for arbitrary solutions $v = v(t)$ and $w = w(t)$ of Eqs. (2.7) belonging to $B(u_0, \eta) \cap \Omega$, we have

$$\lim_{t\to\infty} (v(t) - w(t)) = 0. \tag{2.8}$$

When the limit (2.8) is uniform with respect to $B(u_0, \eta) \cap \Omega$, solutions of Eq. (2.7) are said to be uniformly locally attractive (or equivalently, that solutions of (2.7) are locally asymptotically stable).

Definition 2.57. A solution $v = v(t)$ of Eq. (2.7) is said to be globally attractive if (2.8) holds for each solution $w = w(t)$ of (2.7). If condition (2.8) is satisfied uniformly with respect to the set Ω, solutions of Eq. (2.7) are said to be globally asymptotically stable (or uniformly globally attractive).

Lemma 2.58 ([214]). *Let $D \subset BC$. Then D is relatively compact in BC if the following conditions hold:*

(a) *D is uniformly bounded in BC,*
(b) *The functions belonging to D are almost equicontinuous on $\mathbb{R}_+$, i.e., equicontinuous on every compact of $\mathbb{R}_+$,*
(c) *The functions from D are equiconvergent, that is, given $\epsilon > 0$ there corresponds $T(\epsilon) > 0$ such that $|u(t) - \lim_{t \to \infty} u(t)| < \epsilon$ for any $t \geq T(\epsilon)$ and $u \in D$.*

2.9 Some Ulam stability concepts

Now, we consider the Ulam stability for Eqs. (2.7). Let $\epsilon > 0$ and $\Phi : I \to [0, \infty)$ be a continuous function. We consider the following inequalities:

$$|u(t) - (Nu)(t)| \leq \epsilon; \ t \in I. \tag{2.9}$$

$$|u(t) - (Nu)(t)| \leq \Phi(t); \ t \in I. \tag{2.10}$$

$$|u(t) - (Nu)(t)| \leq \epsilon \Phi(t); \ t \in I. \tag{2.11}$$

Definition 2.59 ([69,345]). The problem (2.7) is Ulam–Hyers stable if there exists a real number $c_f > 0$ such that for each $\epsilon > 0$ and for each solution $u \in C_\gamma$ of the inequality (2.9) there exists a solution $v \in C_\gamma$ of (2.7) with

$$|u(t) - v(t)| \leq \epsilon c_f; \ t \in I.$$

Definition 2.60 ([69,345]). The problem (2.7) is generalized Ulam–Hyers stable if there exists $c_f : C([0, \infty), [0, \infty))$ with $c_f(0) = 0$ such that for each $\epsilon > 0$ and for each solution $u \in C_\gamma$ of the inequality (2.9) there exists a solution $v \in C_\gamma$ of (2.7) with

$$|u(t) - v(t)| \leq c_f(\epsilon); \ t \in I.$$

Definition 2.61 ([69,345]). The problem (2.7) is Ulam–Hyers–Rassias stable with respect to Φ if there exists a real number $c_{f,\Phi} > 0$ such that for each $\epsilon > 0$ and for each solution $u \in C_\gamma$ of the inequality (2.11) there exists a solution $v \in C_\gamma$ of (2.7) with

$$|u(t) - v(t)| \leq \epsilon c_{f,\Phi} \Phi(t); \ t \in I.$$

Definition 2.62 ([69,345]). The problem (2.7) is generalized Ulam–Hyers–Rassias stable with respect to Φ if there exists a real number $c_{f,\Phi} > 0$ such that for each solution $u \in C_\gamma$ of the inequality (2.10) there exists a solution $v \in C_\gamma$ of (2.7) with

$$|u(t) - v(t)| \leq c_{f,\Phi}\Phi(t); \ t \in I.$$

Remark 2.63. It is clear that

(i) Definition 2.59 $\Rightarrow$ Definition 2.60,
(ii) Definition 2.61 $\Rightarrow$ Definition 2.62,
(iii) Definition 2.61 for $\Phi(\cdot) = 1 \Rightarrow$ Definition 2.59.

One can have similar remarks for the inequalities (2.9) and (2.11).

2.10 Some fixed-point theorems

In this section, we give the main fixed-point theorems that will be used in the following chapters.

Definition 2.64 ([125]). Let (M, d) be a metric space. The map $T : M \longrightarrow M$ is said to be Lipschitzian if there exists a constant $k \geq 0$ (called Lipschitz constant) such that

$$d(T(x), T(y)) \leq kd(x, y), \ for \ all \ x, \ y \in M.$$

A Lipschitzian mapping with a Lipschitz constant $k < 1$ is called a contraction.

Definition 2.65 ([355]). A nondecreasing function $\phi : \mathbb{R}_+ \to \mathbb{R}_+$ is called a comparison function if it satisfies one of the following conditions:

(1) For any $t > 0$, we have

$$\lim_{n \to \infty} \phi^{(n)}(t) = 0,$$

where $\phi^{(n)}$ denotes the n-th iteration of ϕ.
(2) The function ϕ is right-continuous and satisfies

$$\forall t > 0 : \phi(t) < t.$$

Theorem 2.66 ([192,307]). *Let (X, d) be a complete metric space and $T : X \to X$ be a mapping such that*

$$d(T(x), T(y)) \leq \phi(d(x, y)),$$

where ϕ is a comparison function. Then T has a unique fixed point in X.

Theorem 2.67 (Banach's fixed-point theorem [253]). *Let C be a nonempty closed subset of a Banach space X. Then any contraction mapping T of C into itself has a unique fixed point.*

Theorem 2.68 (Schauder fixed-point theorem [253]). *Let X be a Banach space, Q be a convex subset of X and $T : Q \longrightarrow Q$ be a compact and continuous map. Then T has at least one fixed point in Q.*

Theorem 2.69 (Schaefer's fixed-point theorem [253]). *Let E be a Banach space and $N : E \to E$ be a completely continuous operator. If the set*

$$D = \{u \in E : u = \lambda N u, \text{ for some } \lambda \in (0, 1)\}$$

is bounded, then N has a fixed point.

Theorem 2.70 (Burton and Kirk fixed-point theorem [195]). *Let X be a Banach space, $A, B : X \to X$ be two operators satisfying*

i) *A is a contraction;*
ii) *B is completely continuous.*

Then either

- *The operator equation $y = A(y) + B(y)$ admits a solution; or*
- *The set $\Omega = \left\{ u \in X : u = \lambda A \left(\frac{u}{\lambda} \right) + \lambda B(u), \text{ for some } \lambda \in [0, 1] \right\}$ is unbounded.*

Theorem 2.71 ([15]). *Let (Ω, d) be a generalized complete metric space and $\Theta : \Omega \to \Omega$ a strictly contractive operator with a Lipschitz constant $L < 1$. If there exists a nonnegative integer k such that $d(\Theta^{k+1} x, \Theta^k x) < \infty$ for some $x \in \Omega$, then the following propositions hold true:*

(A) *The sequence $(\Theta^k x)_{n \in N}$ converges to a fixed point x^* of Θ;*
(B) *x^* is the unique fixed point of Θ in $\Omega^* = \{ y \in \Omega \mid d(\Theta^k x, y) < \infty \}$;*
(C) *If $y \in \Omega^*$, then $d(y, x^*) \le \frac{1}{1-L} d(y, \Theta x)$.*

In the next definition, we will consider a special class of continuous and bounded operators.

Definition 2.72. Let $T : M \subset X \longrightarrow X$ be a bounded operator from a Banach space X into itself. The operator T is called a k-set contraction if there is a number $k \ge 0$ such that

$$\mu(T(A)) \le k\mu(A)$$

for all bounded sets A in M. The bounded operator T is called condensing if $\mu(T(A)) < \mu(A)$ for all bounded sets A in M with $\mu(M) > 0$.

Obviously, every k-set contraction for $0 \le k < 1$ is condensing. Every compact map T is a k-set contraction with $k = 0$.

Theorem 2.73 ([300,301,391]). *Let X be a real Banach space and $B \subset X$ be a bounded, closed and convex set in X. If there exist $u_0 \in B$ and a positive integer n_0 such that $A : B \to B$ be a convex-power condensing operator about u_0 and n_0, then the operator A has at least one fixed point in B.*

Theorem 2.74 (Darbo's fixed-point theorem [153]). *Let M be nonempty, bounded, convex and closed subset of a Banach space X and T : M $\longrightarrow$ M be a continuous operator satisfying $\mu(TA) \le k\mu(A)$ for any nonempty subset A of M and for some constant $k \in [0, 1)$. Then T has at least one fixed point in M.*

Theorem 2.75 (Mönch's fixed-point theorem, [92,313]). *Let D be a bounded, closed and convex subset of a Banach space X such that $0 \in D$, α be the Kuratowski measure of non-compactness and N be a continuous mapping of D into itself. If the implication*

$$[V = \overline{conv}N(V) \quad or \quad V = N(V) \cup \{0\}] \Rightarrow \alpha(V) = 0, \tag{2.12}$$

holds for every subset V of D, then N has a fixed point.

Theorem 2.76 ([323]). *Let Q be a nonempty, closed, convex and equicontinuous subset of a metrizable locally convex vector space X such that $0 \in Q$. Suppose T : Q $\to$ Q is weakly-sequentially continuous. If the implication*

$$\overline{V} = \overline{conv}(\{0\} \cup T(V)) \Rightarrow V \text{ is relatively weakly compact,} \tag{2.13}$$

holds for every subset $V \subset Q$, then the operator T has a fixed point.

Theorem 2.77 (Nonlinear alternative of Leray–Schauder type [253]). *Let X be a Banach space and C a nonempty convex subset of X. Let U, a nonempty open subset of C with $0 \in U$ and T : $\overline{U} \to C$, be a continuous and compact operator.*
   Then either

(a) *T has fixed points, or*
(b) *There exist $u \in \partial U$ and $\lambda \in (0, 1)$ with $u = \lambda T(u)$.*

Theorem 2.78 (Martelli's fixed-point theorem [305]). *Let X be a Banach space and N : X $\to$ $\mathcal{P}_{cl,cv}(X)$ be an u.s.c. and condensing map. If the set $\Omega := \{u \in X : \lambda u \in N(u) \text{ for some } \lambda > 1\}$ is bounded, then N has a fixed point.*

Theorem 2.79 ([140]). *Let $(X, \| \cdot \|_n)$ be a Fréchet space and let A, B : X $\to$ X be two operators such that*

(a) *A is a compact operator;*
(b) *B is a contraction operator with respect to a family of seminorms $\{\| \cdot \|_n\}$;*
(c) *The set $\{x \in X : x = \lambda A(x) + \lambda B\left(\frac{x}{\lambda}\right), \ \lambda \in (0, 1)\}$ is bounded.*

Then the operator equation $A(u) + B(u) = u$ has a solution in X.

Theorem 2.80 (Random fixed-point theorem [270]). *Let K be a nonempty, closed convex bounded subset of the separable Banach space X and let N : $\Omega \times K \to K$ be a compact and continuous random operator. Then the random equation $N(w)u = u$ has a random solution.*

Theorem 2.81 ([252,331]). *Let $(\Omega, \mathcal{F})$ be a measurable space, X be a real separable generalized Banach space and $F : \Omega \times X \to X$ be a continuous random operator, and let $M(w) \in \mathcal{M}_{n \times n}(\mathbb{R}_+)$ be a random variable matrix such that for every $w \in \Omega$, the matrix $M(w)$ converges to 0 and*

$$d(F(w, x_1), F(w, x_2)) \leq M(w)d(x_1, x_2); \; for \; each \; x_1, x_2 \in X \; and \; w \in \Omega,$$

then there exists a random variable $x : \Omega \to X$, which is the unique random fixed point of F.

Theorem 2.82 ([252,331]). *Let $(\Omega, \mathcal{F})$ be a measurable space, X be a real separable generalized Banach space, and $F : \Omega \times X \to X$ be a completely continuous random operator. Then either*

  (i) *the random equation $F(w, x) = x$ has a random solution, i.e., there is a measurable function $x : \Omega \to X$ such that $F(w, x(w)) = x(w)$ for all $w \in \Omega$, or*
  (ii) *the set $M = \{x : \Omega \to X \text{ is measurable} : \lambda(w)F(w, x) = x\}$ is unbounded for some measurable function $\lambda : \Omega \to X$ with $0 < \lambda(w) < 1$ on Ω.*

For more details, see [92,136,250,253,283,429].

Next, we state two multi-valued fixed-point theorems.

Theorem 2.83 (Darbo fixed-point theorem [224]). *Let X be a bounded, closed and convex subset of a Banach space E and let $T : X \to \mathcal{P}_{cl,b}(X)$ be a closed and k-set contraction. Then T has a fixed point.*

Theorem 2.84 (Set-valued version of the Mönch fixed-point theorem [325]). *Let X be Banach space and $K \subset X$ be a closed and convex set. Also, let U be a relatively open subset of K and $N : \overline{U} \to \mathcal{P}_c(K)$. Suppose that N maps compact sets into relatively compact sets, $\text{graph}(N)$ is closed and for some $x_0 \in U$, we have*

$$\text{conv}(x_0 \cup N(M)) \supset M \subset \overline{U} \text{ and } \overline{M} = \overline{U} \text{ imply } \overline{M} \text{ is compact} \qquad (2.14)$$

and

$$x \notin (1 - \lambda)x_0 + \lambda N(x) \quad \forall x \in \overline{U} \backslash U, \; \lambda \in (0, 1). \qquad (2.15)$$

Then there exists $x \in \overline{U}$ with $x \in N(x)$.

Lemma 2.85 (Bohnenblust–Karlin 1950 [188]). *Let X be a Banach space and $K \in \mathcal{P}_{cl,cv}(X)$ and suppose that the operator $G : K \to \mathcal{P}_{cl,cv}(K)$ is upper semicontinuous and the set $G(K)$ is relatively compact in X. Then G has a fixed point in K.*

Lemma 2.86 (Covitz–Nadler [215]). *Let (X, d) be a complete metric space. If $N : X \to \mathcal{P}_{cl}(X)$ is a contraction, then $Fix N \neq \phi$.*

Theorem 2.87 ([224]). *Let $(\Omega, \mathcal{A})$ be a complete σ-finite measure space, X be a separable Banach space, $\mathcal{M}(\Omega, X)$ be the space of all measurable X-valued functions defined on Ω, and* *let $N : \Omega \times X \to \mathcal{P}_{cp,cv}(X)$ be a continuous and condensing multi-valued random operator. If the set $\{u \in \mathcal{M}(\Omega, X) : \lambda u \in N(w)u\}$ is bounded for each $w \in \Omega$ and all $\lambda > 1$, then $N(w)$ has a random fixed point.*

Theorem 2.88 ([320]). *Let $(\Omega, \mathcal{A})$ be a complete σ-finite measure space, E a separable Banach space, and let $N : \Omega \times E \to \mathcal{P}_{cl}(E)$ be a random multi-valued contraction. Then $N(w)$ has a random fixed point.*

Theorem 2.89 ([323]). *Let E be a Banach space with Q a nonempty, bounded, closed, convex, and equicontinuous subset of a metrizable locally convex vector space C such that $0 \in Q$. Suppose $T : Q \to \mathcal{P}_{cl,cv}(Q)$ has weakly sequentially closed graph. If the implication*

$$\overline{V} = \overline{conv}(\{0\} \cup T(V)) \Rightarrow V \text{ is relatively weakly compact,} \tag{2.16}$$

holds for every subset $V \subset Q$, then the operator T has a fixed point.

Theorem 2.90 ([331,334]). *Let (X, d) be a complete generalized metric space and $N : X \to X$ a contractive operator with Lipschitz matrix M. Then N has a unique fixed point x_0 and for each $x \in X$, we have*

$$d(N^k(x), x_0) \leq M^k(M)^{-1}d(x, N(x)); \text{ for all} k \in \mathbb{N}.$$

For $n = 1$, we recover the classical Banach's contraction fixed-point result.

Theorem 2.91 ([331,405]). *Let X be a generalized Banach space, $D \subset E$ be a nonempty closed convex subset of E and $N : D \to D$ be a continuous operator with relatively compact range. Then N has at least a fixed point in D.*

Theorem 2.92 ([330,336]). *Let $(\mathbb{E}, d)$ be a complete generalized metric space and $\mathbb{T} : \mathbb{E} \to \mathbb{E}$ be a contractive operator with Lipschitz matrix $\mathbb{A}$. Then $\mathbb{T}$ has a unique fixed point x_0 and for each $x \in \mathbb{E}$, we have*

$$d(\mathbb{T}^k(x), x_0) \leq \mathbb{A}^k(I - \mathbb{A})^{-1}d(x, \mathbb{T}(x)) \quad \text{for all } k \in \mathbb{N}.$$

Theorem 2.93 ([331]). *Let Ω be a closed, convex, non-empty subset of a generalized Banach spaces $\mathbb{X}$. Suppose that $\mathbb{U}$ and $\mathbb{V}$ map Ω into $\mathbb{X}$ and that*

- $\mathbb{U}x + \mathbb{V}y \in \Omega$ *for all $x, y \in \Omega$;*
- $\mathbb{U}$ *is compact and continuous;*
- $\mathbb{V}$ *is an $\mathbb{A}$-contraction mapping.*

Then the operator equation $\mathbb{U}z + \mathbb{V}z = z$ has at least one solution on Ω.

The following version of the Darbo fixed-point theorem for the DND has an important role in this book.

Theorem 2.94 ([243]). *Let $\mathbb{K}$ be a nonempty, bounded, closed, and convex subset of a Banach space E and let $\mathbb{O} : \mathbb{K} \to \mathbb{K}$ be a continuous operator. Assume that there is $\beta \in \mathcal{A}$ such that*

$$\omega(\mathbb{O}(\mathbb{P})) \leq \beta(\omega(\mathbb{P})), \tag{2.17}$$

for any nonempty subset $\mathbb{P}$ of $\mathbb{K}$. Then $\mathbb{O}$ possesses at least one fixed point in $\mathbb{K}$.

Remark 2.95. Let us note that the above theorem is in form, very similar to the well-known Darbo fixed-point theorem [93]. However, as shown in [243,244] by several examples, both results are essentially different, as Theorem 2.94 performs under more generic circumstances than Darbo fixed-point theorem (DFPT) or its known generalizations.

In 1969, the concepts of Meir–Keeler contraction mapping were introduced by Meir and Keeler.

Definition 2.96 ([310]). Let (X, d) be a metric space. Then a mapping $\mathcal{T}$ on X is said to be a Meir–Keeler contraction (MKC for short) if, for any $\varepsilon > 0$, there exists $\delta > 0$ such that

$$\varepsilon \leq d(x, y) < \varepsilon + \delta \Rightarrow d(\mathcal{T}x, \mathcal{T}y) < \varepsilon, \quad \forall x, y \in X.$$

In [94], the authors defined the notion of Meir–Keeler condensing operators on a Banach space and gave some fixed-point results.

Definition 2.97 ([94]). Let C be a nonempty subset of a Banach space E and μ arbitrary measure of noncompactness on E. We say that an operator $\mathcal{T} : C \to C$ is a Meir–Keeler condensing operator if for any $\varepsilon > 0$, there exists $\delta > 0$ such that

$$\varepsilon \leq \mu(\Omega) < \varepsilon + \delta \Rightarrow \mu(\mathcal{T}\Omega) < \varepsilon,$$

for any bounded subset Ω of C.

The following fixed-point theorem with respect to Meir–Keeler condensing operator, which was introduced by Aghajani et al. [94], plays a key role in the proof of our main results.

Theorem 2.98 ([94]). *Let Ω be a nonempty, bounded, closed, and convex subset of a Banach space E. Also, let μ be an arbitrary measure of noncompactness on E. If $\mathcal{T} : \Omega \to \Omega$ is a continuous and Meir–Keeler condensing operator, then $\mathcal{T}$ has at least one fixed point and the set of all fixed points of $\mathcal{T}$ in Ω is compact.*

Theorem 2.99 (Weissinger's fixed-point theorem [228]). *Assume (E, d) to be a nonempty complete metric space and let $\beta_j \geq 0$ for every $j \in \mathbb{N}$ such that $\sum_{j=0}^{n-1} \beta_j$ converges. Furthermore, let the mapping $\mathbb{T} : E \to E$ satisfy the inequality*

$$d(\mathbb{T}^j u, \mathbb{T}^j v) \leq \beta_j d(u, v),$$

for every $j \in \mathbb{N}$ and every $u, v \in E$. Then $\mathbb{T}$ has a unique fixed point u^*. Moreover, for any $v_0 \in E$, the sequence $\{\mathbb{T}^j v_0\}_{j=1}^{\infty}$ converges to this fixed point u^*.

2.11 Auxiliary lemmas

We state the following generalization of Gronwall's lemma for singular kernel.

Lemma 2.100 ([428]). *Let $v : [0, T] \to [0, +\infty)$ be a real function and $w(\cdot)$ be a nonnegative, locally integrable function on $[0, T]$. Assume that there exist constants $a > 0$ and $0 < \alpha < 1$ such that*

$$v(t) \le w(t) + a \int_0^t (t - s)^{-\alpha} v(s)ds.$$

Then there exists a constant $K = K(\alpha)$ such that

$$v(t) \le w(t) + Ka \int_0^t (t - s)^{-\alpha} w(s)ds, \text{ for every } t \in [0, T].$$

Bainov and Hristova [144] introduced the following integral inequality of Gronwall type for piecewise continuous functions that can be used in the sequel.

Lemma 2.101. *Let, for $t \ge t_0 \ge 0$, the following inequality hold*

$$x(t) \le a(t) + \int_{t_0}^t g(t, s)x(s)ds + \sum_{t_0 < t_k < t} \beta_k(t)x(t_k),$$

where $\beta_k(t)$, ($k \in \mathbb{N}$) are nondecreasing for $t \ge t_0$, $a \in PC([t_0, \infty), \mathbb{R}_+)$, a is nondecreasing, and $g(t, s)$ is a continuous nonnegative function for $t, s \ge t_0$ and nondecreasing with respect to t for any fixed $s \ge t_0$. Then, for $t \ge t_0$,

$$x(t) \le a(t) \prod_{t_0 < t_k < t} (1 + \beta_k(t)) exp\left(\int_{t_0}^t g(t, s)ds\right).$$

Lemma 2.102 ([402]). *We assume that:*

- $v, w \in C(\tilde{I}, \mathbb{R}_+)$
- ψ *belong to both $C^1([a, b], \mathbb{R})$ and $C^1([a, c], \mathbb{R})$ such that ψ is increasing in both cases, and $\psi'(\tau) \ne 0$, $\psi'(u) \ne 0$, for all $(\tau, u) \in \tilde{I}$.*

If there are constants $c > 0$ and $0 < \alpha_1, \alpha_2 < 1$ such that

$$v(\tau, u) \le w(\tau, u) + c \int_a^\tau \int_a^u \psi'(s)\psi'(t)(\psi(\tau) - \psi(s))^{\alpha_1 - 1}(\psi(u) - \psi(t))^{\alpha_2 - 1} v(s, t)dtds,$$

then we have

$$v(\tau, u) \le w(\tau, u)\mathbb{E}_{(\alpha_1, \alpha_2)}\left(c\Gamma(\alpha_1)\Gamma(\alpha_2)(\psi(\tau) - \psi(a))^{\alpha_1}(\psi(u) - \psi(a))^{\alpha_2}\right), \tag{2.18}$$

where $\mathbb{E}_{(\alpha_1, \alpha_2)}(\cdot)$ is the Mittag–Leffler function [249], defined by

$$\mathbb{E}_{(\alpha_1, \alpha_2)}(z) = \sum_{k=0}^\infty \frac{z^k}{\Gamma(\alpha_1 k + 1)\Gamma(\alpha_2 k + 1)}, \quad (z \in \mathbb{R}, \ \alpha_1, \alpha_2 > 0).$$

3

Caputo fractional difference equations in Banach spaces

The results obtained in this chapter are studied and presented as a consequence of the monographs [48,69,281,385,438,439] and the papers [20,21,44,45,54,78,96,103,133,142, 153,156,158–160,200,232,234,280,290,376,394,426]. In [20,21,44,54], Abbas et al. presented some results on the local and global attractivity of solutions for some classes of fractional differential equations involving both the Riemann–Liouville and the Caputo fractional derivatives by employing some fixed-point theorems.

3.1 Implicit fractional q-difference equations in Banach spaces

3.1.1 Introduction

In this section, we discuss the existence of solutions for the following implicit fractional q-difference equation

$$(^cD_q^\alpha u)(t) = f(t, u(t), (^cD_q^\alpha u)(t)), \ t \in I := [0, T], \tag{3.1}$$

with the initial condition

$$u(0) = u_0, \tag{3.2}$$

where $q \in (0, 1), \alpha \in (0, 1], T > 0, f : I \times E \times E \to E$ is a given continuous function, E is a real (or complex) Banach space with norm $\| \cdot \|$, and $^cD_q^\alpha$ is the Caputo fractional q-difference derivative of order α.

3.1.2 Existence results

In this section, we are concerned with the existence results of the problem (3.1)–(3.2).

Definition 3.1. By a solution of the problem (3.1)–(3.2) we mean a continuous function u that satisfies Eq. (3.1) on I and the initial condition (3.2).

The following hypotheses will be used in the sequel.

(3.1.1) The function $t \mapsto f(t, u, v)$ is measurable on I for each $u, v \in E$, and the functions $u \mapsto f(t, u, v)$ and $v \mapsto f(t, u, v)$ are continuous on E for a.e. $t \in I$,

Fractional Difference, Differential Equations, and Inclusions. https://doi.org/10.1016/B978-0-44-323601-3.00010-1
Copyright © 2024 Elsevier Inc. All rights are reserved, including those for text and data mining, AI training, and similar technologies.

(3.1.2) There exists a continuous function $p \in C(I, \mathbb{R}_+)$ such that

$$\|f(t, u, v)\| \leq \frac{p(t)}{1 + \|u\| + \|v\|}; \text{ for a.e. } t \in I, \text{ and each } u, v \in E,$$

(3.1.3) For each bounded and measurable set $B \subset E$ and for each $t \in I$, we have

$$\mu(f(t, B, {}^C D_q^r B)) \leq p(t)\mu(B),$$

where ${}^C D_q^r B = \{{}^C D_q^r w : w \in B\}$.

Set

$$p^* = \sup_{t \in I} p(t), \text{ and } L := \sup_{t \in I} \int_0^T \frac{(t - qs)^{(\alpha-1)}}{\Gamma_q(\alpha)} d_q s.$$

Theorem 3.2. *Assume that the hypotheses (3.1.1)–(3.1.3) hold. If*

$$\ell := Lp^* < 1, \tag{3.3}$$

then the problem (3.1)–(3.2) has at least one solution defined on I.

Proof. By using of Lemma 3.4, we transform the problem (3.1)–(3.2) into a fixed-point problem. Consider the operator $N : C(I) \to C(I)$ defined by

$$(Nu)(t) = u_0 + (I_q^\alpha g)(t); \; t \in I, \tag{3.4}$$

where $g \in C(I)$ such that

$$g(t) = f(t, u(t), g(t)), \text{ or } g(t) = f(t, u_0 + (I_q^\alpha g)(t), g(t)).$$

For any $u \in C(I)$ and each $t \in I$, we have

$$\|(Nu)(t)\| \leq \|u_0\| + \int_0^t \frac{(t - qs)^{(\alpha-1)}}{\Gamma_q(\alpha)} |g(s)| d_q s$$

$$\leq \|u_0\| + \int_0^t \frac{(t - qs)^{(\alpha-1)}}{\Gamma_q(\alpha)} p(s) d_q s$$

$$\leq \|u_0\| + p^* \int_0^t \frac{(t - qs)^{(\alpha-1)}}{\Gamma_q(\alpha)} d_q s$$

$$\leq \|u_0\| + Lp^*$$

$$:= R.$$

Thus

$$\|N(u)\|_\infty \leq R. \tag{3.5}$$

This proves that N transforms the ball $B_R := B(0, R) = \{w \in C : \|w\|_\infty \leq R\}$ into itself.

We will show that the operator $N : B_R \to B_R$ satisfies all the assumptions of Theorem 2.75. The proof will be given in three steps.

Step 1. $N : B_R \to B_R$ *is continuous.*
Let $\{u_n\}_{n\in\mathbb{N}}$ be a sequence such that $u_n \to u$ in B_R. Then for each $t \in I$, we have

$$\|(Nu_n)(t) - (Nu)(t)\| \leq \int_0^t \frac{(t-qs)^{(\alpha-1)}}{\Gamma_q(\alpha)} \|(g_n(s) - g(s))\| d_q s,$$

where $g_n, g \in C(I)$ such that

$$g_n(t) = f(t, u_n(t), g_n(t)), \quad \text{and} \quad g(t) = f(t, u(t), g(t)).$$

Since $u_n \to u$ as $n \to \infty$ and f is continuous, we get

$$g_n(t) \to g(t) \text{ as } n \to \infty, \text{ for each } t \in I.$$

Hence

$$\|N(u_n) - N(u)\|_\infty \leq L\|g_n - g\|_\infty \to 0 \text{ as } n \to \infty.$$

Step 2. $N(B_R)$ *is bounded and equicontinuous.*
Since $N(B_R) \subset B_R$ and B_R is bounded, $N(B_R)$ is bounded.

Next, let $t_1, t_2 \in I$, $t_1 < t_2$ and let $u \in B_R$. Thus we have

$$\|(Nu)(t_2) - (Nu)(t_1)\| \leq \left\| \int_0^{t_2} \frac{(t_2qs)^{(\alpha-1)}}{\Gamma_q(\alpha)} g(s) d_q s - \int_0^{t_1} \frac{(t_1qs)^{(\alpha-1)}}{\Gamma_q(\alpha)} g(s) d_q s \right\|,$$

where $g \in C(I)$ such that

$$g(t) = f(t, u(t), g(t)).$$

Hence we get

$$\begin{aligned}
\|(Nu)(t_2) - (Nu)(t_1)\| &\leq \int_{t_1}^{t_2} \frac{(t_2qs)^{(\alpha-1)}}{\Gamma_q(\alpha)} p(s) d_q s \\
&\quad + \int_0^{t_1} \left| \frac{(t_2qs)^{(\alpha-1)}}{\Gamma_q(\alpha)} - \frac{(t_1qs)^{(\alpha-1)}}{\Gamma_q(\alpha)} \right| p(s) d_q s \\
&\leq p^* \int_{t_1}^{t_2} \frac{(t_2qs)^{(\alpha-1)}}{\Gamma_q(\alpha)} d_q s \\
&\quad + p^* \int_0^{t_1} \left| \frac{(t_2qs)^{(\alpha-1)}}{\Gamma_q(\alpha)} - \frac{(t_1qs)^{(\alpha-1)}}{\Gamma_q(\alpha)} \right| d_q s.
\end{aligned}$$

As $t_1 \longrightarrow t_2$, the right-hand side of the above inequality tends to zero.

Step 3. *The implication (2.12) holds.*
Now let V be a subset of B_R such that $V \subset \overline{N(V)} \cup \{0\}$. Then, V is bounded and equicontinuous and therefore the function $t \to v(t) = \alpha(V(t))$ is continuous on I. By (3.1.3) and the properties of the measure μ, for each $t \in I$, we have

$$v(t) \le \mu((NV)(t) \cup \{0\})$$
$$\le \mu((NV)(t))$$
$$\le \int_0^t \frac{(t - qs)^{(\alpha-1)}}{\Gamma_q(\alpha)} p(s)\mu(V(s)) d_q s$$
$$\le \int_0^t \frac{(t - qs)^{(\alpha-1)}}{\Gamma_q(\alpha)} p(s)v(s) d_q s$$
$$\le L p^* \|v\|_\infty.$$

Thus,

$$\|v\|_\infty \le \ell \|v\|_\infty.$$

From (3.3), we get $\|v\|_\infty = 0$, that is, $v(t) = \mu(V(t)) = 0$, for each $t \in I$, and then $V(t)$ is relatively compact in E. In view of the Ascoli–Arzelà theorem, V is relatively compact in B_R. Applying now Theorem 2.75, we conclude that N has a fixed point which is a solution of the problem (3.1)–(3.2). $\qquad\square$

3.1.3 An example

Let

$$l^1 = \left\{ u = (u_1, u_2, \ldots, u_n, \ldots) : \sum_{n=1}^{\infty} |u_n| < \infty \right\}$$

be the Banach space with the norm

$$\|u\|_{l^1} = \sum_{n=1}^{\infty} |u_n|.$$

Consider the following problem of implicit fractional $\frac{1}{4}$-difference equations

$$\begin{cases} \left({}^c D_{\frac{1}{4}}^{\frac{1}{2}} u_n \right)(t) = f_n \left(t, u(t), ({}^c D_{\frac{1}{4}}^{\frac{1}{2}} u)(t) \right); \ t \in [0, 1], \\ u(0) = (0, 0, \ldots, 0, \ldots), \end{cases} \tag{3.6}$$

where

$$\begin{cases} f_n(t, u, v) = \dfrac{t^{\frac{-1}{4}} (2^{-n} + u_n(t)) \sin t}{64 L (1 + \|u\|_{l^1} + \sqrt{t})(1 + \|u\|_{l^1} + \|v\|_{l^1})}, \ t \in (0, 1], \\ f_n(0, u, v) = 0, \end{cases}$$

with

$$f = (f_1, f_2, \ldots, f_n, \ldots), \text{ and } u = (u_1, u_2, \ldots, u_n, \ldots).$$

For each $t \in (0, 1]$, we have

$$\|f(t, u(t))\|_{l^1} = \sum_{n=1}^{\infty} |f_n(s, u_n(s))|$$

$$\leq \frac{t^{\frac{-1}{4}}|\sin t|}{64L(1 + \|u\|_{l^1} + \sqrt{t})(1 + \|u\|_{l^1} + \|v\|_{l^1})}\left(1 + \|u\|_{l^1}\right)$$

$$\leq \frac{t^{\frac{-1}{4}}|\sin t|}{64L(1 + \|u\|_{l^1} + \|v\|_{l^1})}.$$

Thus the hypothesis (3.1.2) is satisfied with

$$\begin{cases} p(t) = \frac{t^{\frac{-1}{4}}|\sin t|}{64L}; \ t \in (0, 1], \\ p(0) = 0. \end{cases}$$

Thus we have $p^* \leq \frac{1}{64L}$ and then

$$Lp^* = \frac{1}{16} < 1.$$

A simple computation shows that all conditions of Theorem 3.2 are satisfied. Hence the problem (3.6) has at least one solution defined on [0, 1].

3.2 Fractional q-difference equations on the half line

3.2.1 Introduction

In this section, we discuss the existence and the attractivity of solutions for the following functional fractional q-difference equation

$$({}^{c}D_q^{\alpha}u)(t) = f(t, u(t)); \ t \in \mathbb{R}_+ := [0, +\infty), \tag{3.7}$$

with the initial condition

$$u(0) = u_0, \tag{3.8}$$

where $q \in (0, 1)$, $\alpha \in (0, 1]$, $f : \mathbb{R}_+ \times \mathbb{R} \to \mathbb{R}$ is a given continuous function, and ${}^{c}D_q^{\alpha}$ is the Caputo fractional q-difference derivative of order α.

Next, using a generalization of the classical Darbo fixed-point theorem for Fréchet spaces associated with the concept of measure of noncompactness, we discuss the existence of solutions for the problem (3.7)–(3.8) in Fréchet spaces.

Finally, we discuss the existence of bounded solutions for the problem (3.7)–(3.8) on $\mathbb{R}_+$, by applying Schauder's fixed-point theorem associated with the diagonalization process.

3.2.2 Existence and attractivity results

By BC we denote the Banach space of all bounded and continuous functions from $\mathbb{R}_+$ into $\mathbb{R}$ equipped with the norm

$$\|u\|_{BC} := \sup_{t \in \mathbb{R}_+} |u(t)|.$$

Let $\emptyset \neq \Omega \subset BC$ and $G : \Omega \to \Omega$, and consider the solutions of the equation

$$(Gu)(t) = u(t). \tag{3.9}$$

We introduce the following concept of attractivity of solutions for Eq. (3.9).

Definition 3.3. Solutions of Eq. (3.9) are locally attractive if there exists a ball $B(u_0, \eta)$ in the space BC such that, for arbitrary solutions $v = v(t)$ and $w = w(t)$ of Eqs. (3.9) belonging to $B(u_0, \eta) \cap \Omega$, we have

$$\lim_{t \to \infty} (v(t) - w(t)) = 0. \tag{3.10}$$

When the limit (3.10) is uniform with respect to $B(u_0, \eta) \cap \Omega$, solutions of Eq. (3.9) are said to be uniformly locally attractive (or equivalently, that solutions of (3.9) are locally asymptotically stable).

Lemma 3.4. *([214], p. 62). Let $D \subset BC$. Then D is relatively compact in BC if the following conditions hold:*

(a) *D is uniformly bounded in BC;*
(b) *The functions belonging to D are almost equicontinuous on $\mathbb{R}_+$; i.e. equicontinuous on every compact of $\mathbb{R}_+$;*
(c) *The functions from D are equiconvergent, that is, given $\epsilon > 0$ there exists $T(\epsilon) > 0$ such that $|u(t) - \lim_{t \to \infty} u(t)| < \epsilon$ for any $t \geq T(\epsilon)$ and $u \in D$.*

In this section, we are concerned with the existence and the attractivity of solutions of the problem (3.7)–(3.8).

Definition 3.5. By a solution of the problem (3.7)–(3.8) we mean a function $u \in BC$ that satisfies Eq. (3.7) on I and the initial condition (3.8).

The following hypotheses will be used in the sequel:

(3.5.1) The function $f : \mathbb{R}_+ \times \mathbb{R} \to \mathbb{R}$ is continuous.
(3.5.2) There exists a continuous function $p : \mathbb{R}_+ \to \mathbb{R}_+$ such that

$$|f(t, u)| \leq p(t), \text{ for } t \in \mathbb{R}_+, \text{ and each } u \in \mathbb{R},$$

and

$$\lim_{t \to \infty} (I_q^\alpha p)(t) = 0.$$

Set

$$p^* = \sup_{t \in \mathbb{R}_+} (I_q^\alpha p)(t).$$

Now, we present a theorem concerning the existence and the attractivity of solutions of our problem (3.1)–(3.2).

Theorem 3.6. *Assume that the hypotheses* (3.5.1) *and* (3.5.2) *hold. Then the problem* (3.7)–(3.8) *has at least one solution defined on* $\mathbb{R}_+$. *Moreover, solutions of problem* (3.7)–(3.8) *are uniformly locally attractive.*

Proof. Consider the operator N such that, for any $u \in BC$,

$$(Nu)(t) = u_0 + I_q^\alpha f(\cdot, u(\cdot))(t). \tag{3.11}$$

The operator N maps BC into BC. Indeed, the map $N(u)$ is continuous on $\mathbb{R}_+$ for any $u \in BC$, and for each $t \in \mathbb{R}_+$, we have

$$
\begin{aligned}
|(Nu)(t)| &\leq |u_0| + \int_0^t \frac{(t - qs)^{(\alpha-1)}}{\Gamma_q(\alpha)} |f(s, u(s))| d_q s \\
&\leq |u_0| + \int_0^t \frac{(t - qs)^{(\alpha-1)}}{\Gamma_q(\alpha)} p(s) d_q s \\
&\leq |u_0| + p^* \\
&= R.
\end{aligned}
$$

Thus

$$\|N(u)\|_{BC} \leq R. \tag{3.12}$$

Hence $N(u) \in BC$, and the operator N maps the ball

$$B_R := B(0, R) = \{w \in BC : \|w\|_{BC} \leq R\}$$

into itself.

The solutions of the problem (3.7)–(3.8) are the fixed points of the operator N. We shall show that the operator $N : B_R \to B_R$ satisfies all the assumptions of Theorem 2.68. The proof will be given in several steps.

Step 1. *N is continuous.*

Let $\{u_n\}_{n \in \mathbb{N}}$ be a sequence such that $u_n \to u$ in B_R. Then, for each $t \in \mathbb{R}_+$, we have

$$|(Nu_n)(t) - (Nu)(t)| \leq \int_0^t \frac{(t - qs)^{(\alpha-1)}}{\Gamma_q(\alpha)} |f(s, u_n(s)) - f(s, u(s))| d_q s. \tag{3.13}$$

Case 1. If $t \in [0, T]$, $T > 0$, then, since $u_n \to u$ as $n \to \infty$ and f is continuous, by the Lebesgue dominated convergence theorem, Eq. (3.13) implies

$$\|N(u_n) - N(u)\|_{BC} \to 0 \quad \text{as } n \to \infty.$$

Case 2. If $t \in (T, \infty)$, $T > 0$, then from the hypotheses and (3.13), we get

$$|(Nu_n)(t) - (Nu)(t)| \leq 2 \int_0^t \frac{(t - qs)^{(\alpha-1)}}{\Gamma_q(\alpha)} p(s) d_q s. \qquad (3.14)$$

Since $u_n \to u$ as $n \to \infty$ and $(I_q^\alpha p)(t) \to 0$ as $t \to \infty$, then (3.14) gives

$$\|N(u_n) - N(u)\|_{BC} \to 0 \quad \text{as } n \to \infty.$$

Step 2. $N(B_R)$ *is uniformly bounded.*
This is clear since $N(B_R) \subset B_R$ and B_R is bounded.

Step 3. $N(B_R)$ *is equicontinuous on every compact subset* $[0, T]$ *of* $\mathbb{R}_+$; $T > 0$.
Let $t_1, t_2 \in [0, T]$, $t_1 < t_2$, and let $u \in B_R$. Set $p_* = \sup_{t \in [0,T]} p(t)$. Then we have

$$|(Nu)(t_2) - (Nu)(t_1)| \leq \int_0^{t_1} \frac{|(t_2 - qs)^{(\alpha-1)} - (t_1 - qs)^{(\alpha-1)}|}{\Gamma_q(\alpha)} |f(s, u(s))| d_q s$$

$$+ \int_{t_1}^{t_2} \frac{|(t_2 - qs)^{(\alpha-1)}|}{\Gamma_q(\alpha)} |f(s, u(s))| d_q s$$

$$\leq p_* \int_0^{t_1} \frac{|(t_2 - qs)^{(\alpha-1)} - (t_1 - qs)^{(\alpha-1)}|}{\Gamma_q(\alpha)} d_q s$$

$$+ p_* \int_{t_1}^{t_2} \frac{|(t_2 - qs)^{(\alpha-1)}|}{\Gamma_q(\alpha)} d_q s.$$

As $t_1 \longrightarrow t_2$, the right-hand side of the above inequality tends to zero.

Step 4. $N(B_R)$ *is equiconvergent.*
Let $t \in \mathbb{R}_+$ and $u \in B_R$. Then we have

$$|(Nu)(t)| \leq |u_0| + \int_0^t \frac{(t - qs)^{(\alpha-1)}}{\Gamma_q(\alpha)} |f(s, u(s))| d_q s$$

$$\leq |u_0| + \int_0^t \frac{(t - qs)^{(\alpha-1)}}{\Gamma_q(\alpha)} p(s) d_q s$$

$$\leq |u_0| + (I_q^\alpha p)(t).$$

Since $(I_q^\alpha p)(t) \to 0$, as $t \to +\infty$, we get

$$|(Nu)(t)| \to |u_0|, \quad \text{as } t \to +\infty.$$

Hence

$$|(Nu)(t) - (Nu)(+\infty)| \to 0, \quad \text{as } t \to +\infty.$$

As a consequence of Steps 1 to 4, together with the Lemma 3.4, we can conclude that $N : B_R \to B_R$ is continuous and compact. From an application of Schauder's theorem

(Theorem 2.68), we deduce that N has a fixed point u, which is a solution of the problem (3.7)–(3.8) on $\mathbb{R}_+$.

Step 5. *The uniform local attractivity of solutions.*
Let us assume that u_0 is a solution of problem (3.7)–(3.8) with the conditions of this theorem. Taking $u \in B(u_0, 2p^*)$, we have

$$
\begin{aligned}
|(Nu)(t) - u_0(t)| &= |(Nu)(t) - (Nu_0)(t)| \\
&\leq \int_0^t \frac{(t - qs)^{(\alpha-1)}}{\Gamma_q(\alpha)} |f(s, u(s)) - f(s, u_0(s))| d_q s \\
&\leq 2 \int_0^t \frac{(t - qs)^{(\alpha-1)}}{\Gamma_q(\alpha)} p(s) d_q s \\
&\leq 2p^*.
\end{aligned}
$$

Thus we get

$$
\|N(u) - u_0\|_{BC} \leq 2p^*.
$$

Hence we obtain that N is a continuous function such that

$$
N(B(u_0, 2p^*)) \subset B(u_0, 2p^*).
$$

Moreover, if u is a solution of problem (3.7)–(3.8), then

$$
\begin{aligned}
|u(t) - u_0(t)| &= |(Nu)(t) - (Nu_0)(t)| \\
&\leq \int_0^t \frac{(t - qs)^{(\alpha-1)}}{\Gamma_q(\alpha)} |f(s, u(s)) - f(s, u_0(s))| ds \\
&\leq 2(I_q^\alpha p)(t).
\end{aligned}
$$

Thus

$$
|u(t) - u_0(t)| \leq 2(I_q^\alpha p)(t) \to 0 \text{ as } t \to \infty.
$$

Consequently, all solutions of problem (3.7)–(3.8) are uniformly locally attractive. $\qquad\square$

3.2.3 Existence results in Fréchet spaces

Let $X := C(\mathbb{R}_+, E)$ be the Fréchet space of all continuous functions v from $\mathbb{R}_+$ into a Banach space $(E, \|\cdot\|)$, equipped with the family of seminorms

$$
\|v\|_n = \sup_{t \in [0,n]} \|v(t)\|; \ n \in \mathbb{N}^*,
$$

and the distance

$$
d(u, v) = \sum_{n=1}^\infty 2^{-n} \frac{\|u - v\|_n}{1 + \|u - v\|_n}; \ u, v \in X.
$$

Definition 3.7. A nonempty subset $B \subset X$ is said to be bounded if

$$\sup_{v \in B} \|v\|_n < \infty; \text{ for } n \in \mathbb{N}^*.$$

We recall the following definition of the notion of a sequence of measures of noncompactness [229,230].

Definition 3.8. Let $\mathcal{M}_F$ be the family of all nonempty and bounded subsets of a Fréchet space F. A family of functions $\{\mu_n\}_{n \in \mathbb{N}}$ where $\mu_n : \mathcal{M}_F \to [0, \infty)$ is said to be a family of measures of noncompactness in the real Fréchet space F if it satisfies the following conditions for all $B, B_1, B_2 \in \mathcal{M}_F$:

(a) $\{\mu_n\}_{n \in \mathbb{N}}$ is full, that is: $\mu_n(B) = 0$ for $n \in \mathbb{N}$ if and only if B is precompact;
(b) $\mu_n(B_1) \leq \mu_n(B_2)$ for $B_1 \subset B_2$ and $n \in \mathbb{N}$;
(c) $\mu_n(Conv B) = \mu_n(B)$ for $n \in \mathbb{N}$;
(d) If $\{B_i\}_{i=1,\cdots}$ is a sequence of closed sets from $\mathcal{M}_F$ such that $B_{i+1} \subset B_i; i = 1, \cdots$ and if $\lim_{i \to \infty} \mu_n(B_i) = 0$, for each $n \in \mathbb{N}$, then the intersection set $B_\infty := \cap_{i=1}^\infty B_i$ is nonempty.

Example 3.9. [229,322] For $B \in \mathcal{M}_X$, $x \in B$, $n \in \mathbb{N}$ and $\epsilon > 0$, let us denote by $\omega^n(x, \epsilon)$ the modulus of continuity of the function x on the interval $[0, n]$, that is

$$\omega^n(x, \epsilon) = \sup\{\|x(t) - x(s)\| : t, s \in [0, n], |t - s| \leq \epsilon\}.$$

Further, let us put

$$\omega^n(B, \epsilon) = \sup\{\omega^n(x, \epsilon) : x \in B\},$$
$$\omega_0^n(B) = \lim_{\epsilon \to 0^+} \omega^n(B, \epsilon),$$

and

$$\mu_n(B) = \omega_0^n(B) + \sup_{t \in [0,n]} \mu(B(t)),$$

where μ is a measure of noncompactness on the space E.

The family of mappings $\{\mu_n\}_{n \in \mathbb{N}}$ where $\mu_n : \mathcal{M}_X \to [0, \infty)$, satisfies the conditions (a)–(d) from Definition 3.8.

Lemma 3.10. *[189] If Y is a bounded subset of a Banach space F, then for each $\epsilon > 0$, there is a sequence $\{y_k\}_{k=1}^\infty \subset Y$ such that*

$$\mu(Y) \leq 2\mu(\{y_k\}_{k=1}^\infty) + \epsilon.$$

Lemma 3.11. *[313] If $\{u_k\}_{k=0}^\infty \subset L^1([0, n])$ is uniformly integrable, then $\mu_n(\{u_k\}_{k=1}^\infty)$ is measurable for $n \in \mathbb{N}^*$, and*

$$\mu_n\left(\left\{\int_0^t u_k(s)ds\right\}_{k=1}^\infty\right) \leq 2\int_0^t \mu_n(\{u_k(s)\}_{k=1}^\infty)ds,$$

for each $t \in [0, n]$.

Definition 3.12. Let Ω be a nonempty subset of a Fréchet space F, and let $A : \Omega \to F$ be a continuous operator which transforms bounded subsets of onto bounded ones. One says that A satisfies the Darbo condition with constants $(k_n)_{n \in \mathbb{N}}$ with respect to a family of measures of noncompactness $\{\mu_n\}_{n \in \mathbb{N}}$, if

$$\mu_n(A(B)) \leq k_n \mu_n(B)$$

for each bounded set $B \subset \Omega$ and $n \in \mathbb{N}$.

If $k_n < 1$; $n \in \mathbb{N}$ then A is called a contraction with respect to $\{\mu_n\}_{n \in \mathbb{N}}$.

In the sequel we will use the following generalization of the classical Darbo fixed-point theorem for Fréchet spaces.

Theorem 3.13. *[229,230] Let Ω be a nonempty, bounded, closed, and convex subset of a Fréchet space F and let $V : \Omega \to \Omega$ be a continuous mapping. Suppose that V is a contraction with respect to a family of measures of noncompactness $\{\mu_n\}_{n \in \mathbb{N}}$. Then V has at least one fixed point in the set Ω.*

In this section, we are concerned with the existence of solutions of our problem (3.1)–(3.2).

Definition 3.14. By a solution of the problem (3.1)–(3.2) in Fréchet space, we mean a continuous function $u \in X$ that satisfies Eq. (3.1) on $\mathbb{R}_+$ and the initial condition (3.2).

The following hypotheses will be used in the sequel.

(3.14.1) The function $t \mapsto f(t, u)$ is measurable on I for each $u \in E$ and the function $u \mapsto f(t, u)$ is continuous on E for a.e. $t \in \mathbb{R}_+$,

(3.14.2) There exists a continuous function $p : I \to \mathbb{R}_+$ such that

$$\|f(t, u)\| \leq p(t)(1 + \|u\|); \text{ for a.e. } t \in I, \text{ and each } u \in E,$$

(3.14.3) For each bounded and measurable set $B \subset E$ and for each $t \in \mathbb{R}_+$, we have

$$\mu(f(t, B)) \leq p(t)\mu(B),$$

where μ is a measure of noncompactness on the Banach space E.

For $n \in \mathbb{N}^*$, let

$$p_n^* = \sup_{t \in [0,n]} p(t),$$

and define on X the family of measure of noncompactness by

$$\mu_n(D) = \omega_0^n(D) + \sup_{t \in [0,n]} \mu(D(t)),$$

where $D(t) = \{v(t) \in \mathbb{R} : v \in D\}$; $t \in [0, n]$.

Theorem 3.15. *Assume that the hypotheses* (3.14.1)–(3.14.3) *hold.*
  If

$$\frac{4n^\alpha p_n^*}{\Gamma_q(1+\alpha)} < 1; \tag{3.15}$$

for each $n \in \mathbb{N}^*$, then the problem (3.1)–(3.2) has at least one solution in X.

Proof. Consider the operator $N : X \to X$ defined by (3.11). Clearly, the fixed points of the operator N are solution of the problem (3.1)–(3.2).

  For any $n \in \mathbb{N}^*$, we set

$$R_n \geq \frac{|u_0|\Gamma_q(1+\alpha) + p_n^* n^\alpha}{\Gamma_q(1+\alpha) - p_n^* n^\alpha},$$

and we consider the ball

$$B_{R_n} := B(0, R_n) = \{w \in X : \|w\|_n \leq R_n\}.$$

For any $n \in \mathbb{N}^*$, and each $u \in B_{R_n}$ and $t \in [0, n]$, we have

$$\begin{aligned}
\|(Nu)(t)\| &\leq |u_0| + \int_0^t \frac{(tq-s)^{(\alpha-1)}}{\Gamma_q(\alpha)} \|f(s, u(s))\| d_q s \\
&\leq \|u_0\| + \int_0^t \frac{(tq-s)^{(\alpha-1)}}{\Gamma_q(\alpha)} p(s)(1 + \|u(s)\|) d_q s \\
&\leq \|u_0\| + p_n^*(1 + R_n) \int_0^t \frac{(tq-s)^{(\alpha-1)}}{\Gamma_q(\alpha)} d_q s \\
&\leq \|u_0\| + \frac{n^\alpha p_n^*}{\Gamma_q(1+\alpha)}(1 + R_n) \\
&\leq R_n.
\end{aligned}$$

Thus

$$\|N(u)\|_n \leq R_n. \tag{3.16}$$

This proves that N transforms the ball B_{R_n} into itself. We will show that the operator $N : B_{R_n} \to B_{R_n}$ satisfies all the assumptions of Theorem 3.13. The proof will be given in several steps.

  Step 1. $N : B_{R_n} \to B_{R_n}$ *is continuous.*
Let $\{u_k\}_{k \in \mathbb{N}}$ be a sequence such that $u_k \to u$ in B_{R_n}. Then, for each $t \in [0, n]$, we have

$$\|(Nu_k)(t) - (Nu)(t)\| \leq \int_0^t \frac{(tq-s)^{(\alpha-1)}}{\Gamma_q(\alpha)} \|f(s, u_k(s)) - f(s, u(s))\| d_q s.$$

Since $u_k \to u$ as $k \to \infty$, the Lebesgue dominated convergence theorem implies that

$$\|N(u_k) - N(u)\|_n \to 0 \quad \text{as } k \to \infty.$$

Step 2. $N(B_{R_n})$ *is bounded.*
Since $N(B_{R_n}) \subset B_{R_n}$ and B_{R_n} is bounded, then $N(B_{R_n})$ is bounded.

Step 3. *For each bounded and equicontinuous D of B_{R_n}, $\mu_n(N(D)) \leq \ell_n \mu_n(D)$.*
From Lemmas 3.10 and 3.11, for any $D \subset B_{R_n}$ and any $\epsilon > 0$, there exists a sequence $\{u_k\}_{k=0}^{\infty} \subset D$ such that for all $t \in [0, n]$, we have

$$\mu((ND)(t)) = \mu\left(\left\{u_0 + \int_0^t \frac{(t-qs)^{(\alpha-1)}}{\Gamma_q(\alpha)} f(s, u(s))d_qs; \ u \in D\right\}\right)$$

$$\leq 2\mu\left(\left\{\int_0^t \frac{(t-qs)^{(\alpha-1)}}{\Gamma_q(\alpha)} f(s, u_k(s))d_qs\right\}_{k=1}^{\infty}\right) + \epsilon$$

$$\leq 4\int_0^t \frac{(t-qs)^{(\alpha-1)}}{\Gamma_q(\alpha)} \mu(\{f(s, u_k(s))\}_{k=0}^{\infty})d_qs + \epsilon$$

$$\leq 4\int_0^t \frac{(t-qs)^{(\alpha-1)}}{\Gamma_q(\alpha)} p(s)\mu(\{u_k(s)\}_{k=1}^{\infty})d_qs + \epsilon$$

$$\leq \frac{4n^\alpha p_n^*}{\Gamma_q(1+\alpha)} \mu_n(D) + \epsilon.$$

Since $\epsilon > 0$ is arbitrary, then

$$\mu((ND)(t)) \leq \frac{4n^\alpha p_n^*}{\Gamma_q(1+\alpha)} \mu_n(D).$$

Thus

$$\mu_n(N(D)) \leq \frac{4n^\alpha p_n^*}{\Gamma_q(1+\alpha)} \mu_n(D).$$

As a consequence of steps 1 to 3 together with Theorem 3.13, we can conclude that N has at least one fixed point in B_{R_n}, which is a solution of problem (3.1)–(3.2). $\qquad\square$

3.2.4 Existence of bounded solutions

In this section, we are concerned with the existence of bounded solutions of our problem

$$\begin{cases} (^cD_q^\alpha u)(t) = f(t, u(t)); \ t \in \mathbb{R}_+, \\ \\ u(0) = u_0 \in \mathbb{R}, \ u \text{ is bounded on } \mathbb{R}_+, \end{cases} \tag{3.17}$$

Definition 3.16. By a bounded solution of the problem (3.17) we mean a measurable and bounded function u on $\mathbb{R}_+$ such that $u(0) = u_0$, and u satisfies the fractional q-difference equation $(^cD_q^\alpha u)(t) = f(t, u(t))$ on $\mathbb{R}_+$.

The following hypotheses will be used in the sequel.

(3.16.1) The function $t \mapsto f(t, u)$ is measurable on $I_n := [0, n]$; $n \in \mathbb{N}$ for each $u \in \mathbb{R}$, and the function $u \mapsto f(t, u)$ is continuous for a.e. $t \in I_n$,

(3.16.2) There exists a continuous function $p_n : I_n \to \mathbb{R}_+$ such that

$$|f(t, u)| \leq p_n(t), \text{ for a.e. } t \in I_n, \text{ and each } u \in \mathbb{R}.$$

Set

$$p_n^* = \sup_{t \in I_n} p_n(t).$$

Theorem 3.17. *Assume that the hypotheses (3.16.1) and (3.16.2) hold. Then the problem (3.17) has at least one bounded solution defined on $\mathbb{R}_+$.*

Proof. The proof will be given in two parts. Fix $n \in \mathbb{N}$ and consider the problem

$$\begin{cases} (^C D_q^\alpha u)(t) = f(t, u(t)); \ t \in I_n, \\[2mm] u(0) = u_0. \end{cases} \tag{3.18}$$

Part 1. We begin by showing that (3.18) has a solution $u_n \in C(I_n)$ with

$$\|u_n\|_\infty \leq R_n := \frac{n^\alpha p_n^*}{\Gamma_q(1 + \alpha)}.$$

Consider the operator $N : C(I_n) \to C(I_n)$ defined by (3.11). Clearly, the fixed points of the operator N are solution of the problem (3.18).

For any $u \in C(I_n)$, and each $t \in I_n$, we have

$$|(Nu)(t)| \leq |u_0| + \int_0^t \frac{(t - qs)^{(\alpha-1)}}{\Gamma_q(\alpha)} |f(s, u(s))| d_q s$$

$$\leq |u_0| + \int_0^t \frac{(t - qs)^{(\alpha-1)}}{\Gamma_q(\alpha)} p_n(s) d_q s$$

$$\leq |u_0| + p_n^* \int_0^t \frac{(t - qs)^{(\alpha-1)}}{\Gamma_q(\alpha)} d_q s$$

$$\leq \frac{n^\alpha p_n^*}{\Gamma_q(1 + \alpha)}.$$

Thus

$$\|N(u)\|_\infty \leq R_n. \tag{3.19}$$

This proves that N transforms the ball $B_{R_n} := B(0, R_n) = \{w \in C(I_n) : \|w\|_\infty \leq R_n\}$ into itself. We will show that the operator $N : B_{R_n} \to B_{R_n}$ satisfies all the assumptions of Theorem 2.68. The proof will be given in several steps.

Step 1. $N : B_{R_n} \to B_{R_n}$ *is continuous.*

Let $\{u_n\}_{n\in\mathbb{N}}$ be a sequence such that $u_n \to u$ in B_{R_n}. Then, for each $t \in I_n$, we have

$$|(Nu_n)(t) - (Nu)(t)|$$

$$\leq \int_0^t \frac{(t-qs)^{(\alpha-1)}}{\Gamma_q(\alpha)} |f(s, u_n(s)) - f(s, u(s))| d_q s. \tag{3.20}$$

Since $u_n \to u$ as $n \to \infty$ and (3.16.1), the by the Lebesgue dominated convergence theorem, Eq. (3.20) implies

$$\|N(u_n) - N(u)\|_\infty \to 0 \quad \text{as } n \to \infty.$$

Step 2. $N(B_{R_n})$ *is uniformly bounded.*

This is clear since $N(B_{R_n}) \subset B_{R_n}$ and B_{R_n} is bounded.

Step 3. $N(B_{R_n})$ *is equicontinuous.*

Let $t_1, t_2 \in I_n$, $t_1 < t_2$ and let $u \in B_{R_n}$. Thus we have

$$|(Nu)(t_2) - (Nu)(t_1)|$$

$$\leq \int_0^{t_1} \frac{|(t_2-qs)^{(\alpha-1)} - (t_1-qs)^{(\alpha-1)}|}{\Gamma_q(\alpha)} |f(s, u(s))| d_q s$$

$$+ \int_{t_1}^{t_2} \frac{|(t_2-qs)^{(\alpha-1)}|}{\Gamma_q(\alpha)} |f(s, u(s))| d_q s$$

$$\leq p_n^* \int_0^{t_1} \frac{|(t_2-qs)^{(\alpha-1)} - (t_1-qs)^{(\alpha-1)}|}{\Gamma_q(\alpha)} d_q s$$

$$+ p_n^* \int_{t_1}^{t_2} \frac{|(t_2-qs)^{(\alpha-1)}|}{\Gamma_q(\alpha)} d_q s.$$

As $t_1 \longrightarrow t_2$, the right-hand side of the above inequality tends to zero.

As a consequence of steps 1 to 3 together with the Arzelà–Ascoli theorem, we can conclude that N is continuous and compact. From an application of Schauder's theorem (Theorem 2.68), we deduce that N has a fixed point u which is a solution of the problem (3.18).

Part 2. The diagonalization process.

Now, we use the following diagonalization process. For $k \in \mathbb{N}$, let

$$\begin{cases} w_k(t) = u_{n_k}(t); \ t \in [0, n_k], \\ w_k(t) = u_{n_k}(n_k); \ t \in [n_k, \infty). \end{cases}$$

Here $\{n_k\}_{k\in\mathbb{N}^*}$ is a sequence of numbers satisfying

$$0 < n_1 < n_2 < \ldots n_k < \ldots \uparrow \infty.$$

Let $S = \{w_k\}_{k=1}^\infty$. Notice that

$$|w_{n_k}(t)| \leq R_n : \text{for } t \in [0, n_1], \ k \in \mathbb{N}$$

Also, if $k \in \mathbb{N}$ and $t \in [0, n_1]$, we have

$$w_{n_k}(t) = u_0 + \int_0^t \frac{(t-qs)^{(\alpha-1)}}{\Gamma_q(\alpha)} f(s, w_{n_k}(s)) d_q s.$$

Thus, for $k \in \mathbb{N}$ and $t, x \in [0, n_1]$, we have

$$|w_{n_k}(t) - w_{n_k}(x)| \le \int_0^{n_1} \frac{|(t-qs)^{(\alpha-1)} - (x-qs)^{(\alpha-1)}|}{\Gamma_q(\alpha)} |f(s, w_{n_k}(s))| d_q s.$$

Hence

$$|w_{n_k}(t) - w_{n_k}(x)| \le p_1^* \int_0^{n_1} \frac{|(t-qs)^{(\alpha-1)} - (x-qs)^{(\alpha-1)}|}{\Gamma_q(\alpha)} d_q s.$$

The Arzelà–Ascoli Theorem guarantees that there is a subsequence $\mathbb{N}_1^*$ of $\mathbb{N}$ and a function $z_1 \in C([0, n_1], \mathbb{R})$ with $u_{n_k} \to z_1$ as $k \to \infty$ in $C([0, n_1], \mathbb{R})$ through $\mathbb{N}_1^*$. Let $\mathbb{N}_1 = \mathbb{N}_1^* - \{1\}$.

Notice that

$$|w_{n_k}(t)| \le R_n : \text{ for } t \in [0, n_2], \ k \in \mathbb{N}$$

Also, if $k \in \mathbb{N}$ and $t, x \in [0, n_2]$, we have

$$|w_{n_k}(t) - w_{n_k}(x)| \le p_2^* \int_0^{n_2} \frac{|(t-qs)^{(\alpha-1)} - (x-qs)^{(\alpha-1)}|}{\Gamma_q(\alpha)} d_q s.$$

The Arzelà–Ascoli Theorem guarantees that there is a subsequence $\mathbb{N}_2^*$ of $\mathbb{N}_1$ and a function $z_2 \in C([0, n_2], \mathbb{R})$ with $u_{n_k} \to z_2$ as $k \to \infty$ in $C([0, n_2], \mathbb{R})$ through $\mathbb{N}_2^*$. Note that $z_1 = z_2$ on $[0, n_1]$ since $\mathbb{N}_2^* \subset \mathbb{N}_1$. Let $\mathbb{N}_2 = \mathbb{N}_2^* - \{2\}$. Proceed inductively to obtain for $m = 3, 4, \ldots$ a subsequence $\mathbb{N}_m^*$ of $\mathbb{N}_{m-1}$ and a function $z_m \in C([0, n_m], \mathbb{R})$ with $u_{n_k} \to z_m$ as $k \to \infty$ in $C([0, n_m], \mathbb{R})$ through $\mathbb{N}_m^*$. Let $\mathbb{N}_m = \mathbb{N}_m^* - \{m\}$.

Define a function y as follows. Fix $t \in (0, \infty)$ and let $m \in \mathbb{N}$ with $t \le n_m$. Then define $u(t) = z_m(t)$. Thus $u \in C((0, \infty, \mathbb{R}))$, $u(0) = u_0$, and $|u(t)| \le R_n :$ for $t \in [0, \infty)$.

Again fix $t \in (0, \infty)$ and let $m \in \mathbb{N}$ with $t \le n_m$. Then for $n \in \mathbb{N}_m$, we have

$$u_{n_k}(t) = u_0 + \int_0^{n_m} \frac{(t-qs)^{(\alpha-1)}}{\Gamma_q(\alpha)} f(s, w_{n_k}(s)) d_q s.$$

Let $n_k \to \infty$ through $\mathbb{N}_m$ to obtain

$$z_m(t) = u_0 + \int_0^{n_m} \frac{(t-qs)^{(\alpha-1)}}{\Gamma_q(\alpha)} f(s, z_m(s)) d_q s.$$

We can use this method for each $x \in [0, n_m]$ and for each $m \in \mathbb{N}$. Thus

$$(^C D_q^\alpha u)(t) = f(t, u(t)); \text{ for } t \in [0, n_m]$$

for each $m \in \mathbb{N}$ and the constructed function u is a solution of problem (3.17). $\qquad \square$

3.2.5 Some examples

Example 1. Consider the following problem of fractional $\frac{1}{4}$-difference equations

$$\begin{cases} ({}^c D^{\frac{1}{2}}_{\frac{1}{4}} u)(t) = f(t, u(t)); \ t \in \mathbb{R}_+, \\ u(0) = 1, \end{cases} \tag{3.21}$$

where

$$\begin{cases} f(t, u) = \frac{t^{\frac{-1}{4}} \sin t}{(1+\sqrt{t})(1+|u|)}; & t \in (0, \infty), \ u \in \mathbb{R}, \\ f(0, u) = 0; & u \in \mathbb{R}. \end{cases}$$

Clearly, the function f is continuous.

The hypothesis (2.12.2) is satisfied with

$$\begin{cases} p(t) = \frac{t^{\frac{-1}{4}} |\sin t|}{1+\sqrt{t}}; \ t \in (0, \infty), \\ p(0) = 0. \end{cases}$$

All conditions of Theorem 3.17 are satisfied. Hence the problem (3.21) has at least one solution defined on $\mathbb{R}_+$, and solutions of this problem are uniformly locally attractive.

Example 2. Let

$$l^1 = \left\{ u = (u_1, u_2, \ldots, u_k, \ldots) : \sum_{k=1}^{\infty} |u_n| < \infty \right\}$$

be the Banach space with the norm

$$\|u\|_{l^1} = \sum_{k=1}^{\infty} |u_k|,$$

and $F := C(\mathbb{R}_+, l^1)$ be the Fréchet space of all continuous functions v from $\mathbb{R}_+$ into l^1, equipped with the family of seminorms

$$\|v\|_n = \sup_{t \in [0,n]} \|v(t)\|_{l^1}; \ n \in \mathbb{N}^*.$$

Consider the following problem of fractional $\frac{1}{4}$-difference equations

$$\begin{cases} ({}^c D^{\frac{1}{2}}_{\frac{1}{4}} u_k)(t) = f_k(t, u(t)); \ t \in \mathbb{R}_+, \\ u_k(0) = 0; \ t \in \mathbb{R}_+, \ k \in \mathbb{N}, \end{cases} \tag{3.22}$$

where

$$f_k(t, u) = \frac{c_n(2^{-k} + u_k)t^{\frac{5}{4}} \sin t}{64(1 + \sqrt{t})}; \ u \in l^1,$$

for each $t \in [0, n]$; $n \in \mathbb{N}^*$, with

$$c_n = 16n^{-\frac{7}{4}}\Gamma_{\frac{1}{4}}\left(\frac{3}{2}\right); \; n \in \mathbb{N}^*,$$

$$f = (f_1, f_2, \ldots, f_k, \ldots), \; and \; u = (u_1, u_2, \ldots, u_k, \ldots).$$

Since

$$\|f(t, u)\|_{l^1} = \sum_{k=1}^{\infty} |f_k(s, u)| \le \frac{t^{\frac{5}{4}}c_n}{64}(1 + \|u\|_{l^1}); \; t \in \mathbb{R}_+, \; n \in \mathbb{N}^*,$$

then the hypothesis (3.14.2) is satisfied with

$$p(t) = \frac{t^{\frac{5}{4}}c_n}{64}; \; t \in \mathbb{R}_+, \; n \in \mathbb{N}^*.$$

Thus, for any $n \in \mathbb{N}^*$, we have

$$p_n^* = \frac{n^{\frac{5}{4}}c_n}{64}.$$

The condition (3.15) is satisfied. Indeed,

$$\frac{4n^{\frac{1}{2}}p_n^*}{\Gamma_q(1+\alpha)} = 16n^{-\frac{7}{4}}\Gamma_{\frac{1}{4}}\left(\frac{3}{2}\right)\frac{n^{\frac{5}{4}}}{64}\frac{4n^{\frac{1}{2}}}{\Gamma_{\frac{1}{4}}\left(\frac{3}{2}\right)} = \frac{1}{16} < 1.$$

A simple computation shows that all conditions of Theorem 3.15 are satisfied. Hence the problem (3.22) has at least one solution defined on $\mathbb{R}_+$.

Example 3. Consider the following problem of fractional $\frac{1}{4}$-difference equations

$$\begin{cases} (^C D_{\frac{1}{4}}^{\frac{1}{2}} u)(t) = f(t, u(t)); \; t \in \mathbb{R}_+, \\ u(0) = 2, \; u \text{ is bounded on } \mathbb{R}_+, \end{cases} \quad (3.23)$$

where

$$f(t, u) = \frac{e^{t+1}}{1 + |u|}(1 + u); \; t \in \mathbb{R}_+.$$

The hypothesis (3.16.2) is satisfied with $p_n(t) = e^{t+1}$. So, $p_n^* = e^{n+1}$. Simple computations show that all conditions of Theorem 3.17 are satisfied. It follows that the problem (3.23) has at least one bounded solution defined on $\mathbb{R}_+$.

3.3 On the solution of Caputo fractional q-difference equations in Banach spaces

3.3.1 Introduction

This section deals with some existence results for the following fractional q-difference equation

$$({}^{c}D_{q}^{\alpha}u)(t) = f(t, u(t)); \; t \in I := [0, T], \tag{3.24}$$

with the initial condition

$$u(0) = u_{0} \in E, \tag{3.25}$$

where $q \in (0, 1)$, $\alpha \in (0, 1]$, $T > 0$, $f : I \times E \to E$ is a given continuous function, E is a real (or complex) Banach space with norm $\| \cdot \|$, and ${}^{c}D_{q}^{\alpha}$ is the Caputo fractional q-difference derivative of order α.

3.3.2 Existence results

Definition 3.18. By a solution of the problem (3.24)–(3.25) we mean a continuous function u that satisfies Eq. (3.24) on I and the initial condition (3.25).

The following hypotheses will be used in the sequel.

(3.18.1) The function $t \mapsto f(t, u)$ is measurable on I for each $u \in E$, and the function $u \mapsto f(t, u)$ is continuous on E for a.e. $t \in I$,
(3.18.2) There exists a continuous function $p \in C(I, \mathbb{R}_{+})$ such that

$$\|f(t, u)\| \le p(t)(1 + \|u\|); \text{ for a.e. } t \in I, \text{ and each } u \in E,$$

and for some positive integer v, we have

$$p^{*} = \sup_{t \in I} p(t) < \frac{(\Gamma_{q}(1 + v\alpha))^{\frac{1}{v}}}{4T^{\alpha}},$$

(3.18.3) For each bounded and measurable set $B \subset E$ and for each $t \in I$, we have

$$\mu(f(t, B)) \le p(t)\mu(B).$$

Theorem 3.19. *Assume that the hypotheses* (3.18.1)–(3.18.3) *hold. If*

$$\ell := \frac{p^{*}T^{\alpha}}{\Gamma_{q}(1 + \alpha)} < 1, \tag{3.26}$$

then the problem (3.24)–(3.25) *has at least one solution defined on* I.

Proof. By using of Lemma 3.4, we transform the problem (3.24)–(3.25) into a fixed-point problem. Consider the operator $N : C(I) \to C(I)$ defined by

$$(Nu)(t) = u_0 + \int_0^t \frac{(t - qs)^{(\alpha-1)}}{\Gamma_q(\alpha)} f(s, u(s)) d_q s; \quad t \in I. \tag{3.27}$$

Set

$$R > \frac{\|u_0\| + \ell}{1 - \ell},$$

and consider the ball $B_R := \{w \in C(I) : \|w\|_\infty \leq R\}$.

For any $u \in C(I)$ and each $t \in I$, we have

$$\|(Nu)(t)\| \leq \|u_0\| + \int_0^t \frac{(t - qs)^{(\alpha-1)}}{\Gamma_q(\alpha)} \|f(s, u(s))\| d_q s$$

$$\leq \|u_0\| + \int_0^t \frac{(t - qs)^{(\alpha-1)}}{\Gamma_q(\alpha)} p(s)(1 + \|u(s)\|) d_q s$$

$$\leq \|u_0\| + p^*(1 + R) \int_0^t \frac{(t - qs)^{(\alpha-1)}}{\Gamma_q(\alpha)} d_q s$$

$$\leq \|u_0\| + \frac{p^* T^\alpha (1 + R)}{\Gamma_q(1 + \alpha)}$$

$$= \|u_0\| + \ell(1 + R)$$

$$\leq R.$$

Thus

$$\|N(u)\|_\infty \leq R. \tag{3.28}$$

This proves that N transforms the ball B_R into itself.

We will show that the operator $N : B_R \to B_R$ satisfies all the assumptions of Theorem 2.73. The proof will be given in three steps.

Step 1. $N : B_R \to B_R$ *is continuous.*

Let $\{u_n\}_{n \in \mathbb{N}}$ be a sequence such that $u_n \to u$ in B_R. Then, for each $t \in I$, we have

$$\|(Nu_n)(t) - (Nu)(t)\| \leq \int_0^t \frac{(t - qs)^{(\alpha-1)}}{\Gamma_q(\alpha)} \|(f(s, u_n(s)) - f(s, u(s))\| d_q s.$$

Since $u_n \to u$ as $n \to \infty$, then from Lebesgue's dominated convergence theorem and the continuity of the function f, we get

$$f(t, u_n(t)) \to f(t, u(t)) \text{ as } n \to \infty, \text{ for each } t \in I.$$

Hence

$$\|N(u_n) - N(u)\|_\infty \to 0 \text{ as } n \to \infty.$$

Step 2. $N(B_R)$ *is bounded and equicontinuous.*

Since $N(B_R) \subset B_R$ and B_R is bounded, $N(B_R)$ is bounded.

Next, let $t_1, t_2 \in I$, $t_1 < t_2$ and let $u \in B_R$. Thus we have

$$\|(Nu)(t_2) - (Nu)(t_1)\|$$
$$\leq \left\| \int_0^{t_2} \frac{(t_2 - qs)^{(\alpha-1)}}{\Gamma_q(\alpha)} f(s, u(s)) d_q s - \int_0^{t_1} \frac{(t_1 - qs)^{(\alpha-1)}}{\Gamma_q(\alpha)} f(s, u(s)) d_q s \right\|. \tag{3.29}$$

Hence we get

$$\|(Nu)(t_2) - (Nu)(t_1)\| \leq \int_{t_1}^{t_2} \frac{(t_2 - qs)^{(\alpha-1)}}{\Gamma_q(\alpha)} p(s)(1 + \|u(s)\|) d_q s$$
$$+ \int_0^{t_1} \left| \frac{(t_2 - qs)^{(\alpha-1)}}{\Gamma_q(\alpha)} - \frac{(t_1 - qs)^{(\alpha-1)}}{\Gamma_q(\alpha)} \right| p(s)(1 + \|u(s)\|) d_q s$$
$$\leq p^*(1 + R) \int_{t_1}^{t_2} \frac{(t_2 - qs)^{(\alpha-1)}}{\Gamma_q(\alpha)} d_q s$$
$$+ p^*(1 + R) \int_0^{t_1} \left| \frac{(t_2 - qs)^{(\alpha-1)}}{\Gamma_q(\alpha)} - \frac{(t_1 - qs)^{(\alpha-1)}}{\Gamma_q(\alpha)} \right| d_q s.$$

As $t_1 \longrightarrow t_2$, the right-hand side of the above inequality tends to zero.

Step 3. $N : \bar{co} N(B_R) \to \bar{co} N(B_R)$ *is a convex-power condensing operator.*

Set $\Omega = \bar{co} N(B_R)$. It is clear that the operator N maps Ω into itself and $\Omega \subset C(I)$ is equicontinuous. Let $v \in \Omega$. We will prove that there exists a positive integer n_0 such that for any bounded and nonprecompact subset $B \subset \Omega$,

$$\mu_C(N^{(n_0,v)}(B)) \leq \mu_C(B).$$

For any $B \subset \Omega$ and $v \in B$, from the definition of operator $N^{(n,v)}$ and the equicontinuity of Ω, we get that $N(n, v)(B) \subset B_R$ is also equicontinuous. Therefore, from Lemma 2.42, we have

$$\mu_C(N^{(n,v)}(B)) = \max_{t \in I} \mu(N^{(n,v)}(B)(t)); \ n = 1, 2, \cdots \tag{3.30}$$

Let $\epsilon > 0$. By Lemma 2.40, there exists a sequence $\{u_n\}_{n \geq 1} \subset B$ such that

$$\mu(N^{(1,v)}(B)(t)) = \mu(N(B)(t))$$
$$\leq 2\mu \left\{ \int_0^t \frac{(t - qs)^{(\alpha-1)}}{\Gamma_q(\alpha)} f(s, \{u_n(s)\}_{n \geq 1}) d_q s \right\} + \epsilon.$$

Now, by Lemma 2.41 and (3.18.3), we have

$$\mu(N^{(1,v)}(B)(t)) \leq 4 \left\{ \int_0^t \frac{(t-qs)^{(\alpha-1)}}{\Gamma_q(\alpha)} \mu(f(s, \{u_n(s)\}_{n\geq1}))d_qs \right\} + \epsilon$$

$$\leq 4p^* \left\{ \int_0^t \frac{(t-qs)^{(\alpha-1)}}{\Gamma_q(\alpha)} \mu(\{u_n(s)\}_{n\geq1})d_qs \right\} + \epsilon$$

$$\leq 4p^*\mu(B) \left\{ \int_0^t \frac{(t-qs)^{(\alpha-1)}}{\Gamma_q(\alpha)} d_qs \right\} + \epsilon$$

$$\leq \frac{4p^*t^\alpha}{\Gamma_q(1+\alpha)} \mu(B) + \epsilon.$$

As the last inequality is true, for every $\epsilon > 0$, we infer

$$\mu(N^{(1,v)}(B)(t)) \leq \frac{4p^*t^\alpha}{\Gamma_q(1+\alpha)} \mu(B).$$

Again, using Lemma 2.40, for any $\epsilon > 0$, there exists a sequence $\{w_n\}_{n\geq1} \subset \bar{co}\{N^{(1,v)}(B), v\}$ such that

$$\mu(N^{(2,v)}(B)(t)) = \mu(N(\bar{co}\{N^{(1,v)}(B_R), v\})(t))$$

$$\leq 2\mu \left\{ \int_0^t \frac{(t-qs)^{(\alpha-1)}}{\Gamma_q(\alpha)} f(s, \{w_n(s)\}_{n\geq1})d_qs \right\} + \epsilon$$

$$\leq 4 \left\{ \int_0^t \frac{(t-qs)^{(\alpha-1)}}{\Gamma_q(\alpha)} \mu(f(s, \{w_n(s)\}_{n\geq1}))d_qs \right\} + \epsilon$$

$$\leq 4p^* \left\{ \int_0^t \frac{(t-qs)^{(\alpha-1)}}{\Gamma_q(\alpha)} \mu(\{w_n(s)\}_{n\geq1})d_qs \right\} + \epsilon$$

$$\leq 4p^* \left\{ \int_0^t \frac{(t-qs)^{(\alpha-1)}}{\Gamma_q(\alpha)} \mu(\bar{co}\{N^{(1,v)}(B), v\}(s))d_qs \right\} + \epsilon$$

$$\leq 4p^* \left\{ \int_0^t \frac{(t-qs)^{(\alpha-1)}}{\Gamma_q(\alpha)} \mu(N^{(1,v)}(B)(s))d_qs \right\} + \epsilon$$

$$\leq 4p^* \frac{4p^*}{\Gamma_q(1+\alpha)} \mu(B) \left\{ \int_0^t \frac{(t-qs)^{(\alpha-1)}s^\alpha}{\Gamma_q(\alpha)} d_qs \right\} + \epsilon$$

$$\leq \frac{(4p^*)^2}{\Gamma_q(1+\alpha)} \mu(B) \left\{ \int_0^t \frac{(t-qs)^{(\alpha-1)}s^\alpha}{\Gamma_q(\alpha)} d_qs \right\} + \epsilon.$$

On the other hand, we have

$$
\int_0^t \frac{(t - qs)^{(\alpha-1)} s^\alpha}{\Gamma_q(\alpha)} d_q s = \frac{t^{2\alpha}}{\Gamma_q(\alpha)} \int_0^t \left(1 - \frac{s}{tq}\right)^{(\alpha-1)} \left(\frac{s}{tq}\right)^\alpha \frac{d_q s}{tq}
$$

$$
= \frac{t^{2\alpha}}{\Gamma_q(\alpha)} \int_0^1 (1 - x)^{(\alpha-1)} x^\alpha d_q x
$$

$$
= \frac{t^{2\alpha}}{\Gamma_q(\alpha)} \beta_q(\alpha, 1 + \alpha)
$$

$$
= \frac{(t^{2\alpha}}{\Gamma_q(\alpha)} \frac{\Gamma_q(\alpha)\Gamma_q(1 + \alpha)}{\Gamma_q(1 + 2\alpha)}
$$

$$
= t^{2\alpha} \frac{\Gamma_q(1 + \alpha)}{\Gamma_q(1 + 2\alpha)}.
$$

Hence we obtain

$$
\mu(N^{(2,v)}(B)(t)) \le \frac{(4p^*)^2}{\Gamma_q(1 + \alpha)} \mu(B) \left\{ t^{2\alpha} \frac{\Gamma_q(1 + \alpha)}{\Gamma_q(1 + 2\alpha)} \right\} + \epsilon
$$

$$
\le \frac{(4p^*)^2 t^{2\alpha}}{\Gamma_q(1 + 2\alpha)} \mu(B) + \epsilon.
$$

As the last inequality is true, for every $\epsilon > 0$, we get

$$
\mu(N^{(2,v)}(B)(t)) \le \frac{(4p^*)^2 t^{2\alpha}}{\Gamma_q(1 + 2\alpha)} \mu(B).
$$

Repeating the process for $n = 3, 4, \cdots$, for each $t \in I$, we can show by mathematical induction that

$$
\mu(N^{(n,v)}(B)(t)) \le \frac{(4p^*)^n t^{n\alpha}}{\Gamma_q(1 + n\alpha)} \mu(B). \tag{3.31}
$$

By induction, suppose that (3.31) holds for some n and check (3.31) for $n + 1$.

By using Lemma 2.40, for any $\epsilon > 0$, there exists a sequence $\{y_n\}_{n \ge 1} \subset \bar{co}\{N^{(n,v)}(B), v\}$ such that

$$
\mu(N^{(n+1,v)}(B)(t)) = \mu(N(\bar{co}\{N^{(n,v)}(B_R), v\})(t))
$$

$$
\le 2\mu \left\{ \int_0^t \frac{(t - qs)^{(\alpha-1)}}{\Gamma_q(\alpha)} f(s, \{y_n(s)\}_{n \ge 1}) d_q s \right\} + \epsilon
$$

$$
\le 4 \left\{ \int_0^t \frac{(t - qs)^{(\alpha-1)}}{\Gamma_q(\alpha)} \mu(f(s, \{y_n(s)\}_{n \ge 1})) d_q s \right\} + \epsilon
$$

$$
\le 4p^* \left\{ \int_0^t \frac{(t - qs)^{(\alpha-1)}}{\Gamma_q(\alpha)} \mu(\{y_n(s)\}_{n \ge 1}) d_q s \right\} + \epsilon
$$

$$\leq 4p^* \left\{ \int_0^t \frac{(t-qs)^{(\alpha-1)}}{\Gamma_q(\alpha)} \mu(\bar{co}\{y^{(n,v)}(B), v\}(s))d_q s \right\} + \epsilon$$

$$\leq 4p^* \left\{ \int_0^t \frac{(t-qs)^{(\alpha-1)}}{\Gamma_q(\alpha)} \mu(y^{(n,v)}(B)(s))d_q s \right\} + \epsilon$$

$$\leq 4p^* \frac{(4p^*)^n}{\Gamma_q(1+n\alpha)} \left\{ \int_0^t \frac{(t-qs)^{(\alpha-1)} s^{n\alpha}}{\Gamma_q(\alpha)} d_q s \right\} + \epsilon$$

$$\leq \frac{(4p^*)^{n+1} t^{(n+1)\alpha}}{\Gamma_q(1+(n+1)\alpha)} \mu(B) + \epsilon.$$

Thus, as the last inequality is true, for every $\epsilon > 0$, we get

$$\mu(N^{(n+1,v)}(B)(t)) \leq \frac{(4p^*)^{n+1} t^{(n+1)\alpha}}{\Gamma_q(1+(n+1)\alpha)} \mu(B).$$

From (3.30), we get

$$\mu_C(N^{(n,v)}(B)) = \max_{t \in I} \mu(N^{(n,v)}(B)(t)) \leq \frac{(4p^*)^n T^{n\alpha}}{\Gamma_q(1+n\alpha)} \mu(B).$$

Since $p^* < \frac{(\Gamma_q(1+v\alpha))^{\frac{1}{v}}}{4T^\alpha}$, then there exists a positive integer $n_0 = v$ such that

$$\frac{(4p^*)^{n_0} T^{n_0\alpha}}{\Gamma_q(1+n_0\alpha)} < 1.$$

Hence, for any bounded and nonprecompact subset $B \subset \Omega$, we have

$$\mu_C(N^{(n_0,v)}(B)) < \mu_C(B).$$

Therefore $N : \Omega \to \Omega$ is a convex-power condensing operator. Applying now Theorem 2.73, we conclude that N has a fixed point, which is a solution of the problem (3.24)–(3.25).   □

3.3.3 An example

Let

$$l^1 = \left\{ u = (u_1, u_2, \ldots, u_n, \ldots) : \sum_{n=1}^{\infty} |u_n| < \infty \right\}$$

be the Banach space with the norm

$$\|u\|_{l^1} = \sum_{n=1}^{\infty} |u_n|.$$

Consider the following problem of Caputo fractional $\frac{1}{4}$-difference equations

$$\begin{cases} ({}^C D_{\frac{1}{4}}^{\frac{1}{2}} u_n)(t) = f_n(t, u(t)); \ t \in [0, 1], \\ u(0) = (0, 0, \ldots, 0, \ldots), \end{cases} \tag{3.32}$$

where

$$\begin{cases} f_n(t, u) = \dfrac{t^{\frac{-1}{4}} (2^{-n} + u_n(t)) \sin t}{64L(1+\|u\|_{l^1}+\sqrt{t})(1+\|u\|_{l^1})}; \ t \in (0, 1], \\ f_n(0, u, v) = 0, \end{cases}$$

with

$$L > \frac{1}{8\Gamma_{\frac{1}{4}}(\frac{1}{2})}, \ f = (f_1, f_2, \ldots, f_n, \ldots), \text{ and } u = (u_1, u_2, \ldots, u_n, \ldots).$$

For each $t \in (0, 1]$, we have

$$\|f(t, u(t))\|_{l^1} = \sum_{n=1}^{\infty} |f_n(s, u_n(s))|$$

$$\leq \frac{t^{\frac{-1}{4}} |\sin t|}{64L(1 + \|u\|_{l^1} + \sqrt{t})(1 + \|u\|_{l^1})}(1 + \|u\|_{l^1})$$

Thus the hypothesis (3.18.2) is satisfied with

$$\begin{cases} p(t) = \frac{t^{\frac{-1}{4}} |\sin t|}{64L}; \ t \in (0, 1], \\ p(0) = 0. \end{cases}$$

Thus we have $p^* \leq \frac{1}{64L}$ and then

$$\ell = \frac{p^* T^\alpha}{\Gamma_q(1+\alpha)} \leq \frac{1}{64L\Gamma_{\frac{1}{4}}(1 + \frac{1}{2})} < \frac{1}{64} < 1.$$

A simple computation shows that all conditions of Theorem 3.19 are satisfied. Hence the problem (3.32) has at least one solution defined on [0, 1].

3.4 Notes and remarks

The results of Chapter 3 are taken from Abbas et al. [28–30,61,68]. For more details on this subject, we refer the readers to the monographs [19,48], the papers [78,156,158–160,293, 361], and the references therein.

4

Caputo fractional difference inclusions

The outcome of our study in this chapter can be considered as a partial continuation of the problems raised recently in the monographs [48,69,385,394] and the papers [20,21,36, 37,39,50,52,54,77,96,103,179,194,200,280,323–325,351,352,435].

4.1 Fractional q-difference inclusions in Banach spaces

4.1.1 Introduction

In this section, we obtain some existence results using set-valued analysis, the measure of noncompactness, and fixed-point theory for the Caputo fractional q-difference inclusion

$$({}^{c}D_{q}^{\alpha}u)(t) \in F(t, u(t)), \ t \in I := [0, T], \tag{4.1}$$

with the initial condition

$$u(0) = u_0 \in E, \tag{4.2}$$

where $(E, \|\cdot\|)$ is a real or complex Banach space, $q \in (0, 1)$, $\alpha \in (0, 1]$, $T > 0$, $F : I \times E \to \mathcal{P}(E)$ is a multi-valued map, $\mathcal{P}(E) = \{Y \subset E : y \neq \emptyset\}$, and ${}^{c}D_{q}^{\alpha}$ is the Caputo fractional q-difference derivative of order α.

4.1.2 Existence results

Definition 4.1. By a solution of the problem (4.1)–(4.2) we mean a function $u \in C(I)$ that satisfies the initial condition (4.2) and the equation $({}^{C}D_{q}^{\alpha}u)(t) = v(t)$ on I, where $v \in S_{F \circ u}$.

In the sequel, we need the following hypotheses.

(4.1.1) The multi-valued map $F : I \times E \to \mathcal{P}_{cp,cv}(E)$ is Carathéodory,
(4.1.2) There exists a function $p \in L^{\infty}(I, \mathbb{R}_+)$ such that

$$\|F(t, u)\|_{\mathcal{P}} = \sup\{\|v\|_C : v(t) \in F(t, u)\} \leq p(t);$$

for a.e. $t \in I$, and each $u \in E$,
(4.1.3) For each bounded set $B \subset C(I)$ and for each $t \in I$, we have

$$\mu(F(t, B(t)) \leq p(t)\mu(B(t)),$$

where $B(t) = \{u(t) : u \in B\}$,

Fractional Difference, Differential Equations, and Inclusions. https://doi.org/10.1016/B978-0-44-323601-3.00011-3
Copyright © 2024 Elsevier Inc. All rights reserved, including those for text and data mining, AI training, and similar technologies.

(4.1.4) The function $\phi \equiv 0$ is the unique solution in $C(I)$ of the inequality

$$\Phi(t) \le 2p^*(I_q^\alpha \Phi)(t),$$

where p is the function defined in (4.1.2) and

$$p^* = esssup_{t \in I}\, p(t).$$

Remark 4.2. In (4.1.3), μ is the Kuratowski measure of noncompactness on the space E.

Theorem 4.3. *If the hypotheses* (4.1.1)–(4.1.3) *and the condition*

$$L := \frac{p^* T^{(\alpha)}}{\Gamma_q(1+\alpha)} < 1$$

hold, then the problem (4.1)–(4.2) *has at least one solution defined on I.*

Proof. Consider the multi-valued operator $N : C(I) \to \mathcal{P}(C(I))$ defined by:

$$N(u) = \left\{h \in C(I) : h(t) = \mu_0 + \int_0^t \frac{(t-qs)^{(\alpha-1)}}{\Gamma_q(\alpha)} v(s)d_q s; \ v \in S_{Fou}\right\}. \qquad (4.3)$$

Clearly, the fixed points of N are solutions of the problem (4.1)–(4.2). Set

$$R := \|u_0\| + \frac{p^* T^{(\alpha)}}{\Gamma_q(1+\alpha)},$$

and let $B_R := \{u \in C(I) : \|u\|_\infty \le R\}$ be the bounded, closed, and convex ball of $C(I)$. We will show in three steps that the multi-valued operator $N : B_R \to \mathcal{P}_{cl,b}(C(I))$ satisfies all assumptions of Theorem 2.83.

Step 1. $N(B_R) \in \mathcal{P}(B_R)$.
Let $u \in B_R$, and $h \in N(u)$. Then, for each $t \in I$, we have

$$h(t) = u_0 + \int_0^t \frac{(t-qs)^{(\alpha-1)}}{\Gamma_q(\alpha)} v(s)d_q s,$$

for some $v \in S_{Fou}$. On the other hand,

$$\|h(t)\| \le \|u_0\| + \int_0^t \frac{(t-qs)^{(\alpha-1)}}{\Gamma_q(\alpha)} \|v(s)\|d_q s$$

$$\le \|u_0\| + \int_0^t \frac{(t-qs)^{(\alpha-1)}}{\Gamma_q(\alpha)} p(s)d_q s$$

$$\le \|u_0\| + esssup_{t \in I}\, p(t) \int_0^T \frac{(t-qs)^{(\alpha-1)}}{\Gamma_q(\alpha)} d_q s$$

$$= \|u_0\| + \frac{p^* T^{(\alpha)}}{\Gamma_q(1+\alpha)}.$$

Hence $\|h\|_\infty \le R$ and thus $N(B_R) \in \mathcal{P}(B_R)$.

Step 2. $N(u) \in \mathcal{P}_{cl}(B_R)$ *for each* $u \in B_R$.
Let $\{u_n\}_{n\ge 0} \in N(u)$ such that $u_n \longrightarrow \tilde{u}$ in $C(I)$. Then $\tilde{u} \in B_R$ and there exists $f_n(\cdot) \in S_{Fou}$ such that, for each $t \in I$, we have

$$u_n(t) = u_0 + \int_0^t \frac{(t - qs)^{(\alpha-1)}}{\Gamma_q(\alpha)} f_n(s)d_q s.$$

From (4.1.1), and since F has compact values, we may pass to a subsequence if necessary to get that $f_n(\cdot)$ converges to f in $L^1(I)$, and then $f \in S_{Fou}$. Thus, for each $t \in I$, we get

$$u_n(t) \longrightarrow \tilde{u}(t) = u_0 + \int_0^t \frac{(t - qs)^{(\alpha-1)}}{\Gamma_q(\alpha)} f(s)d_q s.$$

Hence $\tilde{u} \in N(u)$.

Step 3. *N satisfies the Darbo condition.*
Let $U \subset B_R$, then, for each $t \in I$, we have

$$\mu((NU)(t)) = \mu(\{(Nu)(t) : u \in U\}).$$

Let $h \in N(u)$. Then there exists $f \in S_{Fou}$ such that, for each $t \in I$, we have

$$h(t) = u_0 + \int_0^t \frac{(t - qs)^{(\alpha-1)}}{\Gamma_q(\alpha)} f(s)d_q s.$$

From Theorem 2.83 and since $U \subset B_R \subset C(I)$, then

$$\mu((NU)(t)) \le 2 \int_0^t \mu\left(\left\{ \frac{(t - qs)^{(\alpha-1)}}{\Gamma_q(\alpha)} f(s) : u \in U \right\}\right) d_q s.$$

Now, since $f \in S_{Fou}$ and $u(s) \in U(s)$, we have

$$\mu(\{(t - qs)^{(\alpha-1)} f(s)\}) = (t - qs)^{(\alpha-1)} p(s) \mu(U(s)).$$

Then

$$\mu((NU)(t)) \le 2 \int_0^t \mu\left(\left\{ \frac{(t - qs)^{(\alpha-1)}}{\Gamma_q(\alpha)} f(s) \right\}\right) d_q s.$$

Thus

$$\mu((NU)(t)) \le 2p^* \int_0^t \frac{(t - qs)^{(\alpha-1)}}{\Gamma_q(\alpha)} \mu(U(s)) d_q s.$$

Hence

$$\mu((NU)(t)) \le \frac{2p^* T^{(\alpha)}}{\Gamma_q(1+\alpha)} \mu(U).$$

Therefore

$$\mu(N(U)) \leq L\mu(U),$$

which implies the N is a L-set-contraction.

As a consequence of Theorem 2.83, we deduce that N has a fixed point that is a solution of the problem (4.1)–(4.2). □

Now, we prove other existence result by applying Theorem 2.84.

Theorem 4.4. *If the hypotheses* (4.1.1)–(4.1.4) *hold, then there exists at least one solution of our problem* (4.1)–(4.2).

Proof. Consider the multi-valued operator $N : C(I) \twoheadrightarrow \mathcal{P}(C(I))$ defined in (4.3). We will show in five steps that the multi-valued operator N satisfies all assumptions of Theorem 2.84.

Step 1. $N(u)$ *is convex for each* $u \in C(I)$.
Let $h_1, h_2 \in N(u)$, then there exist $v_1, v_2 \in S_{Fou}$ such that

$$h_i(t) = \mu_0 + \int_0^t \frac{(t - qs)^{(\alpha-1)}}{\Gamma_q(\alpha)} v_i(s) d_q s; \ t \in I, \ i = 1, 2.$$

Let $0 \leq \lambda \leq 1$. Then, for each $t \in I$, we have

$$(\lambda h_1 + (1 - \lambda)h_2)(t) = \int_0^t \frac{(t - qs)^{(\alpha-1)}}{\Gamma_q(\alpha)} (\lambda v_1(s) + (1 - \lambda)v_2(s)) d_q s.$$

Since S_{Fou} is convex (because F has convex values), we have $\lambda h_1 + (1 - \lambda)h_2 \in N(u)$.

Step 2. *For each compact* $M \subset C(I)$, $N(M)$ *is relatively compact.*
Let (h_n) by any sequence in $N(M)$, where $M \subset C(I)$ is compact. We show that (h_n) has a convergent subsequence from Arzelà–Ascoli compactness criterion in $C(I)$. Since $h_n \in N(M)$ there are $u_n \in M$ and $v_n \in S_{Fou_n}$ such that

$$h_n(t) = \mu_0 + \int_0^t \frac{(t - qs)^{(\alpha-1)}}{\Gamma_q(\alpha)} v_n(s) d_q s.$$

Using Lemma 2.41 and the properties of the measure μ, we have

$$\mu(\{h_n(t)\}) \leq 2 \int_0^t \mu\left(\left\{\frac{(t - qs)^{(\alpha-1)}}{\Gamma_q(\alpha)} v_n(s)\right\}\right) d_q s. \tag{4.4}$$

On the other hand, since M is compact, the set $\{v_n(s) : n \geq 1\}$ is compact. Consequently, $\mu(\{v_n(s) : n \geq 1\}) = 0$ for a.e. $s \in I$. Furthermore,

$$\mu(\{(t - qs)^{(\alpha-1)} v_n(s)\}) = (t - qs)^{(\alpha-1)} \mu(\{v_n(s) : n \geq 1\}) = 0,$$

for a.e. $t, s \in I$. Now (4.4) implies that $\{h_n(t) : n \geq 1\}$ is relatively compact for each $t \in I$. In addition, for each $t_1, t_2 \in I$; with $t_1 < t_2$, we have

$$\|h_n(t_2) - h_n(t_1)\|$$

$$\leq \left\| \int_0^{t_2} \frac{(t_2 - qs)^{(\alpha-1)}}{\Gamma_q(\alpha)} p(s) d_q s - \int_0^{t_1} \frac{(t_1 - qs)^{(\alpha-1)}}{\Gamma_q(\alpha)} p(s) d_q s \right\|$$

$$\leq \int_{t_1}^{t_2} \frac{(t_2 - qs)^{(\alpha-1)}}{\Gamma_q(\alpha)} p(s) d_q s$$

$$+ \int_0^{t_1} \frac{|(t_2 - qs)^{(\alpha-1)} - (t_1 - qs)^{(\alpha-1)}|}{\Gamma_q(\alpha)} p(s) d_q s \qquad (4.5)$$

$$\leq \frac{p^* T^\alpha}{\Gamma_q(1+\alpha)} (t_2 - t_1)^\alpha$$

$$+ p^* \int_0^{t_1} \frac{|(t_2 - qs)^{(\alpha-1)} - (t_1 - qs)^{(\alpha-1)}|}{\Gamma_q(\alpha)} d_q s$$

$$\to 0 \text{ as } t_1 \longrightarrow t_2.$$

This shows that $\{h_n : n \geq 1\}$ is equicontinuous. Consequently, $\{h_n : n \geq 1\}$ is relatively compact in $C(I)$.

Step 3. *The graph of N is closed.*
Let $(u_n, h_n) \in \text{graph}(N)$, $n \geq 1$, with $(\|u_n - u\|, \|h_n - h\|) \to (0.0)$, as $n \to \infty$. We have to show that $(u, h) \in \text{graph}(N)$. $(u_n, h_n) \in \text{graph}(N)$ means that $h_n \in N(u_n)$, which implies that there exists $v_n \in S_{F \circ u_n}$ such that for each $t \in I$,

$$h_n(t) = u_0 + \int_0^t \frac{(t - qs)^{(\alpha-1)}}{\Gamma_q(\alpha)} v_n(s) d_q s.$$

Consider the continuous linear operator $\Theta : L^\infty(I) \to C(I)$,

$$\Theta(v)(t) \mapsto h(t) = u_0 + \int_0^t \frac{(t - qs)^{(\alpha-1)}}{\Gamma_q(\alpha)} v(s) d_q s.$$

Clearly, $\|h_n(t) - h(t)\| \to 0$ as $n \to \infty$. From Lemma 2.29 it follows that $\Theta \circ S_F$ is a closed graph operator. Moreover, $h_n(t) \in \Theta(S_{F \circ u_n})$. Since $u_n \to u$, Lemma 2.29 implies

$$h(t) = u_0 + \int_0^t \frac{(t - qs)^{(\alpha-1)}}{\Gamma_q(\alpha)} v(s) d_q s,$$

for some $v \in S_{F \circ u}$.

Step 4. *M is relatively compact in C(I).*
Let $M \subset \overline{U}$; with $M \subset \text{conv}(\{0\} \cup N(M))$, and let $\overline{M} = \overline{C}$; for some countable set $C \subset M$. The set $N(M)$ is equicontinuous from (4.5). Therefore

$$M \subset \text{conv}(\{0\} \cup N(M)) \implies M \text{ is equicontinuous.}$$

By applying the Arzelà–Ascoli theorem, the set $M(t)$ is relatively compact for each $t \in I$. Since $C \subset M \subset \text{conv}(\{0\} \cup N(M))$, then there exists a countable set $H = \{h_n : n \geq 1\} \subset N(M)$ such that $C \subset \text{conv}(\{0\} \cup H)$. Thus there exist $u_n \in M$ and $v_n \in S_{F \circ u_n}$ such that

$$h_n(t) = u_0 + \int_0^t \frac{(t - qs)^{(\alpha-1)}}{\Gamma_q(\alpha)} v_n(s) d_q s.$$

From Lemma 2.41, we get

$$M \subset \overline{C} \subset \overline{conv}(\{0\} \cup H)) \implies \mu(M(t)) \leq \mu(\overline{C}(t)) \leq \mu(H(t))$$
$$= \mu(\{h_n(t) : n \geq 1\}).$$

Using now the inequality (4.4) in step 2, we obtain

$$\mu(M(t)) \leq 2 \int_0^t \mu\left(\left\{\frac{(t - qs)^{(\alpha-1)}}{\Gamma_q(\alpha)} v_n(s)\right\}\right) d_q s.$$

Since $v_n \in S_{F \circ u_n}$ and $u_n(s) \in M(s)$, we have

$$\mu(M(t)) \leq 2 \int_0^t \mu\left(\left\{\frac{(t - qs)^{(\alpha-1)}}{\Gamma_q(\alpha)} v_n(s) : n \geq 1\right\}\right) d_q s.$$

Also, since $v_n \in S_{F \circ u_n}$ and $u_n(s) \in M(s)$, then, from (4.1.3), we get

$$\mu(\{(t - qs)^{(\alpha-1)} v_n(s); n \geq 1\}) = (t - qs)^{(\alpha-1)} p(s) \mu(M(s)).$$

Hence

$$\mu(M(t)) \leq 2p^* \int_0^t \frac{(t - qs)^{(\alpha-1)}}{\Gamma_q(\alpha)} \mu(M(s)) d_q s.$$

Consequently, from (4.1.4), the function Φ given by $\Phi(t) = \mu(M(t))$ satisfies $\Phi \equiv 0$, that is, $\mu(M(t)) = 0$ for all $t \in I$. Finally, the Arzelà–Ascoli theorem implies that M is relatively compact in $C(I)$.

Step 5. *The priori estimate.*
Let $u \in C(I)$ such that $u \in \lambda N(u)$ for some $0 < \lambda < 1$. Then

$$u(t) = \lambda u_0 + \lambda \int_0^t \frac{(t - qs)^{(\alpha-1)}}{\Gamma_q(\alpha)} v(s) d_q s,$$

for each $t \in I$, where $v \in S_{F \circ u}$. On the other hand,

$$\|u(t)\| \leq \|u_0\| + \int_0^t \frac{(t - qs)^{(\alpha-1)}}{\Gamma_q(\alpha)} \|v(s)\| d_q s$$
$$\leq \|u_0\| + \int_0^t \frac{(t - qs)^{(\alpha-1)}}{\Gamma_q(\alpha)} p(s) d_q s$$
$$\leq \|u_0\| + \frac{p^* T^{(\alpha)}}{\Gamma_q(1+\alpha)}.$$

Then

$$\|u\| \le \|u_0\| + \frac{p^* T^{(\alpha)}}{\Gamma_q(1+\alpha)} := d.$$

Set

$$U = \{u \in C_\gamma : \|u\| < 1 + d\}.$$

Hence the condition (2.15) is satisfied. Finally, Theorem 2.84 implies that N has at least one fixed point $u \in C(I)$, which is a solution of our problem (4.1)–(4.2).   □

4.1.3 An example

Let

$$E = l^1 = \left\{ u = (u_1, u_2, \dots, u_n, \dots), \sum_{n=1}^{\infty} |u_n| < \infty \right\}$$

be the Banach space with the norm

$$\|u\|_E = \sum_{n=1}^{\infty} |u_n|.$$

Consider now the following problem of fractional $\frac{1}{4}$-difference inclusion

$$\begin{cases} (^C D_{\frac{1}{4}}^{\frac{1}{2}} u_n)(t) \in F_n(t, u(t)); \ t \in [0, e], \\ u(0) = (1, 0, \dots, 0, \dots), \end{cases} \tag{4.6}$$

where

$$F_n(t, u(t)) = \frac{t^2 e^{-4-t}}{1 + \|u(t)\|_E} [u_n(t) - 1, u_n(t)]; \ t \in [0, e],$$

with $u = (u_1, u_2, \dots, u_n, \dots)$. Set $\alpha = \frac{1}{2}$, and $F = (F_1, F_2, \dots, F_n, \dots)$.

For each $u \in E$ and $t \in [0, e]$, we have

$$\|F(t, u)\|_{\mathcal{P}} \le c t^2 e^{-t-4}.$$

Hence the hypothesis (4.1.2) is satisfied with $p^* = c e^{-2}$. A simple computation shows that conditions of Theorem 4.4 are satisfied. Hence the problem (4.6) has at least one solution defined on $[0, e]$.

4.2 Weak solutions for Pettis fractional q-difference inclusions

4.2.1 Introduction

In this section we discuss the existence of weak solutions for the following fractional q-difference inclusion

$$({}^cD_q^\alpha u)(t) \in F(t, u(t)), \ t \in I := [0, T], \tag{4.7}$$

with the initial condition

$$u(0) = u_0 \in E, \tag{4.8}$$

where E is a real (or complex) Banach space with norm $\|\cdot\|$ and dual E^*, such that E is the dual of a weakly compactly generated Banach space X, $q \in (0, 1)$, $\alpha \in (0, 1]$, $T > 0$, $F : I \times E \to \mathcal{P}(E)$ is a multi-valued map, $\mathcal{P}(E)$ is the family of all nonempty subsets of E, ${}^cD_q^\alpha$ is the Caputo fractional q-difference derivative of order α.

Next, we consider the following coupled system of fractional q-difference inclusions

$$\begin{cases} ({}^cD_q^\alpha u)(t) \in F(t, v(t)) \\ ({}^cD_q^\alpha v)(t) \in G(t, u(t)) \end{cases} ; \ t \in I, \tag{4.9}$$

with the initial conditions

$$(u(0), v(0)) = (u_0, v_0) \in E \times E, \tag{4.10}$$

$F, G : I \times E \to \mathcal{P}(E)$ are multi-valued maps.

4.2.2 Weak solutions for Caputo Pettis fractional q-difference inclusions

Let us start by defining what we mean by a weak solution of the problem (4.7)–(4.8).

Definition 4.5. By a weak solution of the problem (4.7)–(4.8) we mean a measurable function $u \in C(I)$ that satisfies

$$u(t) = u_0 + \int_0^t \frac{(t - qs)^{(\alpha-1)}}{\Gamma_q(\alpha)} v(s) d_q s,$$

where $v \in S_{Fou}$.

The following hypotheses will be used in the sequel.

(4.5.1) $F : I \times E \to \mathcal{P}_{cp,cl,cv}(E)$ has weakly sequentially closed graph;
(4.5.2) For each continuous $u : I \to E$, there exists a measurable function $v \in S_{Fou}$ a.e. on I and v is Pettis integrable on I;
(4.5.3) There exists $p \in C(I, \mathbb{R}_+)$ such that for all $\varphi \in E^*$, we have

$$\|F(t, u)\|_{\mathcal{P}} = \sup_{v \in S_{Fou}} |\varphi(v)| \le p(t); \text{ for a.e. } t \in I, \text{ and each } u \in E,$$

(4.5.4) For each bounded and measurable set $B \subset E$ and for each $t \in I$, we have

$$\beta(F(t, B) \le p(t)\beta(B).$$

Set

$$p^* = \sup_{t \in I} p(t).$$

Theorem 4.6. *Assume that the hypotheses (4.5.1)–(4.5.4) hold. If*

$$L := \frac{p^* T^\alpha}{\Gamma_q(1 + \alpha)} < 1, \tag{4.11}$$

then the problem (4.7)–(4.8) has at least one weak solution defined on I.

Proof. Consider the multi-valued map $N : C(I) \to \mathcal{P}_{cl}(C(I))$ defined by:

$$(Nu)(t) = \left\{ h \in C(I) : h(t) = u_0 + \int_0^t \frac{(t - qs)^{(\alpha-1)}}{\Gamma_q(\alpha)} v(s)d_q s; \; v \in S_{Fou} \right\}. \tag{4.12}$$

First, notice that the hypotheses imply that for each $u \in C(I)$, there exists a Pettis integrable function $v \in S_{Fou}$, and for each $s \in [0, t]$, the function

$$t \mapsto (t - qs)^{\alpha-1} v(s), \; \text{for a.e. } t \in I,$$

is Pettis integrable. Thus the multi-function N is well defined. Let $R > 0$ be such that

$$R > \frac{p^* T^\alpha}{\Gamma_q(1 + \alpha)},$$

and consider the set

$$Q = \left\{ u \in C(I) : \|u\|_\infty \le R \text{ and } \|u(t_2) - u(t_1)\| \right.$$
$$\left. \le \frac{p^* T^\alpha}{\Gamma_q(1 + \alpha)}(t_2 - t_1)^\alpha + \frac{p^*}{\Gamma_q(\alpha)} \int_0^{t_1} |(t_2 - qs)^{\alpha-1} - (t_1 - qs)^{\alpha-1}| d_q s \right\}.$$

Clearly, the subset Q is closed, convex end equicontinuous. We will show that the operator N satisfies all the assumptions of Theorem 2.89. The proof will be given in several steps.

Step 1. *$N(u)$ is convex for each $u \in Q$.*
For that, let $h_1, h_2 \in N(u)$. Then there exist $v_1, v_2 \in S_{Fou}$ such that, for each $t \in I$, and for any $i = 1, 2$, we have

$$h_i(t) = \frac{\phi}{\Gamma(\gamma)} t^{\gamma-1} + \int_0^t (t - s)^{\alpha-1} \frac{v_i(s)}{\Gamma(\alpha)} ds.$$

Let $0 \leq \lambda \leq 1$. Then, for each $t \in I$, we have

$$[\lambda h_1 + (1 - \lambda)h_2](t) = u_0 + \int_0^t \frac{(t - qs)^{(\alpha-1)}}{\Gamma_q(\alpha)}(\lambda v_1(s) + (1 - \lambda)v_2(s))d_q s.$$

Since $S_{F \circ u}$ is convex (because F has convex values), it follows that

$$\lambda h_1 + (1 - \lambda)h_2 \in N(u).$$

Step 2. *N maps Q into itself.*
Take $h \in N(Q)$. Then there exists $u \in Q$ with $h \in N(u)$, and a Pettis integrable $v : I \to E$ with $v \in S_{F \circ U}$; for a.e. $t \in I$. Assume that $h(t) \neq 0$, then there exists $\varphi \in E^*$ with $\|\varphi\| = 1$ such that $\|h(t)\| = |\varphi(h(t))|$. Then

$$\|h(t)\| = \varphi\left(u_0 + \int_0^t \frac{(t - qs)^{(\alpha-1)}}{\Gamma_q(\alpha)}v(s)d_q s\right).$$

Thus

$$\|h(t)\| \leq \int_0^t \frac{(t - qs)^{(\alpha-1)}}{\Gamma_q(\alpha)}|\varphi(v(s))|d_q s$$

$$\leq \frac{p^*}{\Gamma_q(\alpha)}\int_0^t (t - qs)^{\alpha-1}d_q s$$

$$\leq \frac{p^* T^\alpha}{\Gamma_q(1 + \alpha)}$$

$$\leq R.$$

Next, let $t_1, t_2 \in I$ such that $t_1 < t_2$ and let $h \in N(u)$, with

$$h(t_2) - h(t_1) \neq 0.$$

Then there exists $\varphi \in E^*$ such that

$$\|h(t_2) - h(t_1)\| = |\varphi(h(t_2) - h(t_1))|,$$

and $\|\varphi\| = 1$. Then we have

$$\|h(t_2) - h(t_1)\| = |\varphi(h(t_2) - h(t_1))|$$

$$\leq \varphi\left(\int_0^{t_2} (t_2 - qs)^{\alpha-1}\frac{v(s)}{\Gamma_q(\alpha)}d_q s - \int_0^{t_1} (t_1 - qs)^{\alpha-1}\frac{v(s)}{\Gamma_q(\alpha)}d_q s\right).$$

Thus we get

$$\|h(t_2) - h(t_1)\| \le \int_{t_1}^{t_2} (t_2 - qs)^{\alpha-1} \frac{|\varphi(v(s))|}{\Gamma_q(\alpha)} d_q s$$

$$+ \int_0^{t_1} |(t_2 - qs)^{\alpha-1} - (t_1 - qs)^{\alpha-1}| \frac{|\varphi(v(s))|}{\Gamma_q(\alpha)} d_q s$$

$$\le \int_{t_1}^{t_2} (t_2 - qs)^{\alpha-1} \frac{p(s)}{\Gamma_q(\alpha)} d_q s$$

$$+ \int_0^{t_1} |(t_2 - qs)^{\alpha-1} - (t_1 - qs)^{\alpha-1}| \frac{p(s)}{\Gamma_q(\alpha)} d_q s.$$

Hence we obtain

$$\|h(t_2) - h(t_1)\| \le \frac{p^* T^\alpha}{\Gamma_q(1+\alpha)} (t_2 - t_1)^\alpha$$

$$+ \frac{p^*}{\Gamma_q(\alpha)} \int_0^{t_1} |(t_2 - qs)^{\alpha-1} - (t_1 - qs)^{\alpha-1}| d_q s.$$

This implies that $h \in Q$. Hence $N(Q) \subset Q$.

Step 3. *N has weakly-sequentially closed graph.*
Let (u_n, w_n) be a sequence in $Q \times Q$, with $u_n(t) \to u(t)$ in (E, ω) for each $t \in I$, $w_n(t) \to w(t)$ in (E, ω) for each $(t \in I$, and $w_n \in N(u_n)$ for $n \in \{1, 2, \ldots\}$.
We show that $w \in \Omega(u)$. Since $w_n \in \Omega(u_n)$, there exists $v_n \in S_{F \circ u_n}$ such that

$$w_n(t) = u_0 + \int_0^t \frac{(t - qs)^{(\alpha-1)}}{\Gamma_q(\alpha)} v_n(s) d_q s.$$

We show that there exists $v \in S_{F \circ u}$ such that, for each $t \in I$,

$$w(t) = u_0 + \int_0^t \frac{(t - qs)^{(\alpha-1)}}{\Gamma_q(\alpha)} v(s) d_q s.$$

Since $F(\cdot, \cdot)$ has compact values, there exists a subsequence v_{n_m} such that v_{n_m} is Pettis integrable,

$$v_{n_m}(t) \in F(t, u_n(t)) \text{ a.e. } t \in I,$$
$$v_{n_m}(\cdot) \to v(\cdot) \text{ in } (E, \omega) \text{ as } m \to \infty.$$

As $F(t, \cdot)$ has weakly sequentially closed graph, $v(t) \in F(t, u(t))$. Then by the Lebesgue dominated convergence theorem for the Pettis integral, we obtain

$$\varphi(w_n(t)) \to \varphi\left(u_0 + \int_0^t \frac{(t - qs)^{(\alpha-1)}}{\Gamma_q(\alpha)} v_n(s) d_q s\right),$$

i.e. $w_n(t) \to (Nu)(t)$ in (E, ω). Since this holds, for each $t \in I$, we get $w \in N(u)$.

Step 4. *The Mönch condition (2.16) holds.*

Let V be a subset of Q, such that $\overline{V} = \overline{conv}(\Omega(V) \cup \{0\})$.

Obviously $V(t) \subset \overline{conv}(\Omega(V(t)) \cup \{0\})$ for each $t \in I$. Further, as V is bounded and equicontinuous, the function $t \to v(t) = \beta(V(t))$ is continuous on I. By (4.5.4) and the properties of the measure β, for any $t \in I$, we have

$$v(t) \leq \beta((NV)(t) \cup \{0\})$$

$$\leq \beta((NV)(t))$$

$$\leq \beta\{(Nu)(t) : u \in V\}$$

$$\leq \beta\left\{ \int_0^t (t - qs)^{\alpha-1} \frac{v(s)}{\Gamma_q(\alpha)} d_q s : v(t) \in S_{Fou}, \; u \in V \right\}$$

$$\leq \beta\left\{ \int_0^t (t - qs)^{\alpha-1} \frac{F(s, V(s))}{\Gamma_q(\alpha)} d_q s \right\}$$

$$\leq \int_0^t (t - qs)^{\alpha-1} \frac{\beta(V(s))}{\Gamma_q(\alpha)} d_q s$$

$$\leq \int_0^t (t - qs)^{\alpha-1} \frac{p(s)v(s)}{\Gamma_q(\alpha)} d_q s$$

$$\leq \frac{p^* T^\alpha}{\Gamma_q(1 + \alpha)} \|v\|_\infty$$

$$= L \|v\|_\infty.$$

In particular,

$$\|u\|_\infty \leq L \|v\|_\infty.$$

By (4.11) it follows that $\|v\|_\infty = 0$, that is, $v(t) = \beta(V(t)) = 0$ for each $t \in I$, and then V is weakly relatively compact in $C(I)$. Applying now Theorem 2.89, we conclude that N has a fixed point, which is a weak solution of the problem (4.7)–(4.8). $\square$

4.2.3 Weak solutions for coupled systems of Caputo–Pettis fractional q-difference inclusions

Let us start by defining what we mean by a weak solution of the system (4.9)–(4.10).

Definition 4.7. By a weak solution of the problem (4.9)–(4.10) we mean a coupled measurable functions $(u, v) \in \mathcal{C}$ that satisfies

$$\begin{cases} u(t) = u_0 + \int_0^t \frac{(t-qs)^{(\alpha-1)}}{\Gamma_q(\alpha)} w(s) d_q s, \\ v(t) = v_0 + \int_0^t \frac{(t-qs)^{(\alpha-1)}}{\Gamma_q(\alpha)} z(s) d_q s, \end{cases}$$

where $w \in S_{Fov}$, and $z \in S_{Fou}$

The following hypotheses will be used in the sequel.

(4.7.1) $F, G : I \times E \rightarrow \mathcal{P}_{cp,cl,cv}(E)$ have weakly sequentially closed graph;

(4.7.2) For all continuous functions $u, v : I \rightarrow E$, there exist measurable functions $w \in S_{Fov}$, $z \in S_{Fou}$, a.e. on I and w, z are Pettis integrable on I;

(4.7.3) There exist $p, d \in C(I, \mathbb{R}_+)$ such that for all $\varphi \in E^*$, we have

$$\|F(t, v)\|_{\mathcal{P}} \le p(t), \text{ and } \|G(t, u)\|_{\mathcal{P}} \le d(t); \text{ for a.e. } t \in I, \text{ and each } u, v \in E;$$

(4.7.4) For each bounded and measurable set $B \subset E$ and for each $t \in I$, we have

$$\beta(F(t, B)) \le p(t)\beta(B), \text{ and } \beta(G(t, B)) \le d(t)\beta(B).$$

Set

$$p^* = \sup_{t \in I} p(t), \ d^* = \sup_{t \in I} d(t).$$

Theorem 4.8. *Assume that the hypotheses* (4.7.1)--(4.7.4) *hold. If*

$$\frac{p^* T^\alpha}{\Gamma_q(1 + \alpha)} < 1, \text{ and } \frac{d^* T^\alpha}{\Gamma_q(1 + \alpha)} < 1, \tag{4.13}$$

then the problem (4.9)-(4.10) *has at least one weak solution defined on* I.

Proof. Consider the multi-valued map $N : \mathcal{C} \rightarrow \mathcal{P}_{cl}(\mathcal{C})$ defined by:

$$(N(u, v))(t) = ((N_1 u)(t), (N_2 v)(t)),$$

where $N_1, N_2 : C(I) \rightarrow \mathcal{P}_{cl}(C(I))$ with

$$(N_1 u)(t) = \left\{ h \in C(I) : h(t) = u_0 + \int_0^t \frac{(t - qs)^{(\alpha - 1)}}{\Gamma_q(\alpha)} w(s) d_q s; \ w \in S_{Fov} \right\}, \tag{4.14}$$

and

$$(N_2 v)(t) = \left\{ h \in C(I) : h(t) = v_0 + \int_0^t \frac{(t - qs)^{(\alpha - 1)}}{\Gamma_q(\alpha)} z(s) d_q s; \ z \in S_{Fou} \right\}. \tag{4.15}$$

Notice that, the hypotheses imply that for each $(u, v) \in \mathcal{C}$, there exist Pettis integrable functions $w \in S_{Fov}$, $z \in S_{Fou}$, and for each $s \in [0, t]$, the functions

$$t \mapsto (t - qs)^{\alpha - 1} w(s), \text{ and } t \mapsto (t - qs)^{\alpha - 1} z(s); \text{ for a.e. } t \in I,$$

are Pettis integrable. Thus the multi-function N is well defined. Let $R > 0$ be such that

$$R > \max \left\{ \frac{p^* T^\alpha}{\Gamma_q(1 + \alpha)}, \frac{d^* T^\alpha}{\Gamma_q(1 + \alpha)} \right\}$$

and consider the set

$$\Lambda = \left\{ (u, v) \in \mathcal{C} : \|(u, v)\|_{\mathcal{C}} \leq R \text{ and } \|u(t_2) - u(t_1)\| \leq \frac{p^* T^\alpha}{\Gamma_q(1+\alpha)} (t_2 - t_1)^\alpha \right.$$

$$+ \frac{p^*}{\Gamma_q(\alpha)} \int_0^{t_1} |(t_2 - qs)^{\alpha-1} - (t_1 - qs)^{\alpha-1}| d_q s, \text{ and } \|v(t_2) - v(t_1)\|$$

$$\left. \leq \frac{d^* T^\alpha}{\Gamma_q(1+\alpha)} (t_2 - t_1)^\alpha + \frac{d^*}{\Gamma_q(\alpha)} \int_0^{t_1} |(t_2 - qs)^{\alpha-1} - (t_1 - qs)^{\alpha-1}| d_q s \right\}.$$

Clearly, the subset Λ is closed, convex, and equicontinuous. As in the prove of Theorem 4.3, we can show that $N(u, v)$ is convex for each $(u, v) \in \Lambda$, $N(\Lambda) \subset \Lambda$, N has weakly-sequentially closed graph, and the Mönch condition (2.16) holds. Hence the operator N satisfies all the assumptions of Theorem 2.89. Therefore we conclude that N has a fixed point, which is a weak solution of the problem (4.9)–(4.10). □

4.2.4 Examples

Let

$$E = l^1 = \left\{ u = (u_1, u_2, \ldots, u_n, \ldots) : \sum_{n=1}^\infty |u_n| < \infty \right\}$$

be the Banach space with the norm

$$\|u\|_E = \sum_{n=1}^\infty |u_n|.$$

Example 1. Consider the following problem of fractional $\frac{1}{4}$-difference inclusion

$$\begin{cases} ({}^C D_{\frac{1}{4}}^{\frac{1}{2}} u_n)(t) \in F_n(t, u(t)); \ t \in [0, 1], \\ u(0) = (1, 0, \ldots, 0, \ldots), \end{cases} \tag{4.16}$$

where

$$F_n(t, u(t)) = \frac{ct^2 e^{-4-t}}{1 + \|u(t)\|_E} [u_n(t) - 1, u_n(t)]; \ t \in [0, 1],$$

with

$$u = (u_1, u_2, \ldots, u_n, \ldots), \text{ and } c := \frac{e^4}{4} \Gamma_{\frac{1}{4}} \left(\frac{1}{2}\right).$$

Set

$$F = (F_1, F_2, \ldots, F_n, \ldots).$$

We assume that F is closed and convex valued. Clearly, the function F is continuous.

For each $u \in E$ and $t \in [0, 1]$, we have

$$\|F(t, u(t))\|_{\mathcal{P}} \leq ct^2 \frac{1}{e^{t+4}}.$$

Hence, the hypothesis (4.7.3) is satisfied with $p^* = ce^{-4}$. We will show that condition (4.11) holds with $T = 1$. Indeed,

$$L = \frac{ce^{-4}}{\Gamma_{\frac{1}{4}}(\frac{1}{2})} = \frac{1}{4} < 1.$$

Simple computations show that all conditions of Theorem 4.6 are satisfied. Hence the problem (4.16) has at least one weak solution defined on $[0, 1]$.

Example 2. We consider the following coupled system of fractional $\frac{1}{4}$-difference inclusions

$$\begin{cases} ({}^C D_{\frac{1}{4}}^{\frac{1}{2}} u_n)(t) \in F_n(t, v(t)) \\ ({}^C D_{\frac{1}{4}}^{\frac{1}{2}} v_n)(t) \in G_n(t, u(t)) \\ u(0) = (1, 0, \ldots, 0, \ldots), \ v(0) = (0, 1, 0, \ldots, 0, \ldots) \end{cases} \quad ; \ t \in [0, 1], \qquad (4.17)$$

where

$$F_n(t, v(t)) = \frac{ct^2 e^{-4-t}}{1 + \|u(t)\|_E}[v_n(t) - 1, v_n(t)],$$

$$G_n(t, u(t)) = \frac{ct^2 e^{-4-t}}{1 + \|u(t)\|_E}[u_n(t), 1 + u_n(t)]; \ t \in [0, 1],$$

with

$$u = (u_1, u_2, \ldots, u_n, \ldots), \ v = (v_1, v_2, \ldots, v_n, \ldots), \text{ and } c := \frac{e^4}{4}\Gamma_{\frac{1}{4}}\left(\frac{1}{2}\right).$$

Set

$$F = (F_1, F_2, \ldots, F_n, \ldots), \ G = (G_1, G_2, \ldots, G_n, \ldots).$$

Simple computations show that all conditions of Theorem 4.8 are satisfied. Hence the problem (4.17) has at least one weak solution (u, v) defined on $[0, 1]$.

4.3 Upper and lower solutions for fractional q-difference inclusions

4.3.1 Introduction

In this section we discuss the existence of solutions to the fractional q-difference inclusion

$$({}^c D_q^\alpha u)(t) \in F(t, u(t)), \ t \in I := [0, T], \qquad (4.18)$$

with the boundary condition

$$L(u(0), u(T)) = 0, \tag{4.19}$$

where $q \in (0, 1)$, $\alpha \in (0, 1]$, $T > 0$, $F : I \times \mathbb{R} \to \mathcal{P}(\mathbb{R})$ is a multi-valued map, $\mathcal{P}(\mathbb{R})$ is the family of all nonempty subsets of $\mathbb{R}$, $^C D_q^\alpha$ is the Caputo fractional q-difference derivative of order α, and $L : \mathbb{R}^2 \to \mathbb{R}$ is a given continuous function.

4.3.2 Existence results

We begin by defining what we mean by a solution, an upper solution, and a lower solution to our problem.

Definition 4.9. A function $u \in C(I)$ is said to be a solution of (4.18)–(4.19) if there exists a function $f \in S_{Fou}$ such that $^C D_q^\alpha u(t) = f(t)$ a.e. $t \in I$ and the boundary condition $L(u(0), u(T)) = 0$ is satisfied.

Definition 4.10. A function $w \in C(I)$ is said to be an upper solution of (4.18)–(4.19) if $L(w(0), w(T)) \geq 0$, and there exists a function $v_1 \in S_{Fow}$ such that $^C D_q^\alpha w(t) \geq v_1(t)$ a.e. $t \in I$. Similarly, a function $v \in C(I)$ is said to be a lower solution of (4.18)–(4.19) if $L(v(0), v(T)) \leq 0$, and there exists a function $v_2 \in S_{Fov}$ such that $^C D^\alpha v(t) \leq v_2(t)$ a.e. $t \in I$.

We now present the main result of this paper.

Theorem 4.11. *Assume that the following conditions hold:*

(4.11.1) $F : I \times \mathbb{R} \to \mathcal{P}_{cp,cv}(\mathbb{R})$ *is Carathéodory;*
(4.11.2) *There exist* $v, w \in C(I)$*, which are the lower and upper solutions, respectively, for problem (4.18)–(4.19) such that* $v \leq w$*;*
(4.11.3) *The function* $L(\cdot, \cdot)$ *is continuous on* $[u(0), w(0)] \times [u(T), w(T)]$ *and is nonincreasing in each of its arguments;*
(4.11.4) *There exists* $l \in L^1(I, \mathbb{R}_+)$ *such that*

$$H_d(F(t, u), F(t, \bar{u})) \leq l(t)|u - \bar{u}| \text{ for every } u, \bar{u} \in \mathbb{R},$$

and

$$d(0, F(t, 0)) \leq l(t) \text{ a.e. } t \in I.$$

Then the problem (4.18)-(4.19) has at least one solution u defined on I such that

$$v \leq u \leq w.$$

Proof. Consider the following modified problem

$$^C D_q^\alpha u(t) \in F(t, \tau(u(t))), \text{ for a.e. } t \in I, \tag{4.20}$$

$$u(0) = \tau(u(0)) - L(\bar{u}(0), \bar{u}(T)), \tag{4.21}$$

where

$$\tau(u(t)) = \max\{v(t), \min\{u(t), w(t)\}\},$$

and

$$\overline{u}(t) = \tau(u(t)).$$

A solution to (4.20)–(4.21) is a fixed point of the operator $N : C(I) \to \mathcal{P}(C(I))$ defined by

$$N(u) = \{h \in C(I) \,:\, h(t) = u(0) + (I_q^\alpha v)(t)\},$$

where

$$v \in \{x \in \widetilde{S}^1_{F \circ \tau(u)} \,:\, x(t) \geq v_1(t) \text{ on } A_1 \text{ and } x(t) \leq v_2(t) \text{ on } A_2\},$$

$$S^1_{F \circ \tau(y)} = \{x \in L^1(I) \,:\, x(t) \in F(t, (\tau u)(t)), \text{ a.e. } t \in I\},$$

$$A_1 = \{t \in I : u(t) < v(t) \leq w(t)\}, \quad A_2 = \{t \in I : v(t) \leq w(t) < u(t)\}.$$

Remark 4.12.   (1) For each $u \in C(I)$, the set $\widetilde{S}^1_{F \circ \tau(u)}$ is nonempty. In fact, (4.11.1) implies that there exists $v_3 \in S^1_{F \circ \tau(u)}$, so we set

$$v = v_1 \chi_{A_1} + v_2 \chi_{A_2} + v_3 \chi_{A_3},$$

where

$$A_3 = \{t \in I : v(t) \leq u(t) \leq w(t)\}.$$

Then by decomposability, $x \in \widetilde{S}^1_{F \circ \tau(u)}$.

(2) From the definition of τ, it is clear that $F(\cdot, \tau u(\cdot))$ is an L^1-Carathéodory multi-valued map with compact convex values and there exists $\phi_1 \in C(I, \mathbb{R}_+)$ such that

$$\|F(t, \tau u(t))\|_{\mathcal{P}} \leq \phi_1(t) \text{ for each } u \in \mathbb{R}.$$

(3) Since $\tau(u(t)) = v(t)$ for $t \in A_1$, and $\tau(u(t)) = w(t)$ for $t \in A_2$, in view of (4.11.3), Eq. (4.21) implies that

$$|u(0)| \leq |v(0)| + |L(v(0), v(T)| \leq |v(0)| + |L(u(0), u(T))| = |v(0)| \text{ on } A_1,$$

and

$$u(1) = w(0) - L(w(0), w(T) \leq w(0) - L(u(0), u(T)) = w(0) \text{ on } A_2.$$

Thus

$$|u(0)| \leq \min\{|v(0)|, |w(0)|\}.$$

Now set

$$L := \sup_{t \in I} \int_0^t \frac{(t - qs)^{(\alpha-1)}}{\Gamma_q(\alpha)} d_q s,$$

let

$$R := \min\{|v(0)|, |w(0)|\} + L\|\phi_1\|_\infty,$$

and consider the closed and convex subset of $C(I)$ given by

$$B = \{u \in C(I) : \|u\|_\infty \leq R\}.$$

We will show that the operator $N : B \to \mathcal{P}_{cl,cv}(B)$ satisfies all the assumptions of Theorem 2.85. The proof will be given in steps.

Step 1: *$N(u)$ is convex for each $y \in B$.*
Let h_1, h_2 belong to $N(u)$, then there exist $v_1, v_2 \in \widetilde{S}^1_{F \circ \tau(u)}$ such that, for each $t \in I$ and any $i = 1, 2$, we have

$$h_i(t) = u(0) + (I_q^\alpha v_i)(t).$$

Let $0 \leq d \leq 1$. Then, for each $t \in I$, we have

$$(dh_1 + (1-d)h_2)(t) = u(0) + \int_0^t \frac{(t-qs)^{(\alpha-1)}}{\Gamma_q(\alpha)}[dv_1(s) + (1-d)v_2(s)]d_q s.$$

Since $S_{F \circ \tau(u)}$ is convex (because F has convex values), we have

$$dh_1 + (1-d)h_2 \in N(u).$$

Step 2: *N maps bounded sets into bounded sets in B.*
For each $h \in N(u)$, there exists $v \in \widetilde{S}^1_{F \circ \tau(u)}$ such that

$$h(t) = u(0) + \int_0^t \frac{(t-qs)^{(\alpha-1)}}{\Gamma_q(\alpha)}v(s)d_q s.$$

From conditions (4.11.1)–(4.11.3), for each $t \in I$, we have

$$|h(t)| \leq |u(0)| + \left|\int_0^t \frac{(t-qs)^{(\alpha-1)}}{\Gamma_q(\alpha)}|v(s)|d_q s\right|$$

$$\leq \min\{|v(0)|, |w(0)|\} + \int_0^t \frac{(t-qs)^{(\alpha-1)}}{\Gamma_q(\alpha)}|v(s)|d_q s$$

$$\leq \min\{|v(0)|, |w(0)|\} + L\|\phi_1\|_\infty.$$

Thus

$$\|h\|_\infty \leq R.$$

Step 3: *N maps bounded sets into equicontinuous sets of B.*
Let $t_1, t_2 \in I$ with $t_1 < t_2$, and let $u \in B$ and $h \in N(u)$. Then

$$|h(t_2) - h(t_1)| = \left| \int_0^{t_1} \frac{|(t_2 - qs)^{(\alpha-1)} - (t_1 - qs)^{(\alpha-1)}|}{\Gamma_q(\alpha)} v(s) d_q s \right.$$
$$\left. + \int_{t_1}^{t_2} \frac{(t_2 - qs)^{(\alpha-1)}}{\Gamma_q(\alpha)} v(s) d_q s \right|$$
$$\leq \int_0^{t_1} \frac{|(t_2 - qs)^{(\alpha-1)} - (t_1 - qs)^{(\alpha-1)}|}{\Gamma_q(\alpha)} |v(s)| d_q s$$
$$+ \int_{t_1}^{t_2} \frac{|(t_2 - qs)^{(\alpha-1)}|}{\Gamma_q(\alpha)} |v(s)| d_q s$$
$$\leq \|\phi_1\|_\infty \int_0^{t_1} \frac{|(t_2 - qs)^{(\alpha-1)} - (t_1 - qs)^{(\alpha-1)}|}{\Gamma_q(\alpha)} d_q s$$
$$+ \|\phi_1\|_\infty \int_{t_1}^{t_2} \frac{|(t_2 - qs)^{(\alpha-1)}|}{\Gamma_q(\alpha)} d_q s$$
$$\to 0 \text{ as } t_1 \to t_2.$$

As a consequence of the three steps above, we can conclude from the Arzelà–Ascoli theorem that $N : C(I) \to \mathcal{P}(C(I))$ is continuous and completely continuous.

Step 4: *N has a closed graph.*
Let $u_n \to u_*$, $h_n \in N(u_n)$, and $h_n \to h_*$. We need to show that $h_* \in N(u_*)$. Now $h_n \in N(u_n)$ implies there exists $v_n \in \widetilde{S}^1_{F \circ \tau(u_n)}$ such that, for each $t \in I$,

$$h_n(t) = u(0) + \int_0^t \frac{(t - qs)^{(\alpha-1)}}{\Gamma_q(\alpha)} v_n(s) d_q s.$$

We must show that there exists $v_* \in \widetilde{S}^1_{F \circ \tau(u_*)}$ such that, for each $t \in I$,

$$h_*(t) = u(0) + \int_0^t \frac{(t - qs)^{(\alpha-1)}}{\Gamma_q(\alpha)} v_*(s) d_q s.$$

Since $F(t, \cdot)$ is upper semi-continuous, for every $\epsilon > 0$, there exists a natural number $n_0(\epsilon)$ such that, for every $n \geq n_0(\epsilon)$, we have

$$v_n(t) \in F(t, \tau u_n(t)) \subset F(t, u_*(t)) + \epsilon B(0, 1) \quad \text{a.e. } t \in I.$$

Since $F(\cdot, \cdot)$ has compact values, there exists a subsequence $v_{n_m}(\cdot)$ such that

$$v_{n_m}(\cdot) \to v_*(\cdot) \quad \text{as} \quad m \to \infty,$$

and

$$v_*(t) \in F(t, \tau u_*(t)) \quad \text{a.e. } t \in I.$$

For every $w \in F(t, \tau u_*(t))$, we have

$$|v_{n_m}(t) - v_*(t)| \leq |v_{n_m}(t) - w| + |w - v_*(t)|.$$

Hence

$$|v_{n_m}(t) - v_*(t)| \leq d(v_{n_m}(t), F(t, \tau u_*(t))).$$

We obtain an analogous relation by interchanging the roles of v_{n_m} and v_* to obtain

$$|v_{n_m}(t) - v_*(t)| \leq H_d(F(t, \tau u_n(t)), F(t, \tau u_*(t))) \leq l(t)\|y_n - y_*\|_\infty.$$

Thus

$$|h_{n_m}(t) - h_*(t)| \leq \int_0^t \frac{|(t - qs)^{(\alpha-1)}|}{\Gamma_q(\alpha)} |v_{n_m}(s) - v_*(s)| d_q s$$

$$\leq \|u_{n_m} - u_*\|_\infty \int_0^t \frac{|(t - qs)^{(\alpha-1)}|}{\Gamma_q(\alpha)} l(s) d_q s.$$

Therefore

$$\|h_{n_m} - h_*\|_\infty \leq \|u_{n_m} - u_*\|_\infty \int_0^t \frac{|(t_1 - qs)^{(\alpha-1)}|}{\Gamma_q(\alpha)} l(s) d_q s \to 0 \quad \text{as } m \to \infty,$$

so Lemma 2.28 implies that N is upper semicontinuous.

Step 5: *Every solution u of (4.20)–(4.21) satisfies $v(t) \leq u(t) \leq w(t)$ for all $t \in I$.*
Let u be a solution of (4.20)–(4.21). To prove that $v(t) \leq u(t)$ for all $t \in I$, suppose this is not the case. Then there exist t_1, t_2, with $t_1 < t_2$, such that $v(t_1) = u(t_1)$ and $v(t) > u(t)$ for all $t \in (t_1, t_2)$. In view of the definition of τ,

$$^C D_q^\alpha u(t) \in F(t, v(t)) \text{ for all } t \in (t_1, t_2).$$

Thus there exists $y \in S_{F \circ \tau(v)}$ with $y(t) \geq v_1(t)$ a.e. on (t_1, t_2) such that

$$^C D_q^\alpha u(t) = y(t) \text{ for all } t \in (t_1, t_2).$$

An integration on $(t_1, t]$ with $t \in (t_1, t_2)$ yields

$$u(t) - y(t_1) = \int_{t_1}^t \frac{(t - qs)^{(\alpha-1)}}{\Gamma_q(\alpha)} v(s) d_q.$$

Since v is a lower solution of (4.18)–(4.19),

$$v(t) - v(t_1) \leq \int_{t_1}^t \frac{(t - qs)^{(\alpha-1)}}{\Gamma_q(\alpha)} v_1(s) d_q, \quad t \in (t_1, t_2).$$

From the facts that $u(t_0) = v(t_0)$ and $v(t) \geq v_1(t)$, it follows that

$$v(t) \leq u(t) \text{ for all } t \in (t_1, t_2).$$

This is a contradiction, since $v(t) > u(t)$ for all $t \in (t_1, t_2)$. Consequently,

$$v(t) \le u(t) \text{ for all } t \in I.$$

Similarly, we can prove that

$$u(t) \le w(t) \text{ for all } t \in I.$$

This shows that

$$v(t) \le u(t) \le w(t) \text{ for all } t \in I.$$

Therefore the problem (4.20)–(4.21) has a solution u satisfying $v \le u \le w$.

Step 6: *Every solution of problem (4.20)–(4.21) is solution of (4.18)–(4.19).*
Suppose that u is a solution of the problem (4.20)–(4.21). Then we have

$$^{C}D_q^{\alpha} u(t) \in F(t, \tau(u(t))) \text{ for a.e. } t \in I,$$

and

$$u(0) = \tau(u(0)) - L(\overline{u}(0), \overline{u}(T)).$$

Since, for all $t \in I$, we have

$$v(t) \le u(t) \le w(t),$$

it follows that

$$\tau(u(t)) = u(t).$$

Thus we have

$$^{C}D_q^{\alpha} u(t) \in F(t, u(t)) \text{ for a.e. } t \in I,$$

and

$$L(u(0), u(T)) = 0.$$

We only need to prove that

$$v(0) \le u(0) - L(u(0), u(T)) \le w(1),$$

so suppose that

$$u(0) - L(u(0), u(T)) < u(0).$$

Since $L(v(0), v(T)) \le 0$, we have

$$u(0) \le u(1) - L(v(0), v(T)),$$

and since $L(\cdot, \cdot)$ is nonincreasing with respect to both of its arguments,

$$u(0) \le u(0) - L(v(0), v(T)) \le u(0) - L(u(0), u(T)) < v(0).$$

Hence $u(0) < v(0)$, which is a contradiction. Similarly, we can prove that

$$u(0) - L(u(0), u(T)) \le w(1).$$

Thus u is a solution of (4.18)–(4.19).

This shows that the problem (4.18)–(4.19) has a solution u satisfying $v \le u \le w$ and completes the proof of the theorem.   □

Remark 4.13. In the case where $L(x, y) = ax - by - c$, Theorem 4.11 yields existence results to the problem

$$^{C}D_q^\alpha u(t) \in F(t, u(t)), \quad \text{for a.e. } t \in I, \tag{4.22}$$

$$ay(1) - by(T) = c, \tag{4.23}$$

where $-b < a \le 0 \le b$, $c \in \mathbb{R}$, which includes the anti-periodic problem $b = -a$, $c = 0$, the initial value problem, and the terminal value problem.

4.3.3 An example

Consider now the following problem of Caputo fractional $\frac{1}{4}$-difference inclusion with order $\alpha = \frac{1}{2}$,

$$\begin{cases} \left(^{C}D_{\frac{1}{4}}^{\frac{1}{2}}u\right)(t) \in \frac{7t^2}{27(1+|u(t)|)}[u(t), 33(1+u(t))]; \ t \in [0, 1], \\ u(0) + u(1) = 1. \end{cases} \tag{4.24}$$

Set

$$F(t, u(t)) = \frac{7t^2}{27(1 + |u(t)|)}[u(t), 33(1 + u(t))]; \ t \in [0, 1],$$

and $L(x, y) = -x - y + 1$; $x, y \in \mathbb{R}$.

First, we can see that $F : [0, 1] \times \mathbb{R} \to \mathcal{P}_{cp,cv}(\mathbb{R})$ is Carathéodory.

Next, we can verify that the hypothesis (4.11.2) is satisfied with the functions $v, w \in C([0, 1], \mathbb{R})$, defined by:

$$v(t) = t^{\frac{5}{2}}, \text{ and } w(t) = t^{\frac{3}{2}}.$$

Indeed, $L(v(0), v(1)) = 0 \le 0$ and

$$\left(^{C}D_{\frac{1}{4}}^{\frac{1}{2}}v\right)(t) = \frac{217}{27}t^2 \le \frac{7t^2}{27(1 + |v(t)|)}(31 + 31v(t)) \in F(t, v(t)).$$

Also, $L(w(0), w(1)) = 0 \ge 0$ and

$$\left(^{C}D_{\frac{1}{4}}^{\frac{1}{2}}w\right)(t) = \frac{7}{9}t \ge \frac{7}{9}t^2 = \frac{7t^2}{27(1 + |w(t)|)}(3 + 3w(t)) \in F(t, w(t)).$$

So v and w are lower and upper solutions, respectively, for problem (4.24) with $v \le w$.

Also, hypothesis (4.11.3) is satisfied. Indeed, L is continuous and

$$\frac{\partial L(x, y)}{\partial x} = \frac{\partial L(x, y)}{\partial y} = -1 < 0.$$

Finally, for each $u, \bar{u} \in \mathbb{R}$ and $t \in [0, 1]$, we have

$$H_d(F(t, u), F(t, \bar{u})) \leq \frac{7}{27}t^2 |u - \bar{u}|,$$

and

$$d(0, F(t, 0)) = \|F(t, 0)\|_{\mathcal{P}} \leq \frac{7}{27}t^2.$$

Hence the hypothesis (4.11.4) is satisfied with $l(t) = \frac{7}{27}t^2$.

Consequently, all conditions of Theorem 4.11 are satisfied and we conclude that our problem (4.24) has at least one solution u defined on $[0, 1]$, with $t^2\sqrt{t} \leq u(t) \leq t\sqrt{t}$.

4.4 Notes and remarks

This chapter contains the studies from Abbas et al. [31,47,126]. We refer the reader to the monographs [281,392,439], and the papers [53,81,107,109,232,234,411,441], for more information on the concepts studied in this chapter.

5

Ulam stability for fractional difference equations

We took as motivation the papers [15,48,96,103,232,234,273,345,347–350].

5.1 Existence and Ulam stability for implicit fractional q-difference equations

5.1.1 Introduction

In this section we discuss the existence, uniqueness, and Ulam–Hyers–Rassias stability of solutions for the following implicit fractional q-difference equation

$$({}^{c}D_q^{\alpha}u)(t) = f(t, u(t), ({}^{c}D_q^{\alpha}u)(t)), \ t \in I := [0, T], \tag{5.1}$$

with the initial condition

$$u(0) = u_0, \tag{5.2}$$

where $q \in (0, 1)$, $\alpha \in (0, 1]$, $T > 0$, $f : I \times \mathbb{R} \times \mathbb{R} \to \mathbb{R}$ is a given continuous function, and ${}^{c}D_q^{\alpha}$ is the Caputo fractional q-difference derivative of order α.

5.1.2 Existence results

Definition 5.1. By a solution of the problem (5.1)–(5.2) we mean a continuous function $u \in C(I)$ that satisfies Eq. (5.1) on I and the initial condition (5.2).

The following hypotheses will be used in the sequel.

(5.1.1) The function f satisfies the generalized Lipschitz condition:

$$|f(t, u_1, v_1) - f(t, u_2, v_2)| \le \phi_1(|u_1 - u_2|) + \phi_2(|v_1 - v_2|),$$

for $t \in I$ and $u_1, u_2, v_1, v_2 \in \mathbb{R}$, where ϕ_1 and ϕ_2 are comparison functions.

(5.1.2) There exist functions $p, d, r \in C(I, [0, \infty))$ with $r(t) < 1$ such that

$$|f(t, u, v)| \le p(t) + d(t)|u| + r(t)|v|, \text{ for each } t \in I \text{ and } u, v \in \mathbb{R}.$$

Set

$$p^* = \sup_{t \in I} p(t), \ d^* = \sup_{t \in I} d(t), \ r^* = \sup_{t \in I} r(t).$$

Fractional Difference, Differential Equations, and Inclusions. https://doi.org/10.1016/B978-0-44-323601-3.00012-5
Copyright © 2024 Elsevier Inc. All rights reserved, including those for text and data mining, AI training, and similar technologies.

First, we prove an existence and uniqueness result for the problem (5.1)–(5.2).

Theorem 5.2. *Assume that the hypothesis (5.1.1) holds. Then there exist a unique solution of problem (5.1)–(5.2) on I.*

Proof. By using Lemma 5.6, we transform the problem (5.1)–(5.2) into a fixed-point problem. Consider the operator $N : C(I) \to C(I)$ defined by

$$(Nu)(t) = u_0 + (I_q^\alpha g)(t); \ t \in I, \tag{5.3}$$

where $g \in C(I)$ such that

$$g(t) = f(t, u(t), g(t)), \text{ or } g(t) = f(t, u_0 + (I_q^\alpha g)(t), g(t)).$$

Let $u, v \in C(I)$. Then, for $t \in I$, we have

$$|(Nu)(t) - (Nv)(t)| \le \int_0^t \frac{(t - qs)^{(\alpha-1)}}{\Gamma_q(\alpha)} |g(s) - h(s)| d_q s, \tag{5.4}$$

where $g, h \in C(I)$ such that

$$g(t) = f(t, u(t), g(t)),$$

and

$$h(t) = f(t, v(t), h(t)).$$

From (5.1.1), we obtain

$$|g(t) - h(t)| \le \phi_1(|u(t) - v(t)|) + \phi_2(|g(t) - h(t)|).$$

Thus

$$|g(t) - h(t)| \le (Id - \phi_2)^{-1}\phi_1(|u(t) - v(t)|),$$

where Id is the identity function.

Set

$$L := \sup_{t \in I} \int_0^t \frac{(t - qs)^{(\alpha-1)}}{\Gamma_q(\alpha)} d_q s,$$

and $\phi := L(Id - \phi_2)^{-1}\phi_1$. From (5.4), we get

$$|(Nu)(t) - (Nv)(t)| \le \phi(|u(t) - v(t)|)$$
$$\le \phi(d(u, v)).$$

Hence we get

$$d(N(u), N(v)) \le \phi(d(u, v)).$$

Consequently, from Theorem 2.66, the operator N has a unique fixed point, which is the unique solution of the problem (5.1)–(5.2). $\square$

Theorem 5.3. *Assume that the hypothesis (5.1.2) holds. If*

$$r^* + Ld^* < 1,$$

then the problem (5.1)–(5.2) has at least one solution defined on I.

Proof. Let N be the operator defined in (5.3). Set

$$R \geq \frac{(1 - r^*)|u_0| + Lp^*}{1 - r^* - Ld^*},$$

and consider the closed and convex ball $B_R = \{u \in C(I) : \|u\|_\infty \leq R\}$.

Let $u \in B_R$. Then, for each $t \in I$, we have

$$|(Nu)(t)| \leq |u_0| + \int_0^t \frac{(t - qs)^{(\alpha-1)}}{\Gamma_q(\alpha)} |g(s)| d_q s,$$

where $g \in C(I)$ such that

$$g(t) = f(t, u(t), g(t)).$$

By using (5.1.2), for each $t \in I$, we have

$$\begin{aligned}|g(t)| &\leq p(t) + d(t)|u(t)| + r(t)|g(t)| \\ &\leq p^* + d^*\|u\|_\infty + r^*|g(t)| \\ &\leq p^* + d^*R + r^*|g(t)|.\end{aligned}$$

Thus

$$|g(t)| \leq \frac{p^* + d^*R}{1 - r^*}.$$

Hence

$$\|N(u)\|_\infty \leq |u_0| + \frac{L(p^* + d^*R)}{1 - r^*},$$

which implies that

$$\|N(u)\|_\infty \leq R.$$

This proves that N maps the ball B_R into B_R. We will show that the operator $N : B_R \to B_R$ is continuous and compact. The proof will be given in three steps.

Step 1. N is continuous.

Let $\{u_n\}_{n \in \mathbb{N}}$ be a sequence such that $u_n \to u$ in B_R. Then, for each $t \in I$, we have

$$|(Nu_n)(t) - (Nu)(t)| \leq \int_0^t \frac{(t - qs)^{(\alpha-1)}}{\Gamma_q(\alpha)} |(g_n(s) - g(s))| d_q s$$

where $g_n, g \in C(I)$ such that

$$g_n(t) = f(t, u_n(t), g_n(t)),$$

and

$$g(t) = f(t, u(t), g(t)).$$

Since $u_n \to u$ as $n \to \infty$ and f is continuous function, we get

$$g_n(t) \to g(t) \text{ as } n \to \infty, \text{ for each } t \in I.$$

Hence

$$\|N(u_n) - N(u)\|_\infty \leq \frac{p^* + d^* R}{1 - r^*} \|g_n - g\|_\infty \to 0 \text{ as } n \to \infty.$$

Step 2. $N(B_R)$ is bounded. This is clear since $N(B_R) \subset B_R$ and B_R is bounded.

Step 3. N maps bounded sets into equicontinuous sets in B_R.

Let $t_1, t_2 \in I$, such that $t_1 < t_2$ and let $u \in B_R$. Then we have

$$|(Nu)(t_1) - (Nu)(t_2)| \leq \int_0^{t_1} \frac{|(t_2 - qs)^{(\alpha-1)} - (t_1 - qs)^{(\alpha-1)}|}{\Gamma_q(\alpha)} |g(s)| d_q s$$
$$+ \int_{t_1}^{t_2} \frac{|(t_2 - qs)^{(\alpha-1)}|}{\Gamma_q(\alpha)} |g(s)| d_q s,$$

where $g \in C(I)$ such that $g(t) = f(t, u(t), g(t))$. Hence

$$|(Nu)(t_1) - (Nu)(t_2)| \leq \frac{p^* + d^* R}{1 - r^*} \int_0^{t_1} \frac{|(t_2 - qs)^{(\alpha-1)} - (t_1 - qs)^{(\alpha-1)}|}{\Gamma_q(\alpha)} d_q s$$
$$+ \frac{p^* + d^* R}{1 - r^*} \int_{t_1}^{t_2} \frac{|(t_2 - qs)^{(\alpha-1)}|}{\Gamma_q(\alpha)} d_q s.$$

As $t_1 \to t_2$, the right-hand side of the above inequality tends to zero.

As a consequence of the above three steps with the Arzelà–Ascoli theorem, we can conclude that $N : B_R \to B_R$ is continuous and compact.

From an application of Theorem 2.66, we deduce that N has at least one fixed point, which is a solution of problem (5.1)–(5.2).   □

5.2 Ulam stability results

In this section we are concerned with the generalized Ulam–Hyers–Rassias stability results of the problem (5.1)–(5.2).

Set $\Phi^* = \sup_{t \in I} \Phi(t)$ and

$$p_i^* = \sup_{t \in I} p_i(t), \ i \in \{1, 2, 3\}.$$

Theorem 5.4. *Assume that the following hypotheses hold.*

(5.4.1) *There exist functions p_1, p_2, $p_3 \in C(I, [0, \infty))$ with $p_3(t) < 1$ such that*

$$(1 + |u| + |v|)|f(t, u, v)| \le p_1(t)\Phi(t) + p_2(t)\Phi(t)|u| + p_3(t)|v|,$$

for each $t \in I$ and $u, v \in \mathbb{R}$,

(5.4.2) *There exists $\lambda_\Phi > 0$ such that for each $t \in I$, we have*

$$(I_q^\alpha \Phi)(t) \le \lambda_\Phi \Phi(t).$$

If

$$p_3^* + Lp_2^* \Phi^* < 1,$$

then the problem (5.1)–(5.2) has at least one solution and it is generalized Ulam–Hyers–Rassias stable.

Proof. Consider the operator N defined in (5.3). We can see that Hypothesis (5.4.1) implies (5.1.2) with $p \equiv p_1\Phi$, $d \equiv p_2\Phi$ and $r \equiv p_3$.

Let u be a solution of the inequality (5.16) and assume that v is a solution of problem (5.1)–(5.2). Thus we have

$$v(t) = u_0 + (I_q^\alpha h)(t),$$

where $h \in C(I)$ such that $h(t) = f(t, v(t), h(t))$.

From the inequality (5.16) for each $t \in I$, we have

$$|u(t) - u_0 - (I_q^\alpha g)(t)| \le (I_q^\alpha \Phi)(t),$$

where $g \in C(I)$ such that $g(t) = f(t, u(t), g(t))$.

From the hypotheses (5.4.1) and (5.4.2), for each $t \in I$, we get

$$|u(t) - v(t)| \le |u(t) - u_0 - (I_q^\alpha g)(t) + (I_q^\alpha (g - h))(t)|$$

$$\le (I_q^\alpha \Phi)(t) + \int_0^t \frac{(t - qs)^{(\alpha-1)}}{\Gamma_q(\alpha)}(|(g(s)| + |h(s))|)d_q s$$

$$\le (I_q^\alpha \Phi)(t) + \frac{p_1^* + p_2^*}{1 - p_3^*}(I_q^\alpha \Phi)(t)$$

$$\le \lambda_\phi \Phi(t) + 2\lambda_\phi \frac{p_1^* + p_2^*}{1 - p_3^*}\Phi(t)$$

$$\le \left[1 + 2\frac{p_1^* + p_2^*}{1 - p_3^*}\right]\lambda_\phi \Phi(t)$$

$$:= c_{f,\Phi}\Phi(t).$$

Hence the problem (5.1)–(5.2) is generalized Ulam–Hyers–Rassias stable. □

5.2.1 Examples

Example 1. Consider the following problem of implicit fractional $\frac{1}{4}$-difference equations

$$\begin{cases} (^{C}D_{\frac{1}{4}}^{\frac{1}{2}}u)(t) = f(t, u(t), (^{C}D_{\frac{1}{4}}^{\frac{1}{2}}u)(t)); \ t \in [0, 1], \\ u(0) = 1, \end{cases} \tag{5.5}$$

where

$$f(t, u(t), (^{C}D_{\frac{1}{4}}^{\frac{1}{2}}u)(t)) = \frac{t^2}{1 + |u(t)| + |^{C}D_{\frac{1}{4}}^{\frac{1}{2}}u(t)|} \left(e^{-7} + \frac{1}{e^{t+5}} \right) u(t); \ t \in [0, 1].$$

The hypothesis (5.1.1) is satisfied with

$$\phi_1(t) = \phi_2(t) = t^2 \left(e^{-7} + \frac{1}{e^{t+5}} \right) t.$$

Hence Theorem 5.2 implies that our problem (5.5) has a unique solution defined on $[0, 1]$.

Example 2. Consider now the following problem of implicit fractional $\frac{1}{4}$-difference equations

$$\begin{cases} (^{C}D_{\frac{1}{4}}^{\frac{1}{2}}u)(t) = f(t, u(t), (^{C}D_{\frac{1}{4}}^{\frac{1}{2}}u)(t)); \ t \in [0, 1], \\ u(0) = 2, \end{cases} \tag{5.6}$$

where

$$\begin{cases} f(t, x, y) = \frac{t^2}{1+|x|+|y|} \left(e^{-7} + \frac{1}{e^{t+5}} \right)(t^2 + xt^2 + y); \ t \in (0, 1], \\ f(0, x, y) = 0. \end{cases}$$

The hypothesis (5.4.1) is satisfied with $\Phi(t) = t^2$ and $p_i(t) = \left(e^{-7} + \frac{1}{e^{t+5}} \right) t; \ i \in \{1, 2, 3\}$. Hence Theorem 5.3 implies that our problem (5.6) has at least a solution defined on $[0, 1]$.

Also, the hypothesis (5.4.2) is satisfied. Indeed, for each $t \in (0, 1]$, there exists a real number $0 < \epsilon < 1$ such that $\epsilon < t \leq 1$, and

$$\begin{aligned} (I_q^\alpha \Phi)(t) &\leq \frac{t^2}{\epsilon^2(1 + q + q^2)} \\ &\leq \frac{1}{\epsilon^2} \Phi(t) \\ &= \lambda_\Phi \Phi(t). \end{aligned}$$

Consequently, Theorem 5.4 implies that the problem (5.6) is generalized Ulam–Hyers–Rassias stable.

5.3 Implicit fractional q-difference equations: analysis and stability

5.3.1 Introduction

In this section we discuss the existence and Ulam–Hyers–Rassias stability of random solutions for the following random implicit fractional q-difference equation

$$(^cD_q^\alpha u)(t, w) = f(t, u(t, w), (^cD_q^\alpha u)(t, w), w); \; t \in I := [0, T], \; w \in \Omega, \qquad (5.7)$$

with the initial condition

$$u(0, w) = u_0(w); \; w \in \Omega, \qquad (5.8)$$

where $q \in (0, 1)$, $\alpha \in (0, 1]$, $T > 0$, $(\Omega, \mathcal{A})$ is a measurable space, $u_0 : \Omega \to \mathbb{R}$ is a measurable function, $f : I \times \mathbb{R}^2 \times \Omega \to \mathbb{R}$ is a given function, and $^cD_q^\alpha$ is the Caputo fractional q-difference derivative of order α.

Next, under Carathéodory and certain monotonicity conditions, we prove some existence of solutions and extremal solutions for the following implicit fractional q-difference equations in Banach algebras

$$^cD_q^\alpha \left(\frac{u(t)}{h(t, u(t))} \right) = f \left(t, u(t), ^cD_q^\alpha \left(\frac{u(t)}{h(t, u(t))} \right) \right); \; t \in I := [0, T], \qquad (5.9)$$

with the initial condition

$$u(0) = u_0 \in \mathbb{R}, \qquad (5.10)$$

where $h : I \times \mathbb{R} \to \mathbb{R}^*$, $f : I \times \mathbb{R}^2 \to \mathbb{R}$ are given functions, and $\mathbb{R}^* = \mathbb{R} - \{0\}$.

Finally, we discuss the existence of random solutions for the following implicit random fractional q-difference equation in Banach algebras

$$^cD_q^\alpha \left(\frac{u(t, w)}{h(t, u(t, w), w)} \right) = f \left(t, u(t, w), ^cD_q^\alpha \left(\frac{u(t, w)}{h(t, u(t, w), w)} \right), w \right); \; t \in I := [0, T], \; w \in \Omega,$$
$$(5.11)$$

with the initial condition

$$u(0, w) = u_0(w) \in \mathbb{R}; \; w \in \Omega, \qquad (5.12)$$

where $(\Omega, \mathcal{A})$ is a measurable space and $h : I \times \mathbb{R} \times \Omega \to \mathbb{R}^*$, $f : I \times \mathbb{R}^2 \times \Omega \to \mathbb{R}$ are given functions.

5.3.2 Random solutions and stability results

In this section we are concerned with the existence of random solutions and the Ulam–Hyers–Rassias stability of the problem (5.7)–(5.8).

Definition 5.5. A random solution of the problem (5.7)–(5.8) is a measurable function $u : \Omega \to C(I)$ that satisfies Eq. (5.7) on I and the initial condition (5.8).

Lemma 5.6. *Let $f : I \times \mathbb{R}^2 \times \Omega \to \mathbb{R}$ be a random Carathéodory. Then the problem (5.7)–(5.8) is equivalent to the problem of obtaining the solutions of the integral equation*

$$g(t, w) = f(t, u_0(w) + (I_q^\alpha g)(t, w), g(t, w), w),$$

and if $g(\cdot, w) \in C(I)$ is the solution of this equation, then

$$u(t, w) = u_0(w) + (I_q^\alpha g)(t, w).$$

In the sequel we employ the following random fixed-point theorem.

Theorem 5.7. *(Itoh [270]) Let X be a non-empty, closed convex bounded subset of the separable Banach space E and let $N : \Omega \times X \to X$ be a compact and continuous random operator. Then the random equation $N(w)u = u$ has a random solution.*

The following hypotheses will be used in the sequel.

(5.7.1) The function f is random Carathéodory on $I \times \mathbb{R}^2 \times \Omega$,
(5.7.2) There exists a measurable and bounded function $p : \Omega \to L^\infty(I, \mathbb{R}_+)$ such that

$$|f(t, u, v, w)| \le p(t, w); \text{ for a.e. } t \in I, \text{ and each } u, v \in \mathbb{R}, \ w \in \Omega.$$

Set

$$p^* = \sup_{w \in \Omega} \|p(w)\|_{L^\infty}, \quad u_0^* = \sup_{w \in \Omega} |u_0(w)|,$$

and

$$L := \frac{T^{(\alpha)}}{\Gamma_q(1 + \alpha)}.$$

Theorem 5.8. *Assume that the hypotheses (5.7.1) and (5.7.2) hold. Then the problem (5.7)–(5.8) has at least one random solution defined on $I \times \Omega$.*

Proof. Define a mapping $N : \Omega \times C(I) \to C(I)$ by:

$$(N(w)u)(t) = u_0(w) + \int_0^t \frac{(t - qs)^{(\alpha-1)}}{\Gamma_q(\alpha)} g(s, w) d_q s, \tag{5.13}$$

where $g : \Omega \to C(I)$ such that

$$g(t, w) = f(t, u(t, w), g(t, w), w).$$

The map u_0 is measurable for all $w \in \Omega$. As the indefinite integral is continuous on I, then $N(w)$ defines a mapping $N : \Omega \times C(I) \to C(I)$. Thus u is a random solution for the problem (5.7)–(5.8) if and only if $u = N(w)u$.

Next, for any $u \in C(I)$, and each $t \in I$ and $w \in \omega$, we have

$$|(N(w)u)(t)| \leq |u_0(w)| + \int_0^t \frac{(t - qs)^{(\alpha-1)}}{\Gamma_q(\alpha)} |g(s, w)| d_q s,$$

where $g : \Omega \to C(I)$ such that

$$g(t, w) = f(t, u(t, w), g(t, w), w).$$

By using (5.7.2), we have

$$|g(t, w)| \leq p(t, w)$$
$$\leq p^*.$$

Thus

$$|g(t, w)| \leq p^*.$$

Hence

$$\|(Nw)(u)\|_\infty \leq u_0^* + Lp^* := R.$$

This proves that $N(w)(B_R) \subset B_R$, where $B_R := B(0, R) = \{u \in C(I) : \|u\|_\infty \leq R\}$. We will show that the operator $N : \Omega \times B_R \to B_R$ satisfies all the assumptions of Theorem 5.7. The proof will be given in several steps.

Step 1. $N(w)$ is a random operator on $\Omega \times B_R$ into B_R.
Since $f(t, u, v, w)$ is random Carathéodory, the map $w \to f(t, u, v, w)$ is measurable. Similarly, the product $\frac{(t-qs)^{(\alpha-1)}}{\Gamma_q(\alpha)} f(s, u(s, w), ({}^C D_q^\alpha u)(t, w), w)$ of a continuous and a measurable function is again measurable. Further, the integral is a limit of a finite sum of measurable functions, therefore, the map

$$w \mapsto u_0(w) + \int_0^t \frac{(t - qs)^{(\alpha-1)}}{\Gamma_q(\alpha)} f(s, u(s, w), ({}^C D_q^\alpha u)(t, w), w) d_q s,$$

is measurable. As a result, $N(w)$ is a random operator on $\Omega \times B_R$ into B_R.

Step 2. $N(w)$ *is continuous.*
Let $\{u_n\}_{n \in \mathbb{N}}$ be a sequence such that $u_n \to u$ in B_R. Then, for each $t \in I$, and $w \in \Omega$, we have

$$|(N(w)u_n)(t) - (N(w)u)(t)| \leq \int_0^t |f(s, u_n(s, w), ({}^C D_q^\alpha u_n)(t, w), w)$$
$$- f(s, u(s, w), ({}^C D_q^\alpha u)(t, w), w)| \frac{(t - qs)^{(\alpha-1)} d_q s}{\Gamma_q(\alpha)}. \qquad (5.14)$$

Since $u_n \to u$ as $n \to \infty$ and f is random Carathéodory, then by the Lebesgue dominated convergence theorem, Eq. (5.14) implies

$$\|N(w)u_n - N(w)u\|_\infty \to 0 \quad \text{as } n \to \infty.$$

Step 3. $N(w)B_R$ *is uniformly bounded.*
This is clear since $N(w)B_R \subset B_R$ and B_R is bounded.

Step 4. $N(w)B_R$ *is equicontinuous.*
Let $t_1, t_2 \in I, t_1 < t_2$ and let $u \in B_R$. Then, for each $w \in \Omega$, we have

$$|(Nw)u(t_1) - (Nw)u(t_2)| \leq \int_0^{t_1} \frac{|(t_2 - qs)^{(\alpha-1)} - (t_1 - qs)^{(\alpha-1)}|}{\Gamma_q(\alpha)} |g(s, w)| d_q s$$

$$+ \int_{t_1}^{t_2} \frac{|(t_2 - qs)^{(\alpha-1)}|}{\Gamma_q(\alpha)} |g(s)| d_q s,$$

where $g(w) \in C(I)$ such that $g(t, w) = f(t, u(t, w), g(t, w), w)$. Hence

$$|(Nw)u(t_1) - (Nw)u(t_2)| \leq p^* \int_0^{t_1} \frac{|(t_2 - qs)^{(\alpha-1)} - (t_1 - qs)^{(\alpha-1)}|}{\Gamma_q(\alpha)} d_q s$$

$$+ p^* \int_{t_1}^{t_2} \frac{|(t_2 - qs)^{(\alpha-1)}|}{\Gamma_q(\alpha)} d_q s.$$

As $t_1 \to t_2$, the right-hand side of the above inequality tends to zero.

As a consequence of steps 1 to 4 together with the Arzelà–Ascoli theorem, we can conclude that $N : \Omega \times B_R \to B_R$ is continuous and compact. From an application of Theorem 5.7, we deduce that the operator equation $N(w)u = u$ has a random solution. This implies that the problem (5.7)–(5.8) has a random solution.   □

Now, we are concerned with the generalized Ulam–Hyers–Rassias stability of our problem (5.7)–(5.8).

Let $\epsilon > 0$ and $\Phi : I \times \Omega \to \mathbb{R}_+$ be a measurable function. We consider the following inequalities

$$|(^C D_q^\alpha u)(t, w) - f(t, u(t, w), (^C D_q^\alpha u)(t, w), w)| \leq \epsilon; \; t \in I. \tag{5.15}$$

$$|(^C D_q^\alpha u)(t, w) - f(t, u(t, w), (^C D_q^\alpha u)(t, w), w)| \leq \Phi(t, w); \; t \in I. \tag{5.16}$$

$$|(^C D_q^\alpha u)(t, w) - f(t, u(t, w), (^C D_q^\alpha u)(t, w), w)| \leq \epsilon \Phi(t, w); \; t \in I. \tag{5.17}$$

Definition 5.9 ([70,345]). The problem (5.7)–(5.8) is Ulam–Hyers stable if there exists a real number $c_f > 0$ such that for each $\epsilon > 0$ and for each solution $u(w) \in C(I)$ of the inequality (5.15) there exists a random solution $v(w) \in C(I)$ of (5.7)–(5.8) with

$$|u(t, w) - v(t, w)| \leq \epsilon c_f; \; t \in I, \; w \in \Omega.$$

Definition 5.10 ([70,345]). The problem (5.7)–(5.8) is generalized Ulam–Hyers stable if there exists $c_f : C(\mathbb{R}_+, \mathbb{R}_+)$ with $c_f(0) = 0$ such that for each $\epsilon > 0$ and for each solution $u(w) \in C(I)$ of the inequality (5.15) there exists a random solution $v(w) \in C(I)$ of (5.7)–(5.8) with

$$|u(t, w) - v(t, w)| \leq c_f(\epsilon); \; t \in I, \; w \in \Omega.$$

Definition 5.11 ([70,345]). The problem (5.7)–(5.8) is Ulam–Hyers–Rassias stable with respect to Φ if there exists a real number $c_{f,\Phi} > 0$ such that for each $\epsilon > 0$ and for each solution $u(w) \in C(I)$ of the inequality (5.17) there exists a random solution $v(w) \in C(I)$ of (5.7)–(5.8) with

$$|u(t, w) - v(t, w)| \leq \epsilon c_{f,\Phi} \Phi(t, w); \ t \in I, \ w \in \Omega.$$

Definition 5.12 ([70,345]). The problem (5.7)–(5.8) is generalized Ulam–Hyers–Rassias stable with respect to Φ if there exists a real number $c_{f,\Phi} > 0$ such that for each solution $u(w) \in C(I)$ of the inequality (5.16) there exists a random solution $v(w) \in C(I)$ of (5.7)–(5.8) with

$$|u(t, w) - v(t, w)| \leq c_{f,\Phi} \Phi(t, w); \ t \in I, \ w \in \Omega.$$

Remark 5.13. It is clear that

  (i) Definition 5.9 $\Rightarrow$ Definition 5.10,
 (ii) Definition 5.11 $\Rightarrow$ Definition 5.12,
(iii) Definition 5.11 for $\Phi(\cdot, w) = 1 \Rightarrow$ Definition 5.9.

One can have similar remarks for the inequalities (5.15) and (5.17).

Theorem 5.14. *Assume that the hypotheses* (5.7.1)–(5.7.2) *and the following hypotheses hold.*

(5.14.1) *There exists $\lambda_\Phi > 0$ such that for each $t \in I$, and $w \in \Omega$, we have*

$$(I_q^\alpha \Phi)(t, w) \leq \lambda_\Phi \Phi(t, w),$$

(5.14.2) *There exists $p_1, d_1 \in C(I, \mathbb{R}_+)$ such that for each $t \in I$, and $w \in \Omega$, we have such that*

$$p(t, w) \leq p_1(t)\Phi(t, w), \ and \ d(t, w) \leq d_1(t)\Phi(t, w).$$

Then the problem (5.7)–(5.8) is generalized Ulam–Hyers–Rassias stable.

Proof. Consider the operator N defined in (5.13). Assume that v is a random solution of problem (5.7)–(5.8), then we have

$$v(t, w) = u_0(w) + \int_0^t \frac{(t - qs)^{(\alpha-1)}}{\Gamma_q(\alpha)} h(s) d_q s,$$

where $h(w) \in C(I)$ such that

$$h(t, w) = f(t, v(t, w), h(t, w), w).$$

If u is a random solution of the inequality (5.16), then for each $t \in I$, and $w \in \Omega$, we have

$$\left| u(t, w) - u_0(w) - \int_0^t \frac{(t - qs)^{(\alpha-1)}}{\Gamma_q(\alpha)} g(s) d_q s \right| \leq (I_q^\alpha \Phi)(t, w),$$

where $g(w) \in C(I)$ such that

$$g(t, w) = f(t, u(t, w), g(t, w), w).$$

Set

$$p_1^* = \sup_{t \in I} p_1(t), \ and \ d_1^* = \sup_{t \in I} d_1(t).$$

From hypotheses (5.14.1) and (5.14.2), for each $t \in I$, and $w \in \Omega$, we get

$$|u(t, w) - v(t, w)| \le \left| u(t, w) - u_0(w) - \int_0^t \frac{(t - qs)^{(\alpha - 1)}}{\Gamma_q(\alpha)} g(s) d_q s \right|$$

$$+ \int_0^t \frac{(t - qs)^{(\alpha - 1)}}{\Gamma_q(\alpha)} |g(s) - h(s)| d_q s$$

$$\le (I_q^\alpha \Phi)(t, w) + \int_0^t \frac{(t - qs)^{(\alpha - 1)}}{\Gamma_q(\alpha)} 2(p_1^* + d_1^*) \Phi(s, w) d_q s$$

$$\le (I_q^\alpha \Phi)(t) + 2(p_1^* + d_1^*)(I_q^\alpha \Phi)(t, w)$$

$$\le [1 + 2(p_1^* + d_1^*)] \lambda_\phi \Phi(t, w)$$

$$:= c_{f,\Phi} \Phi(t, w).$$

Hence the problem (5.7)–(5.8) is generalized Ulam–Hyers–Rassias stable. $\qquad \square$

5.3.3 Existence of solutions and extremal solutions

In this section we are concerned with the existence of solutions and extremal solutions of the problem (5.9)–(5.10).

Definition 5.15. By a solution of the problem (5.9)–(5.10) we mean a continuous function $u \in C(I)$ such that the function $t \mapsto \left(\frac{u(t)}{h(t, u(t))} \right)$ is continuous, and u satisfies Eq. (5.9) on I and the initial condition (5.10).

Lemma 5.16. *Let* $f : I \times \mathbb{R}^2 \to \mathbb{R}, h : I \times \mathbb{R} \to \mathbb{R}^*$ *such that* $f(\cdot, u, v) \in C(I)$, *for each* $u, v \in \mathbb{R}$, *and the function* $t \mapsto \left(\frac{u(t)}{h(t, u(t))} \right)$ *is continuous. Then the problem (5.7)–(5.8) is equivalent to the problem of obtaining the solutions of the integral equation*

$$g(t) = f(t, h(t, u(t))(\mu_0 + (I_q^\alpha g)(t)), g(t)),$$

where

$$\mu_0 = \frac{u_0}{h(0, u(0))},$$

and if $g(\cdot) \in C(I)$, *is the solution of this equation, then*

$$u(t) = h(t, u(t)) \left(\mu_0 + (I_q^\alpha g)(t) \right).$$

The nonlinear alternative of Schaefer type proved by Dhage [224] is embodied in the following theorem.

Theorem 5.17. *(Dhage [224]) Let X be a Banach algebra and let $A, B : X \to X$ be two operators satisfying:*

(a) *A is Lipschitz with a Lipschitz constant α;*
(b) *B is compact and continuous;*
(c) *$\alpha M < 1$, where $M = \|B(X)\| := \sup\{\|Bz\| : z \in X\}$.*

Then either

(i) *the equation $\lambda[Au\,Bu] = u$ has a solution for $0 < \lambda < 1$, or*
(ii) *the set $\mathcal{E} = \{u \in X : \lambda[Au\,Bu] = u, \ 0 < \lambda < 1\}$ is unbounded.*

We use the following fixed-point theorem of Dhage [224] for proving the existence of extremal solutions for our problem under certain monotonicity conditions.

Theorem 5.18. *(Dhage [224]) Let K be a cone in a Banach algebra X and let $v, w \in X$. Suppose that $A, B : [v, w] \to K$ are two operators such that*

(a) *A is completely continuous;*
(b) *B is totally bounded;*
(c) *$Au\,Bz \in [v, w]$ for all $u, z \in [v, w]$;*
(d) *A and B are nondecreasing.*

Furthermore, if the cone K is positive and normal, then the operator equation $Au\,Bu = u$ has a least and a greatest positive solution in $[v, w]$.

Theorem 5.19. *(Dhage [224]) Let K be a cone in a Banach algebra X and let $v, w \in X$. Suppose that $A, B : [v, w] \to K$ are two operators such that*

(a) *A is Lipschitz with a Lipschitz constant α;*
(b) *B is totally bounded;*
(c) *$Au\,Bz \in [v, w]$ for all $u, z \in [v, w]$;*
(d) *A and B are nondecreasing.*

Furthermore, if the cone K is positive and normal, then the operator equation $Au\,Bu = u$ has least and a greatest positive solution in $[v, w]$, whenever $\alpha M < 1$, where $M = \|B([v, w])\| := \sup\{\|Bu\| : u \in [v, w]\}$.

Remark 5.20. Note that hypothesis (c) of Theorems 5.18 and 5.19 holds if the operators A and B are positive monotonically increasing and there exist elements v and w in X such that $v \le Av\,Bv$ and $Aw\,Bw \le w$.

Theorem 5.21. *Assume that following hypotheses hold.*

(5.21.1) *The function h is continuous on $I \times \mathbb{R}$.*

(5.21.2) *There exists a function* $\alpha \in C(I, \mathbb{R}_+)$ *such that*

$$|h(t, u) - h(t, \overline{u})| \leq \alpha(t)|u - \overline{u}|, \quad \text{for all } t \in I, \text{ for all } u, \overline{u} \in \mathbb{R}.$$

(5.21.3) *The function* f *is Carathéodory and there exists* $K \in L^\infty(I, \mathbb{R}_+)$ *such that*

$$|f(t, u, v)| \leq K(t), \quad \text{a.e. } t \in I, \text{ for all } u, v \in \mathbb{R}.$$

Let

$$K^* := \|K\|_{L^\infty}, \text{ and } L := \frac{T^{(\alpha)}}{\Gamma_q(1 + \alpha)}.$$

If

$$\|\alpha\|_\infty(|\mu_0| + LK^*) < 1, \tag{5.18}$$

then there exists at least one solution of the problem (5.9)–(5.10) defined on I.

Proof. Define the operators A and B on $C(I)$ by

$$(Au)(t) = h(t, u(t)); \quad t \in I, \tag{5.19}$$

$$(Bu)(t) = \mu_0 + (I_q^\alpha g)(t); \quad t \in I, \tag{5.20}$$

where $g \in C(I)$, such that

$$g(t) = f(t, u(t), g(t)) = f(t, h(t, u(t))(\mu_0 + (I_q^\alpha g)(t)), g(t)).$$

Clearly A and B define the operators $A, B : C(I) \to C(I)$. Now solving (5.9)–(5.10) is equivalent to solving the operator equation

$$(Au)(t)(Bu)(t) = u(t); \quad t \in I. \tag{5.21}$$

We show that operators A and B satisfy all the assumptions of Theorem 5.17. First, we will show that A is a Lipschitz. Let $u_1, u_2 \in X$. Then by (5.21.2),

$$|Au_1(t) - Au_2(t)| = |f(t, u_1(t)) - f(t, u_2(t))|$$
$$\leq \alpha(t)|u_1(t) - u_2(t)|$$
$$\leq \|\alpha\|_\infty\|u_1 - u_2\|_\infty.$$

Taking the maximum over t, in the above inequality yields

$$\|Au_1 - Au_2\|_\infty \leq \|\alpha\|_\infty\|u_1 - u_2\|_\infty,$$

and so A is a Lipschitz with a Lipschitz constant $\|\alpha\|_\infty$.

Next, we show that B is a compact operator on $C(I)$. Let $\{u_n\}$ be a sequence in $C(I)$. From (5.21.3) it follows that

$$\|Bu_n\|_\infty \leq |\mu_0| + LK^*.$$

As a result $\{Bu_n : n \in \mathbb{N}\}$ is a uniformly bounded set in $C(I)$. Let $t_1, t_2 \in I$. Then

$$|Bu_n(t_1) - Bu_n(t_2)| \leq K^* \int_0^{t_1} \frac{|(t_2 - qs)^{(\alpha-1)} - (t_1 - qs)^{(\alpha-1)}|}{\Gamma_q(\alpha)} d_q s$$

$$+ K^* \int_{t_1}^{t_2} \frac{|(t_2 - qs)^{(\alpha-1)}|}{\Gamma_q(\alpha)} d_q s$$

$$\to 0, \quad \text{as } t_1 \to t_2.$$

From this we conclude that $\{Bu_n : n \in \mathbb{N}\}$ is an equicontinuous set in $C(I)$. Hence $B : C(I) \to C(I)$ is compact by Arzelà–Ascoli theorem. Moreover,

$$M = \|B(X)\|$$

$$\leq |\mu_0| + \int_0^t \frac{(t - qs)^{(\alpha-1)}}{\Gamma_q(\alpha)} |g(s)| d_q s$$

$$\leq |\mu_0| + Lh^*,$$

and so

$$\alpha M \leq \|\alpha\|_\infty (|\mu_0| + LK^*) < 1,$$

by assumption (5.18). To finish, it remain to show that either the conclusion (i) or the conclusion (ii) of Theorem 5.17 holds. We now will show that the conclusion (ii) is not possible. Let $u \in C(I)$ be any solution to (5.9)–(5.10). Then, for any $\lambda \in (0, 1)$, we have

$$u(t) = \lambda[h(t, u(t))] \left(\mu_0 + \int_0^t \frac{(t - qs)^{(\alpha-1)}}{\Gamma_q(\alpha)} g(s) d_q s \right); \; t \in I.$$

Therefore

$$|u(t)| \leq |f(t, u(t))| \left((|\mu_0| + \int_0^t \frac{(t - qs)^{(\alpha-1)}}{\Gamma_q(\alpha)} |g(s)| d_q s \right)$$

$$\leq (|f(t, u(t)) - f(t, 0)| + |f(t, 0)|) \left((|\mu_0| + \int_0^t \frac{(t - qs)^{(\alpha-1)}}{\Gamma_q(\alpha)} K(s) d_q s \right)$$

$$\leq (\|\alpha\|_\infty |u(t)| + f^*)(|\mu_0| + LK^*)$$

$$\leq (\|\alpha\|_\infty \|u\|_\infty + f^*)(|\mu_0| + LK^*),$$

where $f^* = \sup_{t \in I} |f(t, 0)|$. Consequently

$$\|u\|_\infty \leq \frac{f^*(|\mu_0| + LK^*)}{1 - \|\alpha\|_\infty (|\mu_0| + LK^*)} := \ell.$$

Thus the conclusion (ii) of Theorem 5.17 does not hold. Therefore the IVP (5.9)–(5.10) has a solution on I. $\qquad \square$

Now, we give some results about the existence of extremal solutions.

We equip the space $C(I)$ with the order relation $\leq$ with the help of the cone defined by

$$K = \{u \in C(I): u(t) \geq 0, \quad \forall t \in I\}.$$

Thus $u \leq \bar{u}$ if and only if $u(t) \leq \bar{u}(t)$ for each $t \in I$. It is well-known that the cone K is positive and normal in $C(I)$ ([262]).

If $\underline{u}, \bar{u} \in C(I)$ and $\underline{u} \leq \bar{u}$, we put

$$[\underline{u}, \bar{u}] = \{u \in C(I) : \underline{u} \leq u \leq \bar{u}\}.$$

Definition 5.22. A function $\beta : I \times \mathbb{R} \to \mathbb{R}$ is called *Chandrabhan* if

(i) the function $t \to \beta(t, u)$ is measurable for each $u \in \mathbb{R}$,
(ii) the function $u \to \beta(t, u)$ is nondecreasing for almost each $t \in I$.

Definition 5.23. A function $u(\cdot, \cdot) \in C(I)$ is said to be a lower solution of (5.9)–(5.10) if we have

$$^c D_q^\alpha \left(\frac{u(t)}{h(t, u(t))} \right) \leq f \left(t, u(t), {}^c D_q^\alpha \left(\frac{u(t)}{h(t, u(t))} \right) \right); \quad t \in I,$$

$$u(0) \leq u_0.$$

Similarly the function $u(\cdot, \cdot) \in C(I)$ is said to be an upper solution of (5.9)–(5.10) if we have

$$^c D_q^\alpha \left(\frac{u(t)}{h(t, u(t))} \right) \geq f \left(t, u(t), {}^c D_q^\alpha \left(\frac{u(t)}{h(t, u(t))} \right) \right); \quad t \in I,$$

$$u(0) \geq u_0.$$

Definition 5.24. A solution u_M of the problem (5.9)–(5.10) is said to be maximal if for any other solution u to the problem (5.9)–(5.10) one has $u(t) \leq u_M(t)$, for all $t \in I$. Again, a solution u_m of the problem (5.9)–(5.10) is said to be minimal if $u_m(t) \leq u(t)$, for all $t \in I$ where u is any solution of the problem (5.9)–(5.10) on I.

The following hypotheses will be used in the sequel.

(5.24.1) $h : I \times \mathbb{R}_+ \to \mathbb{R}_+^*$, $f : I \times \mathbb{R}_+ \times \mathbb{R}_+ \to \mathbb{R}_+$, and $\mu_0 \geq 0$.
(5.24.2) The functions h and f are Chandrabhan.
(5.24.3) There exists a function $\tilde{K} \in L^\infty(I, \mathbb{R}_+)$ such that

$$|f(t, u, v)| \leq \tilde{K}(t), \quad \text{a.e. } t \in I, \text{ for all } u, v \in \mathbb{R}.$$

(5.24.4) The problem (5.9)–(5.10) has a lower solution $\underline{u}$ and an upper solution $\bar{u}$ with $\underline{u} \leq \bar{u}$.

Let

$$\tilde{K}^* = \|\tilde{K}\|_{L^\infty}.$$

Theorem 5.25. *Assume that hypotheses* (5.21.2), (5.24.1)–(5.24.4) *hold. If*

$$\|\alpha\|_\infty(|\mu_0| + L\tilde{K}^*) < 1,$$

then the problem (5.9)–(5.10) *has a minimal and a maximal positive solution on I.*

Proof. Consider a closed interval $[\underline{u}, \overline{u}]$ in $C(I)$, which is well defined in view of hypothesis (5.24.1). Define the operators $A, B : [\underline{u}, \overline{u}] \to C(I)$ by (5.19) and (5.20), respectively. Clearly A and B define the operators $A, B : [\underline{u}, \overline{u}] \to K$.

Now solving (5.9)–(5.10) is equivalent to solving the operator equation

$$Au(t)\, Bu(t) = u(t), \quad t \in I. \tag{5.22}$$

We show that the operators A and B satisfy all the assumptions of Theorem 5.19. As in Theorem 5.21, we can prove that A is Lipschitz with a Lipschitz constant $\|\alpha\|_\infty$ and B is completely continuous operator on $[\underline{u}, \overline{u}]$.

Now hypothesis (5.24.2) implies that A and B are nondecreasing on $[\underline{u}, \overline{u}]$. To see this, let $u_1, u_2 \in [\underline{u}, \overline{u}]$ be such that $u_1 \le u_2$. Then by (5.24.2), we get

$$(Au_1)(t) = h(t, u_1(t)) \le h(t, u_2(t)) = (Au_2)(t), \quad \forall t \in I,$$

and

$$\begin{aligned}
(Bu_1)(t) &= \mu_0 + \int_0^t \frac{(t-qs)^{(\alpha)}}{\Gamma_q(\alpha)} f(s, u_1(s), ({}^C D_q^\alpha u_1)(s)) d_q s \\
&\le \mu_0 + \int_0^t \frac{(t-qs)^{(\alpha)}}{\Gamma_q(\alpha)} f(s, u_2(s), ({}^C D_q^\alpha u_2)(s)) d_q s \\
&= (Bu_2)(t), \quad \forall t \in I.
\end{aligned}$$

So A and B are nondecreasing operators on $[\underline{u}, \overline{u}]$. Again, the hypothesis (5.24.4) imply

$$\begin{aligned}
\underline{u}(t) &= (h(t, \underline{u}(t)))(\mu_0 + \int_0^t \frac{(t-qs)^{(\alpha)}}{\Gamma_q(\alpha)} f(s, \underline{u}(s), ({}^C D_q^\alpha \underline{u})(s)) d_q s) \\
&\le (h(t, \underline{u}(t)))(\mu_0 + \int_0^t \frac{(t-qs)^{(\alpha)}}{\Gamma_q(\alpha)} f(s, v(s), ({}^C D_q^\alpha v)(s)) d_q s) \\
&\le (h(t, \underline{u}(t)))(\mu_0 + \int_0^t \frac{(t-qs)^{(\alpha)}}{\Gamma_q(\alpha)} f(s, \overline{u}(s), ({}^C D_q^\alpha \overline{u})(s)) d_q s) \\
&\le \overline{u}(t),
\end{aligned}$$

for all $t \in I$ and $v \in [\underline{u}, \overline{u}]$. As a result

$$\underline{u}(t) \le (Av)(t)(Bv)(t) \le \overline{u}(t), \quad \forall t \in I \text{ and } v \in [\underline{u}, \overline{u}].$$

Hence $Av\, Bv \in [\underline{u}, \overline{u}]$, for all $v \in [\underline{u}, \overline{u}]$.

Notice for any $u \in [\underline{u}, \overline{u}]$,

$$M = \|B([\underline{u}, \overline{u}])\|$$

$$\leq |\mu_0| + \int_0^t \frac{(t - qs)^{(\alpha)}}{\Gamma_q(\alpha)} f(s, u(s), ({}^C D_q^\alpha u)(s)) d_q s)$$

$$\leq |\mu_0| + L\tilde{K}^*,$$

and so

$$\alpha M \leq \|\alpha\|_\infty (|\mu_0| + L\tilde{K}^*) < 1.$$

Thus all conditions of Theorem 5.19 are satisfied, and so the operator equation (5.20) has a least and a greatest solution in $[\underline{u}, \overline{u}]$. This further implies that the problem (5.9)–(5.10) has a minimal and a maximal positive solution on I. $\qquad\square$

Theorem 5.26. *Assume that hypotheses (5.21.1), (5.24.1)–(5.24.4) hold. Then the problem (5.9)–(5.10) has a minimal and a maximal positive solution on I.*

Proof. Consider the order interval $[\underline{u}, \overline{u}]$ in $C(I)$ and define the operators A and B on $[\underline{u}, \overline{u}]$ by (5.19) and (5.20), respectively. Then the problem (5.9)–(5.10) is transformed into an operator equation $(Au)(t)(Bu)(t) = u(t)$ for all $t \in I$ in the Banach algebra $C(I)$. Notice that (5.24.1) implies $A, B : [\underline{u}, \overline{u}] \to K$. Since the cone K in $C(I)$ is normal, then $[\underline{u}, \overline{u}]$ is a norm bounded set in $C(I)$.

Next we show that A is completely continuous on $[\underline{u}, \overline{u}]$. Now the cone K in $C(I)$ is normal, so the order interval $[\underline{u}, \overline{u}]$ is norm-bounded. Hence there exists a constant $\rho > 0$ such that $|u| \leq \rho$ for all $u \in [\underline{u}, \overline{u}]$. Since h is continuous on the compact set $I \times [-\rho, \rho]$, then it attains its maximum, say M. Therefore, for any subset S of $[\underline{u}, \overline{u}]$, we have

$$\|A(S)\| = \sup\{|Au| : u \in S\}$$

$$= \sup\{\sup_{t \in I} |h(t, u(t))| : u \in S\}$$

$$\leq \sup\{\sup_{t \in I} |h(t, u(t))| : u \in [-\rho, \rho]\}$$

$$\leq M.$$

This shows that $A(S)$ is a uniformly bounded subset of $C(I)$.

We note that the function $h(t, u)$ is uniformly continuous on $I \times [-\rho, \rho]$. Therefore, for any $t_1, t_2 \in I$, we have

$$|h(t_1, u) - h(t_2, u)| \to 0 \quad \text{as } t_1 \to t_2,$$

for all $u \in [-\rho, \rho]$. Similarly, for any $u_1, u_2 \in [-\rho, \rho]$,

$$|h(t, u_1) - h(t, u_2)| \to 0 \quad \text{as } u_1 \to u_2,$$

for all $t \in I$. Hence any $t_1, t_2 \in I$, and for any $u \in S$, one has

$$
\begin{aligned}
|Au(t_1) - Au(t_2)| &= |h(t_1, u(t_1)) - h(t_2, u(t_2))| \\
&\leq |h(t_1, u(t_1)) - h(t_2, u(t_1))| \\
&\quad + |h(t_2, u(t_1)) - h(t_2, u(t_2))| \\
&\to 0 \quad \text{as } t_1 \to t_2.
\end{aligned}
$$

Hence $A(S)$ is an equicontinuous set in K.

From the Arzelà–Ascoli theorem we conclude that A is a completely continuous operator on $[\underline{u}, \overline{u}]$.

Next it can be shown as in the proof of Theorem 5.25 that B is a compact operator on $[\underline{u}, \overline{u}]$. Now an application of Theorem 5.18 yields that the problem (5.9)–(5.10) has a minimal and maximal positive solution on I. $\qquad \square$

5.3.4 Random solutions and extremal random solutions

In this section we are concerned with the existence of solutions of the problem (5.11)–(5.12).

Definition 5.27. By a random solution of the problem (5.11)–(5.12) we mean a measurable function $u : \Omega \to C(I)$ such that the function $t \mapsto \left(\frac{u(t,w)}{h(t,u(t,w),w)} \right)$ is continuous for any $w \in \Omega$, and u satisfies Eq. (5.11) on I and the initial condition (5.12).

Lemma 5.28. *Let $f : I \times \mathbb{R}^2 \times \Omega \to \mathbb{R}$, $h : I \times \mathbb{R} \times \Omega \to \mathbb{R}^*$ such that $f(\cdot, u, v, w) \in C(I)$, for any $w \in \Omega$, and each $u, v \in \mathbb{R}$, and the function $t \mapsto \left(\frac{u(t,w)}{h(t,u(t,w),w)} \right)$ is continuous. Then the problem (5.7)–(5.8) is equivalent to the problem of obtaining the solutions of the integral equation*

$$
g(t, w) = f(t, h(t, u(t, w), w), (\mu_0(w) + (I_q^\alpha g)(t, w)), g(t, w), w),
$$

where

$$
\mu_0(w) = \frac{u_0(w)}{h(0, u(0, w), w)},
$$

and if $g(\cdot, w) \in C(I)$, is the solution of this equation, then

$$
u(t, w) = h(t, u(t, w), w) \left(\mu_0(w) + (I_q^\alpha g)(t, w) \right).
$$

We use the following fixed-point theorem by Dhage for proving the existence of random solutions for our problem.

Theorem 5.29. *(Dhage [224]) Let S be a closed, convex, and bounded subset of a separable Banach algebra X and let $A(w), B(w) : \Omega \times S \to X$ be two random operators satisfying for each $w \in \Omega$*

(a) *$A(w)$ is Lipschitz with the Lipschitz constant $k(w)$;*

(b) $B(w)$ *is continuous and compact;*

(c) $A(w)u\,B(w)u \in S$, *for each* $u \in S$.

Then the random equation $A(w)u\,B(w)u = u$ *has a random solution and the set of all such solutions is compact whenever* $k(w)M(w) < 1$, *for each* $w \in \Omega$, *where*

$$M(w) = \|B(w)(S)\| := \sup\{\|B(w)u\| : u \in S\}.$$

We use the following fixed-point theorem of Dhage [225] for proving the existence of extremal random solutions for our problem under certain monotonicity conditions.

Theorem 5.30. *(Dhage [225]) Let* K *be a cone in a Banach algebra* X *and let* $v, w \in X$. *Suppose that* $A, B : [v, w] \to K$ *are two operators such that*

(a) A *is completely continuous;*

(b) B *is totally bounded;*

(c) $Au\,Bz \in [v, w]$ *for all* $u, z \in [v, w]$;

(d) A *and* B *are nondecreasing.*

Furthermore, if the cone K *is positive and normal, then the operator equation* $Au\,Bu = u$ *has a least and a greatest positive solution in* $[v, w]$.

Theorem 5.31. *(Dhage [225]) Let* K *be a cone in a Banach algebra* X *and let* $v, w \in X$. *Suppose that* $A, B : [v, w] \to K$ *are two operators such that*

(a) A *is Lipschitz with a Lipschitz constant* α;

(b) B *is totally bounded;*

(c) $Au\,Bz \in [v, w]$ *for all* $u, z \in [v, w]$;

(d) A *and* B *are nondecreasing.*

Furthermore, if the cone K *is positive and normal, then the operator equation* $Au\,Bu = u$ *has least and a greatest positive solution in* $[v, w]$, *whenever* $\alpha M < 1$, *where* $M = \|B([v, w])\| := \sup\{\|Bu\| : u \in [v, w]\}$.

Remark 5.32. Note that hypothesis (c) of Theorems 5.30 and 5.31 holds if the operators A and B are positive monotone increasing and there exist elements v and w in X such that $v \le Av\,Bv$ and $Aw\,Bw \le w$.

Theorem 5.33. *Assume that following hypotheses hold.*

(5.33.1) *The function* $\mu_0 : \Omega \to \mathbb{R}$ *is measurable.*

(5.33.2) *The function* h *is random Carathéodory and there exists a measurable function* $\alpha : \Omega \to L^\infty(I, \mathbb{R}_+)$, *such that*

$$|h(t, u, w) - h(t, \overline{u}, w)| \le \alpha(t, w)|u - \overline{u}|;\ \textit{for any } w \in \Omega,\ t \in I,\ \textit{and } u, \overline{u} \in \mathbb{R}.$$

(5.33.3) *The function f is random Carathéodory, and there exists $P : \Omega \to L^\infty(I, \mathbb{R}_+)$ such that*

$$|f(t, u, v, w)| \leq P(t, w): \text{ for any } w \in \omega, \text{ and all } t \in I, \text{ and } u, v \in \mathbb{R}.$$

Set

$$\mu_0^* = \sup_{w \in \Omega} |\mu(w)|, \ h^* = \sup_{w \in \Omega} \|h(\cdot, 0, w)\|_{L^\infty},$$

$$\alpha^* = \sup_{w \in \Omega} \|\alpha(w)\|_{L^\infty}, \ P^* := \sup_{w \in \Omega} \|P(w)\|_{L^\infty},$$

and

$$L := \frac{T^{(\alpha)}}{\Gamma_q(1 + \alpha)}.$$

If

$$\alpha^*(\mu_0^* + LK^*) < 1, \tag{5.23}$$

then the problem (5.11)–(5.12) has at least one random solution defined on $I \times \Omega$, and the set of all solutions is compact in B_η.

Proof. Define the operators $A, B : \Omega \times C(I) \to C(I)$ by

$$(A(w)u)(t) = h(t, u(t, w), w); \quad t \in I, \tag{5.24}$$

$$(B(w)u)(t) = \mu_0(w) + (I_q^\alpha g)(t, w); \quad t \in I, \tag{5.25}$$

where $g(w) \in C(I)$ such that

$$g(t, w) = f(t, u(t, w), g(t, w), w)$$
$$= f(t, h(t, u(t, w), w)(\mu_0(w) + (I_q^\alpha g)(t, w)), g(t, w), w),$$

for all $t \in I$, and each $w \in \Omega$.

The hypothesis (5.33.2) implies that the mapping A is well defined and the function $A(w)u$ is continuous and bounded on I. Again, since the function μ is measurable, then the function $B(w)u$ is also measurable. Therefore $A(w)$ and $B(w)$ define the operators $A(w), B(w) : B_\eta \to C(I)$. Thus u is a random solution for the problem (5.11)–(5.12) if and only if

$$u = A(w)(u).B(w)u. \tag{5.26}$$

Consider the ball $B_R := \{u \in C(I); \|u\|_\infty \leq R\}$, where

$$R > \frac{h^*(\mu_0^* + LP^*)}{1 - \alpha^*(\mu_0^* + LP^*)}.$$

We will show that $A(w)$ and $B(w)$ satisfy all the requirements of Theorem 5.17 on B_R. The proof will be given in four steps.

Step 1. $A(w)$ and $B(w)$ are random operators on $\Omega \times B_R$ into B_R.

Since $h(t, u, w)$ is random Carathéodory, the map $w \to f(t, u, w)$ is measurable. Thus $A(w)$ is a random operator on $\Omega \times B_R$ into B_R.

Similarly, the product $\frac{(t-qs)^{(\alpha-1)}}{\Gamma_q(\alpha)} f(s, u, v, w)$ of a continuous and a measurable function is again measurable. Furthermore, the integral is a limit of a finite sum of measurable functions, therefore, the map

$$w \mapsto \mu_0(w) + \int_0^t \frac{(t-qs)^{(\alpha-1)}}{\Gamma_q(\alpha)} f(t, s, u(s, w), v(s, w), w) d_q s$$

is measurable. As a result, $B(w)$ is a random operator on $\Omega \times B_R$ into B_R.

Step 2. $A(w)$ is a Lipschitz operator on B_R.

Let $u, v \in C(I)$. Then by hypothesis (5.33.2), for any $w \in \omega$, and each $t \in I$, we have

$$|(A(w)u)(t) - (A(w)v)(t)| \le |h(t, u(t, w), w) - h(t, v(t, w), w)|$$
$$\le \alpha(w)\|u - v\|_\infty$$
$$\le \alpha^*\|u - v\|_\infty.$$

Thus, for all $u, v \in C(I)$, we get

$$\|A(w)u - A(w)v\|_\infty \le \alpha^*\|u - v\|_\infty.$$

This shows that $A(w)$ is a Lipschitz on B_R with the Lipschitz constant $\ell = \alpha^*$.

Step 3. $B(w)$ is a compact operator on B_R.

Let $\{u_n\}$ be a sequence in $C(I)$. From (5.33.3) it follows that

$$\|B(w)u_n\|_\infty \le |\mu_0(w)| + LP^*.$$

As a result $\{B(w)u_n : n \in \mathbb{N}\}$ is a uniformly bounded set in $C(I)$. Let $t_1, t_2 \in I$. Then

$$|B(w)u_n(t_1) - B(w)u_n(t_2)| \le P^* \int_0^{t_1} \frac{|(t_2 - qs)^{(\alpha-1)} - (t_1 - qs)^{(\alpha-1)}|}{\Gamma_q(\alpha)} d_q s$$
$$+ P^* \int_{t_1}^{t_2} \frac{|(t_2 - qs)^{(\alpha-1)}|}{\Gamma_q(\alpha)} d_q s$$
$$\to 0, \quad \text{as } t_1 \to t_2.$$

From this we conclude that $\{B(w)u_n : n \in \mathbb{N}\}$ is an equicontinuous set in $C(I)$. Hence from the Arzelà–Ascoli theorem, $B(w) : C(I) \to C(I)$ is compact.

Step 4. $A(w)uB(w)u \in B_R$ for all $u \in B_R$.

Let $u \in B_R$ be arbitrary, then for all $t \in I$ and any $w \in \Omega$, we have

$$|(A(w)u)(t)(B(w)u)(t)| \le |(A(w)u)(t)| + |(B(w)v)(t)|$$
$$\le (|h(t, u(t, w), w) - h(t, 0, w)| + |h(t, 0, w)|)$$

$$\times \left[|\mu_0(w)| + P^* \int_0^t \frac{(t - qs)^{(\alpha-1)}}{\Gamma_q(\alpha)} d_q s \right]$$
$$\leq (\alpha^* R + h^*)(\mu^* + L P^*) < R.$$

Hence we obtain that $A(w)u B(w)u \in B_R$ for any $w \in \Omega$ and all $u \in B_R$.

As a consequence of Steps 1 to 4 together with Theorem 5.17, we deduce that $A(w)B(w)$ has a fixed point in B_R which is a random solution of our problem (5.11)–(5.12).

Moreover, we have

$$M(w) = \|B(w)B_R\| = \sup_{u \in B_R} \|B(w)u\| \leq \mu^* + L P^*.$$

Therefore the assumption (5.23) implies that $M(w)k(w) = \alpha^*(\mu^* + LK^*) < 1$. Consequently, the set of all solutions of the problem (5.11)–(5.12) is compact in B_R.   □

Now, we prove some extremal random solutions results.

We equip the space $C(I)$ with the order relation $\leq$ with the help of the cone defined by

$$K = \{u : \Omega \to \in C(I) : u(t, w) \geq 0, \text{ for any } w \in \Omega, \text{ and } \textit{all } t \in I\}.$$

Thus $u \leq \bar{u}$ if and only if $u(t) \leq \bar{u}(t)$ for each $t \in I$. It is well-known that the cone K is positive and normal in $C(I)$ ([262]).

If $\underline{u}, \bar{u} \in C(I)$ and $\underline{u} \leq \bar{u}$, we put

$$[\underline{u}, \overline{u}] = \{u \in C(I) : \underline{u} \leq u \leq \bar{u}\}.$$

Definition 5.34. A function $\beta : I \times \mathbb{R} \to \mathbb{R}$ is called random Chandrabhan if

(i) The map $(t, w) \to \beta(t, u, w)$ is jointly measurable for all $u \in \mathbb{R}$, and
(ii) The map $u \to \beta(t, u, w)$ is nondecreasing for almost each $t \in I$ and any $w \in \Omega$.

Definition 5.35. A function $u(\cdot, \cdot, w) \in C(I)$ is said to be a lower random solution of (5.11)–(5.12) if for any $w \in \Omega$, we have

$$^c D_q^\alpha \left(\frac{u(t, w)}{h(t, u(t, w), w)} \right) \leq f \left(t, u(t, w), {}^c D_q^\alpha \left(\frac{u(t, w)}{h(t, u(t, w), w)} \right), w \right); \ t \in I,$$
$$u(0, w) \leq u_0(w).$$

Similarly the function $u(\cdot, \cdot, w) \in C(I)$ is said to be an upper random solution of (5.11)–(5.12) if for any $w \in \Omega$, we have

$$^c D_q^\alpha \left(\frac{u(t, w)}{h(t, u(t, w), w)} \right) \geq f \left(t, u(t, w), {}^c D_q^\alpha \left(\frac{u(t, w)}{h(t, u(t, w), w)} \right), w \right); \ t \in I,$$
$$u(0, w) \geq u_0(w).$$

Definition 5.36. A random solution u_M of the problem (5.11)–(5.12) is said to be maximal if for any other random solution u to the problem (5.11)–(5.12) one has $u(t, w) \leq u_M(t, w)$, for any $w \in \Omega$, and all $t \in I$. Again, a random solution u_m of the problem (5.7)–(5.8) is said to be minimal if $u_m(t, w) \leq u(t, w)$, for any $w \in \Omega$, and all $t \in I$, where u is any random solution of the problem (5.11)–(5.12) on I.

The following hypotheses will be used in the sequel.

(5.36.1) $h : I \times \mathbb{R}_+ \times \Omega \to \mathbb{R}_+^*$, $f : I \times \mathbb{R}_+ \times \mathbb{R}_+ \times \Omega \to \mathbb{R}_+$, and $\mu_0(w) \geq 0$, for any $w \in \Omega$.
(5.36.2) The functions h and f are random Chandrabhan.
(5.36.3) There exists a function $\tilde{P} \in L^\infty(I, \mathbb{R}_+)$ such that

$$|f(t, u, v, w)| \leq \tilde{P}(t, w): \quad \text{a.e. } t \in I, \text{ for any } w \in \Omega, \text{and all } u, v \in \mathbb{R}.$$

(5.36.4) The problem (5.1)–(5.8) has a lower random solution $\underline{u}$ and an upper random solution $\overline{u}$ with $\underline{u} \leq \overline{u}$.

Let

$$\tilde{P}^* = \sup_{w \in \Omega} \|\tilde{P}(w)\|_{L^\infty}.$$

Theorem 5.37. *Assume that hypotheses* (5.33.2), (5.36.1)–(5.36.4) *hold. If*

$$\alpha^*(\mu_0^* + L\tilde{P}^*) < 1,$$

then the problem (5.11)–(5.12) *has a minimal and a maximal positive random solution on* I.

Proof. Consider a closed interval $[\underline{u}, \overline{u}]$ in $C(I)$, which is well defined in view of hypothesis (5.36.1). Define the operators $A, B : [\underline{u}, \overline{u}] \to C(I)$ by (5.24) and (5.25), respectively. Clearly, A and B define the operators $A(w), B(w) : [\underline{u}, \overline{u}] \to K$.

Now solving (5.11)–(5.12) is equivalent to solving the operator equation

$$A(w)u(t)\, B(w)u(t) = u(t, w): \quad t \in I, \ w \in \Omega. \tag{5.27}$$

We show that the operators A and B satisfy all the assumptions of Theorem 5.31. As in Theorem 5.33 we can prove that $A(w)$ is Lipschitz with a Lipschitz constant α^* and $B(w)$ is a completely continuous operator on $[\underline{u}, \overline{u}]$.

Now the hypothesis (5.36.2) implies that $A(w)$ and $B(w)$ are nondecreasing on $[\underline{u}, \overline{u}]$. To see this, let $u_1, u_2 \in [\underline{u}, \overline{u}]$ be such that $u_1 \leq u_2$. Then by (5.36.2), we get

$$(A(w)u_1)(t) = h(t, u_1(t, w), w) \leq h(t, u_2(t, w), w)$$

$$= (A(w)u_2)(t), \forall t \in I, \ w \in \Omega,$$

and

$$(B(w)u_1)(t) = \mu_0(w) + \int_0^t \frac{(t-qs)^{(\alpha-1)}}{\Gamma_q(\alpha)} f(s, u_1(s, w), (^C D_q^\alpha u_1)(s, w), w) d_q s$$

$$\leq \mu_0(w) + \int_0^t \frac{(t-qs)^{(\alpha-1)}}{\Gamma_q(\alpha)} f(s, u_2(s, w), (^C D_q^\alpha u_2)(s, w), w) d_q s$$

$$= (B(w)u_2)(t), \quad \forall t \in I, \ w \in \Omega.$$

So A and B are nondecreasing random operators on $[\underline{u}, \overline{u}]$. Again the hypothesis (5.36.4) imply

$$\underline{u}(t, w) = (h(t, \underline{u}(t, w), w))(\mu_0(w)$$

$$+ \int_0^t \frac{(t-qs)^{(\alpha-1)}}{\Gamma_q(\alpha)} f(s, \underline{u}(s, w), (^C D_q^\alpha \underline{u})(s, w), w) d_q s)$$

$$\leq (h(t, \underline{u}(t, w), w))(\mu_0(w)$$

$$+ \int_0^t \frac{(t-qs)^{(\alpha-1)}}{\Gamma_q(\alpha)} f(s, v(s, w), (^C D_q^\alpha v)(s, w), w) d_q s)$$

$$\leq (h(t, \underline{u}(t, w), w))(\mu_0(w)$$

$$+ \int_0^t \frac{(t-qs)^{(\alpha-1)}}{\Gamma_q(\alpha)} f(s, \overline{u}(s, w), (^C D_q^\alpha \overline{u})(s, w), w) d_q s)$$

$$\leq \overline{u}(t, w),$$

for any $w \in \Omega$ and all $t \in I$ and $v \in [\underline{u}, \overline{u}]$. As a result

$$\underline{u}(t, w) \leq (A(w)v)(t)(B(w)v)(t) \leq \overline{u}(t, w), \text{ for any } w \in \Omega, t \in I \text{ and } v \in [\underline{u}, \overline{u}].$$

Hence $A(w)v \, B(w)v \in [\underline{u}, \overline{u}]$, for all $v \in [\underline{u}, \overline{u}]$.

Notice for any $u \in [\underline{u}, \overline{u}]$,

$$M = \|B([\underline{u}, \overline{u}])\|$$

$$\leq |\mu_0(w)| + \int_0^t \frac{(t-qs)^{(\alpha-1)}}{\Gamma_q(\alpha)} f(s, u(s, w), (^C D_q^\alpha u)(s, w), w) d_q s)$$

$$\leq \mu_0^* + L\tilde{K}^*,$$

and so

$$\alpha^* M \leq \alpha^* (|mu_0^* + L\tilde{P}^*) < 1.$$

Thus the operators A and B satisfy all the conditions of Theorem 5.31 and so the operator equation (5.25) has a least and a greatest random solution in $[\underline{u}, \overline{u}]$. Hence the problem (5.11)–(5.12) has a minimal and a maximal positive random solution on $I \times \omega$. $\square$

Theorem 5.38. *Assume that hypotheses* (5.33.1), (5.36.1)–(5.36.4) *hold. Then the problem* (5.11)–(5.12) *has a minimal and a maximal positive random solution on* $I \times \Omega$.

Proof. Consider the order interval $[\underline{u}, \overline{u}]$ in $C(I)$ and define the operators A and B on $[\underline{u}, \overline{u}]$ by (5.24) and (5.25), respectively. Then the problem (5.11)–(5.12) is transformed into an operator equation $(A(w)u)(t)(B(w)u)(t) = u(t)$ for any $w \in \Omega$, and all $t \in I$ in the Banach algebra $C(I)$. Notice that (5.36.1) implies $A, B : [\underline{u}, \overline{u}] \to K$. Since the cone K in $C(I)$ is normal, then $[\underline{u}, \overline{u}]$ is a norm bounded set in $C(I)$.

Next we show that $A(w)$ is completely continuous on $[\underline{u}, \overline{u}]$; for any $w \in \Omega$. Now the cone K in $C(I)$ is normal, so the order interval $[\underline{u}, \overline{u}]$ is norm-bounded. Hence there exists a constant $\rho > 0$ such that $|u| \le \rho$ for all $u \in [\underline{u}, \overline{u}]$. Since h is continuous on the compact set $I \times [-\rho, \rho]$, then it attains its maximum, say M. Therefore, for any subset S of $[\underline{u}, \overline{u}]$, we have

$$
\begin{aligned}
\|A(w)(S)\| &= \sup\{|A(w)u| : u \in S\} \\
&= \sup\{\sup_{t \in I} |h(t, u(t, w), w)| : u \in S\} \\
&\le \sup_{w \in \Omega} \sup_{t \in I} \{\sup |h(t, u, w)| : u \in [-\rho, \rho]\} \\
&\le M.
\end{aligned}
$$

This shows that $A(w)(S)$ is a uniformly bounded subset of $C(I)$ for any $w \in \Omega$.

We note that the function $h(t, u, w)$ is uniformly continuous on $I \times [-\rho, \rho] \times \Omega$. Therefore, for any $t_1, t_2 \in I$, and any $w \in \Omega$, we have

$$
|h(t_1, u, w) - h(t_2, u, w)| \to 0 \quad \text{as } t_1 \to t_2,
$$

for all $u \in [-\rho, \rho]$. Similarly, for any $u_1, u_2 \in [-\rho, \rho]$, and $w \in \Omega$,

$$
|h(t, u_1, w) - h(t, u_2, w)| \to 0 \quad \text{as } u_1 \to u_2,
$$

for any $w \in \Omega$, and all $t \in I$. Hence for any $t_1, t_2 \in I$, one has

$$
\begin{aligned}
|A(w)u(t_1) - A(w)u(t_2)| &= |h(t_1, u(t_1, w), w) - h(t_2, u(t_2, w), w)| \\
&\le |h(t_1, u(t_1, w), w) - h(t_2, u(t_1, w), w)| \\
&\quad + |h(t_2, u(t_1, w), w) - h(t_2, u(t_2, w), w)| \\
&\to 0 \quad \text{as } t_1 \to t_2.
\end{aligned}
$$

This shows that $A(w)(S)$ is an equicontinuous set in K.

Now from the Arzelà–Ascoli theorem, $A(w)$ is completely continuous on $[\underline{u}, \overline{u}]$.

Next it can be shown as in the proof of Theorem 5.37 that $B(w)$ is a compact operator on $[\underline{u}, \overline{u}]$. Now an application of Theorem 5.30 yields that the problem (5.11)–(5.12) has a minimal and maximal positive random solution. $\qquad \square$

5.3.5 Some examples

Let $E = \mathbb{R}$, $\Omega = (-\infty, 0)$ be equipped with the usual σ-algebra consisting of Lebesgue measurable subsets of $(-\infty, 0)$.

Example 1. For a given a measurable function $u : \Omega \to C([0, 1])$, we consider the following problem of random implicit fractional $\frac{1}{4}$-difference equations

$$\begin{cases} ({}^{C}D_{\frac{1}{4}}^{\frac{1}{2}}u)(t, w) = f(t, u(t, w), ({}^{C}D_{\frac{1}{4}}^{\frac{1}{2}}u)(t, w)); \ t \in [0, 1], \ w \in \Omega, \\ u(0, w) = u_0(w); \ w \in \Omega, \end{cases} \qquad (5.28)$$

where

$$\begin{cases} f(t, u, v, w) = \dfrac{ct^{\frac{-1}{4}} \sin t}{64(1 + w^2)(1 + \sqrt{t})(1 + |u| + |v|)}; \ t \in (0, 1] \quad u, v \in \mathbb{R}, \ w \in \Omega \\ f(0, u, v, w) = 0; \qquad\qquad\qquad\qquad\qquad\qquad\qquad\quad u, v \in \mathbb{R}, \ w \in \Omega. \end{cases}$$

$u_0(w) = \sin w + \cos w$, and $c = \frac{9\sqrt{\pi}}{16}$.

The hypothesis (5.33.2) is satisfied with

$$\begin{cases} p(t, w) = \dfrac{ct^{\frac{-1}{4}} |\sin t|}{64(1 + w^2)(1 + \sqrt{t})}; \quad t \in (0, 1], \\ p(0, w) = 0. \end{cases}$$

Hence Theorem 5.8 implies that the problem (5.28) has at least one random solution defined on $[0, 1] \times \Omega$. Also, the hypothesis (5.33.3) is satisfied with

$$\Phi(t, w) = \frac{e^3}{1 + w^2}.$$

Consequently, Theorem 5.14 implies that the problem (5.28) is generalized Ulam–Hyers–Rassias stable.

Example 2. Consider the following problem of implicit fractional $\frac{1}{4}$-difference equations

$$\begin{cases} {}^{C}D_{\frac{1}{4}}^{\frac{1}{2}}\left(\dfrac{u(t)}{h(t,u(t))}\right) = f\left(t, u(t), {}^{C}D_{\frac{1}{4}}^{\frac{1}{2}}\left(\dfrac{u(t)}{h(t,u(t))}\right)\right); \ t \in [0, 1], \\ u(0) = e^{-10}, \end{cases} \qquad (5.29)$$

where

$$h(t, x) = \frac{1}{e^{t+10}(1 + |x|)}; \ t \in [0, 1],$$

and

$$f(t, x, y) = \frac{1}{Le^{t+8}(1 + x^2 + y^2)}; \ t \in [0, 1].$$

We have $h : [0, 1] \times \mathbb{R} \to \mathbb{R}_+^*$, $f : [0, 1] \times \mathbb{R} \times \mathbb{R} \to \mathbb{R}_+$. Clearly, the function h satisfies (5.33.1) and (5.33.2) with $\alpha(t) = \dfrac{1}{e^t+10}$ and $\|\alpha\|_\infty = \frac{1}{e^{10}}$. Also, the function f satisfies (5.33.3) with $K(t) = \frac{1}{Le^t+8}$ and $K^* = \dfrac{1}{Le^8}$. We will show that condition (5.18) holds. Indeed, we have $|\mu_0| = 1 + e^{-10}$ and

$$\|\alpha\|_\infty(|\mu_0| + LK^*) \le \frac{1}{e^{10}}\left(1 + e^{-10} + \frac{1}{e^8}\right) < 1.$$

Hence by Theorem 5.21, the problem (5.29) has a solution defined on $[0, 1]$.

Example 3. We consider now the following problem of implicit random fractional $\frac{1}{4}$-difference equations

$$\begin{cases} {}^c D^{\frac{1}{2}}_{\frac{1}{4}}\left(\dfrac{u(t,w)}{h(t,u(t,w),w)}\right) = f\left(t, u(t, w), {}^c D^{\frac{1}{2}}_{\frac{1}{4}}\left(\dfrac{u(t,w)}{h(t,u(t,w),w)}\right), w\right); t \in [0, 1], \\ u(0, w) = \dfrac{e^{-10}}{1-w} \end{cases} \tag{5.30}$$

where

$$h(t, x, w) = \frac{1}{(1 + w^2)e^{t+10}(1 + |x|)}; \ t \in [0, 1], \ w \in \Omega,$$

and

$$f(t, x, y, w) = \frac{1}{Le^{t+8}(1 + w^2)(1 + x^2 + y^2)}; \ t \in [0, 1], \ w \in \Omega.$$

We have $h : [0, 1] \times \mathbb{R} \times \Omega \to \mathbb{R}^{*+}$ and $f : [0, 1] \times \mathbb{R} \times \mathbb{R} \times \Omega \to \mathbb{R}_+$. Clearly, the function h satisfies (5.33.1) and (5.33.2) with $\alpha(t, w) = \dfrac{1}{(1 + w^2)e^{t+10}}$ and $\alpha^* = \frac{1}{e^{10}}$. Also, the function f satisfies (5.33.3) with $P(t, w) = \frac{1}{L(1+w^2)e^{t+8}}$ and $P^* = \dfrac{1}{Le^8}$. We will show that condition (5.23) holds. Indeed, we have $\mu_0^* = 1 + e^{-10}$ and

$$\alpha^*(\mu_0^* + LP^*) \le \frac{1}{e^{10}}\left(1 + e^{-10} + \frac{1}{e^8}\right) < 1.$$

Hence by Theorem 5.33, the problem (5.30) has a random solution defined on $[0, 1] \times \Omega$.

5.4 Uniqueness and Ulam stability for implicit fractional q-difference equations via Picard operators theory

5.4.1 Introduction

By using the theory of weakly Picard operators, in this section, we discuss the uniqueness and Ulam–Hyers–Rassias stability for the following implicit fractional q-difference equation

$$({}^c D^\alpha_q u)(t) = f(t, u(t), ({}^c D^\alpha_q u)(t)), \ t \in I := [0, T], \tag{5.31}$$

with the initial condition

$$u(0) = u_0, \tag{5.32}$$

where $q \in (0, 1)$, $\alpha \in (0, 1]$, $T > 0$, $f : I \times \mathbb{R} \times \mathbb{R} \to \mathbb{R}$ is a given continuous function, and $^{c}D_q^{\alpha}$ is the Caputo fractional q-difference derivative of order α.

Next, we discuss the uniqueness and Ulam–Hyers–Rassias stability for the following implicit coupled fractional q-difference system

$$\begin{cases} (^{c}D_{q_1}^{\alpha_1} u_1)(t) = f_1(t, u_1(t), u_2(t), (^{c}D_q^{\alpha_1} u_1)(t)), \\ (^{c}D_{q_2}^{\alpha_2} u_2)(t) = f_1(t, u_1(t), u_2(t), (^{c}D_q^{\alpha_2} u_2)(t)), \end{cases} \quad t \in I := [0, T], \tag{5.33}$$

with the initial conditions

$$\begin{cases} u_1(0) = \phi_1 \\ u_2(0) = \phi_2, \end{cases} \tag{5.34}$$

where $T > 0$, $q_i \in (0, 1)$, $\alpha_i \in (0, 1]$; $i = 1, 2$, $f_i : I \times \mathbb{R} \times \mathbb{R} \to \mathbb{R}$ are given continuous functions.

5.4.2 Implicit Caputo fractional q-difference equations

In this section we are concerned with the existence and uniqueness of solutions of the problem (5.31)–(5.32).

Definition 5.39. By a solution of the problem (5.31)–(5.32) we mean a continuous function $u \in C(I)$ that satisfies Eq. (5.31) on I and the initial condition (5.32).

The following hypotheses will be used in the sequel.

(5.39.1) The function f satisfies the Lipschitz condition:

$$|f(t, u_1, v_1) - f(t, u_2, v_2)| \leq \ell_1 |u_1 - u_2| + \ell_2 |v_1 - v_2|,$$

for $t \in I$ and $u_1, u_2, v_1, v_2 \in \mathbb{R}$, where $\ell_1, \ell_2 \in (0, \infty)$, with $\ell_2 < 1$.

(5.39.2) $\Phi \in L^1(I, [0, \infty))$ and there exists $\lambda_\Phi > 0$ such that, for each $x \in I$, we have

$$(I_q^\alpha \Phi)(t) \leq \lambda_\Phi \Phi(t).$$

First, we present a uniqueness and Ulam stability results for our problem (5.31)–(5.32).

Theorem 5.40. *Assume that the hypothesis* (5.39.1) *holds. If*

$$L_f := \frac{T^\alpha \ell_1}{(1 - \ell_2)\Gamma_q(1 + \alpha)} < 1,$$

then there exist a unique solution of problem (5.31)–(5.32) *on I. Moreover, the operator N is k-weakly Picard operator with the positive constant $k_N = \frac{1}{1 - L_f}$ and the fixed-point equation $u = N(u)$ is Ulam–Hyers stable.*

Proof. Consider the operator $N : C(I) \to C(I)$ defined in (5.3). Let $u, v \in C(I)$. Then, for $t \in I$, we have

$$|(Nu)(t) - (Nv)(t)| \leq \int_0^t \frac{(t - qs)^{(\alpha-1)}}{\Gamma_q(\alpha)} |g(s) - h(s)| d_q s, \qquad (5.35)$$

where $g, h \in C(I)$ such that

$$g(t) = f(t, u(t), g(t)),$$

and

$$h(t) = f(t, v(t), h(t)).$$

From (5.39.1), we obtain

$$|g(t) - h(t)| \leq \ell_1 |u(t) - v(t| + \ell_2 |g(t) - h(t|.$$

Thus

$$|g(t) - h(t)| \leq \frac{\ell_1}{1 - \ell_2} |u(t) - v(t)|.$$

Thus, from (5.35), we get

$$
\begin{aligned}
|(Nu)(t) - (Nv)(t)| &\leq \int_0^t \frac{(t - qs)^{(\alpha-1)}}{\Gamma_q(\alpha)} \frac{\ell_1}{1 - \ell_2} |u(s) - v(s)| d_q s \\
&\leq \frac{T^\alpha \ell_1}{(1 - \ell_2)\Gamma_q(1 + \alpha)} \|u - v\|_\infty \\
&= L_f \|u - v\|_\infty.
\end{aligned}
$$

Hence we get

$$\|N(u)N(v)\|_\infty \leq L_f \|u - v\|_\infty.$$

Hence, since $L_f < 1$, then N is a contraction. Consequently, by Banach's contraction principle, N has a unique fixed point, which is the unique solution of the problem (5.31)–(5.32). Moreover, Lemma 5.6 implies that the operator N is k-weakly Picard operator with the positive constant $k_N = \frac{1}{1-L_f}$, and the fixed-point equation $u = N(u)$ is Ulam–Hyers stable.   □

Now, we present conditions for the generalized Ulam–Hyers–Rassias stability of problem (5.31)–(5.32).

Theorem 5.41. *Assume that the hypotheses (5.39.1) and (5.39.2) hold. If $L_f < 1$, then the fixed-point equation $u = N(u)$ is generalized Ulam–Hyers–Rassias stable.*

Proof. Let N be the operator defined in (5.3). Let $u \in C(I)$ be a solution of the inequality $|u(t) - (Nu)(t)| \leq \Phi(t)$; $t \in I$. By Theorem 5.40 there exists a unique solution v of the fixed point equation $u = N(u)$. Then we have

$$v(t) = (Nv)(t) = u_0 + \int_0^t \frac{(t - qs)^{(\alpha-1)}}{\Gamma_q(\alpha)} h(s) d_q s,$$

where $h \in C(I)$ with

$$h(t) = f(t, v(t), h(t)).$$

Thus, for each $t \in I$, it follows that

$$\begin{aligned}
|u(t) - v(t)| &= |u(t) - (Nv)(t)| \\
&\leq |u(t) - (Nu)(t)| + |(Nu)(t) - (Nv)(t)| \\
&\leq \Phi(t) + \int_0^t \frac{(t - qs)^{(\alpha-1)}}{\Gamma_q(\alpha)} |g(s) - h(s)| d_q s,
\end{aligned}$$

where $g \in C(I)$ such that $g(t) = f(t, u(t), g(t))$.

Thus

$$\begin{aligned}
|u(t) - v(t)| &\leq \Phi(t) + \int_0^t \frac{(t - qs)^{(\alpha-1)}}{\Gamma_q(\alpha)} |g(s) - h(s)| d_q s \\
&\leq \Phi(t) + \frac{\ell_1}{1 - \ell_2} \int_0^t \frac{(t - qs)^{(\alpha-1)}}{\Gamma_q(\alpha)} |u(s) - v(s)| d_q s.
\end{aligned}$$

From Lemma 2.100, there exists a constant $\delta = \delta(\alpha)$ such that

$$\begin{aligned}
|u(t) - v(t)| &\leq \Phi(x) + \frac{\ell_1 \delta}{1 - \ell_2} \int_0^t \frac{(t - qs)^{(\alpha-1)}}{\Gamma_q(\alpha)} \Phi(s) d_q s \\
&= \Phi(t) + \frac{\ell_1 \delta}{1 - \ell_2} (I_q^\alpha \Phi)(t).
\end{aligned}$$

Hence by (5.39.2) for each $t \in I$, we get

$$|u(t) - v(t)| \leq \left(1 + \frac{\ell_1 \delta \lambda_\Phi}{1 - \ell_2}\right) \Phi(t)$$

$$:= c_{N,\Phi} \Phi(t).$$

Finally, the fixed-point equation $u = N(u)$ is generalized Ulam–Hyers–Rassias stable.   □

In the sequel we will use of the following theorem.

Theorem 5.42. *[226] Let (Ω, d) be a generalized complete metric space and $\Theta : \Omega \to \Omega$ a strictly contractive operator with a Lipschitz constant $L < 1$. If there exists a nonnegative integer k such that $d(\Theta^{k+1}x, \Theta^k x) < \infty$ for some $x \in \Omega$, then the following propositions hold true:*

(A) *The sequence $(\Theta^k x)_{n \in N}$ converges to a fixed point x^* of Θ;*

(B) *x^* is the unique fixed point of Θ in $\Omega^* = \{y \in \Omega \mid d(\Theta^k x, y) < \infty\}$;*

(C) *If $y \in \Omega^*$, then $d(y, x^*) \leq \frac{1}{1-L} d(y, \Theta x)$.*

Now, we consider $C(I)$ as a metric space, with the metric

$$d(u, v) = \sup_{t \in I} \frac{|u(t) - v(t)|}{\Phi(t)}.$$

Theorem 5.43. *Assume that (5.39.2) and the following hypothesis hold.*

(5.43.1) *There exist a constant $0 < c < 1$ and a function $\varphi \in C(I, \mathbb{R}_+)$, such that for each $t \in I$, and all $u, v \in \mathbb{R}$, we have*

$$|f(t, u_1, v_1) - f(t, u_2, v_2)| \leq \varphi(t)\Phi(t)|u_1 - u_2| + c|v_1 - v_2|.$$

If

$$L_\Phi := \frac{\varphi^* \lambda_\phi}{1 - c} < 1, \tag{5.36}$$

where $\varphi^* = \sup\limits_{t \in I} \varphi(t)$, then there exists a unique solution u_0 of problem (5.31)–(5.32), and the problem (5.31)–(5.32) is generalized Ulam–Hyers–Rassias stable. Furthermore, we have

$$|u(t) - u_0(t)| \leq \frac{\Phi(t)}{1 - L_\Phi}.$$

Proof. Let N be the operator defined in (5.3). Apply Theorem 5.42, for any $u, v \in C(I)$, and each $t \in I$, we have

$$|(Nu)(t) - (Nv)(t)| \leq \int_0^t \frac{(t - qs)^{(\alpha-1)}}{\Gamma_q(\alpha)} |g(s) - h(s)| d_q s,$$

where $g, h \in C(I)$ such that $g(t) = f(t, u(t), g(t))$, and $h(t) = f(t, v(t), h(t))$.

From (5.43.1), we have

$$|g(t) - h(t)| \leq \varphi(t)\Phi(t)|u(t) - v(t| + c|g(t) - h(t|,$$

which gives

$$|g(t) - h(t)| \leq \frac{\varphi(t)\Phi(t)}{1 - c}|u(t) - v(t)|.$$

Thus

$$|(Nu)(t) - (Nv)(t)| \leq \frac{\varphi^*}{1 - c}(I_q^\alpha \Phi)(t)\|u - v\|_\infty.$$

From (5.39.2) we get

$$|(Nu)(t) - (Nv)(t)| \leq \frac{\varphi^* \lambda_\phi}{1 - c}\Phi(t)\|u - v\|_\infty.$$

Hence we get

$$d(N(u), N(v)) = \sup_{t \in I} \frac{|(Nu)(t) - (Nv)(t)|}{\Phi(t)} \le L_\Phi \|u - v\|_\infty,$$

from which we conclude the theorem. □

5.4.3 Implicit coupled Caputo fractional q-difference systems

Now, we are concerned with the existence, uniqueness and Ulam stability of the problem (5.33)–(5.34). By $\mathcal{C} := C(I) \times C(I)$ we denote the Banach space with the norm

$$\|(u, v)\|_{\mathcal{C}} = \|u\|_\infty + \|v\|_\infty.$$

Lemma 5.44. *The solutions of the system (5.33)–(5.34) are the fixed points of the operator $G : \mathcal{C} \to \mathcal{C}$ defined by*

$$(G(u_1, u_2))(t) = ((G_1 u_1)(t), G_2 u_2)(t)); \ t \in I, \tag{5.37}$$

where $G_i : C(I) \to C(I)$; $i = 1, 2$, are defined by

$$(G_i(u_1, u_2))(t) = \phi_i + (I_{q_i}^{\alpha_i} g_i)(t); \ t \in I, \tag{5.38}$$

where $g_i \in C(I)$; $i = 1, 2$, such that

$$g_i(t) = f(t, u_1(t), u_2(t), g_i(t)), \ or \ g_i(t) = f(t, \phi_i + (I_{q_i}^{\alpha_i} g_i)(t), g_i(t)).$$

Now, we will give four types of Ulam stability of the fixed-point equation $(u_1, u_2) = G(u_1, u_2)$. Let ϵ be a positive real number and $\Phi : I \to \mathbb{R}_+$ be a continuous function.

Definition 5.45. The fixed-point equation $(u_1, u_2) = G(u_1, u_2)$ is said to be Ulam–Hyers stable if there exists a real number $c_G > 0$ such that for each $\epsilon > 0$ and for each solution $(u_1, u_2) \in \mathcal{C}$ of the inequality $|u_i(t) - (G_i u_i)(t)| \le \epsilon$; $i = 1, 2$, $t \in I$, there exists a solution $(v_1, v_2) \in \mathcal{C}$ of the equation $(u_1, u_2) = G(u_1, u_2)$ with

$$|u_i(t) - v_i(t)| \le \epsilon c_G; \ t \in I.$$

Definition 5.46. The fixed-point equation $(u_1, u_2) = G(u_1, u_2)$ is said to be generalized Ulam–Hyers stable if there exists $\theta_G \in C([0, \infty), [0, \infty))$, $\theta_G(0) = 0$ such that for each $\epsilon > 0$ and for each solution $(u_1, u_2) \in \mathcal{C}$ of the inequality $|u_i(t) - (G_i u_i)(t)| \le \epsilon$; $t \in I$, there exists a solution $(v_1, v_2) \in \mathcal{C}$ of the equation $(u_1, u_2) = G(u_1, u_2)$ with

$$|u_i(t) - v_i(t)| \le \theta_G(\epsilon); \ t \in I.$$

Definition 5.47. The fixed-point equation $(u_1, u_2) = G(u_1, u_2)$ is said to be Ulam–Hyers–Rassias stable with respect to Φ if there exists a real number $c_{G,\Phi} > 0$ such that for each

$\epsilon > 0$ and for each solution $(u_1, u_2) \in C$ of the inequality $|u_i(t) - (G_i u_i)(t)| \leq \epsilon \Phi(t)$; $t \in I$, there exists a solution $(v_1, v_2) \in C$ of the equation $(u_1, u_2) = G(u_1, u_2)$ with

$$|u_i(t) - v_i(t)| \leq c_{G,\Phi} \Phi(t); \ t \in I.$$

Definition 5.48. The fixed-point equation $u = N(u)$ is said to be generalized Ulam–Hyers–Rassias stable with respect to Φ if there exists a real number $c_{N,\Phi} > 0$ such that for each solution $(u_1, u_2) \in C$ of the inequality $|u_i(t) - (G_i u_i)(t)| \leq \Phi(t)$; $t \in I$, there exists a solution $(v_1, v_2) \in C$ of the equation $(u_1, u_2) = G(u_1, u_2)$ with

$$|u_i(t) - v_i(t)| \leq c_{G,\Phi} \Phi(t); \ t \in I.$$

Definition 5.49. By a solution of the problem (5.33)–(5.34) we mean a pair of continuous functions $(u_1, u_2) \in C$ that satisfy Eqs. (5.33) on I and the initial conditions (5.34).

The following hypotheses will be used in the sequel.

(5.49.1) The functions f_i; $i = 1, 2$ satisfy the Lipschitz conditions:

$$|f_i(t, u_1, v_1, w_1) - f_i(t, u_2, v_2, w_2)| \leq \ell_{1i}|u_1 - u_2| + \ell_{2i}|v_1 - v_2| + \ell_{3i}|w_1 - w_2|,$$

for $t \in I$ and $u_1, u_2, v_1, v_2, w_1, w_2 \in \mathbb{R}$, where $\ell_{1i}, \ell_{2i} \in (0, \infty)$; $i = 1, 2$, with $\ell_{3i} < 1$.

(5.49.2) $\Phi \in L^1(I, [0, \infty))$ and there exists $\Lambda_\Phi > 0$ such that, for each $x \in I$, we have

$$(I_{q_i}^{\alpha_I} \Phi)(t) \leq \Lambda_\Phi \Phi(t); \ i = 1, 2.$$

First, we present a uniqueness and Ulam stability results for our problem (5.33)–(5.34).

Theorem 5.50. *Assume that the hypothesis* (5.49.1) *holds. If*

$$L_{f_1, f_2} = \max \left\{ \frac{T^{(\alpha_i)} \ell_{1i}}{(1 - \ell_{3i}) \Gamma_{q_i}(\alpha_i + 1)}, \frac{T^{(\alpha_i)} \ell_{2i}}{(1 - \ell_{3i}) \Gamma_{q_i}(\alpha_i + 1)} \right\} < 1,$$

then there exist a unique solution of problem (5.33)–(5.34) *on I. Moreover, the operator G is k-weakly Picard operator with the positive constant $k_G = \frac{1}{1 - L_{f_1, f_2}}$ and the fixed-point equation $(u_1, u_2) = G(u_1, u_2)$ is Ulam–Hyers stable.*

Proof. Consider the operator $G : C \to C$ defined in (5.37) and let $(u_1, u_2), (v_1, v_2) \in C$. Then, for $t \in I$, and any $i = 1, 2$, we have

$$|(G_i(u_1, u_2))(t) - (G_i(v_1, v_2))(t)| \leq \int_0^t \frac{(t - q_i s)^{(\alpha_i - 1)}}{\Gamma_{q_i}(\alpha_i)} |g_i(s) - h_i(s)| d_{q_i} s, \quad (5.39)$$

where $g_i, h_i \in C(I)$ such that

$$g_i(t) = f_i(t, u_1(t), u_2(t), g_i(t)), \ and \ h_i(t) = f_i(t, v_1(t), v_2(t), h_i(t)).$$

From (5.49.1) we obtain

$$|g_i(t) - h_i(t)| \le \ell_{1i}|u_1(t) - v_1(t)| + \ell_{2i}|u_2(t) - v_2(t)| + \ell_{3i}|g_i(t) - h_i(t)|.$$

Thus

$$|g_i(t) - h_i(t)| \le \frac{\ell_{1i}}{1 - \ell_{3i}}|u_1(t) - v_1(t)| + \frac{\ell_{2i}}{1 - \ell_{3i}}|u_2(t) - v_2(t)|.$$

Thus we get

$$|(G_i(u_1, u_2))(t) - (G_i(v_1, v_2))(t)|$$

$$\le \int_0^t \frac{(t - q_i s)^{(\alpha_i - 1)}}{\Gamma_{q_i}(\alpha_i)} \left(\frac{\ell_{1i}}{1 - \ell_{3i}}|u_1(s) - v_1(s)| + \frac{\ell_{2i}}{1 - \ell_{3i}}|u_2(s) - v_2(s)| \right) d_{q_i} s$$

$$\le \frac{T^{(\alpha_i)}}{\Gamma_{q_i}(\alpha_i + 1)} \left(\frac{\ell_{1i}}{1 - \ell_{3i}}|u_1(s) - v_1(s)| + \frac{\ell_{2i}}{1 - \ell_{3i}}|u_2(s) - v_2(s)| \right) d_{q_i} s$$

$$\le L_{f_1, f_2}(\|u_1 - v_1\|_\infty + \|u_2 - v_2\|_\infty)$$

$$= L_{f_1, f_2}\|(u_1 - v_1, u_2 - v_2)\|_{\mathcal{C}}.$$

Hence we get

$$\|G(u_1, u_2) - G(v_1, v_2)\|_{\mathcal{C}} \le L_{f_1, f_2}\|(u_1, u_2) - (v_1, v_2)\|_{\mathcal{C}}$$

Hence, since $L_{f_1, f_2} < 1$, then G is a contraction. Consequently, by Banach's contraction principle, G has a unique fixed point, which is the unique solution of the problem (5.33)–(5.34). Moreover, Lemma 5.6 implies that the operator G is k_G-weakly Picard operator with the positive constant $k_G = \frac{1}{1 - L_{f_1, f_2}}$, and the fixed-point equation $(u_1, u_2) = G(u_1, u_2)$ is Ulam–Hyers stable. □

As in Theorem 5.41, we can conclude the following result:

Theorem 5.51. *Assume that the hypotheses (5.49.1) and (5.49.2) hold. If $L_{f_1, f_2} < 1$, then the fixed-point equation $(u_1, u_2) = G(u_1, u_2)$ is generalized Ulam–Hyers–Rassias stable.*

Now, we prove an other result by applying Theorem 5.42. We consider $\mathcal{C}$ as a metric space with the metric

$$d((u_1, u_2), (v_1, v_2)) = \sup_{t \in I} \frac{|u_1(t) - v_1(t)| + |u_2(t) - v_2(t)|}{\Phi(t)}.$$

Theorem 5.52. *Assume that (5.49.2) and the following hypothesis hold.*

(5.52.1) *There exists a constant $0 < c^* < 1$ and a function $\psi \in C(I, \mathbb{R}_+)$, such that for each $t \in I$, and all $u_1, u_2, v_1, v_2, w_1, w_2 \in \mathbb{R}$, we have*

$$|f_i(t, u_1, v_1, w_1) - f_i(t, u_2, v_2, w_2)| \le \psi(t)\Phi(t)(|u_1 - u_2| + |v_1 - v_2|) + c^*|w_1 - w_2|.$$

If

$$L_\Phi := \frac{2\psi^* \lambda_\phi}{1 - c^*} \Phi(t) < 1, \tag{5.40}$$

where $\psi^* = \sup_{t \in I} \psi(t)$, then there exists a unique solution (v_1, v_2) of problem (5.33)–(5.34), and the problem (5.33)–(5.34) is generalized Ulam–Hyers–Rassias stable. Furthermore, we have

$$\|(u_1 - v_1, u_2 - v_2)\|_C \leq \frac{\Phi(t)}{1 - L_\Phi}.$$

Proof. For any $u_1, u_2, v_1, v_2 \in C(I)$, and each $t \in I$, and any $i = 1, 2$, we have

$$|(G_i(u_1, u_2))(t) - (G_i(v_1, v_2))(t)| \leq \int_0^t \frac{(t - q_i s)^{(\alpha_i - 1)}}{\Gamma_{q_i}(\alpha_i)} |g_i(s) - h_i(s)| d_q s,$$

where $g_i, h_i \in C(I)$ such that $g_i(t) = f_i(t, u_1(t), u_2(t), g_i(t))$, and $h_i(t) = f_i(t, v_1(t), v_2(t), h_i(t))$. From (5.52.1), we have

$$|g_i(t) - h_i(t)| \leq \psi(t)\Phi(t)(|u_1(t) - v_1(t)| + |u_2(t) - v_2(t)|), + c^*|g_i(t) - h_i(t)|,$$

which gives

$$|g_i(t) - h_i(t)| \leq \frac{\psi(t)\Phi(t)}{1 - c^*}(|u_1(t) - v_1(t))| + |u_2(t) - v_2(t)|).$$

Thus we get

$$|(G_i(u_1, u_2))(t) - (G_i(v_1, v_2))(t)| \leq \frac{\psi^*}{1 - c^*}(I_{q_i}^{\alpha_i} \Phi)(t)(\|u_1 - v_1\|_\infty + \|u_2 - v_2\|_\infty).$$

From (5.49.2) we get

$$|(G_i(u_1, u_2))(t) - (G_i(v_1, v_2))(t)| \leq \frac{\psi^* \lambda_\phi}{1 - c^*} \Phi(t)\|(u_1 - v_1, u_2 - v_2)\|_C.$$

Hence we get

$$d(G(u_1, u_2), G(v_1, v_2)) = \sup_{t \in I} \frac{\sum_{i=1}^{2} |(G_i(u_1, u_2))(t) - (G_i(v_1, v_2))(t)|}{\Phi(t)}$$

$$\leq L_\Phi \|(u_1 - v_1, u_2 - v_2)\|_C,$$

from which we conclude the theorem. $\square$

5.4.4 An example

Consider the following problem of implicit fractional $\frac{1}{4}$-difference equations

$$
\begin{cases}
({}^C D_{\frac{1}{4}}^{\frac{1}{2}} u)(t) = f(t, u(t), ({}^C D_{\frac{1}{4}}^{\frac{1}{2}} u)(t)); \ t \in [0, 1], \\
u(0) = 1,
\end{cases}
\tag{5.41}
$$

where

$$
f(t, u(t), ({}^C D_{\frac{1}{4}}^{\frac{1}{2}} u)(t)) = \frac{t^2 e^{-t-5}}{1 + |u(t)| + |{}^c D_{\frac{1}{4}}^{\frac{1}{2}} u(t)|}; \ t \in [0, 1].
$$

We can see that the solutions of the problem (5.41) are solutions of the fixed-point equation $u = A(u)$, where $A : C([0, 1], \mathbb{R}) \to C([0, 1], \mathbb{R})$ is the operator defined by

$$
(Au)(t) = 1 + \int_0^t \frac{(t - \frac{s}{4})^{-\frac{1}{2}}}{\Gamma_{\frac{1}{4}}(\frac{1}{2})} |g(s) - h(s)| d_{\frac{1}{4}} s; \ t \in [0, 1].
$$

The hypothesis (5.49.1) is satisfied with

$$
\ell_1 = \ell_2 = e^{-5}.
$$

Also, we can easily see that for $T = 1$, we get

$$
L_f := \frac{T^\alpha \ell_1}{(1 - \ell_2)\Gamma_{\frac{1}{4}}(\frac{3}{2})} < 1.
$$

Hence Theorem 5.40 implies that problem (5.41) has a unique solution defined on $[0, 1]$. Moreover, the operator A is k-weakly Picard operator with the positive constant $k_A = \frac{1}{1-L_f}$ and the fixed-point equation $u = A(u)$ is Ulam–Hyers stable.

Also, the hypothesis (5.49.2) is satisfied with $\Phi(t) = t^2$. Indeed, for each $t \in (0, 1]$, there exists a real number $0 < \epsilon < 1$ such that $\epsilon < t \leq 1$ and

$$
(I_q^\alpha \Phi)(t) \leq \frac{t^2}{\epsilon^2(1 + q + q^2)}
$$

$$
\leq \frac{1}{\epsilon^2} \Phi(t)
$$

$$
= \lambda_\Phi \Phi(t).
$$

Consequently, Theorem 5.41 implies that the fixed-point equation $u = A(u)$ is generalized Ulam–Hyers–Rassias stable.

5.5 Notes and remarks

The results of Chapter 5 are taken from the papers [49,60]. The monographs [48,70] and the papers [8–10,12,69,159,160,272] include additional pertinent results and investigations.

Impulsive fractional difference equations

We explored and demonstrated the results obtained in this chapter by taking into consideration the previously stated publications in the preceding chapters, the monographs of Abbas et al. [48,70], Benchohra et al. [165], Graef et al. [251], and papers such as [32,69,78, 133,142,145,146,153,155,170,212,242,309,379,380,410,411,420,421,427,433]. Impulsive differential equations have gained importance in recent years in some mathematical models of real phenomena, especially in biological or medical domains, and in control theory, see for example the monographs [70,165,251], and the papers [20,52,76,339,340]. In [21,36,41–43,182,183,185–187,191,265] the authors initially offered to study some classes of impulsive differential equations with noninstantaneous impulses.

6.1 Impulsive implicit Caputo fractional q-difference equations

6.1.1 Introduction

In this section we first discuss the existence of solutions for the following problem of implicit fractional q-difference equations

$$\begin{cases} ({}^C_q D^r_{t_k} u)(t) = f(t, u(t), ({}^C_q D^r_{t_k} u)(t)); \ t \in J_k, \ k = 0, \dots, m, \\ u(t^+_k) = u(t^-_k) + L_k(u(t^-_k)); \ k = 1, \dots, m, \\ u(0) = u_0 \in \mathbb{R}, \end{cases} \tag{6.1}$$

where $J_0 = [0, t_1]$, $J_k := (t_k, t_{k+1}]$; $k = 1, \dots, m$, $0 = t_0 < t_1 < \cdots < t_m < t_{m+1} = T$, $f : J_k \times \mathbb{R} \times \mathbb{R} \to \mathbb{R}$; $k = 1, \dots, m$, $L_k : \mathbb{R} \to \mathbb{R}$; $k = 1, \dots, m$ are given continuous functions, and ${}^c_q D^r_{t_k}$ is the Caputo fractional q-difference derivative of order $r \in (0, 1]$.

Recently, in [69,133,142,153,170], the authors applied the measure of noncompactness to some classes of functional Riemann-Liouville or Caputo fractional differential equations in Banach spaces. Motivated by the above papers, we next discuss the existence of solutions for the problem (6.1), when $u_0 \in E$, $f : J_k \times E \times E \to E$; $k = 1, \dots, m$, $L_k : E \to E$; $k = 1, \dots, m$ are given continuous functions, E is a real (or complex) Banach space with norm $\| \cdot \|$.

Fractional Difference, Differential Equations, and Inclusions. https://doi.org/10.1016/B978-0-44-323601-3.00013-7
Copyright © 2024 Elsevier Inc. All rights reserved, including those for text and data mining, AI training, and similar technologies.

6.1.2 Existence results in the scalar case

In this section we present some results concerning the existence of solutions for the problem (6.1).

Definition 6.1. By a solution of the problem (6.1) we mean a function $u \in PC$ that satisfies the condition $u(0) = u_0$, and the equation $(^C_q D^r_{t_k} u)(t) = f(t, u(t), (^C_q D^r_{t_k} u)(t))$ on J_k; $k = 0, \ldots, m$.

The following hypotheses will be used in the sequel:

(6.1.1) The function $f : J_k \mapsto f(t, u, v)$; $k = 0, \ldots, m$, is continuous;

(6.1.2) The functions f and L_k; $k = 1, \ldots, m$, satisfy the generalized Lipschitz conditions:

$$|f(t, u_1, v_1) - f(t, u_2, v_2)| \leq \phi_1(|u_1 - u_2|) + \phi_2(|v_1 - v_2|),$$

and

$$|L_k(u_1) - L_k(u_2)| \leq \phi_3(|u - u_2|),$$

for $t \in I$ and $u, v \in \mathbb{R}$, where ϕ_i; $i = 1, 2, 3$; are comparison functions;

(6.1.3) There exists a continuous function $\ell \in C(J_k, \mathbb{R}_+)$; $k = 0, \ldots, m$, such that

$$|f(t, u, v)| \leq \ell(t)(1 + |u| + |v|); \text{ for a.e. } t \in J_k, \text{ and each } u, v \in \mathbb{R},$$

with

$$\ell^* = \sup_{t \in I} \ell(t) < 1;$$

(6.1.4) There exists a constant $l > 0$ such that

$$|L_k(u)| \leq l(1 + |u|); \text{ for each } u \in \mathbb{R}.$$

Theorem 6.2. *Assume that the hypotheses* (6.1.1) *and* (6.1.2) *hold. Then the problem* (6.1) *has a unique solution defined on* I.

Proof. Consider the Banach space $C(I) := C(I, \mathbb{R})$ as a complete metric space of continuous functions from I into $\mathbb{R}$ equipped with the usual metric

$$d(u, v) := \max_{t \in I} |u(t) - v(t)|.$$

Transform the problem (6.1) into a fixed-point equation. Consider the operator $N : PC \to PC$ defined by:

$$\begin{cases} (Nu)(t) = u_0 + (_qI_0^r g)(t); & \text{if } t \in J_0, \\[2mm] (Nu)(t) = u_0 + \displaystyle\sum_{i=1}^{k} L_i(u(t_i^-)) \\[2mm] \quad + \displaystyle\sum_{i=1}^{k} \int_{t_{i-1}}^{t_i} \frac{(t_i - qs)^{(r-1)}}{\Gamma_q(r)} g(s) d_q s \\[2mm] \quad + \displaystyle\int_{t_k}^{t} \frac{(t - qs)^{(r-1)}}{\Gamma_q(r)} g(s) d_q s; & \text{if } t \in J_k, \ k = 1, \ldots, m, \end{cases} \tag{6.2}$$

where $g(\cdot) \in C(J_k)$; $k = 0, \ldots, m$, with

$$g(t) = f(t, u_0 + (_qI_0^r g)(t), g(t)).$$

Clearly, the fixed-points of the operator N are solutions of the problem (6.1).

Let $u \in PC$ and $t \in J_0$. Then

$$|(Nu)(t) - (Nv)(t)| = |(_qI_0^r(g - h))(t)|,$$

where $g, h \in C(J_k)$; $k = 0, \ldots, m$, with

$$g(t) = f(t, u_0 + (_qI_0^r g)(t), g(t)), \text{ and } h(t) = f(t, u_0 + (_qI_0^r h)(t), h(t)).$$

Thus, for each $u, v \in C(I)$ and $t \in J_0$, we have

$$|(Nu)(t) - (Nv)(t)| = \int_0^t \frac{|t - qs|^{(r-1)}}{\Gamma_q(r)} |g(s) - h(s)| d_q s.$$

From (6.1.2) we have

$$|g(t) - h(t)| \le \phi_1(|u(t) - v(t)|) + \phi_2(|g(t) - h(t)|).$$

Thus

$$|g(t) - h(t)| \le (Id - \phi_2)^{-1}\phi_1(|u(t) - v(t)|).$$

Hence

$$|(Nu)(t) - (Nv)(t)| \le \int_0^t \frac{|t - qs|^{(r-1)}}{\Gamma_q(r)} (Id - \phi_2)^{-1}\phi_1(|u(s) - v(s)|) d_q s$$

$$\le \frac{T^r}{\Gamma_q(1+r)}(Id - \phi_2)^{-1}\phi_1(d(u,v))$$

$$= \phi(d(u,v)),$$

where Id is the identity function and ϕ is the comparison function defined by

$$\phi(t) = \frac{T^r}{\Gamma_q(1+r)}(Id - \phi_2)^{-1}\phi_1(t); \ t \in J_0.$$

So we get

$$d(N(u), N(v)) \leq \phi(d(u, v)).$$

Next, for each $u, v \in C(I)$ $t \in J_k : k = 1, \ldots, m$, we get

$$\begin{aligned}
|(Nu)(t) - (Nv)(t)| &\leq \frac{T^r}{\Gamma_q(1+r)} (Id - \phi_2)^{-1} \phi_1(d(u, v)) + m\phi_3(d(u, v)) \\
&\leq (\frac{T^r}{\Gamma_q(1+r)} (Id - \phi_2)^{-1} \phi_1 + m\phi_3)(d(u, v)) \\
&= \phi(d(u, v)),
\end{aligned}$$

where ϕ is the comparison function defined by

$$\phi(t) = \frac{T^r}{\Gamma_q(1+r)} (Id - \phi_2)^{-1} \phi_1(t) + m\phi_3(t); \ t \in J_k : k = 1, \ldots, m, .$$

So we get

$$d(N(u), N(v)) \leq \phi(d(u, v)).$$

Consequently, from Theorem 2.66, the operator N has a unique fixed point, which is the unique solution of our problem (6.1) on I. $\qquad\square$

Theorem 6.3. *Assume that the hypotheses* (6.1.1), (6.1.3) *and* (6.1.4) *hold. If*

$$ml + \frac{2T^r \ell^*}{(1 - \ell^*)\Gamma_q(1+r)} < 1,$$

then the problem (6.1) has at least one solution defined on I.

Proof. Consider the operator $N : PC \to PC$ defined in (6.2). Let $R > 0$ such that

$$R \geq \frac{|u_0| + ml + \frac{2T^r \ell^*}{(1-\ell^*)\Gamma_q(1+r)}}{1 - ml - \frac{2T^r \ell^*}{(1-\ell^*)\Gamma_q(1+r)}},$$

and consider the ball $B_R := B(0, R) = \{w \in \|w\|_{PC} \leq R\}$. We will show that the operator $N : B_R \to B_R$ satisfies all the assumptions of Theorem 2.68. The proof will be given in several steps.

Step 1. $N : B_R \to B_R$ *is continuous.*

Let $\{u_n\}_{n \in \mathbb{N}}$ be a sequence such that $u_n \to u$ in B_R. Then, for each $t \in J_0$, we have

$$|(Nu_n)(t) - (Nu)(t)| \leq \int_0^t \frac{(t - qs)^{(r-1)}}{\Gamma_q(r)} |g_n(s) - g(s)| d_q s, \tag{6.3}$$

where $g, g_n \in C(J_0)$ with

$$g(t) = f(t, u_0 + (_q I_0^r g)(t), g(t)),$$

and
$$g_n(t) = f(t, u_0 + ({}_qI_0^r g_n)(t), g_n(t)).$$

Since $u_n \to u$ as $n \to \infty$ and f is continuous, then by the Lebesgue dominated convergence theorem, (6.3) implies

$$\|N(u_n) - N(u)\|_{PC} \to 0 \quad \text{as } n \to \infty.$$

Also, for each $t \in J_k : k = 1, \ldots, m$, we have

$$
\begin{aligned}
|(Nu_n)(t) - (Nu)(t)| &\le \sum_{i=1}^k \|L_i(u_n(t_i^-)) - L_i(u(t_i^-))\| \\
&+ \sum_{i=1}^k \int_{t_{i-1}}^{t_i} \frac{(t_i - qs)^{(r-1)}}{\Gamma_q(r)} |g_n(s) - g(s)| d_q s \\
&+ \int_{t_k}^t \frac{(t - qs)^{(r-1)}}{\Gamma_q(r)} |g_n(s) - g(s)| d_q s.
\end{aligned}
\tag{6.4}
$$

Again, by the Lebesgue dominated convergence theorem, (6.4) implies the continuity of our operator N.

Step 2. $N(B_R)$ *is bounded.*
Let $u \in B_R$, and $t \in J_0$. Then

$$|((Nu)(t)| = \left| u_0 + \int_0^t \frac{(t - qs)^{(r-1)}}{\Gamma_q(r)} g(s) d_q s \right|,$$

where $g(\cdot) \in C(I)$ with

$$g(t) = f(t, u_0 + ({}_qI_0^r g)(t), g(t)).$$

Thus

$$
\begin{aligned}
|(Nu)(t)| &\le |u_0| + \int_0^t \frac{(t - qs)^{(r-1)}}{\Gamma_q(r)} |g(s)| d_q s \\
&\le |u_0| + \frac{T^r \ell^*(1 + R)}{(1 - \ell^*)\Gamma_q(1 + r)} \\
&\le R.
\end{aligned}
$$

Next, if $u \in B_R$, and $t \in J_k : k = 1, \ldots, m$, we have

$$
\begin{aligned}
|(Nu)(t)| &\le |u_0| + \sum_{i=1}^k \|L_i(u(t_i^-))\| \\
&+ \sum_{i=1}^k \int_{t_{i-1}}^{t_i} \frac{(t_i - qs)^{(r-1)}}{\Gamma_q(r)} |g(s)| d_q s
\end{aligned}
$$

$$+ \int_{t_k}^t \frac{(t - qs)^{(r-1)}}{\Gamma_q(r)} |g(s)| d_q s$$

$$\leq |u_0| + ml(1 + R) + \frac{2T^r \ell^*(1 + R)}{(1 - \ell^*)\Gamma_q(1 + r)}$$

$$\leq R.$$

Hence, for any $u \in B_R$, and each $t \in J$, we get

$$\|N(u)\|_{PC} \leq R.$$

This proves that N transforms the ball $B_R := B(0, R) = \{w \in \|w\|_{PC} \leq R\}$ into itself.

Step 3. $N(B_R)$ *is equicontinuous.*
Let $x_1, x_2 \in J_0$ such that $0 \leq x_1 < x_2 \leq t_1$ and let $u \in B_R$. Then

$$|(Nu)(x_2) - (Nu)(x_1)|$$

$$\leq \left| \int_0^{x_2} \frac{(x_2 - qs)^{(r-1)}}{\Gamma_q(r)} g(s) d_q s - \int_0^{x_1} \frac{(x_1 - qs)^{(r-1)}}{\Gamma_q(r)} g(s) d_q s \right|,$$

where $g \in C(J_0)$ with

$$g(t) = f(t, u_0 + (q I_0^r g)(t), g(t)).$$

Thus

$$|(Nu)(x_2) - (Nu)(x_1)|$$

$$\leq \int_{x_1}^{x_2} \frac{(x_2 - qs)^{(r-1)}}{\Gamma_q(r)} |g(s)| d_q s$$

$$+ \int_0^{x_1} \frac{|(x_2 - qs)^{(r-1)} - (x_1 - qs)^{(r-1)}|}{\Gamma_q(r)} |g(s)| d_q s$$

$$\leq \frac{\ell^*(1 + R)(x_2 - x_1)^r}{(1 - \ell^*)\Gamma_q(1 + r)}$$

$$+ \frac{\ell^*(1 + R)}{1 - \ell^*} \int_0^{x_1} \frac{|(x_2 - qs)^{(r-1)} - (x_1 - qs)^{(r-1)}|}{\Gamma_q(r)} d_q s.$$

As $x_1 \longrightarrow x_2$, the right-hand side of the above inequality tends to zero.
 Also, if we let $x_1, x_2 \in J_k$; $k = 1, \ldots, m$ such that $t_k \leq x_1 < x_2 \leq t_{k+1}$ and let $u \in B_R$, we obtain

$$|(Nu)(x_2) - (Nu)(x_1)|$$

$$\leq \frac{2\ell^*(1 + R)}{(1 - \ell^*)\Gamma_q(1 + r)} |x_2 - x_1|^r$$

$$+ \frac{2\ell^*(1 + R)}{1 - \ell^*} \int_0^{x_1} \left| \frac{(x_2 - qs)^{(r-1)}}{\Gamma_q(r)} - \frac{(x_1 - qs)^{(r-1)}}{\Gamma_q(r)} \right| d_q s.$$

Again, as $x_1 \longrightarrow x_2$, the right-hand side of the above inequality tends to zero. Hence $N(B_R)$ is equicontinuous.

As a consequence of the above three steps, together with the Arzelà–Ascoli theorem, we can conclude that $N : B_R \to B_R$ is continuous and compact. From an application of Theorem 2.66, we deduce that N has a fixed point u, which is a solution of problem (6.1). $\qquad\square$

Now, we use Schaefer's fixed-point theorem to prove the following results.

Theorem 6.4. *Assume that the hypotheses* (6.1.1), (6.1.3), (6.1.4) *and the conditions*

$$ml + \frac{2T^r \ell^*}{(1 - \ell^*)\Gamma_q(1+r)} < 1,$$

hold. Then the problem (6.1) *has at least one solution defined on I.*

Proof. We consider the operator $N : PC \to PC$ defined in (6.2). As in the prove of Theorem 6.3, we can show that $N : PC \to PC$ is continuous and compact. Now it remains to show that the set

$$\mathcal{E} = \{u \in X : u = \lambda N(u); \text{ for some } \lambda \in (0, 1)\}$$

is bounded.

Let $u \in \mathcal{E}$, then $u = \lambda N(u)$, for some $\lambda \in (0, 1)$. Thus, for each $t \in J_0$, we have

$$|u(t)| \le \left| u_0 + \int_1^t \frac{(t - qs)^{(r-1)}}{\Gamma_q(r)} g(s) d_q s \right|,$$

where $g(\cdot) \in C(I)$ with

$$g(t) = f(t, u_0 + (_q I_1^r g)(t), g(t)).$$

Thus

$$|u(t)| \le |u_0| + \int_0^t \frac{(t - qs)^{(r-1)}}{\Gamma_q(r)} |g(s)| d_q s$$

$$\le |u_0| + \int_0^t \frac{(t - qs)^{(r-1)}}{(1 - \ell^*)\Gamma_q(r)} \ell^* (1 + |u(s)|) d_q s$$

$$\le |u_0| + \frac{T^r \ell^*}{(1 - \ell^*)\Gamma_q(1+r)} + \int_0^t \frac{(t - qs)^{(r-1)}}{(1 - \ell^*)\Gamma_q(r)} \ell^* |u(s)| d_q s.$$

We can apply a version of Gronwall lemma to obtain that $|u(t)| \le M_1$, with $M_1 > 0$.

Next, for each $t \in J_k : k = 1, \ldots, m$, we have

$$|u(t)| \leq |u_0| + \sum_{i=1}^{k} \|L_i(u(t_i^-))\|$$

$$+ \sum_{i=1}^{k} \int_{t_{i-1}}^{t_i} \frac{(t_i - qs)^{(r-1)}}{\Gamma_q(r)} |g(s)| d_q s$$

$$+ \int_{t_k}^{t} \frac{(t - qs)^{(r-1)}}{\Gamma_q(r)} |g(s)| d_q s$$

$$\leq |u_0| + ml(1 + |u(t)|) + \frac{2T^r \ell^*}{(1 - \ell^*)\Gamma_q(1 + r)}$$

$$+ \int_{0}^{t} \frac{(t - qs)^{(r-1)}}{(1 - \ell^*)\Gamma_q(r)} \ell^* 2 |u(s)| d_q s.$$

This implies that, for each $t \in J_k : k = 1, \ldots, m$, we get

$$|u(t)| \leq \frac{|u_0| + ml}{1 - ml} + \frac{2T^r \ell^*}{(1 - ml)(1 - \ell^*)\Gamma_q(1 + r)}$$

$$+ \frac{2\ell^*}{1 - ml} \int_{0}^{t} \frac{(t - qs)^{(r-1)}}{(1 - \ell^*)\Gamma_q(r)} |u(s)| d_q s.$$

Also, by applying a version of Gronwall lemma, we can obtain $|u(t)| \leq M_2$, with $M_2 > 0$.

Hence the set $\mathcal{E}$ is bounded. As a consequence of Schaefer's fixed-point theorem (Theorem 2.69), we deduce that N has a fixed point, which is a solution of our problem (6.1).   □

6.1.3 Existence results in Banach spaces

In this section we present some results concerning the existence of solutions for the problem (6.1) in Banach spaces.

Theorem 6.5. *Assume that the following hypotheses hold.*

(6.5.1) *The function f is continuous,*

(6.5.2) *There exists a continuous function $p \in C(J_k, \mathbb{R}_+);\ k = 0, \ldots, m$ such that*

$$\|f(t, u, v)\| \leq p(t)(1 + |u| + |v|);\ \text{for a.e. } t \in J_k,\ \text{and each } u, v \in E,$$

with $p^* = \sup_{t \in J} p(t) < 1$,

(6.5.3) *For each bounded and measurable set $B \subset E$ and for each $t \in J_k;\ k = 0, \ldots, m$, we have*

$$\mu(f(t, B, {}^C_q D^r_{t_k} B)) \leq p(t)\mu(B);\ t \in J_k,\ k = 0, \ldots, m,$$

where ${}^C_q D^r_{t_k} B = \{{}^C_q D^r_{t_k} w : w \in B\}$,

(6.5.4) *There exists a constant L > 0 such that*

$$|L_k(u)| \leq L(1+|u|); \ \textit{for each } u \in E,$$

and, for each bounded and measurable set $B \subset E$ and for each $t \in J_k$; $k = 0, \ldots, m$, we have

$$\mu(L_k(B)) \leq l\mu(B).$$

If

$$\rho := mL + \frac{2p^* T^r}{\Gamma_q(1+r)} < 1, \tag{6.5}$$

then the problem (6.1) has at least one solution defined on I.

Proof. Consider the operator $N : PC \to PC$ defined in (6.2). Let $R > 0$ such that

$$R \geq \frac{|u_0| + mL + \frac{2T^r p^*}{(1-p^*)\Gamma_q(1+r)}}{1 - mL - \frac{2T^r p^*}{(1-p^*)\Gamma_q(1+r)}}.$$

Let $u \in PC$ and $t \in J_0$. Then

$$\|((Nu)(t)\| = \left\| u_0 + \int_0^t \frac{(t-qs)^{(r-1)}}{\Gamma_q(r)} g(s) d_q s \right\|,$$

where $g(\cdot) \in C(I)$ with

$$g(t) = f(t, u_0 + (_q I_0^r g)(t), g(t)).$$

Thus

$$\|(Nu)(t)\| \leq \|u_0\| + \int_0^t \frac{(t-qs)^{(r-1)}}{\Gamma_q(r)} \|g(s)\| d_q s$$

$$\leq \|u_0\| + \frac{T^r p^*(1+R)}{(1-p^*)\Gamma_q(1+r)}.$$

On the other hand, if $u \in PC$ and $t \in J_k$: $k = 1, \ldots, m$, we have

$$\|(Nu)(t)\| \leq \|u_0\| + \sum_{i=1}^k \|L_i(u(t_i^-))\|$$

$$+ \sum_{i=1}^k \int_{t_{i-1}}^{t_i} \frac{(t_i - qs)^{(r-1)}}{\Gamma_q(r)} \|g(s)\| d_q s$$

$$+ \int_{t_k}^t \frac{(t-qs)^{(r-1)}}{\Gamma_q(r)} \|g(s)\| d_q s$$

$$\leq \|u_0\| + mL + \frac{2T^r p^*(1+R)}{(1-p^*)\Gamma_q(1+r)}.$$

Hence, for any $u \in PC$ and each $t \in J$, we get

$$\|N(u)\|_{PC} \le \|u_0\| + mL + \frac{2T^r p^*(1+R)}{(1-p^*)\Gamma_q(1+r)} \le R.$$

This proves that N transforms the ball $B_R := B(0,R) = \{w \in \|w\|_{PC} \le R\}$ into itself.

We will show that the operator $N : B_R \to B_R$ satisfies all the assumptions of Theorem 2.75. The proof will be given in three steps.

Step 1. $N : B_R \to B_R$ *is continuous.*
Let $\{u_n\}_{n \in \mathbb{N}}$ be a sequence such that $u_n \to u$ in B_R. Then, for each $t \in J_0$, we have

$$\|(Nu_n)(t) - (Nu)(t)\| \le \int_0^t \frac{(t-qs)^{(r-1)}}{\Gamma_q(r)} \|g_n(s) - g(s)\| d_q s, \tag{6.6}$$

where $g, g_n \in C(J_0)$ with

$$g(t) = f(t, u_0 + ({}_qI_0^r g)(t), g(t)),$$

and

$$g_n(t) = f(t, u_0 + ({}_qI_0^r g_n)(t), g_n(t)).$$

Since $u_n \to u$ as $n \to \infty$ and f is continuous, then by the Lebesgue dominated convergence theorem, (6.6) implies

$$\|N(u_n) - N(u)\|_{PC} \to 0 \quad \text{as } n \to \infty.$$

Also, for each $t \in J_k : k = 1, \ldots, m$, we have

$$\begin{aligned}
\|(Nu_n)(t) - (Nu)(t)\| \le &\sum_{i=1}^k \|L_i(u_n(t_i^-)) - L_i(u(t_i^-))\| \\
&+ \sum_{i=1}^k \int_{t_{i-1}}^{t_i} \frac{(t_i - qs)^{(r-1)}}{\Gamma_q(r)} \|g_n(s) - g(s)\| d_q s \\
&+ \int_{t_k}^t \frac{(t-qs)^{(r-1)}}{\Gamma_q(r)} \|g_n(s) - g(s)\| d_q s.
\end{aligned} \tag{6.7}$$

Again, by the Lebesgue dominated convergence theorem, (6.7) implies the continuity of our operator N.

Step 2. $N(B_R)$ *is bounded and equicontinuous.*
Since $N(B_R) \subset B_R$ and B_R is bounded, then $N(B_R)$ is bounded.

Next, let $x_1, x_2 \in J_0$ such that $0 \le x_1 < x_2 \le t_1$ and let $u \in B_R$. Then

$$\begin{aligned}
&\|(Nu)(x_2) - (Nu)(x_1)\| \\
&\le \left\| \int_0^{x_2} \frac{(x_2 - qs)^{(r-1)}}{\Gamma_q(r)} g(s) d_q s - \int_0^{x_1} \frac{(x_1 - qs)^{(r-1)}}{\Gamma_q(r)} g(s) d_q s \right\|,
\end{aligned}$$

where $g \in C(J_0)$ with

$$g(t) = f(t, u_0 + (_q I_0^r g)(t), g(t)).$$

Thus

$$\|(Nu)(x_2) - (Nu)(x_1)\|$$
$$\leq \int_{x_1}^{x_2} \frac{(x_2 - qs)^{(r-1)}}{\Gamma_q(r)} \|g(s)\| d_q s$$
$$+ \int_0^{x_1} \frac{|(x_2 - qs)^{(r-1)} - (x_1 - qs)^{(r-1)}|}{\Gamma_q(r)} \|g(s)\| d_q s$$
$$\leq \frac{p^*(1+R)(x_2 - x_1)^r}{(1-p^*)\Gamma_q(1+r)}$$
$$+ \frac{p^*(1+R)}{1-p^*} \int_0^{x_1} \frac{|(x_2 - qs)^{(r-1)} - (x_1 - qs)^{(r-1)}|}{\Gamma_q(r)} d_q s.$$

As $x_1 \longrightarrow x_2$, the right-hand side of the above inequality tends to zero.

Also, if we let $x_1, x_2 \in J_k$; $k = 1, \ldots, m$ such that $t_k \leq x_1 < x_2 \leq t_{k+1}$ and let $u \in B_R$, we obtain

$$\|(Nu)(x_2) - (Nu)(x_1)\|$$
$$\leq \frac{2p^*(1+R)}{(1-p^*)\Gamma_q(1+r)}|x_2 - x_1|^r$$
$$+ \frac{2p^*(1+R)}{1-p^*} \int_0^{x_1} \left| \frac{(x_2 - qs)^{(r-1)}}{\Gamma_q(r)} - \frac{(x_1 - qs)^{(r-1)}}{\Gamma_q(r)} \right| d_q s.$$

Again, as $x_1 \longrightarrow x_2$, the right-hand side of the above inequality tends to zero. Hence $N(B_R)$ is bounded and equicontinuous.

Step 3. *The implication (2.12) holds.*
Now let V be a subset of B_R such that $V \subset \overline{N(V)} \cup \{0\}$, V is bounded and equicontinuous and therefore the function $t \to v(t) = \mu(V(t))$ is continuous on J. By (6.5.3) and the properties of the measure μ, for each $t \in J_0$, we have

$$v(t) \leq \mu((NV)(t) \cup \{0\})$$
$$\leq \mu((NV)(t))$$
$$\leq \int_0^t \frac{(t - qs)^{(r-1)} p(s)}{\Gamma_q(r)} v(s) d_q s$$
$$\leq \int_0^t \frac{(t - qs)^{(r-1)} p(s)}{\Gamma_q(r)} \mu(V(s)) d_q s$$
$$\leq \frac{p^* T^r}{\Gamma_q(1+r)} \|v\|_{PC}.$$

Thus

$$\|v\|_{PC} \leq \rho \|v\|_{PC}.$$

Also, for each $t \in J_k$; $k = 1, \ldots, m$, we get

$$
\begin{aligned}
v(t) &\leq \mu((NV)(t) \cup \{0\}) \\
&\leq \mu((NV)(t)) \\
&\leq \sum_{i=1}^{k} l^* \mu(V(s)) + \sum_{i=1}^{k} \int_{t_{i-1}}^{t_i} \frac{(t_i - qs)^{(r-1)} p(s)}{\Gamma_q(r)} \mu(V(s)) d_q s \\
&\quad + \int_{t_k}^{t} \frac{(t - qs)^{(r-1)} p(s)}{\Gamma_q(r)} \mu(V(s)) d_q s \\
&\leq L \sum_{i=1}^{k} v(t) + \sum_{i=1}^{k} \int_{t_{i-1}}^{t_i} \frac{(t_i - qs)^{(r-1)} p(s)}{\Gamma_q(r)} v(s) d_q s \\
&\quad + \int_{t_k}^{t} \frac{(t - qs)^{(r-1)} p(s)}{\Gamma_q(r)} v(s) d_q s \\
&\leq \left(mL + \frac{2p^* T^r}{\Gamma_q(1+r)} \right) \|v\|_{PC}.
\end{aligned}
$$

Hence

$$\|v\|_{PC} \leq \rho \|v\|_{PC}.$$

From (6.5), we get $\|v\|_{PC} = 0$, that is $v(t) = \beta(V(t)) = 0$, for each $t \in I$, and then $V(t)$ is relatively compact in E. In view of the Ascoli–Arzelà theorem, V is relatively compact in B_R. Applying now Theorem 2.75, we conclude that N has a fixed point, which is a solution of the problem (6.1). $\qquad\square$

6.1.4 Examples

Example 1. Consider the problem of implicit impulsive q-fractional differential equation of the form

$$
\begin{cases}
({}_q^C D_{t_k}^r u)(t) = f(t, u(t), ({}_q^C D_{t_k}^r u)(t)); \ t \in J_k, \ k = 0, \ldots, m, \\
u(t_k^+) = u(t_k^-) + L_k(u(t_k^-)); \ k = 1, \ldots, m, \\
u(0) = 0,
\end{cases}
\tag{6.8}
$$

where $I = [0, 1]$, $r \in (0, 1]$,

$$f(t, u(t), ({}_q^C D_{t_k}^r u)(t)) = \frac{\Gamma_q(1+r)t^2 \left(e^{-7} + \frac{1}{e^{t+5}} \right)(2^{-n} + u_n(t))}{1 + |u(t)| + |{}_q^C D_{t_k}^r u(t)|}; \ t \in [0, 1],$$

$$L_k(u(t_k^-)) = \frac{1}{(3e^{45})(1 + |u(t_k^-)|)}; \ k = 1, \ldots, m.$$

Clearly, the function f is continuous.

For each $t \in [0, 1]$, we have

$$|f(t, u(t), ({}^{C}_{q}D^{r}_{t_k}u)(t))| \leq \Gamma_q(1+r)t^2 \left(e^{-7} + \frac{1}{e^{t+5}}\right),$$

and

$$|L_k(u)| \leq \frac{1}{3e^5}.$$

Hence the hypothesis (6.5.2) is satisfied with $\ell^* = 2e^{-5}\Gamma_q(1+r)$ and (6.5.4) is satisfied with $l = \frac{1}{3e^4}$.

Simple computations show that all conditions of Theorem 6.3 are satisfied. It follows that the problem (6.8) has at least one solution on $[0, 1]$.

Example 2. Let

$$E = l^1 = \left\{u = (u_1, u_2, \ldots, u_n, \ldots), \sum_{n=1}^{\infty} |u_n| < \infty\right\}$$

be the Banach space with the norm

$$\|u\|_E = \sum_{n=1}^{\infty} |u_n|.$$

Consider the problem of implicit impulsive q-fractional differential equation of the form

$$\begin{cases} ({}^{C}_{q}D^{r}_{t_k}u)(t) = f(t, u(t), ({}^{C}_{q}D^{r}_{t_k}u)(t)); \ t \in J_k, \ k = 0, \ldots, m, \\ u(t_k^+) = u(t_k^-) + L_k(u(t_k^-)); \ k = 1, \ldots, m, \\ u(0) = 0, \end{cases} \qquad (6.9)$$

where $I = [0, 1]$, $r \in (0, 1]$, $u = (u_1, u_2, \ldots, u_n, \ldots)$,

$$f = (f_1, f_2, \ldots, f_n, \ldots),$$

$${}^{C}_{q}D^{r}_{t_k}u = ({}^{C}_{q}D^{r}_{t_k}u_1, {}^{C}_{q}D^{r}_{t_k}u_2, \ldots, {}^{C}_{q}D^{r}_{t_k}u_n, \ldots); \ k = 0, \ldots, m,$$

$$f_n(t, u(t), ({}^{C}_{q}D^{r}_{t_k}u)(t)) = \frac{\Gamma_q(1+r)t^2 \left(e^{-7} + \frac{1}{e^{t+5}}\right)(2^{-n} + u_n(t))}{1 + \|u(t)\|_E + \|{}^{C}_{q}D^{r}_{t_k}u(t)\|_E}; \ t \in [0, 1],$$

$$L_k(u(t_k^-)) = \frac{1}{(3e^{45})(1 + \|u(t_k^-)\|_E)}; \ k = 1, \ldots, m.$$

For each $u \in E$ and $t \in [0, 1]$, we have

$$\|f(t, u(t), ({}^{C}_{q}D^{r}_{t_k}u)(t))\|_E \leq \Gamma_q(1+r)t^2 \left(e^{-7} + \frac{1}{e^{t+5}}\right),$$

and

$$\|L_k(u)\|_E \leq \frac{1}{3e^5}.$$

Hence the hypothesis (6.5.2) is satisfied with $p^* = 2e^{-5}\Gamma_q(1+r)$ and (6.5.4) is satisfied with $L = \frac{1}{3e^4}$.

We will show that condition (6.5) holds with $T = 1$. Indeed, if we assume, for instance, that the number of impulses $m = 3$ and $r = \frac{1}{2}$, then we have

$$L := mL + \frac{2p^*T^r}{\Gamma_q(1+r)} = e^{-5} + \frac{2e^{-5}\Gamma_q(1+r)}{\Gamma_q(\frac{3}{2})} = 3e^{-5} < 1.$$

Simple computations show that all conditions of Theorem 6.5 are satisfied. It follows that the problem (6.9) has at least one solution on [0, 1].

6.2 Implicit Caputo fractional q-difference equations with noninstantaneous impulses

6.2.1 Introduction

In this section we first discuss the existence of solutions for the following problem of implicit fractional q-difference equations with noninstantaneous impulses.

$$\begin{cases} (^C_q D^r_{s_k} u)(t) = f(t, u(t), (^C_q D^r_{s_k} u)(t)); \ t \in I_k, \ k = 0, \dots, m, \\ u(t) = g_k(t, u(t_k^-)); \ \text{if } t \in J_k, \ k = 1, \dots, m, \\ u(s_k) + Q(u) = u_k; \ k = 0, \dots, m, \end{cases} \tag{6.10}$$

where $I_0 := [0, t_1]$, $J_k := (t_k, s_k]$, $I_k := (s_k, t_{k+1}]$; $u_k \in \mathbb{R}$, $k = 1, \dots, m$, $f : I_k \times \mathbb{R} \times \mathbb{R} \to \mathbb{R}$, $g_k : J_k \times \mathbb{R} \to \mathbb{R}$, $Q : PC \to \mathbb{R}$ are given functions such that $g_k(t, u(t_k^-))|_{t=s_k} = u_k - Q(u) \in \mathbb{R}$; $k = 1, \dots, m$, $0 = s_0 < t_1 \leq s_1 < t_2 \leq s_2 < \cdots \leq s_{m-1} < t_m \leq s_m < t_{m+1} = T$, and $^C_q D^r_{s_k}$ is the Caputo fractional q-difference derivative of order $r \in (0, 1]$.

Next we discuss the existence of solutions for problem (6.10), when $u_k \in E$, $f : I_k \times E \times E \to E$, $g_k : J_k \times E \to E$, $Q : PC \to E$ are given functions such that $g_k(t, u(t_k^-))|_{t=s_k} = u_k \in E$; $k = 1, \dots, m$, and E is a real (or complex) Banach space with norm $\| \cdot \|$.

6.2.2 Existence results in the scalar case

In this section we present some results concerning the existence of solutions for problem (6.10).

Definition 6.6. By a solution of problem (6.10) we mean a function $u \in PC$ that satisfies the condition $u(s_k) = u_k$; $k = 0, \dots, m$, and the equations $(^C_q D^r_{t_k} u)(t) = f(t, u(t), (^C_q D^r_{t_k} u)(t))$ on I_k; $k = 0, \dots, m$, and $u(t) = g_k(t, u(t_k^-))$ on J_k; $k = 1, \dots, m$.

The following hypotheses will be used in the sequel:

(6.6.1) The functions Q, f, g_k; $k = 1, \ldots, m$, are continuous.

(6.6.2) The functions Q, f and g_k; $k = 1, \ldots, m$, satisfy the Lipschitz conditions:

   (i) $|f(t, u_1, v_1) - f(t, u_2, v_2)| \le \phi_1 |u_1 - u_2| + \phi_2 |v_1 - v_2|$;

   (ii) $|Q(u) - Q(v)| \le \phi_3 \|u - v\|_{PC}$;

   (iii) $|g_k(t, u_1) - g_k(t, u_2)| \le \phi_4 |u_1 - u_2|$, for $t \in I$ and $u, v \in PC$, $u_i, v_i \in \mathbb{R}$; $i = 1, 2$, where ϕ_i; $i = 1, \ldots, 4$; are positive constants with $\phi_4 < 1$.

(6.6.3) There exist continuous functions $l_k \in C(I_k, \mathbb{R}_+)$; $k = 0, \ldots, m$, $c_k \in C(J_k, \mathbb{R}_+)$; $k = 1, \ldots, m$, and a constant $L > 0$, such that

$$|f(t, u, v)| \le l_k(t)(1 + |u| + |v|); \text{ for a.e. } t \in I_k, \text{ and each } u, v \in \mathbb{R},$$

$$|g_k(t, u)| \le c_k(t)(1 + |u|); \text{ for each } u \in \mathbb{R},$$

with

$$c^* = \max_{k=1,\ldots,m} \left\{ \sup_{t \in J_k} \{c_k(t)\} \right\} < 1,$$

and

$$|Q(u)| \le L(1 + \|u\|_{PC}); \text{ for each } u \in PC.$$

Remark 6.7. From (6.6.2), we have that

$$|f(t, u, v| \le |f(t, 0, 0)| + \phi_1 |u| + \phi_2 |v|,$$

$$|Q(u)| \le |Q(0)| + \phi_3 \|u\|_{PC},$$

and

$$|g_k(t, u)| \le |g_k(t, 0)| + \phi_4 |u|.$$

So under some additional conditions, we can see that (6.6.2) imply (6.6.3).

The first result is based on the Banach contraction mapping principle.

Theorem 6.8. *Assume that hypotheses* (6.6.1) *and* (6.6.2) *hold. By denoting*

$$\rho = \max_{k \in \{1,2,\ldots,m\}} (t_{k+1} - s_k),$$

if

$$\phi_1 \frac{\rho^r}{\Gamma_q(r+1)} + \phi_2 < 1, \tag{6.11}$$

and

$$\phi := \phi_3 \left(1 + \frac{\rho^r \phi_1}{\Gamma_q(1+r)\left(1 - \phi_1 \frac{\rho^r}{\Gamma_q(r+1)} - \phi_2\right)} \right) < 1, \tag{6.12}$$

then problem (6.10) *has a unique solution defined on I.*

Proof. Consider the Banach space PC as a complete metric space of continuous functions from I into $\mathbb{R}$ equipped with the usual metric

$$d(u, v) := \max_{t \in I} |u(t) - v(t)|.$$

Now we consider operator $N : PC \to PC$ defined by:

$$\begin{cases} (Nu)(t) = u_k - Q(u) + ({}_q I^r_{s_k} g)(t); \ t \in I_k, \ k = 0, \ldots, m, \\ (Nu)(t) = g_k(t, u(t_k^-)); \ t \in J_k, \ k = 1, \ldots, m, \end{cases} \tag{6.13}$$

where $g \in C(I_k); \ k = 0, \ldots, m$, is the unique solution of

$$g(t) = f(t, u, g(t)). \tag{6.14}$$

Clearly, the fixed points of operator N are the solutions of problem (6.10).

First, we must verify that operator N is well defined, i.e., g is uniquely determined by Eq. (6.14). To this end, we define the following operator $H_k : C(I_k) \to C(I_k)$, as follows:

$$H_k(g(t)) = f(t, u_k - Q(u) + ({}_q I^r_{s_k} g)(t), g(t)) \tag{6.15}$$

So given $g_1, \ g_2 \in C(I)$, using (6.6.2), we have that, for all $t \in I_k$, the following inequalities are fulfilled

$$\begin{aligned} |H_k(g_2(t)) - H_k(g_1(t))| &\leq |f(t, u_k - Q(u) + ({}_q I^r_{s_k} g_2)(t), g_2(t)) \\ &\quad - f(t, u_k - Q(u) + ({}_q I^r_{s_k} g_1)(t), g_1(t))| \\ &\leq \phi_1 |{}_q I^r_{s_k} (g_1 - g_2)(t)| + \phi_2 |(g_1 - g_2)(t)| \\ &\leq \phi_1 \frac{(t_{k+1} - s_k)^r}{\Gamma_q(r+1)} d(g_1, g_2) + \phi_2 d(g_1, g_2) \\ &\leq \left(\phi_1 \frac{\rho^r}{\Gamma_q(r+1)} + \phi_2 \right) d(g_1, g_2). \end{aligned}$$

As a consequence, we have that Eq. (6.14) has a unique solution for any $t \in I_k$.

The continuity of f and Q implies that $g \in C(I_k)$ for all $k \in \{1, 2, \ldots, m\}$.

Let $u \in PC$ and $t \in I_k, k = 0, \ldots, m$. Then

$$|(Nu)(t) - (Nv)(t)| = |Q(u) - Q(v)| + |({}_q I^r_{s_k} (g - h))(t)|,$$

where $g, h \in C(I_k); \ k = 0, \ldots, m$, are the unique solutions of the equations

$$g(t) = f(t, u_k - Q(u) + ({}_q I^r_{s_k} g)(t), g(t)),$$

and

$$h(t) = f(t, u_k - Q(v) + ({}_q I^r_{s_k} h)(t), h(t)).$$

Thus, for each $u, v \in C(I_k)$ and $t \in I_k$, we have

$$|(Nu)(t) - (Nv)(t)| \leq |Q(u) - Q(v)| + \int_{s_k}^{t} \frac{|t - qs|^{(r-1)}}{\Gamma_q(r)} |g(s) - h(s)| d_q s.$$

From (6.6.2) we have, for all $t \in I_k$:

$$|g(t) - h(t)| \leq \phi_1 |Q(u) - Q(v)| + \phi_1 {}_q I_{s_k}^{r} |(g - h)(t)| + \phi_2 |g(t) - h(t)|$$

$$\leq \phi_1 \phi_3 \|u - v\|_{PC} + \phi_1 \frac{\rho^r}{\Gamma_q(r+1)} |g(t) - h(t)| + \phi_2 |g(t) - h(t)|$$

Thus

$$|g(t) - h(t)| \leq \frac{\phi_1 \phi_3}{1 - \phi_1 \frac{\rho^r}{\Gamma_q(r+1)} - \phi_2} \|u - v\|_{PC}, \quad t \in I_k.$$

Hence

$$|(Nu)(t) - (Nv)(t)|$$

$$\leq |Q(u) - Q(v)| + \int_{s_k}^{t} \frac{|t - qs|^{(r-1)}}{\Gamma_q(r)} \frac{\phi_1 \phi_3}{1 - \phi_1 \frac{\rho^r}{\Gamma_q(r+1)} - \phi_2} \|u - v\|_{PC} d_q s$$

$$\leq \phi_3 \, d(u, v) + \frac{\rho^r}{\Gamma_q(1+r)} \frac{\phi_1 \phi_3}{1 - \phi_1 \frac{\rho^r}{\Gamma_q(r+1)} - \phi_2} d(u, v)$$

$$=: \phi \, d(u, v).$$

So we deduce that

$$d(N(u), N(v)) \leq \phi \, d(u, v).$$

Next, for each $u, v \in C(J_k)$ and $t \in J_k: k = 1, \dots, m$, we get

$$|(Nu)(t) - (Nv)(t)| \leq |g_k(t, u(t_k^-)) - g_k(t, v(t_k^-))|$$
$$\leq \phi_4 \, d(u, v).$$

So we arrive at

$$d(N(u), N(v)) \leq \phi_4 \, d(u, v).$$

Consequently, from the Banach contraction principle, the operator N has a unique fixed point, which is the unique solution of our problem (6.10) on I. $\qquad\square$

The next result is based on Schauder's fixed-point theorem. Set

$$l^* = \max_{k=0,\dots,m} \left\{ \sup_{t \in I_k} \{l_k(t)\} \right\}.$$

Theorem 6.9. *Assume that the hypotheses* (6.6.1), (6.6.2)(*i*), (*ii*) *and* (6.6.3) *hold. If*

$$l^* \left(1 + \frac{\rho^r}{\Gamma_q(1+r)} \right) < 1,$$

and

$$L \left(1 + \frac{\rho^r l^*}{\Gamma_q(1+r)} \frac{1}{1 - l^* \left(1 + \frac{\rho^r}{\Gamma_q(1+r)} \right)} \right) < 1,$$

then problem (6.10) *has at least one solution defined on I.*

Proof. Consider the operator $N : PC \to PC$ defined in (6.13). Let $R > 0$ be such that

$$R = \max_{k=1,2,\dots,m} \left\{ \frac{c^*}{1-c^*}, \frac{|u_k| + L + \frac{\rho^r l^*}{\Gamma_q(1+r)} \frac{1+|u_k|+L\,(1+R)}{1-l^* \left(1 + \frac{\rho^r}{\Gamma_q(1+r)} \right)}}{1 - L \left(1 + \frac{\rho^r l^*}{\Gamma_q(1+r)} \frac{1}{1-l^* \left(1 + \frac{\rho^r}{\Gamma_q(1+r)} \right)} \right)} \right\},$$

and consider the ball $B_R := B(0, R) = \{w \in PC, \; \|w\|_{PC} \le R\}$.

We will show that operator $N : B_R \to B_R$ satisfies all the assumptions of Theorem 2.68. The proof will be given in several steps.

Step 1. $N : PC \to PC$ *is continuous.*
Let $\{u_n\}_{n \in \mathbb{N}} \subset PC$ be a sequence such that $u_n \to u$ in PC. Then, for each $t \in I_k$; $k = 0, \dots, m$, we have

$$|(Nu_n)(t) - (Nu)(t)| \le |Q(u_n) - Q(u)| + \int_{s_k}^{t} \frac{(t-qs)^{(r-1)}}{\Gamma_q(r)} |g_n(s) - g(s)| d_q s, \qquad (6.16)$$

where $g, g_n \in C(I_k)$ are the unique solutions of the following equations

$$g(t) = f(t, u_k - Q(u) + ({}_q I_{s_k}^r g)(t), g(t)),$$

and

$$g_n(t) = f(t, u_k - Q(u_n) + ({}_q I_{s_k}^r g_n)(t), g_n(t)).$$

The uniqueness of such functions is deduced from (6.6.2)(*i*), (*ii*) and (6.11), as in the proof of Theorem 6.8.

Since $\|u_n - u\|_{PC} \to 0$ as $n \to \infty$ and f and Q are continuous, then the Lebesgue dominated convergence theorem, (6.14) and (6.16) imply that

$$\|N(u_n) - N(u)\|_{PC} \to 0 \quad \text{as } n \to \infty.$$

Also, for each $t \in J_k$; $k = 1, \dots, m$, we have

$$|(Nu_n)(t) - (Nu)(t)| \le |g_k(t, u_n(t_k^-)) - g_k(t, u(t_k^-))|.$$

Using again that $\|u_n - u\|_{PC} \to 0$ as $n \to \infty$ and the continuity of functions g_k, we deduce the continuity of operator N on PC.

Step 2. $N(B_R) \subset B_R$.
Let $u \in B_R$, and $t \in I_k$; $k = 0, \ldots, m$; Then

$$|(Nu)(t)| = \left| u_k - Q(u) + \int_{s_k}^{t} \frac{(t - qs)^{(r-1)}}{\Gamma_q(r)} g(s) d_q s \right|.$$

In this case, using (6.6.3), we know that $g \in C(I)$ satisfies

$$|g(t)| \leq l_k(t)(1 + |u_k - Q(u) +_q I_{s_k}^r g(t)| + |g(t)|)$$

$$\leq l^* \left(1 + |u_k| + |Q(u)| + \frac{\rho^r}{\Gamma_q(1+r)} \|g\|_\infty + \|g\|_\infty \right).$$

As a consequence, using (6.6.3) again, we deduce that

$$\|g\|_\infty \leq l^* \frac{1 + |u_k| + L(1 + R)}{1 - l^* \left(1 + \frac{\rho^r}{\Gamma_q(1+r)} \right)}.$$

Thus

$$|(Nu)(t)| \leq |u_k| + |Q(u)| + \int_{s_k}^{t} \frac{(t - qs)^{(r-1)}}{\Gamma_q(r)} |g(s)| d_q s$$

$$\leq |u_k| + L(1 + R) + \frac{\rho^r l^*}{\Gamma_q(1+r)} \frac{1 + |u_k| + L(1 + R)}{1 - l^* \left(1 + \frac{\rho^r}{\Gamma_q(1+r)} \right)}$$

$$\leq R.$$

Next, if $u \in B_R$, and $t \in J_k$; $k = 1, \ldots, m$, we have

$$|(Nu)(t)| \leq c^*(1 + R) \leq R.$$

Hence, for any $u \in B_R$ and each $t \in I$, we get

$$\|N(u)\|_{PC} \leq R.$$

This proves that N transforms the ball $B_R := B(0, R) = \{w \in \|w\|_{PC} \leq R\}$ into itself.

Step 3. *$N(B_R)$ is equicontinuous.*
Let $x_1, x_2 \in I_k$; $k = 0, \ldots, m$ such that $s_k \leq x_1 < x_2 \leq t_{k+1}$ and let $u \in B_R$. Then

$$|(Nu)(x_2) - (Nu)(x_1)|$$

$$\leq \left| \int_{s_k}^{x_2} \frac{(x_2 - qs)^{(r-1)}}{\Gamma_q(r)} g(s) d_q s - \int_{s_k}^{x_1} \frac{(x_1 - qs)^{(r-1)}}{\Gamma_q(r)} g(s) d_q s \right|.$$

Thus

$$|(Nu)(x_2) - (Nu)(x_1)|$$

$$\leq \int_{x_1}^{x_2} \frac{(x_2 - qs)^{(r-1)}}{\Gamma_q(r)} |g(s)| d_q s$$

$$+ \int_{s_k}^{x_1} \frac{|(x_2 - qs)^{(r-1)} - (x_1 - qs)^{(r-1)}|}{\Gamma_q(r)} |g(s)| d_q s$$

$$\leq l^* \frac{1 + |u_k| + L(1+R)}{1 - l^*\left(1 + \frac{\rho^r}{\Gamma_q(1+r)}\right)} \frac{(x_2 - x_1)^r}{\Gamma_q(1+r)}$$

$$+ l^* \frac{1 + |u_k| + L(1+R)}{1 - l^*\left(1 + \frac{\rho^r}{\Gamma_q(1+r)}\right)} \int_0^{x_1} \frac{|(x_2 - qs)^{(r-1)} - (x_1 - qs)^{(r-1)}|}{\Gamma_q(r)} d_q s.$$

As $x_1 \longrightarrow x_2$, the right-hand side of the above inequality tends to zero.

Also, if we let $x_1, x_2 \in J_k$; $k = 1, \ldots, m$ such that $t_k \leq x_1 < x_2 \leq s_k$ and let $u \in B_R$, we obtain

$$|(Nu)(x_2) - (Nu)(x_1)| \leq |g_k(x_2, u(t_k^-)) - g_k(x_1, u(t_k^-))|.$$

From the continuity of g_k, again, as $x_1 \longrightarrow x_2$, the right-hand side of the above inequality tends to zero.

Hence, $N(B_R)$ is equicontinuous.

As a consequence of the above three steps, together with the Arzelà–Ascoli theorem, we can conclude that $N : B_R \to B_R$ is continuous and compact. As a direct application of Theorem 2.68, we deduce that N has a fixed point u, which is a solution of problem (6.10). □

Remark 6.10. We can use Schaefer's fixed-point theorem (Theorem 2.69), to prove Theorem 6.9. Indeed, we can show that the operator $N : PC \to PC$ is continuous and compact and the set $\mathcal{E} = \{u \in X : u = \lambda N(u); \text{ for some } \lambda \in (0, 1)\}$ is bounded.

6.2.3 Existence results in Banach spaces

In this section we present some results concerning the existence of solutions for problem (6.10) in Banach spaces.

Theorem 6.11. *Assume that the following hypotheses hold.*

(6.11.1) *The functions Q, f and g_k; $k = 0, \ldots, m$, are continuous.*
(6.11.2) *The functions Q, f and g_k; $k = 1, \ldots, m$, satisfy the Lipschitz conditions:*

$$\|f(t, u_1, v_1) - f(t, u_2, v_2)\| \leq \phi_1 \|u_1 - u_2\| + \phi_2 \|v_1 - v_2\|,$$
$$\|Q(u) - Q(v)\| \leq \phi_3 \|u - v\|_{PC},$$

for $t \in I$ and $u, v \in PC$, $u_i, v_i \in E$; $i = 1, 2$, where ϕ_i; $i = 1, \ldots, 3$; are positive constants.

(6.11.3) *There exist a constant $L > 0$, and continuous functions $p_k \in C(J_k, \mathbb{R}_+)$; $k = 0, \ldots, m$, and $c_k \in C(J_k, \mathbb{R}_+)$; $k = 1, \ldots, m$, such that*

$$\|f(t, u, v)\| \leq p_k(t)(1 + \|u\| + \|v\|); \ for \ a.e. \ t \in I_k, \ and \ each \ u, v \in E,$$
$$\|g_k(t, u)\| \leq c_k(t)(1 + \|u\|); \ for \ a.e. \ t \in J_k, \ and \ each \ u \in E,$$

with $c^* = \max\limits_{k=1,\ldots,m} \left\{ \sup\limits_{t \in J_k} \{l(t)\} \right\} < 1$, and

$$\|Q(u)\| \leq L(1 + \|u\|_{PC}); \ for \ each \ u \in PC.$$

(6.11.4) *For each bounded set $D \subset PC$, we have*

$$\mu(Q(D)) \leq L \sup\limits_{t \in I} \mu(D(t)),$$

where $D(t) = \{u(t) : u \in D\}$; $t \in I$, and for each bounded and measurable set $B \subset E$, we have

$$\mu(f(t, B, {}^C_q D^r_{t_k} B)) \leq p_k(t)\mu(B); \ t \in I_k, \ k = 0, \ldots, m,$$

where ${}^C_q D^r_{t_k} B = \{{}^C_q D^r_{t_k} w : w \in B\}$, and

$$\mu(g_k(t, B)) \leq c_k(t)\mu(B); \ t \in J_k, \ k = 0, \ldots, m.$$

Set

$$p^* = \max\limits_{k=0,\ldots,m} \left\{ \sup\limits_{t \in J} \{p_k(t)\} \right\}.$$

If the conditions

$$\phi_1 \frac{\rho^r}{\Gamma_q(r+1)} + \phi_2 < 1, \tag{6.17}$$

$$p^* \left(1 + \frac{\rho^r}{\Gamma_q(1+r)}\right) < 1,$$

$$L \left(1 + \frac{\rho^r p^*}{\Gamma_q(1+r)} \frac{1}{1 - p^* \left(1 + \frac{\rho^r}{\Gamma_q(1+r)}\right)}\right) < 1,$$

and

$$L + \frac{p^* \rho^r}{\Gamma_q(1+r)} < 1, \tag{6.18}$$

are satisfied, then problem (6.10) has at least one solution defined on I.

Proof. Consider the operator $N : PC \to PC$ defined in (6.13). Let $R > 0$ such that

$$R = \max_{k=1,2,\ldots,m} \left\{ \frac{c^*}{1-c^*}, \frac{|u_k| + L + \frac{\rho^r p^*}{\Gamma_q(1+r)} \frac{1+|u_k|+L(1+R)}{1-p^*\left(1+\frac{\rho^r}{\Gamma_q(1+r)}\right)}}{1-L\left(1+\frac{\rho^r p^*}{\Gamma_q(1+r)} \frac{1}{1-p^*\left(1+\frac{\rho^r}{\Gamma_q(1+r)}\right)}\right)} \right\},$$

and consider the ball $B_R := B(0, R) = \{w \in PC, \|w\|_{PC} \le R\}$.

Let $u \in PC$ and $t \in I_k$. Then

$$\|(Nu)(t)\| = \left\| u_k - Q(u) + \int_{s_k}^t \frac{(t-qs)^{(r-1)}}{\Gamma_q(r)} g(s)d_q s \right\|,$$

where $g \in C(I_k)$ is the unique solution of the equation

$$g(t) = f(t, u_k - Q(u) + ({}_q I_{s_k}^r g)(t), g(t)).$$

The uniqueness of such functions is deduced from (6.11.2) and condition (6.17), as in the proof of Theorem 6.8.

We will show that the operator $N : B_R \to B_R$ satisfies all the assumptions of Theorem 2.75.

As in the prove of Theorem 6.9, we can show that $N(B_R) \subset B_R$, $N : B_R \to B_R$ is continuous and $N(B_R)$ is bounded and equicontinuous. We still have to prove that the implication (2.12) holds.

Set

$$v := \max \left\{ c^*, L + \frac{p^* \rho^r}{\Gamma_q(1+r)} \right\}.$$

Let V be a subset of B_R such that $V \subset \overline{N(V)} \cup \{0\}$, V is bounded and equicontinuous and therefore the function $t \mapsto v(t) = \mu(V(t))$ is continuous on J. By (6.11.4) and the properties of the measure μ, for each $t \in I_k$, we have

$$v(t) \le \mu((NV)(t) \cup \{0\})$$
$$\le \mu((NV)(t))$$
$$= \mu(u_k - Q(v) + ({}_q I_{s_k}^r g)(t)),$$

where $g \in C(I_k)$; $k = 0, \ldots, m$, is the unique solution of the equation

$$g(t) = f(t, u_k - Q(v) + ({}_q I_{s_k}^r g)(t), g(t)).$$

Then we obtain

$$v(t) \le \mu \left(Lv(t) + \int_{s_k}^t \frac{(t-qs)^{(r-1)} p(s)}{\Gamma_q(r)} v(s)d_q s \right)$$

$$\leq L\mu(V(t)) + \int_{s_k}^t \frac{(t-qs)^{(r-1)}p(s)}{\Gamma_q(r)} \mu(V(s)) d_q s$$

$$\leq \left(L + \frac{p^* \rho^r}{\Gamma_q(1+r)} \right) \|v\|_{PC}.$$

Thus, for each $t \in I_k$, we get

$$v(t) \leq v\|v\|_{PC}.$$

Also, for each $t \in J_k$; $k = 1, \ldots, m$, we obtain

$$v(t) \leq \mu((NV)(t) \cup \{0\})$$
$$\leq \mu((NV)(t))$$
$$\leq c^* \mu(V(s))$$
$$\leq c^* \|v\|_{PC}.$$

Thus we obtain

$$\|v\|_{PC} \leq v\|v\|_{PC}.$$

Hence we get $\|v\|_{PC} = 0$, that is $v(t) = \beta(V(t)) = 0$, for each $t \in I$, and then $V(t)$ is relatively compact in E. In view of the Ascoli–Arzelà theorem, V is relatively compact in B_R. Applying now Theorem 2.75, we conclude that N has a fixed point, which is a solution of problem (6.10). $\qquad\square$

6.2.4 Examples

Example 1. Consider the problem of implicit impulsive q-fractional differential equation of the form

$$\begin{cases} ({}^C_q D^r_{t_k} u)(t) = f(t, u(t), ({}^C_q D^r_{t_k} u)(t)); \ t \in I_k, \ k = 0, 1, 2, \\ u(t) = g_k(t, u(t_k^-)); \ t \in J_k, \ k = 1, 2, \\ u(0) = Q(u), \end{cases} \qquad (6.19)$$

where $I = [0, 1]$, $I_0 = [0, \frac{1}{5}]$, $J_1 = (\frac{1}{5}, \frac{2}{5}]$, $I_1 = (\frac{2}{5}, \frac{3}{5}]$, $J_2 = (\frac{3}{5}, \frac{4}{5}]$, $I_2 = (\frac{4}{5}, 1]$, $r \in (0, 1]$,

$$f(t, u(t), ({}^C_q D^r_{t_k} u)(t)) = \frac{\Gamma_q(1+r)t^2(1+u(t)) + {}^C_q D^r_{t_k} u(t)}{e^{t+5}(2+|u(t)|)}; \ t \in [0, 1],$$

$$Q(u) = \frac{1 + \|u\|_{PC}}{3e^5},$$

and

$$g_k(t, u(t_k^-)) = \frac{1 + u(t_k^-)}{3e^{t+5}}; \ k = 1, \ldots, m.$$

Clearly, the function f is continuous.

For each $t \in I_k$, we have

$$|f(t, u(t), (^C_q D^r_{t_k} u)(t))| \leq \Gamma_q (1+r) t^2 e^{-(t+5)} (1 + |u(t)| + |(^C_q D^r_{t_k} u)(t)|).$$

$$|Q(u)| \leq \frac{1}{3e^5}$$

and

$$|g_k(t, u)| \leq \frac{1}{3e^5} (1 + |u|).$$

Hence the hypotheses (6.11.2) and (6.11.3) is satisfied with $\phi_1 = \phi_2 = l^* = e^{-5}\Gamma_q(1+r)$, and $c^* = L = \frac{1}{3e^5}$.

Also, we can verify that conditions (6.11),

$$l^* \left(1 + \frac{\rho^r}{\Gamma_q (1+r)}\right) < 1,$$

and

$$L \left(1 + \frac{\rho^r l^*}{\Gamma_q(1+r)} \frac{1}{1 - l^*\left(1 + \frac{\rho^r}{\Gamma_q(1+r)}\right)}\right) < 1,$$

are satisfied. Hence all conditions of Theorem 6.9 are satisfied. It follows that the problem (6.19) has at least one solution.

Example 2. Let

$$E = l^1 = \left\{u = (u_1, u_2, \ldots, u_n, \ldots), \sum_{n=1}^{\infty} |u_n| < \infty\right\}$$

be the Banach space with the norm

$$\|u\|_E = \sum_{n=1}^{\infty} |u_n|.$$

Consider the problem of implicit impulsive q-fractional differential equation of the form

$$\begin{cases} (^C_q D^r_{t_k} u)(t) = f(t, u(t), (^C_q D^r_{t_k} u)(t)); \ t \in J_k, \ k = 0, \ldots, m, \\ u(t) = g_k(t, u(t)); \ k = 1, \ldots, m, \\ u(0) = Q(u), \end{cases} \qquad (6.20)$$

where $I = [0, 1]$, $r \in (0, 1]$, $u = (u_1, u_2, \ldots, u_n, \ldots)$,

$$f = (f_1, f_2, \ldots, f_n, \ldots),$$

$$^C_q D^r_{t_k} u = (^C_q D^r_{t_k} u_1, ^C_q D^r_{t_k} u_2, \ldots, ^C_q D^r_{t_k} u_n, \ldots); \ k = 0, \ldots, m,$$

$$f_n(t, u(t), (^C_q D^r_{t_k} u)(t))) = \Gamma_q(1+r)t^2 \left(e^{-7} + \frac{1}{e^{t+5}}\right)(2^{-n} + u_n(t) + ^C_q D^r_{t_k} u_n(t)); \ t \in [0, 1],$$

$$g_k(t, u(t_k^-)) = \frac{1 + \|u(t_k^-)\|_E}{3e^{t+5}}; \ k = 1, \ldots, m,$$

and

$$Q(u) = \frac{1 + \|u\|_{PC}}{3e^5}.$$

For each $u \in E$ and $t \in [0, 1]$, we have

$$\|f(t, u(t), (^C_q D^r_{t_k} u)(t))\|_E \leq \Gamma_q(1+r)t^2 \left(e^{-7} + \frac{1}{e^{t+5}}\right)(1 + \|u\|_E + \|^C_q D^r_{t_k} u\|_E),$$

$$\|g_k(t, u)\|_E \leq \frac{1 + \|u\|_E}{3e^5},$$

and

$$\|Q(u)\|_E \leq \frac{1 + \|u\|_{PC}}{3e^5}.$$

Hence the hypotheses (6.11.3), (6.11.4) are satisfied with $\phi_1 = \phi_2 = p^* = 2e^{-5}\Gamma_q(1+r)$, and $L = c^* = \frac{1}{3e^5}$.

We assume, for instance, that the number of impulses $m = 3$ and $r = \frac{1}{2}$. Then by simple computations, we can show that all conditions of Theorem 6.11 are satisfied. Consequently, problem (6.20) has at least one solution on $[0, 1]$.

6.3 Instantaneous and noninstantaneous impulsive integro-differential equations in Banach spaces

6.3.1 Introduction

In this section we first discuss the existence of mild solutions for the following nonlocal problem of impulsive integro-differential equations

$$\begin{cases} u'(t) = Au(t) + \int_0^t \Upsilon(t-s)u(s)ds + f(t, u(t)); \ t \in I_k, \ k = 0, \ldots, m, \\ u(t_k^+) = u(t_k^-) + L_k(u(t_k^-)); \ k = 1, \ldots, m, \\ u(0) + g(u) = u_0 \in E, \end{cases} \tag{6.21}$$

where $I_0 = [0, t_1]$, $I_k := (t_k, t_{k+1}]$; $k = 1, \ldots, m$, $0 = t_0 < t_1 < \cdots < t_m < t_{m+1} = T$, $f : I_k \times E \to E$; $k = 1, \ldots, m$, $L_k : E \to E$; $k = 1, \ldots, m$, $g : PC \to E$ are given functions, E is a real (or complex) Banach space with norm $\| \cdot \|$, $u'(t) := \frac{du}{dt}$, $A : D(A) \subset E \to E$ generates a C_0-semigroup on the Banach space E, $\Upsilon(t)$ is a closed linear operator on E with $D(A) \subset D(\Upsilon)$.

Next, we discuss the existence of mild solutions for the following nonlocal problem of noninstantaneous impulsive integro-differential equations

$$\begin{cases} u'(t) = Au(t) + \int_0^t \Upsilon(t-s)u(s)ds + f(t, u(t)); \ t \in I_k, \ k = 0, \dots, m, \\ u(t) = g_k(t, u(t_k^-)); \ t \in J_k, \ k = 1, \dots, m, \\ u(s_k) + g(u) = u_k \in E; \ k = 0, \dots, m, \end{cases} \tag{6.22}$$

where $I_0 := [0, t_1]$, $J_k := (t_k, s_k]$, $I_k := (s_k, t_{k+1}]$; $k = 1, \dots, m$, $f : I_k \times E \to E$, $g_k : J_k \times E \to E$ are given functions such that $g_k(t, u(t_k^-))|_{t=s_k} = u_k \in E$; $k = 1, \dots, m$, $g : \mathcal{PC} \to E$ is a given function, and $0 = s_0 < t_1 \le s_1 < t_2 \le s_2 < \dots \le s_{m-1} < t_m \le s_m < t_{m+1} = T$.

6.3.2 Mild solutions with instantaneous impulses

In this section we are concerned with the existence results of the problem(6.21).

Definition 6.12. [227] A resolvent operator for the Cauchy problem

$$\begin{cases} u'(t) = Au(t) + \int_0^t \Upsilon(t-s)u(s)ds; \ t \in [0, \infty), \\ u(0) = u_0 \in E, \end{cases} \tag{6.23}$$

is a bounded linear operator-valued function $R(t) \in B(E)$; $t \ge 0$, verifying the following conditions:

(i) $R(0) = I$ (the identity map of E) and $\|R(t)\| \le Ne^{vt}$ for some constants $N > 0$, and $v \in \mathbb{R}$.

(ii) For each $u \in E$, $R(t)u$ is strongly continuous for $t \ge 0$.

(iii) $R(t)$ is bounded for $t \ge 0$. For $u \in D(A)$, $R(\cdot)u \in C(\mathbb{R}_+, D(A)) \cap C^1(\mathbb{R}_+, E)$ and

$$R'(t)u = AR(t)u + \int_0^t \Upsilon(t-s)R(s)uds$$

$$= R(t)Au + \int_0^t R(t-s)\Upsilon(s)uds; \ t \in [0, \infty).$$

Theorem 6.13 ([227,254]). *Assume that the following hypotheses hold.*

(6.13.1) *The operator A is the infinitesimal generator of a uniformly continuous semigroup $(S(t))_{t\ge0}$.*

(6.13.2) *For all $t \ge 0$, $\Upsilon(t)$ is a closed linear operator from $D(A)$ to E and $\Upsilon(t) \in B(E)$. For any $u \in E$, the map $t \mapsto \Upsilon(t)u$ is bounded differentiable and the derivative $t \mapsto \Upsilon'(t)u$ is bounded uniformly continuous on $\mathbb{R}_+$.*

Then there exists a unique uniformly continuous resolvent operator for the Cauchy problem (6.23).

Definition 6.14 ([254]). By a mild solution of the problem (6.21) we mean a function $u \in PC$ that satisfies

$$u(t) = R(t)[u_0 - g(u)] + \int_0^t R(t-s)f(s,u(s))ds + \sum_{0<t_i<t} R(t-t_i)L_i(u(t_i)); \; t \in I.$$

The following hypotheses will be used in the sequel.

(6.14.1) The function $t \mapsto f(t,u)$ is measurable on I for each $u \in E$ and the function $u \mapsto f(t,u)$ is continuous on E for a.e. $t \in I_k$,

(6.14.2) There exists a function $p \in L^\infty(I)$, such that

$$\|f(t,u)\| \leq p(t)(1+\|u\|); \text{ for a.e. } t \in I_k, \text{ and each } u \in E,$$

(6.14.3) There exist positive constants q^*, l_k^*; $k=0,\ldots,m$ such that

$$\|g(u)\| \leq q^*(1+\|u\|_{PC}); \text{ for each } u \in PC,$$

and

$$\|L_k(u)\| \leq l_k^*(1+\|u\|); \; k=0,\ldots,m, \text{ for a.e. } t \in I, \text{ and each } u \in E,$$

(6.14.4) For each bounded set $B \subset E$, we have

$$\mu(f(t,B)) \leq p(t)\mu(B), \; \mu(L_k(B)) \leq l_k^*\mu(B); \; k=0,\ldots,m,$$

and for each bounded set $B_1 \subset PC$, we have

$$\mu(g(B_1)) \leq q^* \sup_{t \in I} \mu(B_1(t)),$$

where $B_1(t) = \{u(t) : u \in B_1\}; \; t \in I.$

Set

$$p^* = \|p\|_{L^\infty}, \text{ and } M = \sup_{t \in I} \|R(t)\|_{B(E)}.$$

Theorem 6.15. *Assume that the hypotheses* (6.13.1), (6.13.2), (6.14.1)–(6.14.4) *hold. If*

$$\ell := M\left(q^* + Tp^* + \sum_{k=0}^m l_k^*\right) < 1, \tag{6.24}$$

then the problem (6.21) has at least one mild solution defined on I.

Proof. Transform the problem (6.21) into a fixed-point problem. Consider the operator $N : PC \to PC$ defined by

$$(Nu)(t) = R(t)[u_0 - g(u)] + \int_0^t R(t-s)f(s,u(s))ds + \sum_{0<t_i<t} R(t-t_i)L_i(u(t_i)); \; t \in I. \tag{6.25}$$

Let $\rho > 0$, such that

$$\rho \geq \frac{M\left(\|u_0\| + q^* + Tp^* + \sum\limits_{k=0}^{m} l_k^*\right)}{1 - M\left(q^* + Tp^* + \sum\limits_{k=0}^{m} l_k^*\right)},$$

and consider the ball $B_\rho := B(0, \rho) = \{w \in PC : \|w\|_{PC} \leq \rho\}$.

For any $u \in B_\rho$ and each $t \in I$, we have

$$\|(Nu)(t)\| \leq \|R(t)\|_{B(E)}[\|u_0\| + \|g(u)\|]$$

$$+ \int_0^t \|R(t-s)\|_{B(E)} \|f(s, u(s))\| ds$$

$$+ \sum_{0 < t_i < t} \|R(t - t_i)\|_{B(E)} \|L_i(u(t_i))\|$$

$$\leq M\left[\|u_0\| + (1 + \rho)\left(q^* + Tp^* + \sum_{k=0}^{m} l_k^*\right)\right]$$

$$:\leq \rho.$$

Thus

$$\|N(u)\|_{PC} \leq \rho. \tag{6.26}$$

This proves that N transforms the ball B_ρ into itself. We will show that the operator $N : B_\rho \to B_\rho$ satisfies all the assumptions of Theorem 2.75. The proof will be given in three steps.

Step 1. $N : B_\rho \to B_\rho$ *is continuous.*

Let $\{u_n\}_{n \in \mathbb{N}}$ be a sequence such that $u_n \to u$ as $n \to \infty$ in B_ρ. Then, for each $t \in I$, we have

$$\|(Nu_n)(t) - (Nu)(t)\| \leq \|R(t)\|_{B(E)} \|g(u_n) - g(u)\|$$

$$+ \int_0^t \|R(t-s)\|_{B(E)} \|f(s, u_n(s)) - f(s, u(s))\| ds$$

$$+ \sum_{0 < t_i < t} \|R(t - t_i)\|_{B(E)} \|L_i(u_n(t_i)) - L_i(u(t_i))\|$$

$$\leq M\|g(u_n) - g(u)\|$$

$$+ M \int_0^T \|f(s, u_n(s)) - f(s, u(s))\| ds$$

$$+ M \sum_{0 < t_i < t} \|L_i(u_n(t_i)) - L_i(u(t_i))\|.$$

Since $u_n \to u$ as $n \to \infty$ and f, g, L_i are continuous, the Lebesgue dominated convergence theorem implies that

$$\|N(u_n) - N(u)\|_{PC} \to 0 \text{ as } n \to \infty.$$

Step 2. *$N(B_\rho)$ is bounded and equicontinuous.*
Since $N(B_\rho) \subset B_\rho$ and B_ρ is bounded, then $N(B_\rho)$ is bounded.
Next, let $t, \tau \in I, \tau < t$ and let $u \in B_\rho$. Thus we have

$$\|(Nu)(t) - (Nu)(\tau)\| \leq \|R(t) - R(\tau)\|_{B(E)}(\|u_0\| + \|g(u)\|)$$
$$+ \int_0^\tau \|R(t-s) - R(\tau-s)\|_{B(E)}\|f(s, u(s))\|ds$$
$$+ \int_\tau^t \|R(t-s)\|_{B(E)}\|f(s, u(s))\|ds$$
$$+ \sum_{0<t_i<t} \|R(t-t_i) - R(\tau-t_i)\|_{B(E)}\|L_i(u(t_i))\|.$$

Hence we get

$$\|(Nu)(t) - (Nu)(\tau)\| \leq (\|u_0\| + q^*(1+\rho))\|R(t) - R(\tau)\|_{B(E)}$$
$$+ p^*(1+\rho)\int_0^\tau \|R(t-s) - R(\tau-s)\|_{B(E)}ds$$
$$+ Mp^*(1+\rho)(t-\tau)$$
$$+ \sum_{0<t_i<t} l_i^*(1+\rho)\|R(t-t_i) - R(\tau-t_i)\|_{B(E)}.$$

As the resolvent operator $R(\cdot)$ is uniformly continuous, the right-hand side of the above inequality tends to zero as $\tau \longrightarrow t$.

Step 3. *The implication (2.12) holds.*
Now let V be a subset of B_ρ such that $V \subset \overline{N(V)} \cup \{0\}$. V is bounded and equicontinuous and therefore the function $t \to v(t) = \mu(V(t))$ is continuous on I. By (6.14.3) and the properties of the measure μ, for each $t \in I$, we have

$$v(t) \leq \mu((NV)(t) \cup \{0\})$$
$$\leq \mu((NV)(t))$$
$$\leq \|R(t)\|_{B(E)}q^* \sup_{t\in I}\mu(V(t)) + \int_0^t \|R(t-s)\|_{B(E)}p(s)\mu(V(s))ds$$
$$+ \sum_{k=0}^m \|R(t-t_k)\|_{B(E)}l_k(t)\mu(V(t))$$

$$\leq Mq^*\|v\|_\infty + Mp^* \int_0^t v(s)ds + M \sum_{k=0}^m l_k^* v(t)$$

$$\leq M\left(q^* + Tp^* + \sum_{k=0}^m l_k^*\right)\|v\|_\infty.$$

Hence

$$\|v\|_\infty \leq \ell\|v\|_\infty.$$

From (6.24), we get $\|v\|_\infty = 0$, that is, $v(t) = \mu(V(t)) = 0$, for each $t \in I$, and then $V(t)$ is relatively compact in E. In view of the Ascoli–Arzelà theorem, V is relatively compact in B_ρ. Applying now Theorem 2.75, we conclude that N has a fixed point, which is a mild solution of our problem (6.21). □

6.3.3 Mild solutions with noninstantaneous impulses

In this section we are concerned with the existence results of the problem (6.22).

Denote by

$$\mathcal{PC} = \{y : I \to E : y \in C([0, t_1] \cup (t_k, s_k] \cup (s_k, t_{k+1}], E), \ k = 1, \ldots, m$$
$$\text{and there exist } y(t_k^-), \ y(t_k^+), \ y(s_k^-) \text{ and } y(s_k^+) \ k = 1, \ldots, m$$
$$\text{with } y(t_k^-) = y(t_k) \text{ and } y(s_k^-) = y(s_k)\},$$

the Banach space equipped with the standard supremum norm.

Definition 6.16. [254] By a mild solution of the problem (6.22) we mean a function $u \in \mathcal{PC}$ that satisfies

$$\begin{cases} u(t) = R(t)[u_k - g(u)] + \int_{s_k}^t R(t-s)f(s, u(s))ds; \ t \in I_k, \ k = 0, \ldots, m, \\ u(t) = g_k(t, u(t_k^-)); \ t \in J_k, \ k = 1, \ldots, m, \end{cases}$$

The following hypotheses will be used in the sequel.

(6.16.1) The functions $t \mapsto f(t, u)$ and $t \mapsto g_k(t, u)$ are measurable on I_k, J_k, respectively, for each $u \in E$, and the functions $u \mapsto f(t, u)$ and $u \mapsto g_k(t, u)$ are continuous on E for a.e. t in I_k, J_k, respectively.

(6.16.2) There exist functions $p, l_k \in L^\infty(I)$; $k = 0, \ldots, m$, such that

$$\|f(t, u)\| \leq p(t)(1 + \|u\|); \text{ for a.e. } t \in I_k, \text{ and each } u \in E,$$

and

$$\|g_k(t, u)\| \leq l_k(t)(1 + \|u\|); \ k = 1, \ldots, m, \text{ for a.e. } t \in J_k, \text{ and each } u \in E.$$

(6.16.3) There exists a positive constant q^*, such that

$$\|g(u)\| \leq q^*(1 + \|u\|_{\mathcal{PC}}); \text{ for a.e. } t \in I, \text{ and each } u \in \mathcal{PC}.$$

(6.16.4) For each bounded set $B \subset E$ and for each $t \in I$, we have

$$\mu(f(t, B)) \le p(t)\mu(B), \ \mu(g_k(t, B)) \le l_k(t)\mu(B); \ k = 0, \dots, m,$$

and for each bounded set $B_0 \subset \mathcal{PC}$, we have

$$\mu(g(B_0)) \le q^* \sup_{t \in I} \mu(B_0(t)),$$

where $B_0(t) = \{u(t) : u \in B_0\}; \ t \in I$.

Set

$$p^* = \|p\|_{L^\infty}, \ l^* = \max_{k=0,\dots,m} \|l_k\|_{L^\infty}, \ M = \sup_{t \in I} \|R(t)\|_{B(E)}.$$

Theorem 6.17. *Assume that the hypotheses* (6.13.1), (6.13.2), (6.16.1)–(6.16.4) *hold. If*

$$\ell := \max\{l^*, M(q^* + Tp^*)\} < 1, \tag{6.27}$$

then the problem (6.22) has at least one mild solution defined on I.

Proof. Transform the problem (6.22) into a fixed-point problem. Consider the operator $N : \mathcal{PC} \to \mathcal{PC}$ defined by

$$\begin{cases} (Nu)(t) = R(t)[u_k - g(u)] + \int_{s_k}^t R(t-s)f(s, u(s))ds; \ t \in I_k, \ k = 0, \dots, m, \\ (Nu)(t) = g_k(t, u(t_k^-)); \ t \in J_k, \ k = 1, \dots, m. \end{cases} \tag{6.28}$$

Let $L > 0$ such that

$$L \ge \frac{M(\|u_k\| + q^* + Tp^*)}{1 - M(q^* + Tp^*)}.$$

For any $u \in \mathcal{PC}$ and each $t \in I_k$, we have

$$\|(Nu)(t)\| \le \|R(t)\|_{B(E)}[\|u_k\| + \|g(u)\|]$$
$$+ \int_{s_k}^t \|R(t-s)\|_{B(E)} \|f(s, u(s))\| ds$$
$$\le M[\|u_k\| + (1 + L)(q^* + Tp^*)]$$
$$\le L.$$

Thus

$$\|N(u)\|_{\mathcal{PC}} \le L. \tag{6.29}$$

Next, for each $t \in J_k; \ k = 1, \dots, m$, it is clear that

$$\|(Nu)(t)\|_E \le l^*.$$

Hence

$$\|N(u)\|_{\mathcal{PC}} \leq \max\left\{L, l^*\right\} := \rho.$$

This proves that N transforms the ball $B_\rho := \{w \in \mathcal{PC} : \|w\|_{\mathcal{PC}} \leq \rho\}$ into itself.

We will show that the operator $N : B_\rho \to B_\rho$ satisfies all the assumptions of Theorem 2.75. The proof will be given in three steps.

Step 1. $N : B_\rho \to B_\rho$ *is continuous.*

Let $\{u_n\}_{n\in\mathbb{N}}$ be a sequence such that $u_n \to u$ as $n \to \infty$ in B_ρ. Then, for each $t \in J_k$; $k = 1, \ldots, m$, we have

$$\|(Nu_n)(t) - (Nu)(t)\| \leq \|R(t)\|_{B(E)}\|g_k(t, u_n(t_k^-)) - g_k(t, u(t_k^-))\|,$$

and for each $t \in I_k$; $k = 0, \ldots, m$, we have

$$\|(Nu_n)(t) - (Nu)(t)\| \leq \|R(t)\|_{B(E)}\|g(u_n) - g(u)\|$$
$$+ \int_{s_k}^{t} \|R(t-s)\|_{B(E)}\|f(s, u_n(s)) - f(s, u(s))\|ds$$
$$\leq M\|g(u_n) - g(u)\|$$
$$+ M\int_{s_k}^{T} \|f(s, u_n(s)) - f(s, u(s))\|ds.$$

Since $u_n \to u$ as $n \to \infty$ and f, g, g_k are continuous, the Lebesgue dominated convergence theorem implies that

$$\|N(u_n) - N(u)\|_{\mathcal{PC}} \to 0 \text{ as } n \to \infty.$$

Step 2. $N(B_\rho)$ *is bounded and equicontinuous.*

Since $N(B_R) \subset B_\rho$ and B_ρ is bounded, then $N(B_\rho)$ is bounded.

Next, let $t, \tau \in I_k$, $\tau < t$ and let $u \in B_\rho$. Thus we have

$$\|(Nu)(t) - (Nu)(\tau)\| \leq \|R(t) - R(\tau)\|_{B(E)}(\|u_0\| + \|g(u)\|)$$
$$+ \int_{0}^{\tau} \|R(t-s) - R(\tau-s)\|_{B(E)}\|f(s, u(s))\|ds$$
$$+ \int_{\tau}^{t} \|R(t-s)\|_{B(E)}\|f(s, u(s))\|ds.$$

Hence we get

$$\|(Nu)(t) - (Nu)(\tau)\| \leq (\|u_0\| + q^*(1 + \rho))\|R(t) - R(\tau)\|_{B(E)}$$
$$+ p^*(1 + \rho)\int_{0}^{\tau} \|R(t-s) - R(\tau-s)\|_{B(E)}ds.$$

As $\tau \longrightarrow t$, the right-hand side of the above inequality tends to zero.

Step 3. *The implication (2.12) holds.*
Now let V be a subset of B_R such that $V \subset \overline{N(V)} \cup \{0\}$. V is bounded and equicontinuous and therefore the function $t \to v(t) = \mu(V(t))$ is continuous on I. By (6.16.3) and the properties of the measure μ, for each $t \in I_k$, we have

$$v(t) \leq \mu((NV)(t) \cup \{0\})$$
$$\leq \mu((NV)(t))$$
$$\leq l^* \|v\|_\infty$$
$$\leq \ell \|v\|_\infty.$$

Next, for each $t \in I_k$, we have

$$v(t) \leq \mu((NV)(t) \cup \{0\})$$
$$\leq \mu((NV)(t))$$
$$\leq \|R(t)\|_{B(E)} q^* \sup_{t \in I} \mu(V(t)) + \int_0^t \|R(t-s)\|_{B(E)} p(s) \mu(V(s)) ds$$
$$\leq M q^* \|v\|_{\mathcal{PC}} + M p^* \int_0^t \|v(s)\| ds$$
$$\leq M(q^* + T p^*) \|v\|_\infty$$
$$\leq \ell \|v\|_\infty.$$

Thus, for each $t \in I$, we get

$$v(t) \leq \ell \|v\|_\infty.$$

Hence

$$\|v\|_\infty \leq \ell \|v\|_\infty$$

From (6.27), we get $\|v\|_\infty = 0$, that is, $v(t) = \mu(V(t)) = 0$, for each $t \in I$, and then $V(t)$ is relatively compact in E. In view of the Ascoli–Arzelà theorem, V is relatively compact in B_ρ. Applying now Theorem 2.75, we conclude that N has a fixed point, which is a mild solution of our problem (6.22). □

6.3.4 Examples

Let

$$H := L^2([0, \pi]) = \left\{ u : [0, \pi] \to \mathbb{R} : \int_0^\pi |u(x)|^2 dx < \infty \right\},$$

be the Hilbert space with the scalar product $< u, v > = \int_0^\pi u(x)v(x)dx$. It is known that H is a Banach space with the norm

$$\|u\|_2 = \left(\int_0^\pi |u(x)|^2 dx \right)^{\frac{1}{2}}.$$

Example 1. Consider the following problem of impulsive integro-differential equations

$$
\begin{cases}
\frac{\partial}{\partial t} z(t, x) = \frac{\partial^2}{\partial x^2} z(t, x) + Q(t, z(t, x)) \\
\quad + \int_0^t b(t - s) \frac{\partial^2}{\partial x^2} z(s, x) ds; \ t \in [0, 1] \cup (1, 2], \ x \in [0, \pi], \\
z(1^+, x) = z(1^-, x) + L_1(z(1^-, x)); \ x \in [0, \pi], \\
z(t, 0) = z(t, \pi) = 0; \ t \in [0, 1] \cup (1, 2], \\
z(0, x) + g(z) = 1 + x^2; \ x \in [0, \pi], \ z \in PC,
\end{cases}
\tag{6.30}
$$

where $t \in I = [0, 2]$, $PC := PC([0, 2], H)$,

$$
Q(t, z(t, x)) = \frac{ct^2}{1 + \|z\|_2} \left(e^{-7} + \frac{1}{e^{t+x+5}} \right) (1 + z(t, x)); \ t \in [0, 1] \cup (1, 2],
$$

$$
L_1(z(1^-), x) = \frac{z(1^-, x)}{3e^4(1 + \|z(1^-, x)\|_2)},
$$

and

$$
g(z) = \int_0^\pi K(x, y) \frac{e^{-y}}{1 + \|z\|_{PC}} dy,
$$

with $\int_0^\pi \int_0^\pi K^2(x, y) dx dy < \infty$.

We define the strongly elliptic operator $A : D(A) \subset H \to H$ by:

$$
Au = \mathcal{A}(x, D)u = \sum_{|\mu| \le 2m} a_\mu(x) D^\mu u,
$$

where $a_\mu \in C^{2m}([0, \pi])$, and $D(A) = H^{2m}([0, \pi]) \cap H_0^m([0, \pi])$.

It is well known (see [327]) that A generates a uniformly continuous semigroup $T(t)$; $t \ge 0$ in the Hilbert space H.

For $x \in [0, \pi]$, we have

$$
u(t)(x) = z(t, x); \quad t \in [0, 1] \cup (1, 2],
$$
$$
f(t, u(t))(x) = Q(t, z(t, x)); \quad t \in [0, 1] \cup (1, 2],
$$
$$
\Upsilon(t) = b(t)A
$$
$$
u_0(x) = 1 + x^2; \ x \in [0, \pi].
$$

Thus under the above definitions of f, u_0, and A, the system (6.30) can be represented by the problem (6.22). Furthermore, more appropriate conditions on Q ensure the hypotheses (6.13.1), (6.13.1), (6.16.1)–(6.16.4). Consequently, Theorem 6.15 implies that the problem (6.30) has at least one mild solution on $[0, 2]$.

Example 2. Consider now the following problem of impulsive integro-differential equations

$$
\begin{cases}
\frac{\partial}{\partial t} z(t,x) = \frac{\partial^2}{\partial x^2} z(t,x) + Q(t, z(t,x)) \\
\quad + \int_0^t b(t-s) \frac{\partial^2}{\partial x^2} z(s,x)\, ds;\ t \in [0,1] \cup (2,3],\ x \in [0,\pi], \\[4pt]
z(t,x) = g_1(t, z(1^-,x));\ t \in (1,2],\ x \in [0,\pi], \\[4pt]
z(t,0) = z(t,\pi) = 0;\ t \in [0,1] \cup (2,3], \\[4pt]
z(0,x) + g(z) = 1 + e^x;\ x \in [0,\pi], \\[4pt]
z(2,x) + g(z) = 2 + e^x;\ x \in [0,\pi],
\end{cases}
\tag{6.31}
$$

where $t \in [0,3]$, $\mathcal{PC} := \mathcal{PC}([0,3], H)$,

$$
Q(t, z(t,x)) = \frac{ct^2}{1 + \|z\|_2}\left(e^{-7} + \frac{1}{e^{t+x+5}}\right)(1 + z(t,x));\ t \in [0,1] \cup (2,3],
$$

$$
g_1(t, z(1^-,x)) = \frac{z(1^-,x)}{(3e^4)(1 + \|z(1^-,x)\|_2)};\ t \in (1,2],\ x \in [0,\pi],
$$

and

$$
g(z) = \int_0^\pi K(x,y)\frac{e^{-y}}{1 + \|z\|_{\mathcal{PC}}}\, dy,
$$

with $\int_0^\pi \int_0^\pi K^2(x,y)\, dx\, dy < \infty$.

Again, as the above example, simple computations show that all conditions of Theorem 6.17 are satisfied. It follows that the problem (6.31) has at least one mild solution on $[0,3]$.

6.4 Notes and remarks

This chapter contains the studies from Abbas et al. [40,71,72]. One can see the monographs [165,251] and the papers [143,165,198,217,275,289,308,381,386], for additional details and results on the subject.

7

Coupled fractional difference systems

We explored and demonstrated the results obtained in this chapter by taking into consideration the previously stated publications in the preceding chapters and the papers [4,7,11,12,15,16,55,56,115,134,232,234,271,273,278].

7.1 Implicit coupled Caputo fractional q-difference systems

7.1.1 Introduction

In this section we discuss the existence and Ulam–Hyers–Rassias stability of solutions for the following coupled implicit fractional q-difference system

$$\begin{cases} ({}^{c}D_q^{\alpha_1} u_1)(t) = f_1(t, u_1(t), u_2(t), ({}^{c}D_q^{\alpha_1} u_1)(t)), \\ ({}^{c}D_q^{\alpha_2} u_2)(t) = f_2(t, u_1(t), u_2(t), ({}^{c}D_q^{\alpha_2} u_2)(t)), \end{cases} \; ; \; t \in I := [0, T], \tag{7.1}$$

with the initial conditions

$$(u_1(0), u_2(0)) = (u_{01}, u_{02}), \tag{7.2}$$

where $q \in (0, 1)$, $T > 0$, $\alpha_i \in (0, 1]$, $f_i : I \times \mathbb{R} \times \mathbb{R} \times \mathbb{R} \to \mathbb{R}$; $i = 1, 2$, are given continuous functions, and ${}^{c}D_q^{\alpha_i}$ is the Caputo fractional q-difference derivative of order α_i; $i = 1, 2$.

Next, we discuss the existence and uniqueness of solutions for the problem (7.1)–(7.2) in generalized Banach spaces, where $f_i : I \times \mathbb{R}^{3m} \to \mathbb{R}^m$; $i = 1, 2$, are given continuous functions, $\mathbb{R}^m$; $m \in \mathbb{N}^*$ is the Euclidean Banach space with a suitable norm $\| \cdot \|$.

7.1.2 Existence and Ulam stability results

In this section we are concerned with the existence stability of solutions of the system (7.1)–(7.2). We denote by $\mathcal{C} := C(I) \times C(I)$ the Banach space with the norm

$$\|(u, v)\|_{\mathcal{C}} = \|u\|_\infty + \|v\|_\infty.$$

Definition 7.1. By a solution of the problem (7.1)–(7.2) we mean a coupled functions $(u, v) \in \mathcal{C}$ that satisfies the system

$$\begin{cases} ({}^{c}D_q^{\alpha_1} u)(t) = f_1(t, u(t), v(t), ({}^{c}D_q^{\alpha_1} u)(t)), \\ ({}^{c}D_q^{\alpha_2} v)(t) = f_2(t, u(t), v(t), ({}^{c}D_q^{\alpha_2} v)(t)), \end{cases}$$

Fractional Difference, Differential Equations, and Inclusions. https://doi.org/10.1016/B978-0-44-323601-3.00014-9

Copyright © 2024 Elsevier Inc. All rights reserved, including those for text and data mining, AI training, and similar technologies.

on I and the initial condition $(u(0), v(0)) = (u_{01}, u_{02})$.

Now, we consider the Ulam stability for the system (7.1)–(7.2). Let $\epsilon > 0$ and $\Phi : I \to \mathbb{R}_+$ be a continuous function. We consider the following inequalities

$$\begin{cases} |({}^c D_q^{\alpha_1} u)(t) - f_1(t, u(t), v(t), ({}^c D_q^{\alpha_1} u)(t))| \le \epsilon \\ |({}^c D_q^{\alpha_2} v)(t) - f_2(t, u(t), v(t), ({}^c D_q^{\alpha_2} v)(t))| \le \epsilon \end{cases} \; ; t \in I. \qquad (7.3)$$

$$\begin{cases} |({}^c D_q^{\alpha_1} u)(t) - f_1(t, u(t), v(t), ({}^c D_q^{\alpha_1} u)(t))| \le \Phi(t) \\ |({}^c D_q^{\alpha_2} v)(t) - f_2(t, u(t), v(t), ({}^c D_q^{\alpha_2} v)(t))| \le \Phi(t) \end{cases} \; ; t \in I. \qquad (7.4)$$

$$\begin{cases} |({}^c D_q^{\alpha_1} u)(t) - f_1(t, u(t), v(t), ({}^c D_q^{\alpha_1} u)(t))| \le \epsilon \Phi(t) \\ |({}^c D_q^{\alpha_2} v)(t) - f_2(t, u(t), v(t), ({}^c D_q^{\alpha_2} v)(t))| \le \epsilon \Phi(t) \end{cases} \; ; t \in I. \qquad (7.5)$$

Set

$$|(u(t), v(t))| := |u(t)| + |v(t)|.$$

Definition 7.2. [70,345] The system (7.1)–(7.2) is Ulam–Hyers stable if there exists a real number $c_{f_1, f_2} > 0$ such that for each $\epsilon > 0$ and for each solution $(u_1, v_1) \in C$ of the inequalities (7.3) there exists a solution $(u, v) \in C(I)$ of (7.1)–(7.2) with

$$|(u_1(t) - u(t), v_1(t) - v(t))| \le \epsilon c_{f_1, f_2}; \; t \in I.$$

Definition 7.3. [70,345] The system (7.1)–(7.2) is generalized Ulam–Hyers stable if there exists $c_{f_1, f_2} : C(\mathbb{R}_+, \mathbb{R}_+)$ with $c_{f_i}(0) = 0; \; i = 1, 2$ such that for each $\epsilon > 0$ and for each solution $(u_1, v_1) \in C$ of the inequalities (7.3) there exists a solution $(u, v) \in C$ of (7.1)–(7.2) with

$$|(u_1(t) - u(t), v_1(t) - v(t))| \le c_{f_1, f_2}(\epsilon); \; t \in I.$$

Definition 7.4. [70,345] The system (7.1)–(7.2) is Ulam–Hyers–Rassias stable with respect to Φ if there exists a real number $c_{f_1, f_2, \Phi} > 0$ such that for each $\epsilon > 0$ and for each solution $(u_1, v_1) \in C$ of the inequalities (7.5) there exists a solution $(u, v) \in C$ of (7.1)–(7.2) with

$$|(u_1(t) - u(t), v_1(t) - v(t))| \le \epsilon c_{f_1, f_2, \Phi} \Phi(t); \; t \in I.$$

Definition 7.5. [70,345] The system (7.1)–(7.2) is generalized Ulam–Hyers–Rassias stable with respect to Φ if there exists a real number $c_{f_1, f_2, \Phi} > 0$ such that for each solution $(u_1, v_1) \in C$ of the inequalities (7.4) there exists a solution $(u, v) \in C$ of (7.1)–(7.2) with

$$|(u_1(t) - u(t), v_1(t) - v(t))| \le c_{f_1, f_2, \Phi} \Phi(t); \; t \in I.$$

Remark 7.6. It is clear that

  (i) Definition 7.2 $\Rightarrow$ Definition 7.3,

  (ii) Definition 7.4 $\Rightarrow$ Definition 7.5,

  (iii) Definition 7.4 for $\Phi(\cdot) = 1 \Rightarrow$ Definition 7.2.

One can have similar remarks for the inequalities (7.3) and (7.5).

The following hypotheses will be used in the sequel:

(7.6.1) There exist functions $p_i, d_i, r_i \in C(I, [0, \infty))$; $i = 1, 2$, with $r_i(t) < 1$ such that

$$|f_i(t, u, v, w)| \le p_i(t) + d_i(t) \min(|u|, |v|) + r_i(t)|w|;$$

for each $t \in I$ and $u, v, w \in \mathbb{R}$,

(7.6.2) There exists $\lambda_\Phi > 0$ such that for each $t \in I$, we have

$$(I_q^{\alpha_i} \Phi)(t) \le \lambda_\Phi \Phi(t); \ i = 1, 2.$$

Set

$$L_i := \frac{T^{\alpha_i}}{\Gamma_q(1 + \alpha_i)}, \quad p_i^* = \sup_{t \in I} p_i(t), \quad d_i^* = \sup_{t \in I} d_i(t), \quad r_i^* = \sup_{t \in I} r_i(t), \quad \Phi^* = \sup_{t \in I} \Phi(t).$$

Theorem 7.7. *Assume that the hypothesis* (7.6.1) *holds. If*

$$r_1^* + r_2^* - r_1^* r_2^* + (1 - r_2^*)L_1 d_1^* + (1 - r_1^*)L_2 d_2^* < 1, \tag{7.6}$$

then the system (7.1)–(7.2) has at least one solution defined on I. Moreover, if the hypothesis (7.6.2) holds, then the system (7.1)–(7.2) is generalized Ulam–Hyers–Rassias stable.

Proof. Define the operators $N_i : C(I) \to C(I)$; $i = 1, 2$ by

$$(N_1 u)(t) = u_{01} + (I_q^{\alpha_1} g_1)(t); \ t \in I, \tag{7.7}$$

and

$$(N_2 v)(t) = u_{02} + (I_q^{\alpha_2} g_2)(t); \ t \in I, \tag{7.8}$$

where $g_i \in C(I)$ such that

$$g_i(t) = f(t, u(t), v(t), g_i(t)).$$

Consider the continuous operator $N : \mathcal{C} \to \mathcal{C}$ defined by

$$(N(u, v))(t) = ((N_1 u)(t), (N_2 v)(t)). \tag{7.9}$$

Set

$$R \ge \frac{(1 - r_1^*)(1 - r_2^*)(|u_{01}| + |u_{02}|) + (1 - r_2^*)L_1 p_1^* + (1 - r_1^*)L_2 p_2^*}{1 - r_1^* - r_2^* + r_1^* r_2^* - (1 - r_2^*)L_1 d_1^* - (1 - r_1^*)L_2 d_2^*},$$

and consider the closed and convex ball $B_R = \{u \in C : \|(u, v)\|_C \leq R\}$.

Let $u \in B_R$. Then, for each $t \in I$, we have

$$|(N_1 u)(t)| \leq |u_{01}| + \int_0^t \frac{(t - qs)^{(\alpha_1 - 1)}}{\Gamma_q(\alpha_1)} |g_1(s)| d_q s,$$

and

$$|(N_2 v)(t)| \leq |u_{02}| + \int_0^t \frac{(t - qs)^{(\alpha_2 - 1)}}{\Gamma_q(\alpha_2)} |g_2(s)| d_q s.$$

By using (7.6.1), for each $t \in I$, we have

$$|g_i(t)| \leq p_i(t) + d_i(t) \min(|u(t)|, |v(t)| + r_i(t)|g_i(t)|$$
$$\leq p_i^* + d_i^* R + r_i^* |g_i(t)|.$$

Thus

$$|g_i(t)| \leq \frac{p_i^* + d_i^* R}{1 - r_i^*}.$$

Hence

$$\|N_1(u)\|_\infty \leq |u_{01}| + \frac{L_1(p_1^* + d_1^* R)}{1 - r_1^*},$$

and

$$\|N_2(v)\|_\infty \leq |u_{02}| + \frac{L_2(p_2^* + d_2^* R)}{1 - r_2^*}.$$

This implies that

$$\|N(u, v)\|_C = \|N_1(u)\|_\infty + \|N_2(v)\|_\infty$$
$$\leq |u_{01}| + |u_{02}| + \sum_{i=1}^2 \frac{L_i(p_i^* + d_i^* R)}{1 - r_i^*}$$
$$\leq R.$$

This proves that N maps the ball B_R into B_R. We will show that the operator $N : B_R \to B_R$ is continuous and compact. The proof will be given in several steps.

Step 1. N is continuous.

Let $\{u_n\}_{n \in \mathbb{N}}$ and $\{v_n\}_{n \in \mathbb{N}}$ two sequences such that $(u_n, v_n) \to (u, v)$ in B_R. Then, for each $t \in I$, we have

$$|(N_1 u_n)(t) - (N_1 u)(t)| \leq \int_0^t \frac{(t - qs)^{(\alpha_1 - 1)}}{\Gamma_q(\alpha_1)} |(g_{1n}(s) - g_1(s))| d_q s,$$

and

$$|(N_2 v_n)(t) - (N_2 v)(t)| \le \int_0^t \frac{(t - qs)^{(\alpha_2 - 1)}}{\Gamma_q(\alpha_2)} |(g_{2n}(s) - g_2(s))| d_q s,$$

where $g_{in}, g_i \in C(I)$; $i = 1, 2$, such that

$$g_{in}(t) = f_i(t, u_n(t), v_n(t), g_{in}(t)),$$

and

$$g_i(t) = f_i(t, u(t), v(t), g_i(t)).$$

Since $(u_n, v_n) \to u$ as $n \to \infty$ and f_i are continuous functions, we get

$$g_{in}(t) \to g_i(t) \text{ as } n \to \infty, \text{ for each } t \in I.$$

Thus

$$\|N_1(u_n) - N_1(u)\|_\infty \le \frac{p_1^* + d_1^* R}{1 - r_1^*} \|g_{1n} - g_1\|_\infty \to 0 \text{ as } n \to \infty,$$

and

$$\|N_2(v_n) - N_2(v)\|_\infty \le \frac{p_2^* + d_2^* R}{1 - r_2^*} \|g_{2n} - g_2\|_\infty \to 0 \text{ as } n \to \infty.$$

Hence

$$\|N(u_n, v_n) - N(u, v)\|_C \to 0 \text{ as } n \to \infty.$$

Step 2. $N(B_R)$ is bounded. This is clear since $N(B_R) \subset B_R$ and B_R is bounded.

Step 3. N maps bounded sets into equicontinuous sets in B_R.
Let $t_1, t_2 \in I$, such that $t_1 < t_2$ and let $(u, v) \in B_R$. Then we have

$$|(N_i u)(t_1) - (N_i u)(t_2)| \le \int_0^{t_1} \frac{|(t_2 - qs)^{(\alpha_i - 1)} - (t_1 - qs)^{(\alpha_i - 1)}|}{\Gamma_q(\alpha_i)} |g_i(s)| d_q s$$
$$+ \int_{t_1}^{t_2} \frac{|(t_2 - qs)^{(\alpha_i - 1)}|}{\Gamma_q(\alpha_i)} |g_i(s)| d_q s,$$

where $g_i \in C(I)$ such that $g_i(t) = f(t, u(t), v(t), g_i(t))$. Hence

$$|(N_1 u)(t_1) - (N_1 u)(t_2)| \le \frac{p_1^* + d_1^* R}{1 - r_1^*} \int_0^{t_1} \frac{|(t_2 - qs)^{(\alpha_1 - 1)} - (t_1 - qs)^{(\alpha_1 - 1)}|}{\Gamma_q(\alpha_1)} d_q s$$
$$+ \frac{p_1^* + d_1^* R}{1 - r_1*} \int_{t_1}^{t_2} \frac{|(t_2 - qs)^{(\alpha_1 - 1)}|}{\Gamma_q(\alpha_1)} d_q s$$
$$\to 0 \text{ as } t_1 \to t_2,$$

and

$$|(N_2v)(t_1) - (N_2v)(t_2)| \leq \frac{p_2^* + d_2^* R}{1 - r_2^*} \int_0^{t_1} \frac{|(t_2 - qs)^{(\alpha_2 - 1)} - (t_1 - qs)^{(\alpha_2 - 1)}|}{\Gamma_q(\alpha_2)} d_q s$$

$$+ \frac{p_2^* + d_2^* R}{1 - r_{2*}} \int_{t_1}^{t_2} \frac{|(t_2 - qs)^{(\alpha_2 - 1)}|}{\Gamma_q(\alpha_2)} d_q s$$

$$\to 0 \text{ as } t_1 \to t_2.$$

As a consequence of the above three steps with the Arzelà–Ascoli theorem, we can conclude that $N : B_R \to B_R$ is continuous and compact. From an application of Theorem 2.68, we deduce that N has at least a fixed point (u, v), which is a solution of our system (7.1)–(7.2).

Step 4. Generalized Ulam–Hyers–Rassias stability.
Let (u_1, v_1) be a solution of the inequality (7.4), and let us assume that (u, v) is a solution of the system (7.1)–(7.2). Thus we have

$$(u(t), v(t)) = \left(u_{01} + (I_q^{\alpha_1} g_1)(t), u_{02} + (I_q^{\alpha_2} g_2)(t)\right),$$

where $g_i \in C(I)$; $i = 1, 2$, such that $g_i(t) = f(t, u(t), v(t), g_i(t))$.
From the inequality (7.4), for each $t \in I$, we have

$$|u_1(t) - u_{01} - (I_q^{\alpha_1} g_1)(t)| \leq (I_q^{\alpha_1} \Phi)(t),$$

and

$$|v_1(t) - u_{02} - (I_q^{\alpha_2} g_2)(t)| \leq (I_q^{\alpha_2} \Phi)(t).$$

From the hypotheses (7.6.1) and (7.6.2), for each $t \in I$, we have

$$|u(t) - u_1(t)| \leq |u(t) - u_{01} - (I_q^{\alpha_1} g_1)(t) + (I_q^{\alpha_1}(g_1 - g_2))(t)|$$

$$\leq (I_q^{\alpha_1} \Phi)(t) + \int_0^t \frac{(t - qs)^{(\alpha_1 - 1)}}{\Gamma_q(\alpha_1)} (|(g_1(s)| + |g_2(s))|)d_q s$$

$$\leq (I_q^{\alpha_1} \Phi)(t) + \frac{p_1^* + d_1^*}{1 - r_1^*}(I_q^{\alpha_1} \Phi)(t)$$

$$\leq \lambda_\phi \Phi(t) + 2\lambda_\phi \frac{p_1^* + d_1^*}{1 - r_1^*} \Phi(t)$$

$$\leq \left[1 + 2\frac{p_1^* + d_1^*}{1 - r_1^*}\right] \lambda_\phi \Phi(t)$$

$$:= c_{f_1, \Phi} \Phi(t).$$

Also, we get

$$|v(t) - v_1(t)| \leq \left[1 + 2\frac{p_2^* + d_2^*}{1 - r_2^*}\right] \lambda_\phi \Phi(t)$$

$$:= c_{f_2,\Phi} \Phi(t).$$

Thus

$$|(u(t), v(t)) - (u_1(t), v_1(t))| = |u(t) - u_1(t)| + |v(t) - v_1(t)|$$

$$\leq \left[\lambda_\phi \sum_{i=1}^{2}\left(1 + 2\frac{p_i^* + d_i^*}{1 - r_i^*}\right)\right] \Phi(t)$$

$$:= c_{f_1,f_2,\Phi} \Phi(t).$$

Hence the problem (7.1)–(7.2) is generalized Ulam–Hyers–Rassias stable.   □

7.1.3 Results in generalized Banach spaces

Now, we are concerned with the existence and uniqueness results of the coupled system (7.1)–(7.2), in generalized Banach spaces.

Let C be the Banach space of all continuous functions v from I into $\mathbb{R}^m$ with the supremum (uniform) norm

$$\|v\|_C := \sup_{t \in I} \|v(t)\|.$$

By $L^\infty(I, \mathbb{R}_+)$, we denote the Banach space of measurable functions from I into $\mathbb{R}_+$ which are essentially bounded.

Let $x, y \in \mathbb{R}^m$ with $x = (x_1, x_2, \ldots, x_m)$, $y = (y_1, y_2, \ldots, y_m)$.

By $x \leq y$ we mean $x_i \leq y_i$; $i = 1, \ldots, m$. Also

$$|x| = (|x_1|, |x_2|, \ldots, |x_m|),$$
$$\max(x, y) = (\max(x_1, y_1), \max(x_2, y_2), \ldots, \max(x_m, y_m)),$$

and

$$\mathbb{R}_+^m = \{x \in \mathbb{R}^m : x_i \in \mathbb{R}_+, \ i = 1, \ldots, m\}.$$

If $c \in \mathbb{R}$, then $x \leq c$ means $x_i \leq c$; $i = 1, \ldots, m$.

Definition 7.8. Let X be a nonempty set. By a vector-valued metric on X we mean a map $d : X \times X \to \mathbb{R}^m$ with the following properties:

   (i) $d(x, y) \geq 0$ for all $x, y \in X$, and if $d(x, y) = 0$, then $x = y$;
   (ii) $d(x, y) = d(y, x)$ for all $x, y \in X$;
   (iii) $d(x, z) \leq d(x, y) + d(y, z)$ for all $x, y, z \in X$.

We call the pair (X, d) a generalized metric space with

$$d(x, y) := \begin{pmatrix} d_1(x, y) \\ d_2(x, y) \\ \vdots \\ d_m(x, y) \end{pmatrix}.$$

Notice that d is a generalized metric space on X if and only if $d_i; \ i = 1, \ldots, m$ are metrics on X.

Definition 7.9. [118,404] A square matrix of real numbers is said to be convergent to zero if and only if its spectral radius $\rho(M)$ is strictly less than 1. In other words, this means that all the eigenvalues of M are in the open-unit disc, i.e., $|\lambda| < 1$, for every $\lambda \in \mathbb{C}$ with $det(M - \lambda I) = 0$, where I denotes the unit matrix of $M_{m \times m}(\mathbb{R})$.

Example 7.10. The matrix $A \in M_{2 \times 2}(\mathbb{R})$ defined by

$$A = \begin{pmatrix} a & b \\ c & d \end{pmatrix},$$

converges to zero in the following cases:

(1) $b = c = 0, a, d > 0$ and $\max\{a, d\} < 1$.
(2) $c = 0, a, d > 0, a + d < 1$ and $-1 < b < 0$.
(3) $a + b = c + d = 0, a > 1, c > 0$ and $|a - c| < 1$.

Definition 7.11. Let (X, d) be a generalized metric space. An operator $N : X \to X$ is said to be contractive if there exists a matrix M convergent to zero such that

$$d(N(x), N(y)) \le Md(x, y); \ for \ all \ x, y \in X.$$

In the sequel we will make use of the following fixed-point theorems in generalized Banach spaces:

Theorem 7.12. *[331,334] Let (X, d) be a complete generalized metric space and $N : X \to X$ a contractive operator with Lipschitz matrix M. Then N has a unique fixed point x_0 and, for each $x \in X$, we have*

$$d(N^k(x), x_0) \le M^k(M)^{-1}d(x, N(x)); \ for \ all \ k \in \mathbb{N}.$$

For $n = 1$, we recover the classical Banach's contraction fixed-point result.

Theorem 7.13. *[331] Let X be a generalized Banach space, $D \subset E$ be a nonempty closed convex subset of E and $N : D \to D$ be a continuous operator with relatively compact range. Then N has at least one fixed point in D.*

The following hypotheses will be used in the sequel.

(7.13.1) There exist continuous functions $p_i, d_i, l_i : I \to \mathbb{R}_+$; $i = 1, 2$ such that $l_i < 1$, and
$$\|f_i(t, u_1, v_1, w_1) - f_i(t, u_2, v_2, w_2)\|$$

$$\leq p_i(t)\|u_1 - u_2\| + d_i(t)\|v_1 - v_2\| + l_i(t)\|w_1 - w_2\|;$$

for a.e. $t \in I$, and each $u_i, v_i, w_i \in \mathbb{R}^m$, $i = 1, 2$.

(7.13.2) There exist continuous functions $K_i, P_i, D_i, L_i : I \to \mathbb{R}_+$; $i = 1, 2$ such that
$$\|f_i(t, u, v, w)\| \leq K_i(t) + P_i(t)\|u\| + D_i(t)\|v\| + L_i(t)\|w\|;$$

for a.e. $t \in I$, and each $u, v, w \in \mathbb{R}^m$, $i = 1, 2$.

Set

$$p_i^* := \sup_{t \in I} p_i(t), \ d_i^* := \sup_{t \in I} d_i(t), \ l_i^* := \sup_{t \in I} l_i(t), \ K_i^* := \sup_{t \in I} K_i(t),$$

$$P_i^* := \sup_{t \in I} P_i(t), \ D_i^* := \sup_{t \in I} D_i(t), \ L_i^* := \sup_{t \in I} L_i(t),$$

and

$$\ell_i := \frac{T^{\alpha_i}}{\Gamma_q(1 + \alpha_i)}; \ i = 1, 2.$$

The space $C^2 := C \times C$ is a generalized Banach space with the norm

$$\|(u_1, u_2)\|_{C^2} := (\|u_1\|_C, \|u_2\|_C).$$

Definition 7.14. By a solution of the problem (7.1)–(7.2) we mean a coupled continuous functions $(u, v) \in C^2$ satisfying the initial condition (7.2) and the system (7.1) on I.

First, we prove an existence and uniqueness result for the coupled system (7.1)–(7.2) by using Banach's fixed-point theorem type in generalized Banach spaces.

Theorem 7.15. *Assume that the hypothesis* (7.13.1) *holds. If the matrix*

$$M := \begin{pmatrix} \frac{\ell_1 p_1^*}{1 - l_1^*} & \frac{\ell_1 d_1^*}{1 - l_1^*} \\ \frac{\ell_2 p_2^*}{1 - l_2^*} & \frac{\ell_2 d_2^*}{1 - l_2^*} \end{pmatrix}$$

converges to 0, then the coupled system (7.1)–(7.2) *has a unique solution.*

Proof. We can define the operators $N_1, N_2 : C^2 \to C$ by

$$(N_i(u_1, u_2))(t) = u_{0i} + (I_q^{\alpha_i} g_i)(t); \ i = 1, 2, \ t \in I, \tag{7.10}$$

where $g_i(\cdot) \in C$, with

$$g_i(t) = f_i(t, u_1(t), u_2(t), g_i(t))$$
$$= f_i(t, u_{01} + (I_q^{\alpha_1} g_1)(t), u_{02} + (I_q^{\alpha_2} g_2)(t), g_i(t)); \ i = 1, 2.$$

Consider the operator $N : C^2 \to C^2$ defined by

$$(N(u_1, u_2))(t) = ((N_1(u_1, u_2))(t), (N_2(u_1, u_2))(t)). \tag{7.11}$$

Clearly, the fixed points of the operator N are solutions of the coupled system (7.1)–(7.2). We show that N satisfies all conditions of Theorem 7.12.

For each $(u_1, u_2), (v_1, v_2) \in C^2$ and $t \in I$, we have

$$\|(N_i(u_1, u_2))(t) - (N_i(v_1, v_2))(t)\| \le \int_0^t \frac{(t - qs)^{(\alpha_i - 1)}}{\Gamma_q(\alpha_i)} \|g_i(s) - h_i(s)\| d_q s, \tag{7.12}$$

where $g_i(\cdot), h_i(\cdot) \in C$; $i = 1, 2$, with

$$\begin{aligned}
g_i(t) &= f_i(t, u_1(t), u_2(t), g_i(t)) \\
&= f_i(t, u_{01} + (I_q^{\alpha_1} g_1)(t), u_{02} + (I_q^{\alpha_2} g_2)(t), g_i(t)),
\end{aligned}$$

and

$$\begin{aligned}
h_i(t) &= f_i(t, v_1(t), v_2(t), h_i(t)) \\
&= f_i(t, u_{01} + (I_q^{\alpha_1} h_1)(t), u_{02} + (I_q^{\alpha_2} h_2)(t), h_i(t)).
\end{aligned}$$

From hypothesis (7.13.1), we have

$$\|g_i(t) - h_i(t)\| = p_i(t)\|u_1(t) - v_1(t)\| + d_i(t)\|u_2 - v_2\| + l_i(t)\|g_i - h_i\|.$$

Then

$$\|g_i(t) - h_i(t)\| = p_i(t)\|u_1(t) - v_1(t)\| + d_i(t)\|u_2(t) - v_2(t)\| + l_i(t)\|g_i(t) - h_i(t)\|.$$

Thus

$$\|g_i - h_i\|_C = p_i^* \|u_1 - v_1\|_C + i_1^* \|u_2 - v_2\|_C + l_i^* \|g_i - h_i\|_C.$$

This implies that

$$(1 - l_i^*)\|g_i - h_i\|_C = p_i^* \|u_1 - v_1\|_C + d_i^* \|u_2 - v_2\|_C.$$

Hence

$$\|g_i - h_i\|_C = \frac{p_i^*}{1 - l_i^*}\|u_1 - v_1\|_C + \frac{d_i^*}{1 - l_i^*}\|u_2 - v_2\|_C.$$

From (7.12), we get

$$\begin{aligned}
&\|(N_i(u_1, u_2)) - (N_i(v_1, v_2))\|_C \\
&\le \int_0^t \frac{(t - qs)^{(\alpha_i - 1)}}{\Gamma_q(\alpha_i - 1)} \|g_i(s) - h_i(s)\| d_q s
\end{aligned}$$

$$\leq \frac{\ell_i p_i^*}{1 - l_i^*} \|u_1 - v_1\|_C + \frac{\ell_i d_i^*}{1 - l_i^*} \|u_2 - v_2\|_C.$$

Consequently,

$$d((N(u_1, u_2)), (N(v_1, v_2))) \leq Md((u_1, u_2), (v_1, v_2)),$$

where

$$d((u_1, u_2), (v_1, v_2)) = \begin{pmatrix} \|u_1 - v_1\|_C \\ \|u_2 - v_2\|_C \end{pmatrix}.$$

Since the matrix M converges to zero, then Theorem 7.12 implies that coupled system (7.1)–(7.2) has a unique solution. $\qquad\square$

Now, we prove an existence result for the coupled system (7.1)–(7.2) by using Schauder fixed-point theorem type in generalized Banach space.

Theorem 7.16. *Assume that the hypothesis* (7.13.2) *holds. Then the coupled system* (7.1)–(7.2) *has at least one solution.*

Proof. Let $N : C^2 \to C^2$ be the operator defined in (7.9). We show that N satisfies all conditions of Theorem 7.13. The proof will be given in several steps.

Step 1. *N is continuous.*
Let $\{(u_{1n}, u_{2n})\}_n$ be a sequence such that $(u_{1n}, u_{2n}) \to (u_1, u_2) \in C^2$ as $n \to \infty$. For any $i = 1, 2$ and each $t \in I$, we have

$$\|(N_i(u_{1n}, u_{2n}))(t) - (N_i(u_1, u_2))(t)\| \leq \int_0^t \frac{(t - qs)^{(\alpha_i - 1)}}{\Gamma_q(\alpha_i - 1)} \|g_{in}(s) - g_i(s)\| d_q s,$$

where $g_i(\cdot), g_{in}(\cdot) \in C(I)$; $i = 1, 2$, with

$$\begin{aligned} g_i(t) &= f_i(t, u_1(t), u_2(t), g_i(t)) \\ &= f_i(t, u_{01} + (I_q^{\alpha_1} g_1)(t), u_{02} + (I_q^{\alpha_2} g_2)(t), g_i(t)), \end{aligned}$$

and

$$\begin{aligned} g_{in}(t) &= f_i(t, u_{1n}(t), u_{2n}(t), g_{in}(t)) \\ &= f_i(t, u_{01} + (I_q^{\alpha_1} g_{1n})(t), u_{02} + (I_q^{\alpha_2} g_{2n})(t), g_{in}(t)). \end{aligned}$$

From (7.13.2), we have

$$\|g_{in} - g_i\|_C \leq \frac{P_i^*}{1 - L_i^*} \|u_{1n} - u_1\|_C + \frac{D_i^*}{1 - L_i^*} \|u_{2n} - u_2\|_C.$$

Thus

$$\|(N_i(u_{1n}, u_{2n}))(t) - (N_i(u_1, u_2))(t)\|$$

$$\leq \frac{\ell_i P_i^*}{1 - L_i^*}\|u_{1n} - u_1\|_C + \frac{\ell_i D_i^*}{1 - L_i^*}\|u_{2n} - u_2\|_C$$

$$\to 0 \text{ as } n \to \infty.$$

Hence we get

$$\|N_i(u_{1n}, u_{2n}) - N_i(u_1, u_2)\|_C \to 0 \text{ as } n \to \infty.$$

Consequently,

$$\|N(u_{1n}, u_{2n}) - N(u_1, u_2)\|_{C^2}$$
$$:= (\|N_1(u_{1n}, u_{2n}) - N_1(u_1, u_2)\|_C, \|N_2(u_{1n}, u_{2n}) - N_2(u_1, u_2)\|_C)$$
$$\to (0, 0) \text{ as } n \to \infty.$$

Finally, N is continuous.

Step 2. *N maps bounded sets into bounded sets in C^2.*
Set

$$h_i^* := \sup_{t \in I} h_i(t), \quad k_i^* := \sup_{t \in I} k_i(t), \quad l_i^* := \sup_{t \in I} l_i(t).$$

Let $R > 0$ and set

$$B_R := \{(\mu, \nu) \in C^2 : \|\mu\|_C \leq R, \|\nu\|_C \leq R\}.$$

For any $i = 1, 2$ and each $(u, v) \in B_R$ and $t \in I$, we have

$$\|(N_i(u_1, u_2))(t)\| \leq \int_0^t \frac{(t - qs)^{(\alpha_i - 1)}}{\Gamma_q(\alpha_i - 1)}\|g_i(s)\|d_q s,$$

where $g_i(\cdot), \in C(I); \ i = 1, 2$, with

$$g_i(t) = f_i(t, u_1(t), u_2(t), g_i(t))$$
$$= f_i(t, u_{01} + (I_q^{\alpha_1} g_1)(t), u_{02} + (I_q^{\alpha_2} g_2)(t), g_i(t)).$$

Since

$$\|g_i\|_C \leq \frac{P_i^*}{1 - L_i^*}\|u_1\|_C + \frac{D_i^*}{1 - L_i^*}\|u_2\|_C,$$

we get

$$\|N_i(u_1, u_2)\|_C \leq \frac{\ell_i P_i^*}{1 - L_i^*}\|u_1\|_C + \frac{\ell_i D_i^*}{1 - L_i^*}\|u_2\|_C.$$

Thus

$$\|N_i(u_1, u_2)\|_C \leq \frac{R\ell_i P_i^*}{1 - L_i^*} + \frac{R\ell_i D_i^*}{1 - L_i^*} := M_i.$$

Hence

$$\|(N(u, v))\|_{C^2} \leq (M_1, M_2) := M.$$

Step 3. *N maps bounded sets into equicontinuous sets in* C^2.
Let B_R be the ball defined in Step 2. For each $t_1, t_2 \in I$ with $t_1 \leq t_2$ and any $(u, v) \in B_R$ and $i = 1, 2$, we have

$$\|(N_i(u_1, u_2))(t_1) - (N_i(u_1, u_2))(t_2)\|$$
$$\leq \int_0^{t_1} \frac{(t_1 - qs)^{(\alpha_i - 1)}}{\Gamma_q(\alpha_i - 1)} \|g_i(s)\| d_q s - \int_0^{t_2} \frac{(t_2 - qs)^{(\alpha_i - 1)}}{\Gamma_q(\alpha_i - 1)} \|g_i(s)\| d_q s,$$

where $g_i(\cdot), \in C(I)$; $i = 1, 2$, with

$$g_i(t) = f_i(t, u_1(t), u_2(t), g_i(t))$$
$$= f_i(t, u_{01} + (I_q^{\alpha_1} g_1)(t), u_{02} + (I_q^{\alpha_2} g_2)(t), g_i(t)).$$

Thus

$$\|(N_i(u_1, u_2))(t_1) - (N_i(u_1, u_2))(t_2)\|$$
$$\leq \left(\frac{RP_i^*}{1 - L_i^*} + \frac{RD_i^*}{1 - L_i^*} \right) \int_0^{t_1} \frac{|(t_2 - qs)^{(\alpha_i - 1)} - (t_1 - qs)^{(\alpha_i - 1)}|}{\Gamma_q(\alpha_1)} d_q s$$
$$+ \left(\frac{RP_i^*}{1 - L_i^*} + \frac{RD_i^*}{1 - L_i^*} \right) \int_{t_1}^{t_2} \frac{|(t_2 - qs)^{(\alpha_i - 1)}|}{\Gamma_q(\alpha_i)} d_q s$$
$$\to 0 \text{ as } t_1 \to t_2.$$

Hence

$$\|(N(u_1, u_2))(t_1) - (N(u_1, u_2))(t_2)\|$$
$$= (\|(N_1(u_1, u_2))(t_1) - (N_1(u_1, u_2))(t_2)\|, \|(N_2(u_1, u_2))(t_1) - (N_2(u_1, u_2))(t_2)\|)$$
$$\to (0, 0) \text{ as } t_1 \to t_2.$$

As a consequence of steps 1 to 3 together with Theorem 7.13, we can conclude that N has at least one fixed point in B_R, which is a solution of our coupled system (7.1)–(7.2).   □

7.1.4 Examples

Example 1. Consider the following coupled system of implicit fractional $\frac{1}{4}$-difference equation:

$$\begin{cases} ({}^{c}D_{\frac{1}{4}}^{\frac{1}{2}}u)(t) = f_1(t, u(t), v(t), ({}^{c}D_{\frac{1}{4}}^{\frac{1}{2}}u)(t)) \\ ({}^{c}D_{\frac{1}{4}}^{\frac{1}{2}}v)(t) = f_2(t, u(t), v(t), ({}^{c}D_{\frac{1}{4}}^{\frac{1}{2}}v)(t)) \quad ; t \in [0, 1], \\ (u(0), v(0)) = (1, 2) \end{cases} \tag{7.13}$$

where

$$\begin{cases} f_1(t, x, y, z) = \frac{ct^2}{1+|x|+|y|+|z|}\left(e^{-7} + \frac{1}{e^{t+5}}\right)(t^2 + xt^2 + z) \\ f_2(t, x, y, z) = \frac{ct^2}{e^{t+5}(1+|x|+|y|+|z|)}(e^t + ty + z) \end{cases} \quad ; t \in (0, 1],$$

and $c > 0$. The hypothesis (7.13.1) is satisfied with

$$p_1(t) = d_1(t) = \left(e^{-7} + \frac{1}{e^{t+5}}\right)ct^4,$$

$$r_1(t) = \left(e^{-7} + \frac{1}{e^{t+5}}\right)ct^2, \quad p_2(t) = \frac{ct^2}{e^{t+4}}, \quad d_2 = \frac{ct^3}{e^{t+5}}, \quad r_2(t) = \frac{ct^2}{e^{t+5}}.$$

Also, the condition (7.6) is satisfied with a good choice of the constant c. Hence Theorem 7.7 implies that our system (7.13) has at least a solution defined on [0, 1].

On the other hand, hypothesis (7.13.2) is satisfied with $\Phi(t) = t^2$. Indeed, for each $t \in (0, 1]$, there exists a real number $0 < \epsilon < 1$ such that $\epsilon < t \le 1$, and

$$(I_q^\alpha \Phi)(t) \le \frac{t^2}{\epsilon^2(1 + q + q^2)}$$

$$\le \frac{1}{\epsilon^2}\Phi(t)$$

$$= \lambda_\Phi \Phi(t).$$

Consequently, the problem (7.13) is generalized Ulam–Hyers–Rassias stable.

Example 2. Consider now the following coupled system of implicit fractional $\frac{1}{4}$-difference equation:

$$\begin{cases} ({}^{c}D_{\frac{1}{4}}^{\frac{1}{2}}u)(t) = f_1(t, u(t), v(t), ({}^{c}D_{\frac{1}{4}}^{\frac{1}{2}}u)(t)) \\ ({}^{c}D_{\frac{1}{4}}^{\frac{1}{2}}v)(t) = f_2(t, u(t), v(t), ({}^{c}D_{\frac{1}{4}}^{\frac{1}{2}}v)(t)) \quad ; t \in [0, 1], \\ (u(0), v(0)) = (1, 2) \end{cases} \tag{7.14}$$

where

$$\begin{cases} f_1(t,x,y,z) = \frac{ct^2}{1+|x|+|y|+|z|}\left(e^{-7} + \frac{1}{e^{t+5}}\right)(xt^2+z) \\ f_2(t,x,y,z) = \frac{ct^2}{e^{t+5}(1+|x|+|y|+|z|)}(ty+z) \end{cases} \quad ; t \in (0,1],$$

and $c > 0$. Hypothesis (7.13.1) is satisfied with

$$p_1(t) = \left(e^{-7} + \frac{1}{e^{t+5}}\right)ct^4, \ d_1(t) = 0,$$

$$l_1(t) = \left(e^{-7} + \frac{1}{e^{t+5}}\right)ct^2, \ p_2(t) = 0, \ d_2 = \frac{ct^3}{e^{t+5}}, \ l_2(t) = \frac{ct^2}{e^{t+5}}.$$

Also, with a good choice of the constant c, the matrix

$$M := \begin{pmatrix} \frac{l_1 p_1^*}{1-l_1^*} & 0 \\ 0 & \frac{l_2 d_2^*}{1-l_2^*} \end{pmatrix}$$

converges to 0. Hence Theorem 7.15 implies that the system (7.14) has a unique solution defined on $[0,1]$.

7.2 Existence and oscillation for coupled fractional q-difference systems

7.2.1 Introduction

In this section we discuss the existence of solutions and the oscillation for the following coupled fractional q-difference system

$$\begin{cases} ({}^cD_q^{\alpha_1}u_1)(t) = f_1(t,u_2(t)), \\ ({}^cD_q^{\alpha_2}u_2)(t) = f_2(t,u_1(t)), \\ (u_1(0),u_2(0)) = (u_{01},u_{02}) \in \mathbb{R} \times \mathbb{R}, \ u \text{ and } v \text{ are bounded on } \mathbb{R}_+, \end{cases} \quad ; t \in \mathbb{R}_+, \quad (7.15)$$

where $q \in (0,1)$, $\alpha_i \in (0,1]$, $\mathbb{R}_+ := [0,+\infty)$, $f_i : I \times \mathbb{R} \to \mathbb{R}$; $i = 1,2$ are given functions, and ${}^cD_q^{\alpha_i}$ is the Caputo fractional q-difference derivative of order α_i; $i = 1,2$.

7.2.2 Existence of bounded solutions

In this section we are concerned with the existence of solutions of the coupled system (7.15).

Definition 7.17. By a solution of the coupled system (7.15) we mean a bounded coupled functions $(u_1, u_2) \in C$ that satisfies the system

$$\begin{cases} (^cD_q^{\alpha_1} u_1)(t) = f_1(t, u_2(t)), \\ (^cD_q^{\alpha_2} u_2)(t) = f_2(t, u_1(t)), \end{cases}$$

on $\mathbb{R}_+ \times \mathbb{R}_+$ and the initial conditions $(u_1(0), u_2(0)) = (u_{01}, u_{02})$.

Let $I_n := [0, n]$; $n \in \mathbb{N}$. We denote by $X_n := C(I_n) \times C(I_n)$ the Banach space with the norm

$$\|(u, v)\|_{X_n} = \|u\|_\infty + \|v\|_\infty.$$

The following hypotheses will be used in the sequel.

(7.17.1) The functions $t \mapsto f_1(t, v)$ and $t \mapsto f_2(t, u)$ are measurable on $I_n := [0, n]$; $n \in \mathbb{N}$ for each $u \in \mathbb{R}$, and the functions $u \mapsto f_1(t, v)$ and $v \mapsto f_2(t, u)$ is continuous for a.e. $t \in I_n$,

(7.17.2) There exist continuous functions $p_{in} : I_n \to \mathbb{R}_+$; $n = 1, 2$ such that

$$|f_i(t, u_i)| \leq p_{in}(t); \text{ for a.e. } t \in I_n, \text{ and each } u_i \in \mathbb{R}.$$

Set

$$p_{in}^* = \sup_{t \in I_n} p_{in}(t); \ i = 1, 2.$$

Theorem 7.18. *Assume that the hypotheses (7.17.1) and (7.17.2) hold. Then the problem (7.15) has at least one bounded solution defined on $\mathbb{R}_+$.*

Proof. The proof will be given in two parts. Fix $n \in \mathbb{N}$ and consider the problem

$$\begin{cases} (^cD_q^{\alpha_1} u_1)(t) = f_1(t, u_2(t)) \\ (^cD_q^{\alpha_2} u_2)(t) = f_2(t, u_1(t)) \quad ; t \in I_n := [0, n]. \\ (u_1(0), u_2(0)) = (u_{01}, u_{02}) \end{cases} \tag{7.16}$$

Part 1. We begin by showing that (7.16) has a solution $(u_{1n}, u_{2n}) \in X_n$ with

$$\|(u_{1n}, u_{2n})\|_{X_n} \leq R_n := \|u_{01}\| + \|u_{02}\| + \sum_{i=1}^{2} \frac{n^\alpha p_{in}^*}{\Gamma_q(1 + \alpha_i)}.$$

Consider the operators $N_i : C(I_n) \to C(I_n)$; $i = 1, 2$, and $N : X_n \to X_n$ defined by

$$(N_1 u_1)(t) = u_{01} + (I_q^{\alpha_1} f_1(\cdot, u_2(\cdot)))(t); \ t \in I, \tag{7.17}$$

$$(N_2) u_2(t) = u_{02} + (I_q^{\alpha_2} f_2(\cdot, u_1(\cdot)))(t); \ t \in I, \tag{7.18}$$

and

$$(N(u_1, u_2))(t) = ((N_1 u_1)(t), (N_2 u_2)(t)). \tag{7.19}$$

Clearly, the fixed points of the operator N are solution of our coupled system (7.16).

For any $n \in \mathbb{N}^*$, we consider the ball

$$B_{R_n} := B(0, R_n) = \{w = (w_1, w_2) \in X_n : \|w\|_{X_n} \le R_n\}.$$

We shall show that the operator $N : B_{R_n} \to B_{R_n}$ satisfies all the assumptions of Theorem 2.68. The proof will be given in several steps.

Step 1. $N : B_{R_n} \to B_{R_n}$ *is continuous.*
Let $\{u_k\}_{k \in \mathbb{N}}$ be a sequence such that $u_k = (u_{1k}, u_{2u}) \to u = (u_1, u_2)$ in B_{R_n}. Then, for each $t \in I_n$, we have

$$\|(N_1 u_{1k})(t) - (N_1 u_1)(t)\| \le \int_0^t \frac{(tq - s)^{(\alpha_1 - 1)}}{\Gamma_q(\alpha_1)} \|f_1(s, u_{2k}(s)) - f_1(s, u_2(s))\| d_q s,$$

and

$$\|(N_2 u_{2k})(t) - (N_2 u_2)(t)\| \le \int_0^t \frac{(tq - s)^{(\alpha_2 - 1)}}{\Gamma_q(\alpha_2)} \|f_2(s, u_{1k}(s)) - f_1(s, u_1(s))\| d_q s.$$

Since $u_{ik} \to u_i$ as $k \to \infty$, the Lebesgue dominated convergence theorem implies that

$$\|N_i(u_{ik}) - N_i(u_i)\|_\infty \to 0 \quad \text{as } k \to \infty.$$

Hence

$$\|N(u_k) - N(u)\|_{X_n} \to 0 \quad \text{as } k \to \infty.$$

Step 2. $N(B_{R_n})$ *is uniformly bounded.*
For any $u = (u_1, u_2) \in X_n$, and each $t \in I_n$, we have

$$\begin{aligned}
|(N_1(u_1))(t)| &\le |u_{01}| + \int_0^t \frac{(t - qs)^{(\alpha_1 - 1)}}{\Gamma_q(\alpha_1)} |f_1(s, u_2(s))| d_q s \\
&\le |u_{01}| + \int_0^t \frac{(t - qs)^{(\alpha_1 - 1)}}{\Gamma_q(\alpha)} p_{1n}(s) d_q s \\
&\le |u_{01}| + p_{1n}^* \int_0^t \frac{(t - qs)^{(\alpha_1 - 1)}}{\Gamma_q(\alpha_1)} d_q s \\
&\le |u_{01}| + \frac{n^\alpha p_{1n}^*}{\Gamma_q(1 + \alpha_1)}.
\end{aligned}$$

Also,

$$|(N_2(u_2)(t)| \leq |u_{02}| + \int_0^t \frac{(t-qs)^{(\alpha_2-1)}}{\Gamma_q(\alpha_2)} |f_2(s, u_1(s))| d_q s$$

$$\leq |u_{02}| + \frac{n^\alpha p_{2n}^*}{\Gamma_q(1+\alpha_2)}.$$

Thus we get

$$\|(Nu)(t)\| = \|(N_1u_1)(t)\| + \|(N_2u_2)(t)\|$$

$$\leq \|u_{01}\| + \frac{n^{\alpha_1} p_{1n}^*}{\Gamma_q(1+\alpha_1)} + \|u_{02}\| + \frac{n^{\alpha_2} p_{2n}^*}{\Gamma_q(1+\alpha_2)}$$

$$= R_n.$$

Hence

$$\|N(u)\|_{X_n} \leq R_n. \tag{7.20}$$

This proves that $N(B_{R_n}) \subset B_{R_n}$ and then B_{R_n} is bounded.

Step 3. *$N(B_{R_n})$ is equicontinuous.*
Let $t_1, t_2 \in I_n$, $t_1 < t_2$ and let $u = (u_1, u_2) \in B_{R_n}$. Thus we have

$$|(N_1u_1)(t_2) - (N_1u_1)(t_1)|$$

$$\leq \int_0^{t_1} \frac{|(t_2-qs)^{(\alpha-1)} - (t_1-qs)^{(\alpha_1-1)}|}{\Gamma_q(\alpha_1)} |f+1(s, u_2(s))| d_q s$$

$$+ \int_{t_1}^{t_2} \frac{|(t_2-qs)^{(\alpha_1-1)}|}{\Gamma_q(\alpha_1)} |f_1(s, u_2(s))| d_q s$$

$$\leq p_{1n}^* \int_0^{t_1} \frac{|(t_2-qs)^{(\alpha_1-1)} - (t_1-qs)^{(\alpha_1-1)}|}{\Gamma_q(\alpha_1)} d_q s$$

$$+ p_{1n}^* \int_{t_1}^{t_2} \frac{|(t_2-qs)^{(\alpha_1-1)}|}{\Gamma_q(\alpha_1)} d_q s.$$

Also, we get

$$|(N_2u_2)(t_2) - (N_2u_2)(t_1)|$$

$$\leq p_{2n}^* \int_0^{t_1} \frac{|(t_2-qs)^{(\alpha_2-1)} - (t_1-qs)^{(\alpha_2-1)}|}{\Gamma_q(\alpha_2)} d_q s$$

$$+ p_{2n}^* \int_{t_1}^{t_2} \frac{|(t_2-qs)^{(\alpha_1-1)}|}{\Gamma_q(\alpha_2)} d_q s.$$

Thus

$$|(Nu)(t_2) - (Nu)(t_1)| = |(N_1u_1)(t_2) - (N_1u_1)(t_1)| + |(N_2u_2)(t_2) - (N_2u_2)(t_1)|$$

$$\leq p_{1n}^* \int_0^{t_1} \frac{|(t_2 - qs)^{(\alpha_1-1)} - (t_1 - qs)^{(\alpha_1-1)}|}{\Gamma_q(\alpha_1)} d_qs$$

$$+ p_{2n}^* \int_0^{t_1} \frac{|(t_2 - qs)^{(\alpha_2-1)} - (t_1 - qs)^{(\alpha_2-1)}|}{\Gamma_q(\alpha_2)} d_qs$$

$$+ p_{1n}^* \int_{t_1}^{t_2} \frac{|(t_2 - qs)^{(\alpha_1-1)}|}{\Gamma_q(\alpha_1)} d_qs$$

$$+ p_{2n}^* \int_{t_1}^{t_2} \frac{|(t_2 - qs)^{(\alpha_1-1)}|}{\Gamma_q(\alpha_2)} d_qs.$$

As $t_1 \longrightarrow t_2$, the right-hand side of the above inequality tends to zero.

As a consequence of steps 1 to 3 together with the Arzelà–Ascoli theorem, we can conclude that N is continuous and compact. From an application of Theorem 2.68, we deduce that N has a fixed point (u_1, u_2), which is a solution of the problem (7.16).

Part 2. The diagonalization process.

Now, we use the following diagonalization process. For $k \in \mathbb{N}$ and $i = 1, 2$, we let $w_k = (w_{1k}, w_{2k})$ such that

$$\begin{cases} w_{ik}(t) = u_{in_k}(t); \ t \in [0, n_k], \\ w_{ik}(t) = u_{in_k}(n_k); \ t \in [n_k, \infty) \end{cases} ; \ i = 1, 2.$$

Here $\{n_k\}_{k \in \mathbb{N}^*}$ is a sequence of numbers satisfying

$$0 < n_1 < n_2 < \ldots n_k < \ldots \uparrow \infty.$$

Let $S = \{w_k\}_{k=1}^\infty$. Notice that

$$|w_{n_k}(t)| = \sum_{i=1}^2 |w_{in_k}(t)| \leq R_n; \ \text{for } t \in [0, n_1], \ k \in \mathbb{N}.$$

Also, if $k \in \mathbb{N}$ and $t \in [0, n_1]$, we have

$$w_{1n_k}(t) = u_0 + \int_0^t \frac{(t - qs)^{(\alpha_1-1)}}{\Gamma_q(\alpha_1)} f(s, w_{2n_k}(s)) d_qs,$$

and

$$w_{2n_k}(t) = u_0 + \int_0^t \frac{(t - qs)^{(\alpha_2-1)}}{\Gamma_q(\alpha_2)} f(s, w_{1n_k}(s)) d_qs.$$

Thus, for $k \in \mathbb{N}$ and $t, x \in [0, n_1]$, we have

$$|w_{1n_k}(t) - w_{1n_k}(x)| \leq \int_0^{n_1} \frac{|(t - qs)^{(\alpha_1 - 1)} - (x - qs)^{(\alpha_1 - 1)}|}{\Gamma_q(\alpha_1)} |f_1(s, w_{2n_k}(s))| d_q s,$$

and

$$|w_{2n_k}(t) - w_{2n_k}(x)| \leq \int_0^{n_1} \frac{|(t - qs)^{(\alpha_2 - 1)} - (x - qs)^{(\alpha_2 - 1)}|}{\Gamma_q(\alpha_2)} |f_2(s, w_{1n_k}(s))| d_q s.$$

Hence

$$|w_{1n_k}(t) - w_{1n_k}(x)| \leq p_{1n_1}^* \int_0^{n_1} \frac{|(t - qs)^{(\alpha_1 - 1)} - (x - qs)^{(\alpha_1 - 1)}|}{\Gamma_q(\alpha_1)} d_q s,$$

and

$$|w_{2n_k}(t) - w_{2n_k}(x)| \leq p_{2n-1}^* \int_0^{n_1} \frac{|(t - qs)^{(\alpha_2 - 1)} - (x - qs)^{(\alpha_2 - 1)}|}{\Gamma_q(\alpha_2)} d_q s.$$

The Arzelà–Ascoli theorem guarantees that there is a subsequence $\mathbb{N}_1^*$ of $\mathbb{N}$ and a coupled functions $z_1 = (z_{11}, z_{21}) \in X_{n_1}$ with $u_{n_k} \to z_1$ as $k \to \infty$ in X_{n_1} through $\mathbb{N}_1^*$. Let $\mathbb{N}_1 = \mathbb{N}_1^* - \{1\}$.

Notice that

$$|w_{n_k}(t)| \leq R_n : \text{ for } t \in [0, n_2], \ k \in \mathbb{N}.$$

Also, if $k \in \mathbb{N}$ and $t, x \in [0, n_2]$, we have

$$|w_{1n_k}(t) - w_{1n_k}(x)| \leq p_{1n_2}^* \int_0^{n_2} \frac{|(t - qs)^{(\alpha_1 - 1)} - (x - qs)^{(\alpha_1 - 1)}|}{\Gamma_q(\alpha_1)} d_q s,$$

and

$$|w_{2n_k}(t) - w_{2n_k}(x)| \leq p_{2n_2}^* \int_0^{n_2} \frac{|(t - qs)^{(\alpha_2 - 1)} - (x - qs)^{(\alpha_2 - 1)}|}{\Gamma_q(\alpha_2)} d_q s.$$

The Arzelà–Ascoli theorem guarantees that there is a subsequence $\mathbb{N}_2^*$ of $\mathbb{N}_1$ and a function $z_2 = (z_{12}, z_{22}) \in X_{n_2}$ with $u_{n_k} \to z_2$ as $k \to \infty$ in X_{n_2} through $\mathbb{N}_2^*$. Note that $z_1 = z_2$ on $[0, n_1]$ since $\mathbb{N}_2^* \subset \mathbb{N}_1$. Let $\mathbb{N}_2 = \mathbb{N}_2^* - \{2\}$. Proceed inductively to obtain for $m = 3, 4, \ldots$ a subsequence $\mathbb{N}_m^*$ of $\mathbb{N}_{m-1}$ and a function $z_m = (z_{1m}, z_{2m}) \in X_{n_m}$ with $u_{n_k} \to z_m$ as $k \to \infty$ in X_{n_m} through $\mathbb{N}_m^*$. Let $\mathbb{N}_m = \mathbb{N}_m^* - \{m\}$.

Define a function y as follows. Fix $t \in (0, \infty)$ and let $m \in \mathbb{N}$ with $t \leq n_m$. Then define $u(t) = z_m(t)$. Thus $u = (u_1, u_2) \in X_\infty = C[0, \infty) \times C[0, \infty)$, $u(0) = (u_{01}, u_{02})$ and $|u(t)| \leq R_n$: for $t \in [0, \infty)$.

Again fix $t \in (0, \infty)$ and let $m \in \mathbb{N}$ with $t \leq n_m$. Then, for $n \in \mathbb{N}_m$, we have

$$u_{1n_k}(t) = u_{01} + \int_0^{n_m} \frac{(t - qs)^{(\alpha_1 - 1)}}{\Gamma_q(\alpha_1)} f_1(s, w_{2n_k}(s)) d_q s,$$

and

$$u_{2n_k}(t) = u_{02} + \int_0^{n_m} \frac{(t-qs)^{(\alpha_2-1)}}{\Gamma_q(\alpha_2)} f_2(s, w_{1n_k}(s))d_q s.$$

Let $n_k \to \infty$ through $\mathbb{N}_m$ to obtain

$$z_{1m}(t) = u_{01} + \int_0^{n_m} \frac{(t-qs)^{(\alpha_1-1)}}{\Gamma_q(\alpha_1)} f_1(s, z_{2m}(s))d_q s,$$

and

$$z_{2m}(t) = u_{02} + \int_0^{n_m} \frac{(t-qs)^{(\alpha_2-1)}}{\Gamma_q(\alpha_2)} f_2(s, z_{1m}(s))d_q s.$$

We can use this method for each $x \in [0, n_m]$ and for each $m \in \mathbb{N}$. Thus for $t \in [0, n_m]$,

$$(^C D_q^{\alpha_1} u_1(t) = f_1(t, u_2(t)),$$

and

$$(^C D_q^{\alpha_2} u_2(t) = f_2(t, u_1(t)).$$

For each $m \in \mathbb{N}$ and the constructed function u is a solution of coupled system (7.15). $\square$

7.2.3 Oscillation and nonoscillation results

Definition 7.19. [91] A solution u of problem (7.15) is said to be oscillatory if it is neither eventually positive nor eventually negative. Otherwise u is called nonoscillatory.

Definition 7.20. [91] A solution $u = (u_1, u_2)$ of the coupled system (7.15) is said to be strongly (weakly) oscillatory if each (at least one) of its components is oscillatory. Otherwise, it is said to be strongly (weakly) nonoscillatory if each (at least one) of its nontrivial components is nonoscillatory.

The following hypothesis will be used in the sequel.

(7.20.1) There exist continuous functions $q_{in} : I_n \to \mathbb{R}_+$; $n = 1, 2$ and continuous, bounded, and increasing real functions g_i; $n = 1, 2$ such that, for a.e. $t \in I_n$, and each $u_i \in \mathbb{R}$, and $v \in \mathbb{R}^*$,

$$|f_1(t, u_1)| = q_{1n}(t)g_i(u_2), \quad |f_2(t, u_1)| = q_{2n}(t)g_i(u_2), \text{ and } vg_i(v) > 0.$$

Remark 7.21. We can see that (7.20.1) implies (7.17.2) with $p_{in}(t) = Mq_{in}(t)$, where

$$M = \sup_{v \in \mathbb{R}} |g_i(v)|.$$

The following theorem gives sufficient conditions to ensure the nonoscillation of solutions of the coupled system (7.15).

Theorem 7.22. *Assume that (7.17.1) and (7.20.1) hold. If $u = (u_1, u_2)$ is a weakly nonoscillatory solution of (7.15) such that u_1 and u_2 have the same sign, then the first component u_1 is also nonoscillatory.*

Proof. Assume to the contrary that u_1 is an oscillatory, but u_2 is an eventually positive. Then, in view of (7.20.1), there exists $n_m > 0$ such that $q_{1n}(t)g_1(u_2(t)) \geq 0$ for t larger than n_m. Thus, for all $t > n_m$,

$$u_1(t) = u_1(n_m) + \int_{n_m}^{n_m+1} \frac{(t-qs)^{(\alpha_1-1)}}{\Gamma_q(\alpha_1)} q_{1n}(s)g_1(u_2(s))d_q s > 0.$$

Hence $u_1(t) > 0$ for all large t. This is a contradiction.

Analogously, the case when u_2 is an eventually negative is proved similarly. Indeed, if u_2 is an eventually negative, then from (7.20.1), there exists $n_m > 0$, such that for all $t > n_m$,

$$u_1(t) = u_1(n_m) + \int_{n_m}^{n_m+1} \frac{(t-qs)^{(\alpha_1-1)}}{\Gamma_q(\alpha_1)} q_{1n}(s)g_1(u_2(s))d_q s < 0.$$

Thus $u_1(t) < 0$ for all large t. This is again a contradiction. This means that u_1 is nonoscillatory. $\square$

Corollary 7.23. *Assume that (7.17.1) and (7.20.1) hold. If $u = (u_1, u_2)$ is a weakly nonoscillatory solution of (7.15) such that u_1 and u_2 have the same sign, then the second component u_2 is also nonoscillatory.*

Corollary 7.24. *Assume that (7.17.1) and (7.20.1) hold. If $u = (u_1, u_2)$ is a weakly nonoscillatory solution of (7.15) such that u_1 and u_2 have the same sign, then u is a strongly nonoscillatory solution of (7.15).*

The following theorem presents the oscillatory of the coupled system (7.15).

Theorem 7.25. *Assume that (7.17.1) and (7.20.1) hold. If $u = (u_1, u_2)$ is a weakly oscillatory solution of (7.15) such that u_1 and u_2 have the same sign, then the first component u_1 is also oscillatory.*

Proof. Assume to the contrary that u_1 is nonoscillatory. If u_1 is an eventually positive solution, then, in view of (7.20.1), there exists $n_m > 0$ such that $q_{2n}(t)g_2(u_1(t)) \geq 0$ for t larger than n_m. Thus, for all $t > n_m$,

$$u_2(t) = u_2(n_m) + \int_{n_m}^{n_m+1} \frac{(t-qs)^{(\alpha_2-1)}}{\Gamma_q(\alpha_2)} q_{2n}(s)g_2(u_1(s))d_q s > 0.$$

Hence $u_2(t) > 0$ for all large t. This is a contradiction since u_2 is an oscillatory solution.

Analogously, the case when u_1 is an eventually negative is proved similarly. Indeed, if u_1 is an eventually negative, then, from (7.20.1), there exists $n_m > 0$ such that for all $t > n_m$,

$$u_2(t) = u_2(n_m) + \int_{n_m}^{n_m+1} \frac{(t-qs)^{(\alpha_2-1)}}{\Gamma_q(\alpha_2)} q_{2n}(s) g_2(u_1(s)) d_q s < 0.$$

Thus $u_2(t) < 0$ for all large t. This is again a contradiction since u_2 is an oscillatory solution. This means that u_1 is oscillatory.   □

Corollary 7.26. *Assume that (7.17.1) and (7.20.1) hold. If $u = (u_1, u_2)$ is a weakly oscillatory solution of (7.15) such that u_1 and u_2 have the same sign, then u_2 is an oscillatory solution.*

Corollary 7.27. *Assume that (7.17.1) and (7.20.1) hold. If $u = (u_1, u_2)$ is a weakly oscillatory solution of (7.15) such that u_1 and u_2 have the same sign, then u is an oscillatory solution of (7.15),*

7.3 Notes and remarks

This chapter is written by taking into account the paper of Abbas et al. [75]. We refer the reader to the papers [82,83,96,103,129–131,279,345,346,350,358] and the references therein, for some additional and relevant results.

8

Coupled Caputo–Hadamard fractional differential systems in generalized Banach spaces

We took as motivation the monographs of Abbas et al. [48,69,70], Samko et al. [385], Kilbas et al. [281], Zhou [438], and the papers [38,53,257,369].

8.1 Coupled Caputo–Hadamard fractional differential systems with multipoint boundary conditions

8.1.1 Introduction

In this section we discuss the existence and uniqueness of solutions for the following coupled system of Caputo–Hadamard fractional differential equations

$$
\begin{cases}
({}^{Hc}D_1^{\alpha_1}u)(t) = f_1(t, u(t), v(t)) \\[2mm]
({}^{Hc}D_1^{\alpha_2}v)(t) = f_2(t, u(t), v(t))
\end{cases}
; \; t \in I := [1, T],
\tag{8.1}
$$

with the multi-point boundary conditions

$$
\begin{cases}
a_1 u(1) - b_1 u'(1) = d_1 u(\xi_1) \\[2mm]
a_2 u(T) + b_2 u'(T) = d_2 u(\xi_2) \\[2mm]
a_3 v(1) - b_3 v'(1) = d_3 v(\xi_3) \\[2mm]
a_4 v(T) + b_4 v'(T) = d_4 v(\xi_4)
\end{cases}
; \; w \in \Omega,
\tag{8.2}
$$

where $\alpha \in (1, 2]$, $T > 1$, $a_i, b_i, d_i \in \mathbb{R}$, $\xi_i \in (1, T)$; $i = 1, 2, 3, 4$, $f_1, f_2 : I \times \mathbb{R}^m \times \mathbb{R}^m \to \mathbb{R}^m$ are given continuous functions, $\mathbb{R}^m$; $m \in \mathbb{N}^*$ is the Euclidean Banach space with a

Fractional Difference, Differential Equations, and Inclusions. https://doi.org/10.1016/B978-0-44-323601-3.00015-0
Copyright © 2024 Elsevier Inc. All rights reserved, including those for text and data mining, AI training, and similar technologies.

suitable norm $\| \cdot \|$, $^{Hc}D_1^{\alpha_i}$ is the Caputo–Hadamard fractional derivative of order α_i; $i = 1, 2$.

8.1.2 Main results

In this section we are concerned with the existence and uniqueness results of the coupled system (8.1)–(8.2).

Definition 8.1. By a solution of the problem (8.1)–(8.2) we mean a coupled continuous functions $(u, v) \in C \times C$ satisfying the boundary conditions (8.2) and Eqs. (8.1) on I.

The following hypotheses will be used in the sequel.

(8.1.1) There exist continuous functions $p_i, q_i : I \rightarrow \mathbb{R}_+$; $i = 1, 2$ such that

$$\| f_i(t, u_1, v_1) - f_i(t, u_2, v_2) \| \leq p_i(t) \| u_1 - u_2 \| + q_i(t) \| v_1 - v_2 \|;$$

for a.e. $t \in I$, and each $u_i, v_i \in \mathbb{R}^m$, $i = 1, 2$.

(8.1.2) There exist continuous functions $h_i, k_i, l_i : I \rightarrow \mathbb{R}_+$; $i = 1, 2$ such that

$$\| f_i(t, u, v) \| \leq h_i(t) + k_i(t) \| u \| + l_i(t) \| v \|; \text{ for a.e. } t \in I, \text{ and each } u, v \in \mathbb{R}^m.$$

First, we prove an existence and uniqueness result for the coupled system (8.1)–(8.2) by using Banach fixed-point theorem type in generalized Banach spaces. We define the operators $N_1, N_2 : C \rightarrow C$ by

$$(N_1(u, v))(t) = \int_1^T G_1(t, s) f_1(s, u(s)) ds, \tag{8.3}$$

and

$$(N_2(u, v))(t) = \int_1^T G_2(t, s) f_2(s, u(s)) ds, \tag{8.4}$$

where

$G_1(t,s)$

$$
= \begin{cases}
\begin{aligned}
&\frac{1}{s\Gamma(\alpha_1)}(ln\frac{t}{s})^{\alpha_1-1} + \frac{d_1(ln\xi_1)^{\alpha_1-2}(lnt)^{\alpha_1-1}}{s\Delta_1}[\frac{a_2}{\Gamma(\alpha_1)}(ln\frac{T}{s})^{\alpha_1-1} + \frac{b_2}{\Gamma(\alpha_1-1)}(ln\frac{T}{s})^{\alpha_1-2}\\
&-\frac{d_2}{\Gamma(\alpha_1)}(ln\frac{\xi_2}{s})^{\alpha_1-1}] - \frac{d_1(lnt)^{\alpha_1-1}}{s\Delta_1\Gamma(\alpha_1)}(ln\frac{\xi_1}{s})^{\alpha_1-1}[a_2(lnT)^{\alpha_1-2}\\
&+\frac{b_2}{T}(\alpha_1-2)(lnT)^{\alpha_1-3} - d_2(ln\xi_2)^{\alpha_1-2}]\\
&+\frac{d_1(lnt)^{\alpha_1-2}}{s\Delta_1\Gamma(\alpha_1)}(ln\frac{\xi_1}{s})^{\alpha_1-1}[a_2(lnT)^{\alpha_1-1} + \frac{b_2}{T}(\alpha_1-1)(lnT)^{\alpha_1-2} - d_2(ln\xi_2)^{\alpha_1-1}]\\
&-\frac{d_1(ln\xi_1)^{\alpha_1-1}(lnt)^{\alpha_1-2}}{s\Delta_1}[\frac{a_2}{\Gamma(\alpha_1)}(ln\frac{T}{s})^{\alpha_1-1} + \frac{b_2}{\Gamma(\alpha_1-1)}(ln\frac{T}{s})^{\alpha_1-2}\\
&-\frac{d_2}{\Gamma(\alpha_1)}(ln\frac{\xi_2}{s})^{\alpha_1-1}]; \ s\le\xi_1,\ s\le t,\\[8pt]

&\frac{d_1(ln\xi_1)^{\alpha_1-2}(lnt)^{\alpha_1-1}}{s\Delta_1}[\frac{a_2}{\Gamma(\alpha_1)}(ln\frac{T}{s})^{\alpha_1-1} + \frac{b_2}{\Gamma(\alpha_1-1)}(ln\frac{T}{s})^{\alpha_1-2} - \frac{d_2}{\Gamma(\alpha_1)}(ln\frac{\xi_2}{s})^{\alpha_1-1}]\\
&-\frac{d_1(lnt)^{\alpha_1-1}}{s\Delta_1\Gamma(\alpha_1)}(ln\frac{\xi_1}{s})^{\alpha_1-1}[a_2(lnT)^{\alpha_1-2} + \frac{b_2}{T}(\alpha_1-2)(lnT)^{\alpha_1-3} - d_2(ln\xi_2)^{\alpha_1-2}]\\
&+\frac{d_1(lnt)^{\alpha_1-2}}{s\Delta_1\Gamma(\alpha_1)}(ln\frac{\xi_1}{s})^{\alpha_1-1}[a_2(lnT)^{\alpha_1-1} + \frac{b_2}{T}(\alpha_1-1)(lnT)^{\alpha_1-2} - d_2(ln\xi_2)^{\alpha_1-1}]\\
&-\frac{d_1(ln\xi_1)^{\alpha_1-1}(lnt)^{\alpha_1-2}}{s\Delta_1}[\frac{a_2}{\Gamma(\alpha_1)}(ln\frac{T}{s})^{\alpha_1-1} + \frac{b_2}{\Gamma(\alpha_1-1)}(ln\frac{T}{s})^{\alpha_1-2}\\
&-\frac{d_2}{\Gamma(\alpha_1)}(ln\frac{\xi_2}{s})^{\alpha_1-1}]; \ s\le\xi_1,\ t\le s,\\[8pt]

&\frac{1}{s\Gamma(\alpha_1)}(ln\frac{t}{s})^{\alpha_1-1}\\
&+\frac{d_1(ln\xi_1)^{\alpha_1-2}(lnt)^{\alpha_1-1}}{s\Delta_1}[\frac{a_2}{\Gamma(\alpha_1)}(ln\frac{T}{s})^{\alpha_1-1} + \frac{b_2}{\Gamma(\alpha_1-1)}(ln\frac{T}{s})^{\alpha_1-2} - \frac{d_2}{\Gamma(\alpha_1)}(ln\frac{\xi_2}{s})^{\alpha_1-1}]\\
&-\frac{d_1(ln\xi_1)^{\alpha_1-1}(lnt)^{\alpha_1-2}}{s\Delta_1}[\frac{a_2}{\Gamma(\alpha_1)}(ln\frac{T}{s})^{\alpha_1-1} + \frac{b_2}{\Gamma(\alpha_1-1)}(ln\frac{T}{s})^{\alpha_1-2}\\
&-\frac{d_2}{\Gamma(\alpha_1)}(ln\frac{\xi_2}{s})^{\alpha_1-1}]; \ \xi_1\le s\le\xi_2,\ s\le t,\\[8pt]

&\frac{d_1(ln\xi_1)^{\alpha_1-2}(lnt)^{\alpha_1-1}}{s\Delta_1}[\frac{a_2}{\Gamma(\alpha_1)}(ln\frac{T}{s})^{\alpha_1-1} + \frac{b_2}{\Gamma(\alpha_1-1)}(ln\frac{T}{s})^{\alpha_1-2} - \frac{d_2}{\Gamma(\alpha_1)}(ln\frac{\xi_2}{s})^{\alpha_1-1}]\\
&-\frac{d_1(ln\xi_1)^{\alpha_1-1}(lnt)^{\alpha_1-2}}{s\Delta_1}[\frac{a_2}{\Gamma(\alpha_1)}(ln\frac{T}{s})^{\alpha_1-1} + \frac{b_2}{\Gamma(\alpha_1-1)}(ln\frac{T}{s})^{\alpha_1-2}\\
&-\frac{d_2}{\Gamma(\alpha_1)}(ln\frac{\xi_2}{s})^{\alpha_1-1}]; \ \xi_1\le s\le\xi_2,\ t\le s,\\[8pt]

&\frac{1}{s\Gamma(\alpha_1)}(ln\frac{t}{s})^{\alpha_1-1}\\
&+\frac{d_1(ln\xi_1)^{\alpha_1-2}(lnt)^{\alpha_1-1}}{s\Delta_1}[\frac{a_2}{\Gamma(\alpha_1)}(ln\frac{T}{s})^{\alpha_1-1} + \frac{b_2}{\Gamma(\alpha_1-1)}(ln\frac{T}{s})^{\alpha_1-2}]\\
&-\frac{d_1(ln\xi_1)^{\alpha_1-1}(lnt)^{\alpha_1-2}}{s\Delta_1}[\frac{a_2}{\Gamma(\alpha_1)}(ln\frac{T}{s})^{\alpha_1-1} + \frac{b_2}{\Gamma(\alpha_1-1)}(ln\frac{T}{s})^{\alpha_1-2}]; \ \xi_2\le s,\ s\le t,\\[8pt]

&\frac{d_1(ln\xi_1)^{\alpha_1-2}(lnt)^{\alpha_1-1}}{s\Delta_1}[\frac{a_2}{\Gamma(\alpha_1)}(ln\frac{T}{s})^{\alpha_1-1} + \frac{b_2}{\Gamma(\alpha_1-1)}(ln\frac{T}{s})^{\alpha_1-2}]\\
&-\frac{d_1(ln\xi_1)^{\alpha_1-1}(lnt)^{\alpha_1-2}}{s\Delta_1}[\frac{a_2}{\Gamma(\alpha_1)}(ln\frac{T}{s})^{\alpha_1-1} + \frac{b_2}{\Gamma(\alpha_1-1)}(ln\frac{T}{s})^{\alpha_1-2}]; \ \xi_2\le s,\ t\le s,
\end{aligned}
\end{cases}
$$

with

$$
\Delta_1 = d_1(ln\xi_1)^{\alpha_1-1}[a_2(lnT)^{\alpha_1-2} + \frac{b_2}{T}(\alpha_1-2)(lnT)^{\alpha_1-3} - d_2(ln\xi_2)^{\alpha_1-2}]
$$

$$
- d_1(ln\xi_1)^{\alpha_1-2}[a_2(lnT)^{\alpha_1-1} + \frac{b_2}{T}(\alpha_1-1)(lnT)^{\alpha_1-2} - d_2(ln\xi_2)^{\alpha_1-1}] \ne 0,
$$

and

$G_2(t,s)$

$$= \begin{cases}
\begin{aligned}
&\frac{1}{s\Gamma(\alpha_2)}(ln\frac{t}{s})^{\alpha_2-1} \\
&+\frac{d_3(ln\xi_3)^{\alpha_2-2}(lnt)^{\alpha_2-1}}{s\Delta_2}[\frac{a_4}{\Gamma(\alpha_2)}(ln\frac{T}{s})^{\alpha_2-1}+\frac{b_4}{\Gamma(\alpha_2-1)}(ln\frac{T}{s})^{\alpha_2-2}-\frac{d_4}{\Gamma(\alpha_2)}(ln\frac{\xi_4}{s})^{\alpha_2-1}] \\
&-\frac{d_3(lnt)^{\alpha_2-1}}{s\Delta_2\Gamma(\alpha_2)}(ln\frac{\xi_3}{s})^{\alpha_2-1}[a_4(lnT)^{\alpha_2-2}+\frac{b_4}{T}(\alpha_2-2)(lnT)^{\alpha_2-3}-d_4(ln\xi_4)^{\alpha_2-2}] \\
&+\frac{d_3(lnt)^{\alpha_2-2}}{s\Delta_2\Gamma(\alpha_2)}(ln\frac{\xi_3}{s})^{\alpha_2-1}[a_4(lnT)^{\alpha_2-1}+\frac{b_4}{T}(\alpha_2-1)(lnT)^{\alpha_2-2}-d_4(ln\xi_4)^{\alpha_2-1}] \\
&-\frac{d_3(ln\xi_3)^{\alpha_2-1}(lnt)^{\alpha_2-2}}{s\Delta_2}[\frac{a_4}{\Gamma(\alpha_2)}(ln\frac{T}{s})^{\alpha_2-1}+\frac{b_4}{\Gamma(\alpha_2-1)}(ln\frac{T}{s})^{\alpha_2-2} \\
&-\frac{d_4}{\Gamma(\alpha_2)}(ln\frac{\xi_4}{s})^{\alpha_2-1}];\ s\le\xi_3,\ s\le t,
\end{aligned} \\[2em]
\begin{aligned}
&\frac{d_3(ln\xi_3)^{\alpha_2-2}(lnt)^{\alpha_2-1}}{s\Delta_2}[\frac{a_4}{\Gamma(\alpha_2)}(ln\frac{T}{s})^{\alpha_2-1}+\frac{b_4}{\Gamma(\alpha_2-1)}(ln\frac{T}{s})^{\alpha_2-2}-\frac{d_4}{\Gamma(\alpha_2)}(ln\frac{\xi_4}{s})^{\alpha_2-1}] \\
&-\frac{d_3(lnt)^{\alpha_2-1}}{s\Delta_2\Gamma(\alpha_2)}(ln\frac{\xi_3}{s})^{\alpha_2-1}[a_4(lnT)^{\alpha_2-2}+\frac{b_4}{T}(\alpha_2-2)(lnT)^{\alpha_2-3}-d_4(ln\xi_4)^{\alpha_2-2}] \\
&+\frac{d_3(lnt)^{\alpha_2-2}}{s\Delta_2\Gamma(\alpha_2)}(ln\frac{\xi_3}{s})^{\alpha_2-1}[a_4(lnT)^{\alpha_2-1}+\frac{b_4}{T}(\alpha_2-1)(lnT)^{\alpha_2-2}-d_4(ln\xi_4)^{\alpha_2-1}] \\
&-\frac{d_3(ln\xi_3)^{\alpha_2-1}(lnt)^{\alpha_2-2}}{s\Delta_2}[\frac{a_4}{\Gamma(\alpha_2)}(ln\frac{T}{s})^{\alpha_2-1}+\frac{b_4}{\Gamma(\alpha_2-1)}(ln\frac{T}{s})^{\alpha_2-2} \\
&-\frac{d_4}{\Gamma(\alpha_2)}(ln\frac{\xi_4}{s})^{\alpha_2-1}];\ s\le\xi_3,\ t\le s,
\end{aligned} \\[2em]
\begin{aligned}
&\frac{1}{s\Gamma(\alpha_2)}(ln\frac{t}{s})^{\alpha_2-1} \\
&+\frac{d_3(ln\xi_3)^{\alpha_2-2}(lnt)^{\alpha_2-1}}{s\Delta_2}[\frac{a_4}{\Gamma(\alpha_2)}(ln\frac{T}{s})^{\alpha_2-1}+\frac{b_4}{\Gamma(\alpha_2-1)}(ln\frac{T}{s})^{\alpha_2-2}-\frac{d_4}{\Gamma(\alpha_2)}(ln\frac{\xi_4}{s})^{\alpha_2-1}] \\
&-\frac{d_3(ln\xi_3)^{\alpha_2-1}(lnt)^{\alpha_2-2}}{s\Delta_2}[\frac{a_4}{\Gamma(\alpha_2)}(ln\frac{T}{s})^{\alpha_2-1}+\frac{b_4}{\Gamma(\alpha_2-1)}(ln\frac{T}{s})^{\alpha_2-2} \\
&-\frac{d_4}{\Gamma(\alpha_2)}(ln\frac{\xi_4}{s})^{\alpha_2-1}];\ \xi_3\le s\le\xi_4,\ s\le t,
\end{aligned} \\[2em]
\begin{aligned}
&\frac{d_3(ln\xi_3)^{\alpha_2-2}(lnt)^{\alpha_2-1}}{s\Delta_2}[\frac{a_4}{\Gamma(\alpha_2)}(ln\frac{T}{s})^{\alpha_2-1}+\frac{b_4}{\Gamma(\alpha_2-1)}(ln\frac{T}{s})^{\alpha_2-2}-\frac{d_4}{\Gamma(\alpha_2)}(ln\frac{\xi_4}{s})^{\alpha_2-1}] \\
&-\frac{d_3(ln\xi_3)^{\alpha_2-1}(lnt)^{\alpha_2-2}}{s\Delta_2}[\frac{a_4}{\Gamma(\alpha_2)}(ln\frac{T}{s})^{\alpha_2-1}+\frac{b_4}{\Gamma(\alpha_2-1)}(ln\frac{T}{s})^{\alpha_2-2} \\
&-\frac{d_4}{\Gamma(\alpha_2)}(ln\frac{\xi_4}{s})^{\alpha_2-1}];\ \xi_3\le s\le\xi_4,\ t\le s,
\end{aligned} \\[2em]
\begin{aligned}
&\frac{1}{s\Gamma(\alpha_2)}(ln\frac{t}{s})^{\alpha_2-1} \\
&+\frac{d_3(ln\xi_3)^{\alpha_2-2}(lnt)^{\alpha_2-1}}{s\Delta_2}[\frac{a_4}{\Gamma(\alpha_2)}(ln\frac{T}{s})^{\alpha_2-1}+\frac{b_4}{\Gamma(\alpha_2-1)}(ln\frac{T}{s})^{\alpha_2-2}] \\
&-\frac{d_3(ln\xi_3)^{\alpha_2-1}(lnt)^{\alpha_2-2}}{s\Delta_2}[\frac{a_4}{\Gamma(\alpha_2)}(ln\frac{T}{s})^{\alpha_2-1}+\frac{b_4}{\Gamma(\alpha_2-1)}(ln\frac{T}{s})^{\alpha_2-2}];\ \xi_4\le s,\ s\le t,
\end{aligned} \\[2em]
\begin{aligned}
&\frac{d_3(ln\xi_3)^{\alpha_2-2}(lnt)^{\alpha_2-1}}{s\Delta_2}[\frac{a_4}{\Gamma(\alpha_2)}(ln\frac{T}{s})^{\alpha_2-1}+\frac{b_4}{\Gamma(\alpha_2-1)}(ln\frac{T}{s})^{\alpha_2-2}] \\
&-\frac{d_3(ln\xi_3)^{\alpha_2-1}(lnt)^{\alpha_2-2}}{s\Delta_2}[\frac{a_4}{\Gamma(\alpha_2)}(ln\frac{T}{s})^{\alpha_2-1}+\frac{b_4}{\Gamma(\alpha_2-1)}(ln\frac{T}{s})^{\alpha_2-2}];\ \xi_4\le s,\ t\le s,
\end{aligned}
\end{cases}$$

with

$$\Delta_2 = d_3(ln\xi_3)^{\alpha_2-1}[a_4(lnT)^{\alpha_2-2}+\frac{b_4}{T}(\alpha_2-2)(lnT)^{\alpha_2-3}-d_4(ln\xi_4)^{\alpha_2-2}]$$

$$-d_3(ln\xi_3)^{\alpha_2-2}[a_4(lnT)^{\alpha_2-1}+\frac{b_4}{T}(\alpha_2-1)(lnT)^{\alpha_2-2}-d_4(ln\xi_4)^{\alpha_2-1}]\neq 0.$$

Set

$$p_i^* := \sup_{t \in I} p(t), \quad q_i^* := \sup_{t \in I} q(t), \text{ and } G_i^* = \sup_{t \in I} \|G_i(t, \cdot)\|_{BV}; \ i = 1, 2.$$

Theorem 8.2. *Assume that the hypothesis (8.1.1) holds. If the matrix*

$$M := (T - 1) \begin{pmatrix} G_1^* p_1^* & G_1^* q_1^* \\ G_2^* p_2^* & G_2^* q_2^* \end{pmatrix}$$

converges to 0, *then the coupled system (8.1)–(8.2) has a unique solution.*

Proof. Consider the operator $N : \mathcal{C} \to \mathcal{C}$ defined by

$$(N(u, v))(t) = ((N_1(u, v))(t), (N_2(u, v))(t)). \tag{8.5}$$

Clearly, the fixed points of the operator N are solutions of the coupled system (8.1)–(8.2). We show that N satisfies all conditions of Theorem 2.90.

For each $(u_1, v_1), (u_2, v_2) \in \mathcal{C}$ and $t \in I$, we have

$$\|(N_1(u_1, v_1))(t) - (N_1(u_2, v_2))(t)\|$$

$$\leq \int_1^T \|G_1(t, s)\|_{BV} \|f_1(s, u_1(s), v_1(s)) - f_1(s, u_2(s), v_2(s))\| ds$$

$$\leq \int_1^T \|G_1(t, s)\|_{BV} (p_1(s)\|u_1(s) - v_1(s)\| + q_1(s)\|u_2(s) - v_2(s)\|) ds$$

$$\leq G_1^*(T - 1)(p_1^* \|u_1 - v_1\|_\infty + q_1^* \|u_2 - v_2\|_\infty).$$

Then

$$\|(N_1(u_1, v_1)) - (N_1(u_2, v_2))\|_\infty$$
$$\leq G_1^*(T - 1)(p_1^* \|u_1 - v_1\|_\infty + q_1^* \|u_2 - v_2\|_\infty).$$

Also, for each $(u_1, v_1), (u_2, v_2) \in \mathcal{C}$ and $t \in I$, we get

$$\|(N_2(u_1, v_1)) - (N_2(u_2, v_2))\|_\infty$$
$$\leq G_2^*(T - 1)(p_2^* \|u_1 - v_1\|_\infty + q_2^* \|u_2(\cdot) - v_2(\cdot)\|_\infty).$$

Thus

$$d((N(u_1, v_1)), (N(u_2, v_2))) \leq M d((u_1, v_1), (u_2, v_2)),$$

where

$$d((u_1, v_1), (u_2, v_2)) = \begin{pmatrix} \|u_1 - v_1\|_\infty \\ \|u_2 - v_2\|_\infty \end{pmatrix}.$$

Since the matrix M converges to zero, then Theorem 2.90 implies that coupled system (8.1)–(8.2) has a unique solution. $\qquad\square$

Now, we prove an existence result for the coupled system (8.1)–(8.2) by using Schauder fixed-point theorem type in generalized Banach space.

Theorem 8.3. *Assume that the hypothesis (8.1.2) holds. Then the coupled system (8.1)–(8.2) has at least one solution.*

Proof. Let $N : C \to C$ be the operator defined in (8.5). We show that N satisfies all conditions of Theorem 2.68. The proof will be given in several steps.

Step 1. N is continuous.
Let $(u_n, v_n)_n$ be a sequence such that $(u_n, v_n) \to (u, v) \in C$ as $n \to \infty$. For any $i = 1, 2$ and each $t \in I$, we have

$$\|(N_i(u_n, v_n))(t) - (N_i(u, v))(t)\|$$
$$\leq \int_1^T \|G_i(t, s)\|_{BV} \|f_i(s, u_n(s), v_n(s)) - f_i(s, u(s), v(s))\| ds$$
$$\leq G_i^*(T - 1)\|f_i(\cdot, u_n(\cdot), v_n(\cdot)) - f_i(\cdot, u(\cdot), v(\cdot))\|_\infty.$$

Since f_i; $i = 1, 2$ is continuous, then by the Lebesgue dominated convergence theorem, we have

$$\|(N_i(u_n, v_n)) - (N_i(u, v))\|_\infty \to 0 \text{ as } n \to \infty.$$

Hence N is continuous.

Step 2. N maps bounded sets into bounded sets in C.
Set

$$h_i^* := \sup_{t \in I} h_i(t), \quad k_i^* := \sup_{t \in I} k_i(t), \quad l_i^* := \sup_{t \in I} l_i(t).$$

Let $R > 0$ and set

$$B_R := \{(\mu, v) \in C : \|\mu\|_\infty \leq R, \|v\|_\infty \leq R\}.$$

For any $i = 1, 2$ and each $(u, v) \in B_R$ and $t \in I$, we have

$$\|(N_i(u, v))(t)\| \leq \int_1^T \|G_i(t, s)\|_{BV} \|f_i(s, u(s), v(s))\| ds$$
$$\leq \int_1^T \|G_i(t, s)\|_{BV} (h_i(s) + k_i(s)\|u(s\| + l_i(s)\|v(s)\|) ds$$
$$\leq \int_1^T \|G_i(t, s)\|_{BV} (h_i^* + Rk_i^* + Rl_i^*) ds$$
$$\leq (T - 1)G_i^*(h_i^* + Rk_i^* + Rl_i^*)$$
$$:= \ell_i.$$

Thus

$$\|(N_i(u, v))\|_\infty \leq \ell_i.$$

Hence

$$\|(N(u, v))\|_C \leq \max\{\ell_1, \ell_2\} := \ell.$$

Step 3. *N* maps bounded sets into equicontinuous sets in $\mathcal{C}$.

Let B_R be the ball defined in Step 2. For each $t_1, t_2 \in I$ with $t_1 \leq t_2$ and any $(u, v) \in B_R$ and $i = 1, 2$, we have

$$\|(N_i(u, v))(t_1) - (N_i(u, v))(t_2)\|$$

$$\leq \int_1^T \|G_i(t_1, s) - G_i(t_2, s)\|_{BV} \|f_i(s, u(s), v(s))\| ds$$

$$\leq \int_1^T \|G_i(t_1, s) - G_i(t_2, s)\|_{BV} (h_i(s) + k_i(s)\|u\|_\infty + l_i(s)\|v\|_\infty) ds$$

$$\leq (h_i^* + Rk_i^* + Rl_i^*) \int_1^T \|G_i(t_1, s) - G_i(t_2, s)\|_{BV} ds$$

$$\to 0 \text{ as } t_1 \to t_2.$$

As a consequence of steps 1 to 3 together with Theorem 2.68, we can conclude that *N* has at least one fixed point in B_R, which is a solution of our coupled system (8.1)–(8.2).   □

8.1.3 An example

Consider the following coupled system of Caputo–Hadamard fractional differential equations

$$\begin{cases} (^{Hc}D_1^{\frac{3}{2}}u)(t) = f(t, u(t), v(t)); \\ (^{Hc}D_1^{\frac{3}{2}}v)(t) = g(t, u(t), v(t)); \end{cases} ; \ t \in [1, e], \tag{8.6}$$

with the multi-point boundary conditions

$$\begin{cases} u(1) - u'(1) = u(2) \\ 2u(T) + u'(T) = 2u\left(\frac{3}{2}\right) \\ 3v(1) - v'(1) = 3v\left(\frac{5}{4}\right) \\ v(T) + 2v'(T) = v(2), \end{cases} \tag{8.7}$$

where

$$f(t, u, v) = \frac{ct^{\frac{-1}{4}}u(t)\sin t}{64(1 + \sqrt{t})(1 + |u| + |v|)}; \ t \in [1, e],$$

$$g(t, u, v) = \frac{cv(t)\cos t}{64(1 + |u| + |v|)}; \ t \in [1, e],$$

and $c < \frac{1}{\max\{G_1^*, G_2^*\}}$. The hypothesis (8.1.1) is satisfied with

$$p_2(t) = q_1(t) = 0, \; p_1(t) = \frac{c \sin t}{64}, \; q_2(t) = \frac{c \cos t}{64}.$$

Also, the matrix

$$\frac{c(e-1)}{64} \begin{pmatrix} G_1^* & 0 \\ 0 & G_2^* \end{pmatrix}$$

converges to 0. Hence Theorem 8.2 implies that the system (8.6)–(8.7) has a unique solution defined on $[1, e]$.

8.2 Random coupled Caputo–Hadamard fractional differential systems with four-point boundary conditions

8.2.1 Introduction

In this section we discuss the existence and uniqueness of solutions for the following coupled system of Caputo–Hadamard fractional differential equations

$$\begin{cases} (^{Hc}D_1^{\alpha_1} u)(t, w) = f_1(t, u(t, w), v(t, w), w) \\ (^{Hc}D_1^{\alpha_2} v)(t, w) = f_2(t, u(t, w), v(t, w), w) \end{cases} \; ; \; t \in I := [1, T], \; w \in \Omega, \qquad (8.8)$$

with the four-point boundary conditions

$$\begin{cases} a_1 u(1, w) - b_1 u'(1, w) = d_1 u(\xi_1, w) \\ a_2 u(T, w) + b_2 u'(T, w) = d_2 u(\xi_2, w) \\ a_3 v(1, w) - b_3 v'(1, w) = d_3 v(\xi_3, w) \\ a_4 v(T, w) + b_4 v'(T, w) = d_4 v(\xi_4, w) \end{cases} \; ; \; w \in \Omega, \qquad (8.9)$$

where $T > 1$, $a_i, b_i, d_i \in \mathbb{R}$, $\xi_i \in (1, T)$; $i = 1, 2, 3, 4$, (Ω, F, P) is a complete probability space, $f_1, f_2 : I \times \mathbb{R}^m \times \mathbb{R}^m \times \Omega \to \mathbb{R}^m$ are given functions, $\mathbb{R}^m$; $m \in \mathbb{N}^*$ is the Euclidean Banach space with a suitable norm $\| \cdot \|$, $^{Hc}D_1^{\alpha_i}$ is the Caputo–Hadamard fractional derivative of order $\alpha_i \in (1, 2]$; $i = 1, 2$.

8.2.2 Main results

In this section we are concerned with the existence and uniqueness results of the coupled system (8.8)–(8.9).

Definition 8.4. By a random solution of the problem (8.8)–(8.9) we mean a coupled measurable functions $(u, v) \in C(I) \times C(I)$ satisfying the boundary conditions (8.9) and Eqs. (8.8) on I.

The following hypotheses will be used in the sequel.

(8.4.1) The functions f_i; $i = 1, 2$ are Carathéodory.
(8.4.2) There exist continuous functions $p_i, q_i : I \to L^\infty(\Omega, \mathbb{R}_+)$; $i = 1, 2$ such that

$$\|f_i(t, u_1, v_1) - f_i(t, u_2, v_2)\| \le p_i(t, w)\|u_1 - u_2\| + q_i(t, w)\|v_1 - v_2\|;$$

for a.e. $t \in I$, and each $u_i, v_i \in \mathbb{R}^m$, $i = 1, 2$.
(8.4.3) There exist continuous functions $h_i, k_i, l_i : I \to L^\infty(\Omega, \mathbb{R}_+)$; $i = 1, 2$ such that

$$\|f_i(t, u, v)\| \le h_i(t, w) + k_i(t, w)\|u\| + l_i(t, w)\|v\|,$$

for a.e. $t \in I$ and each $u, v \in \mathbb{R}^m$.

First, we prove an existence and uniqueness result for the coupled system (8.8)–(8.9) by using Banach random fixed-point theorem type in generalized Banach spaces. We define the operators $N_1, N_2 : C \times \Omega \to C(I)$ by

$$(N_1(u, v))(t, w) = \int_1^T G_1(t, s) f_1(s, u(s, w), w) ds, \tag{8.10}$$

and

$$(N_2(u, v))(t, w) = \int_1^T G_2(t, s) f_2(s, u(s, w), w) ds, \tag{8.11}$$

where

$G_1(t,s)$

$$
= \begin{cases}
\begin{aligned}
& \frac{1}{s\Gamma(\alpha_1)}(ln\frac{t}{s})^{\alpha_1-1} \\
& + \frac{d_1(ln\xi_1)^{\alpha_1-2}(lnt)^{\alpha_1-1}}{s\Delta_1}[\frac{a_2}{\Gamma(\alpha_1)}(ln\frac{T}{s})^{\alpha_1-1} + \frac{b_2}{\Gamma(\alpha_1-1)}(ln\frac{T}{s})^{\alpha_1-2} - \frac{d_2}{\Gamma(\alpha_1)}(ln\frac{\xi_2}{s})^{\alpha_1-1}] \\
& - \frac{d_1(lnt)^{\alpha_1-1}}{s\Delta_1\Gamma(\alpha_1)}(ln\frac{\xi_1}{s})^{\alpha_1-1}[a_2(lnT)^{\alpha_1-2} + \frac{b_2}{T}(\alpha_1-2)(lnT)^{\alpha_1-3} - d_2(ln\xi_2)^{\alpha_1-2}] \\
& + \frac{d_1(lnt)^{\alpha_1-2}}{s\Delta_1\Gamma(\alpha_1)}(ln\frac{\xi_1}{s})^{\alpha_1-1}[a_2(lnT)^{\alpha_1-1} + \frac{b_2}{T}(\alpha_1-1)(lnT)^{\alpha_1-2} - d_2(ln\xi_2)^{\alpha_1-1}] \\
& - \frac{d_1(ln\xi_1)^{\alpha_1-1}(lnt)^{\alpha_1-2}}{s\Delta_1}[\frac{a_2}{\Gamma(\alpha_1)}(ln\frac{T}{s})^{\alpha_1-1} + \frac{b_2}{\Gamma(\alpha_1-1)}(ln\frac{T}{s})^{\alpha_1-2} \\
& - \frac{d_2}{\Gamma(\alpha_1)}(ln\frac{\xi_2}{s})^{\alpha_1-1}]; \ s \leq \xi_1, \ s \leq t
\end{aligned} \\[4pt]
\begin{aligned}
& \frac{d_1(ln\xi_1)^{\alpha_1-2}(lnt)^{\alpha_1-1}}{s\Delta_1}[\frac{a_2}{\Gamma(\alpha_1)}(ln\frac{T}{s})^{\alpha_1-1} + \frac{b_2}{\Gamma(\alpha_1-1)}(ln\frac{T}{s})^{\alpha_1-2} - \frac{d_2}{\Gamma(\alpha_1)}(ln\frac{\xi_2}{s})^{\alpha_1-1}] \\
& - \frac{d_1(lnt)^{\alpha_1-1}}{s\Delta_1\Gamma(\alpha_1)}(ln\frac{\xi_1}{s})^{\alpha_1-1}[a_2(lnT)^{\alpha_1-2} + \frac{b_2}{T}(\alpha_1-2)(lnT)^{\alpha_1-3} - d_2(ln\xi_2)^{\alpha_1-2}] \\
& + \frac{d_1(lnt)^{\alpha_1-2}}{s\Delta_1\Gamma(\alpha_1)}(ln\frac{\xi_1}{s})^{\alpha_1-1}[a_2(lnT)^{\alpha_1-1} + \frac{b_2}{T}(\alpha_1-1)(lnT)^{\alpha_1-2} - d_2(ln\xi_2)^{\alpha_1-1}] \\
& - \frac{d_1(ln\xi_1)^{\alpha_1-1}(lnt)^{\alpha_1-2}}{s\Delta_1}[\frac{a_2}{\Gamma(\alpha_1)}(ln\frac{T}{s})^{\alpha_1-1} + \frac{b_2}{\Gamma(\alpha_1-1)}(ln\frac{T}{s})^{\alpha_1-2} \\
& - \frac{d_2}{\Gamma(\alpha_1)}(ln\frac{\xi_2}{s})^{\alpha_1-1}]; \ s \leq \xi_1, \ t \leq s,
\end{aligned} \\[4pt]
\begin{aligned}
& \frac{1}{s\Gamma(\alpha_1)}(ln\frac{t}{s})^{\alpha_1-1} \\
& + \frac{d_1(ln\xi_1)^{\alpha_1-2}(lnt)^{\alpha_1-1}}{s\Delta_1}[\frac{a_2}{\Gamma(\alpha_1)}(ln\frac{T}{s})^{\alpha_1-1} + \frac{b_2}{\Gamma(\alpha_1-1)}(ln\frac{T}{s})^{\alpha_1-2} - \frac{d_2}{\Gamma(\alpha_1)}(ln\frac{\xi_2}{s})^{\alpha_1-1}] \\
& - \frac{d_1(ln\xi_1)^{\alpha_1-1}(lnt)^{\alpha_1-2}}{s\Delta_1}[\frac{a_2}{\Gamma(\alpha_1)}(ln\frac{T}{s})^{\alpha_1-1} + \frac{b_2}{\Gamma(\alpha_1-1)}(ln\frac{T}{s})^{\alpha_1-2} \\
& - \frac{d_2}{\Gamma(\alpha_1)}(ln\frac{\xi_2}{s})^{\alpha_1-1}]; \ \xi_1 \leq s \leq \xi_2, \ s \leq t,
\end{aligned} \\[4pt]
\begin{aligned}
& \frac{d_1(ln\xi_1)^{\alpha_1-2}(lnt)^{\alpha_1-1}}{s\Delta_1}[\frac{a_2}{\Gamma(\alpha_1)}(ln\frac{T}{s})^{\alpha_1-1} + \frac{b_2}{\Gamma(\alpha_1-1)}(ln\frac{T}{s})^{\alpha_1-2} - \frac{d_2}{\Gamma(\alpha_1)}(ln\frac{\xi_2}{s})^{\alpha_1-1}] \\
& - \frac{d_1(ln\xi_1)^{\alpha_1-1}(lnt)^{\alpha_1-2}}{s\Delta_1}[\frac{a_2}{\Gamma(\alpha_1)}(ln\frac{T}{s})^{\alpha_1-1} + \frac{b_2}{\Gamma(\alpha_1-1)}(ln\frac{T}{s})^{\alpha_1-2} \\
& - \frac{d_2}{\Gamma(\alpha_1)}(ln\frac{\xi_2}{s})^{\alpha_1-1}]; \ \xi_1 \leq s \leq \xi_2, \ t \leq s,
\end{aligned} \\[4pt]
\begin{aligned}
& \frac{1}{s\Gamma(\alpha_1)}(ln\frac{t}{s})^{\alpha_1-1} \\
& + \frac{d_1(ln\xi_1)^{\alpha_1-2}(lnt)^{\alpha_1-1}}{s\Delta_1}[\frac{a_2}{\Gamma(\alpha_1)}(ln\frac{T}{s})^{\alpha_1-1} + \frac{b_2}{\Gamma(\alpha_1-1)}(ln\frac{T}{s})^{\alpha_1-2}] \\
& - \frac{d_1(ln\xi_1)^{\alpha_1-1}(lnt)^{\alpha_1-2}}{s\Delta_1}[\frac{a_2}{\Gamma(\alpha_1)}(ln\frac{T}{s})^{\alpha_1-1} + \frac{b_2}{\Gamma(\alpha_1-1)}(ln\frac{T}{s})^{\alpha_1-2}]; \ \xi_2 \leq s, \ s \leq t,
\end{aligned} \\[4pt]
\begin{aligned}
& \frac{d_1(ln\xi_1)^{\alpha_1-2}(lnt)^{\alpha_1-1}}{s\Delta_1}[\frac{a_2}{\Gamma(\alpha_1)}(ln\frac{T}{s})^{\alpha_1-1} + \frac{b_2}{\Gamma(\alpha_1-1)}(ln\frac{T}{s})^{\alpha_1-2}] \\
& - \frac{d_1(ln\xi_1)^{\alpha_1-1}(lnt)^{\alpha_1-2}}{s\Delta_1}[\frac{a_2}{\Gamma(\alpha_1)}(ln\frac{T}{s})^{\alpha_1-1} + \frac{b_2}{\Gamma(\alpha_1-1)}(ln\frac{T}{s})^{\alpha_1-2}]; \ \xi_2 \leq s, \ t \leq s,
\end{aligned}
\end{cases}
$$

with

$$
\Delta_1 = d_1(ln\xi_1)^{\alpha_1-1}[a_2(lnT)^{\alpha_1-2} + \frac{b_2}{T}(\alpha_1-2)(lnT)^{\alpha_1-3} - d_2(ln\xi_2)^{\alpha_1-2}]
$$

$$
- d_1(ln\xi_1)^{\alpha_1-2}[a_2(lnT)^{\alpha_1-1} + \frac{b_2}{T}(\alpha_1-1)(lnT)^{\alpha_1-2} - d_2(ln\xi_2)^{\alpha_1-1}] \neq 0,
$$

and

$G_2(t,s)$

$$
=\begin{cases}
\begin{aligned}
&\frac{1}{s\Gamma(\alpha_2)}(ln\frac{t}{s})^{\alpha_2-1}\\
&+\frac{d_3(ln\xi_3)^{\alpha_2-2}(lnt)^{\alpha_2-1}}{s\Delta_2}[\frac{a_4}{\Gamma(\alpha_2)}(ln\frac{T}{s})^{\alpha_2-1}+\frac{b_4}{\Gamma(\alpha_2-1)}(ln\frac{T}{s})^{\alpha_2-2}-\frac{d_4}{\Gamma(\alpha_2)}(ln\frac{\xi_4}{s})^{\alpha_2-1}]\\
&-\frac{d_3(lnt)^{\alpha_2-1}}{s\Delta_2\Gamma(\alpha_2)}(ln\frac{\xi_3}{s})^{\alpha_2-1}[a_4(lnT)^{\alpha_2-2}+\frac{b_4}{T}(\alpha_2-2)(lnT)^{\alpha_2-3}-d_4(ln\xi_4)^{\alpha_2-2}]\\
&+\frac{d_3(lnt)^{\alpha_2-2}}{s\Delta_2\Gamma(\alpha_2)}(ln\frac{\xi_3}{s})^{\alpha_2-1}[a_4(lnT)^{\alpha_2-1}+\frac{b_4}{T}(\alpha_2-1)(lnT)^{\alpha_2-2}-d_4(ln\xi_4)^{\alpha_2-1}]\\
&-\frac{d_3(ln\xi_3)^{\alpha_2-1}(lnt)^{\alpha_2-2}}{s\Delta_2}[\frac{a_4}{\Gamma(\alpha_2)}(ln\frac{T}{s})^{\alpha_2-1}+\frac{b_4}{\Gamma(\alpha_2-1)}(ln\frac{T}{s})^{\alpha_2-2}\\
&-\frac{d_4}{\Gamma(\alpha_2)}(ln\frac{\xi_4}{s})^{\alpha_2-1}];\; s\le\xi_3,\; s\le t,\\[6pt]
&\frac{d_3(ln\xi_3)^{\alpha_2-2}(lnt)^{\alpha_2-1}}{s\Delta_2}[\frac{a_4}{\Gamma(\alpha_2)}(ln\frac{T}{s})^{\alpha_2-1}+\frac{b_4}{\Gamma(\alpha_2-1)}(ln\frac{T}{s})^{\alpha_2-2}-\frac{d_4}{\Gamma(\alpha_2)}(ln\frac{\xi_4}{s})^{\alpha_2-1}]\\
&-\frac{d_3(lnt)^{\alpha_2-1}}{s\Delta_2\Gamma(\alpha_2)}(ln\frac{\xi_3}{s})^{\alpha_2-1}[a_4(lnT)^{\alpha_2-2}+\frac{b_4}{T}(\alpha_2-2)(lnT)^{\alpha_2-3}-d_4(ln\xi_4)^{\alpha_2-2}]\\
&+\frac{d_3(lnt)^{\alpha_2-2}}{s\Delta_2\Gamma(\alpha_2)}(ln\frac{\xi_3}{s})^{\alpha_2-1}[a_4(lnT)^{\alpha_2-1}+\frac{b_4}{T}(\alpha_2-1)(lnT)^{\alpha_2-2}-d_4(ln\xi_4)^{\alpha_2-1}]\\
&-\frac{d_3(ln\xi_3)^{\alpha_2-1}(lnt)^{\alpha_2-2}}{s\Delta_2}[\frac{a_4}{\Gamma(\alpha_2)}(ln\frac{T}{s})^{\alpha_2-1}+\frac{b_4}{\Gamma(\alpha_2-1)}(ln\frac{T}{s})^{\alpha_2-2}\\
&-\frac{d_4}{\Gamma(\alpha_2)}(ln\frac{\xi_4}{s})^{\alpha_2-1}];\; s\le\xi_3,\; t\le s,\\[6pt]
&\frac{1}{s\Gamma(\alpha_2)}(ln\frac{t}{s})^{\alpha_2-1}\\
&+\frac{d_3(ln\xi_3)^{\alpha_2-2}(lnt)^{\alpha_2-1}}{s\Delta_2}[\frac{a_4}{\Gamma(\alpha_2)}(ln\frac{T}{s})^{\alpha_2-1}+\frac{b_4}{\Gamma(\alpha_2-1)}(ln\frac{T}{s})^{\alpha_2-2}-\frac{d_4}{\Gamma(\alpha_2)}(ln\frac{\xi_4}{s})^{\alpha_2-1}]\\
&-\frac{d_3(ln\xi_3)^{\alpha_2-1}(lnt)^{\alpha_2-2}}{s\Delta_2}[\frac{a_4}{\Gamma(\alpha_2)}(ln\frac{T}{s})^{\alpha_2-1}+\frac{b_4}{\Gamma(\alpha_2-1)}(ln\frac{T}{s})^{\alpha_2-2}\\
&-\frac{d_4}{\Gamma(\alpha_2)}(ln\frac{\xi_4}{s})^{\alpha_2-1}];\; \xi_3\le s\le\xi_4,\; s\le t,\\[6pt]
&\frac{d_3(ln\xi_3)^{\alpha_2-2}(lnt)^{\alpha_2-1}}{s\Delta_2}[\frac{a_4}{\Gamma(\alpha_2)}(ln\frac{T}{s})^{\alpha_2-1}+\frac{b_4}{\Gamma(\alpha_2-1)}(ln\frac{T}{s})^{\alpha_2-2}-\frac{d_4}{\Gamma(\alpha_2)}(ln\frac{\xi_4}{s})^{\alpha_2-1}]\\
&-\frac{d_3(ln\xi_3)^{\alpha_2-1}(lnt)^{\alpha_2-2}}{s\Delta_2}[\frac{a_4}{\Gamma(\alpha_2)}(ln\frac{T}{s})^{\alpha_2-1}+\frac{b_4}{\Gamma(\alpha_2-1)}(ln\frac{T}{s})^{\alpha_2-2}\\
&-\frac{d_4}{\Gamma(\alpha_2)}(ln\frac{\xi_4}{s})^{\alpha_2-1}];\; \xi_3\le s\le\xi_4,\; t\le s,\\[6pt]
&\frac{1}{s\Gamma(\alpha_2)}(ln\frac{t}{s})^{\alpha_2-1}\\
&+\frac{d_3(ln\xi_3)^{\alpha_2-2}(lnt)^{\alpha_2-1}}{s\Delta_2}[\frac{a_4}{\Gamma(\alpha_2)}(ln\frac{T}{s})^{\alpha_2-1}+\frac{b_4}{\Gamma(\alpha_2-1)}(ln\frac{T}{s})^{\alpha_2-2}]\\
&-\frac{d_3(ln\xi_3)^{\alpha_2-1}(lnt)^{\alpha_2-2}}{s\Delta_2}[\frac{a_4}{\Gamma(\alpha_2)}(ln\frac{T}{s})^{\alpha_2-1}+\frac{b_4}{\Gamma(\alpha_2-1)}(ln\frac{T}{s})^{\alpha_2-2}];\; \xi_4\le s,\; s\le t,\\[6pt]
&\frac{d_3(ln\xi_3)^{\alpha_2-2}(lnt)^{\alpha_2-1}}{s\Delta_2}[\frac{a_4}{\Gamma(\alpha_2)}(ln\frac{T}{s})^{\alpha_2-1}+\frac{b_4}{\Gamma(\alpha_2-1)}(ln\frac{T}{s})^{\alpha_2-2}]\\
&-\frac{d_3(ln\xi_3)^{\alpha_2-1}(lnt)^{\alpha_2-2}}{s\Delta_2}[\frac{a_4}{\Gamma(\alpha_2)}(ln\frac{T}{s})^{\alpha_2-1}\\
&+\frac{b_4}{\Gamma(\alpha_2-1)}(ln\frac{T}{s})^{\alpha_2-2}];\; \xi_4\le s,\; t\le s,
\end{aligned}
\end{cases}
$$

with

$$
\Delta_2=d_3(ln\xi_3)^{\alpha_2-1}[a_4(lnT)^{\alpha_2-2}+\frac{b_4}{T}(\alpha_2-2)(lnT)^{\alpha_2-3}-d_4(ln\xi_4)^{\alpha_2-2}]
$$
$$
-d_3(ln\xi_3)^{\alpha_2-2}[a_4(lnT)^{\alpha_2-1}+\frac{b_4}{T}(\alpha_2-1)(lnT)^{\alpha_2-2}-d_4(ln\xi_4)^{\alpha_2-1}]\ne0.
$$

Set

$$G_i^* = \sup_{t \in [1,T]} \int_1^t |G_i(t,s)| ds; \; i = 1, 2.$$

Theorem 8.5. *Assume that the hypotheses (8.4.1) and (8.4.2) hold. If, for every $w \in \Omega$, the matrix*

$$M(w) := \begin{pmatrix} G_1^* \|p_1(\cdot, w)\|_\infty & G_1^* \|q_1(\cdot, w)\|_\infty \\ G_2^* \|p_2(\cdot, w)\|_\infty & G_2^* \|q_2(\cdot, w)\|_\infty \end{pmatrix}$$

converges to 0, then the coupled system (8.8)–(8.9) has a unique random solution.

Proof. Consider the operator $N : \mathcal{C} \times \Omega \to \mathcal{C}$ defined by

$$(N(u, v))(t, w) = ((N_1(u, v))(t, w), (N_2(u, v))(t, w)). \tag{8.12}$$

Clearly, the fixed points of the operator N are random solutions of the coupled system (8.8)–(8.9). Let us show that N is a random operator on $\mathcal{C}$. Since f_i; $i = 1, 2$ are Carathéodory functions, then $w \to f_i(t, u, v, w)$ are measurable maps. Thus, we conclude that the maps

$$w \to (N_1(u, v))(t, w) \text{ and } w \to (N_2(u, v))(t, w),$$

are measurable. As a result, N is a random operator on $\mathcal{C} \times \Omega$ into $\mathcal{C}$. We show that N satisfies all conditions of Theorem 2.81.

For any $w \in \Omega$ and each $(u_1, v_1), (u_2, v_2) \in \mathcal{C}$ and $t \in I$, we have

$$\|(N_1(u_1, v_1))(t, w) - (N_1(u_2, v_2))(t, w)\|$$
$$\leq \int_1^T |G_1(t,s)| \|f_1(s, u_1(s, w), v_1(s, w), w) - f_1(s, u_2(s, w), v_2(s, w), w)\| ds$$
$$\leq \int_1^T |G_1(t,s)| (p_1(s, w) \|u_1(s, w) - v_1(s, w)\| + q_1(s, w) \|u_2(s, w) - v_2(s, w)\|) ds$$
$$\leq G_1^* (\|p_1(\cdot, w)\|_\infty \|u_1(\cdot, w) - v_1(\cdot, w)\|_\infty + \|q_1(\cdot, w)\|_\infty \|u_2(\cdot, w) - v_2(\cdot, w)\|_\infty).$$

Then

$$\|(N_1(u_1, v_1))(\cdot, w) - (N_1(u_2, v_2))(\cdot, w)\|_\infty$$
$$\leq G_1^* (\|p_1(\cdot, w)\|_\infty \|u_1(\cdot, w) - v_1(\cdot, w)\|_\infty + \|q_1(\cdot, w)\|_\infty \|u_2(\cdot, w) - v_2(\cdot, w)\|_\infty).$$

Also, for any $w \in \Omega$ and each $(u_1, v_1), (u_2, v_2) \in \mathcal{C}$ and $t \in I$, we get

$$\|(N_2(u_1, v_1))(\cdot, w) - (N_2(u_2, v_2))(\cdot, w)\|_\infty$$
$$\leq G_2^* (\|p_2(\cdot, w)\|_\infty \|u_1(\cdot, w) - v_1(\cdot, w)\|_\infty + \|q_2(\cdot, w)\|_\infty \|u_2(\cdot, w) - v_2(\cdot, w)\|_\infty).$$

Thus

$$d((N(u_1, v_1))(\cdot, w), (N(u_2, v_2))(\cdot, w))$$
$$\leq M(w) d((u_1(\cdot, w), v_1(\cdot, w)), (u_2(\cdot, w), v_2(\cdot, w))),$$

where

$$d((u_1(\cdot, w), v_1(\cdot, w)), (u_2(\cdot, w), v_2(\cdot, w))) = \begin{pmatrix} \|u_1(\cdot, w) - v_1(\cdot, w)\|_\infty \\ \|u_2(\cdot, w) - v_2(\cdot, w)\|_\infty \end{pmatrix}.$$

Since for every $w \in \Omega$, the matrix $M(w)$ converges to zero, then Theorem 2.81 implies that coupled system (8.8)–(8.9) has a unique random solution. □

Set

$$A(w) := (G_1^* \|h_1(\cdot, w)\|_\infty + G_2^* \|h_2(\cdot, w)\|_\infty),$$

and

$$C(w) := \max\{G_1^* \|k_1(\cdot, w)\|_\infty + G_2^* \|k_2(\cdot, w)\|_\infty, G_1^* \|l_1(\cdot, w)\|_\infty + G_2^* \|l_2(\cdot, w)\|_\infty\}.$$

Now, we prove an existence result for the coupled system (8.8)–(8.9) using Leray–Schauder random fixed-point theorem type in generalized Banach space.

Theorem 8.6. *Assume that the hypotheses (8.4.1) and (8.4.3) hold. If $C(w) < 1$, then the coupled system (8.8)–(8.9) has at least a random solution.*

Proof. Let $N : C \times \Omega \to C$ be the operator defined in (8.12). We show that N satisfies all conditions of Theorem 2.82. The proof will be given in several steps.

Step 1. $N(\cdot, \cdot, w)$ is continuous.
Let $(u_n, v_n)_n$ be a sequence such that $(u_n, v_n) \to (u, v) \in C$ as $n \to \infty$. For any $w \in \Omega$ and each $t \in I$, we have

$$\|(N_1(u_n, v_n))(t, w) - (N_1(u, v))(t, w)\|$$
$$\leq \int_1^T |G_1(t, s)| \|f_1(s, u_n(s, w), v_n(s, w), w) - f_1(s, u(s, w), v(s, w), w)\| ds$$
$$\leq G_1^* \|f_1(\cdot, u_n(\cdot, w), v_n(\cdot, w), w) - f_1(\cdot, u(\cdot, w), v(\cdot, w), w)\|_\infty.$$

Since f_1 is Carathéodory, we have

$$\|(N_1(u_n, v_n))(\cdot, w) - (N_1(u, v))(\cdot, w)\|_\infty \to 0 \text{ as } n \to \infty.$$

On the other hand, for any $w \in \Omega$ and each $t \in I$, we obtain

$$\|(N_2(u_n, v_n))(t, w) - (N_2(u, v))(t, w)\|$$
$$\leq \int_1^T |G_2(t, s)| \|f_2(\cdot, u_n(\cdot, w), v_n(\cdot, w), w) - f_2(\cdot, u(\cdot, w), v(\cdot, w), w)\|_\infty$$
$$\leq G_2^* \|f_2(\cdot, u_n(\cdot, w), v_n(\cdot, w), w) - f_2(\cdot, u(\cdot, w), v(\cdot, w), w)\|_\infty.$$

Also the fact that f_2 is Carathéodory, we get

$$\|(N_2(u_n, v_n))(\cdot, w) - (N_2(u, v))(\cdot, w)\|_\infty \to 0 \text{ as } n \to \infty.$$

Hence $N(\cdot, \cdot, w)$ is continuous.

Step 2. $N(\cdot, \cdot, w)$ maps bounded sets into bounded sets in $\mathcal{C}$.
Let $R > 0$ and set

$$B_R := \{(\mu, v) \in \mathcal{C} : \|\mu\|_\infty \leq R, \|v\|_\infty \leq R\}.$$

For any $w \in \Omega$ and each $(u, v) \in B_R$ and $t \in I$, we have

$$\|(N_1(u, v))(t, w)\|$$
$$\leq \int_1^T |G_1(t, s)| \|f_1(s, u(s, w), v(s, w), w)\| ds$$
$$\leq \int_1^T |G_1(t, s)|(h_1(s, w) + k_1(w)\|u(s, w\| + l_1(s, w)\|v(s, w\|) ds$$
$$\leq \int_1^T |G_1(t, s)|(\|h_1(\cdot, w)\|_\infty + R\|k_1(\cdot, w)\|_\infty + R\|l_1(\cdot, w)\|_\infty) ds$$
$$\leq G_1^*(\|h_1(\cdot, w)\|_\infty + R\|k_1(\cdot, w)\|_\infty + R\|l_1(\cdot, w)\|_\infty)$$
$$:= \ell_1(w).$$

Thus

$$\|(N_1(u, v))(\cdot, w)\|_\infty \leq \ell_1(w).$$

Also, for any $w \in \Omega$ and each $(u, v) \in B_R$ and $t \in I$, we get

$$\|(N_2(u, v))(\cdot, w)\|_\infty \leq G_2^*(\|h_2(\cdot, w)\|_\infty + R\|k_2(\cdot, w)\|_\infty + R\|l_2(\cdot, w)\|_\infty)$$
$$:= \ell_2(w).$$

Hence

$$\|(N(u, v))(\cdot, w)\|_\mathcal{C} \leq \max\{\ell_1(w), \ell_2(w)\} := \ell(w).$$

Step 3. $N(\cdot, \cdot, w)$ maps bounded sets into equicontinuous sets in $\mathcal{C}$.
Let B_R be the ball defined in Step 2. For each $t_1, t_2 \in I$ with $t_1 \leq t_2$ and any $(u, v) \in B_R$ and $w \in \Omega$, we have

$$\|(N_1(u, v))(t_1, w) - (N_1(u, v))(t_2, w)\|$$
$$\leq \int_1^T |G_1(t_1, s) - G_1(t_2, s)| \|f_1(s, u(s, w), v(s, w), w)\| ds$$
$$\leq \int_1^T |G_1(t_1, s) - G_1(t_2, s)|(h_1(s, w) + k_1(s, w)\|u(\cdot, w)\|_\infty + l_1(s, w)\|v(\cdot, w)\|_\infty) ds$$
$$\leq (\|h_1(\cdot, w)\|_\infty + R\|k_1(\cdot, w)\|_\infty + R\|l_1(\cdot, w)\|_\infty) \int_1^T |G_1(t_1, s) - G_1(t_2, s)| ds$$
$$\to 0 \text{ as } t_1 \to t_2.$$

Also, we get

$$\|(N_2(u, v))(t_1, w) - (N_2(u, v))(t_2, w)\|$$

$$\leq (\|h_2(\cdot, w)\|_\infty + R\|k_2(\cdot, w)\|_\infty + R\|l_2(\cdot, w)\|_\infty) \int_1^T |G_2(t_1, s) - G_2(t_2, s)| ds$$

$$\to 0 \text{ as } t_1 \to t_2.$$

As a consequence of Steps 1 to 3, with the Arzelà–Ascoli theorem, we conclude that $N(\cdot, \cdot, w)$ maps B_R into a precompact set in $\mathcal{C}$.

Step 4. The set $E(w)$ consisting of $(u(\cdot, w), v(\cdot, w)) \in \mathcal{C}$ such that $(u(\cdot, w), v(\cdot, w)) = \lambda(w)(N((u, v))(\cdot, w)$ for some measurable function $\lambda : \Omega \to (0, 1)$ is bounded in $\mathcal{C}$.
Let $(u(\cdot, w), v(\cdot, w)) \in \mathcal{C}$ such that $(u(\cdot, w), v(\cdot, w)) = \lambda(w)(N((u, v))(\cdot, w)$. Then $u(\cdot, w) = \lambda(w)(N_1((u, v))(\cdot, w)$ and $v(\cdot, w) = \lambda(w)(N_2((u, v))(\cdot, w)$. Thus, for any $w \in \Omega$ and each $t \in I$, we have

$$\|u(t, w)\| \leq \int_1^T |G_1(t, s)| \|f_1(s, u(s, w), v(s, w), w)\| ds$$

$$\leq \int_1^T |G_1(t, s)| (h_1(s, w) + k_1(s, w)\|u(s, w\| + l_1(s, w)\|v(s, w\|) ds$$

$$\leq G_1^*(\|h_1(\cdot, w)\|_\infty + \|k_1(\cdot, w)\|_\infty \|u(\cdot, w\|_\infty + \|l_1(\cdot, w)\|_\infty \|v(\cdot, w\|_\infty).$$

So we get

$$\|u(\cdot, w)\|_\infty \leq G_1^*(\|h_1(\cdot, w)\|_\infty + \|k_1(\cdot, w)\|_\infty \|u(\cdot, w\|_\infty + \|l_1(\cdot, w)\|_\infty \|v(\cdot, w\|_\infty).$$

Also, we get

$$\|v(\cdot, w)\|_\infty \leq G_2^*(\|h_2(\cdot, w)\|_\infty + \|k_2(\cdot, w)\|_\infty \|u(\cdot, w\|_\infty + \|l_2(\cdot, w)\|_\infty \|v(\cdot, w\|)_\infty.$$

We obtain

$$\|u(\cdot, w)\|_\infty + \|v(\cdot, w)\|_\infty \leq A(w) + C(w)(\|u(\cdot, w)\|_\infty + \|v(\cdot, w)\|_\infty).$$

It follows that

$$\|u(\cdot, w)\|_\infty + \|v(\cdot, w)\|_\infty \leq \frac{A(w)}{1 - C(w)}$$

$$:= L(w).$$

Hence we get

$$\|(u(\cdot, w), v(\cdot, w))\|_\mathcal{C} \leq L(w).$$

This shows that the set $E(w)$ is bounded.

As a consequence of steps 1 to 4 together with Theorem 2.82, we can conclude that N has at least one random fixed point in B_R, which is a random solution of the coupled system (8.8)–(8.9). $\qquad\qquad\qquad\qquad\qquad\qquad\qquad\qquad\qquad\qquad\qquad\qquad\qquad\square$

8.2.3 An example

Let $\Omega = (-\infty, 0)$ be equipped with the usual σ-algebra consisting of Lebesgue measurable subsets of $(-\infty, 0)$. Consider the following random coupled system of Caputo–Hadamard fractional differential equations

$$
\begin{cases}
({}^{Hc}D_1^{\frac{3}{2}}u)(t, w) = f(t, u(t, w), v(t, w), w); \\
({}^{Hc}D_1^{\frac{3}{2}}v)(t, w) = g(t, u(t, w), v(t, w), w);
\end{cases} \quad ; w \in \Omega, \ t \in [1, e], \tag{8.13}
$$

with the three-point boundary conditions

$$
\begin{cases}
u(1, w) - u'(1, w) = u(2, w), \\
2u(T, w) + u'(T, w) = 2u(\frac{3}{2}, w), \\
3v(1, w) - v'(1, w) = 3v(\frac{5}{4}, w), \\
v(T, w) + 2v'(T, w) = v(2, w),
\end{cases} \quad ; w \in \Omega, \tag{8.14}
$$

where

$$
f(t, u, v, w) = \frac{ct^{\frac{-1}{4}} w^2 u(t) \sin t}{64(1 + w^2 + \sqrt{t})(1 + |u| + |v|)}; \ t \in [1, e],
$$

$$
g(t, u, v) = \frac{cw^2 v(t) \cos t}{64(1 + |u| + |v|)}; \ w \in \Omega, \ t \in [1, e],
$$

and $c < \frac{1}{\max\{G_1^*, G_2^*\}}$. The hypothesis (8.4.2) is satisfied with

$$
p_2(t, w) = q_1(t, w) = 0, \ p_1(t, w) = \frac{cw^2 \sin t}{64(1 + w^2)}, \ q_2(t, w) = \frac{cw^2 \cos t}{64(1 + w^2)}.
$$

Also, if for every $w \in \Omega$, the matrix

$$
\frac{cw^2}{64(1 + w^2)} \begin{pmatrix} G_1^* & 0 \\ 0 & G_2^* \end{pmatrix}
$$

converges to 0. Hence, Theorem 8.5 implies that the system (8.13)–(8.14) has a unique random solution defined on $[1, e]$.

8.3 Random coupled systems of implicit Caputo–Hadamard fractional differential equations with multi-point boundary conditions

8.3.1 Introduction

In this section we discuss the existence and uniqueness of solutions for the following coupled system of Caputo–Hadamard fractional differential equations

$$\begin{cases} (^{Hc}D_1^{\alpha_1}u)(t,w) = f_1(t,u(t,w),v(t,w),(^{Hc}D_1^{\alpha_1}u)(t,w),w) \\ (^{Hc}D_1^{\alpha_2}v)(t,w) = f_2(t,u(t,w),v(t,w),(^{Hc}D_1^{\alpha_2}v)(t,w),w) \end{cases} ; \qquad (8.15)$$

where $t \in I := [1,T]$, $w \in \Omega$, with the four-point boundary conditions

$$\begin{cases} a_1u(1,w) - b_1u'(1,w) = d_1u(\xi_1,w) \\ a_2u(T,w) + b_2u'(T,w) = d_2u(\xi_2,w) \\ a_3v(1,w) - b_3v'(1,w) = d_3v(\xi_3,w) \\ a_4v(T,w) + b_4v'(T,w) = d_4v(\xi_4,w) \end{cases} ; w \in \Omega, \qquad (8.16)$$

where $\alpha \in (1,2]$, $T > 1$, $a_i, b_i, d_i \in \mathbb{R}$, $\xi_i \in (1,T)$; $i = 1,2,3,4$, (Ω, F, P) is a complete probability space, $f_1, f_2 : I \times \mathbb{R}^m \times \mathbb{R}^m \times \mathbb{R}^m \times \Omega \to \mathbb{R}^m$ are given functions, $\mathbb{R}^m$; $m \in \mathbb{N}^*$ is the Banach space with a suitable norm $\| \cdot \|$, and $^{Hc}D_1^{\alpha_i}$ is the Caputo–Hadamard fractional derivative of order α_i; $i = 1,2$.

8.3.2 Main results

In this section we are concerned with the existence and uniqueness results of the coupled system (8.15)–(8.16).

From [338, Theorem 21], we concluded the following lemma.

Lemma 8.7. *Let $f_i : I \times \mathbb{R}^m \times \mathbb{R}^m \times \mathbb{R}^m \times \Omega \to \mathbb{R}^m$; $i = 1,2$ such that $f_i(\cdot, u,v,x,w) \in C(I)$ for each $u,v,x \in C(I)$ and $w \in \Omega$. Then the coupled system (8.15)–(8.16) is equivalent to the problem of obtaining the solution of the coupled system*

$$\begin{cases} g_1(t,w) = f_1\left(t, \int_1^T G_1(t,s)g_1(s,w)ds, \int_1^T G_2(t,s)g_2(s,w)ds, g_1(t),w\right) \\ g_2(t,w) = f_2\left(t, \int_1^T G_1(t,s)g_1(s,w)ds, \int_1^T G_2(t,s)g_2(s,w)ds, g_2(t),w\right), \end{cases}$$

and if $g_i(\cdot,w) \in C(I)$; $w \in \Omega$ are the solutions of this system, then

$$\begin{cases} u(t,w) = \int_1^T G_1(t,s)g_1(s,w)ds \\ v(t,w) = \int_1^T G_2(t,s)g_2(s,w)ds, \end{cases}$$

where

$G_1(t,s)$

$$= \begin{cases}
\begin{aligned}
&\frac{1}{s\Gamma(\alpha_1)}(ln\tfrac{t}{s})^{\alpha_1-1} + \frac{d_1(ln\xi_1)^{\alpha_1-2}(lnt)^{\alpha_1-1}}{s\Delta_1}[\frac{a_2}{\Gamma(\alpha_1)}(ln\tfrac{T}{s})^{\alpha_1-1} + \frac{b_2}{\Gamma(\alpha_1-1)}(ln\tfrac{T}{s})^{\alpha_1-2}\\
&-\frac{d_2}{\Gamma(\alpha_1)}(ln\tfrac{\xi_2}{s})^{\alpha_1-1}]\\
&-\frac{d_1(lnt)^{\alpha_1-1}}{s\Delta_1\Gamma(\alpha_1)}(ln\tfrac{\xi_1}{s})^{\alpha_1-1}[a_2(lnT)^{\alpha_1-2} + \tfrac{b_2}{T}(\alpha_1-2)(lnT)^{\alpha_1-3} - d_2(ln\xi_2)^{\alpha_1-2}]\\
&+\frac{d_1(lnt)^{\alpha_1-2}}{s\Delta_1\Gamma(\alpha_1)}(ln\tfrac{\xi_1}{s})^{\alpha_1-1}[a_2(lnT)^{\alpha_1-1} + \tfrac{b_2}{T}(\alpha_1-1)(lnT)^{\alpha_1-2} - d_2(ln\xi_2)^{\alpha_1-1}]\\
&-\frac{d_1(ln\xi_1)^{\alpha_1-1}(lnt)^{\alpha_1-2}}{s\Delta_1}[\frac{a_2}{\Gamma(\alpha_1)}(ln\tfrac{T}{s})^{\alpha_1-1} + \frac{b_2}{\Gamma(\alpha_1-1)}(ln\tfrac{T}{s})^{\alpha_1-2}\\
&-\frac{d_2}{\Gamma(\alpha_1)}(ln\tfrac{\xi_2}{s})^{\alpha_1-1}];\ s\le\xi_1,\ s\le t,
\end{aligned}\\[4pt]
\begin{aligned}
&\frac{d_1(ln\xi_1)^{\alpha_1-2}(lnt)^{\alpha_1-1}}{s\Delta_1}[\frac{a_2}{\Gamma(\alpha_1)}(ln\tfrac{T}{s})^{\alpha_1-1} + \frac{b_2}{\Gamma(\alpha_1-1)}(ln\tfrac{T}{s})^{\alpha_1-2} - \frac{d_2}{\Gamma(\alpha_1)}(ln\tfrac{\xi_2}{s})^{\alpha_1-1}]\\
&-\frac{d_1(lnt)^{\alpha_1-1}}{s\Delta_1\Gamma(\alpha_1)}(ln\tfrac{\xi_1}{s})^{\alpha_1-1}[a_2(lnT)^{\alpha_1-2} + \tfrac{b_2}{T}(\alpha_1-2)(lnT)^{\alpha_1-3} - d_2(ln\xi_2)^{\alpha_1-2}]\\
&+\frac{d_1(lnt)^{\alpha_1-2}}{s\Delta_1\Gamma(\alpha_1)}(ln\tfrac{\xi_1}{s})^{\alpha_1-1}[a_2(lnT)^{\alpha_1-1} + \tfrac{b_2}{T}(\alpha_1-1)(lnT)^{\alpha_1-2} - d_2(ln\xi_2)^{\alpha_1-1}]\\
&-\frac{d_1(ln\xi_1)^{\alpha_1-1}(lnt)^{\alpha_1-2}}{s\Delta_1}[\frac{a_2}{\Gamma(\alpha_1)}(ln\tfrac{T}{s})^{\alpha_1-1} + \frac{b_2}{\Gamma(\alpha_1-1)}(ln\tfrac{T}{s})^{\alpha_1-2}\\
&-\frac{d_2}{\Gamma(\alpha_1)}(ln\tfrac{\xi_2}{s})^{\alpha_1-1}];\ s\le\xi_1,\ t\le s,
\end{aligned}\\[4pt]
\begin{aligned}
&\frac{1}{s\Gamma(\alpha_1)}(ln\tfrac{t}{s})^{\alpha_1-1}\\
&+\frac{d_1(ln\xi_1)^{\alpha_1-2}(lnt)^{\alpha_1-1}}{s\Delta_1}[\frac{a_2}{\Gamma(\alpha_1)}(ln\tfrac{T}{s})^{\alpha_1-1} + \frac{b_2}{\Gamma(\alpha_1-1)}(ln\tfrac{T}{s})^{\alpha_1-2} - \frac{d_2}{\Gamma(\alpha_1)}(ln\tfrac{\xi_2}{s})^{\alpha_1-1}]\\
&-\frac{d_1(ln\xi_1)^{\alpha_1-1}(lnt)^{\alpha_1-2}}{s\Delta_1}[\frac{a_2}{\Gamma(\alpha_1)}(ln\tfrac{T}{s})^{\alpha_1-1} + \frac{b_2}{\Gamma(\alpha_1-1)}(ln\tfrac{T}{s})^{\alpha_1-2}\\
&-\frac{d_2}{\Gamma(\alpha_1)}(ln\tfrac{\xi_2}{s})^{\alpha_1-1}];\ \xi_1\le s\le\xi_2,\ s\le t,
\end{aligned}\\[4pt]
\begin{aligned}
&\frac{d_1(ln\xi_1)^{\alpha_1-2}(lnt)^{\alpha_1-1}}{s\Delta_1}[\frac{a_2}{\Gamma(\alpha_1)}(ln\tfrac{T}{s})^{\alpha_1-1} + \frac{b_2}{\Gamma(\alpha_1-1)}(ln\tfrac{T}{s})^{\alpha_1-2} - \frac{d_2}{\Gamma(\alpha_1)}(ln\tfrac{\xi_2}{s})^{\alpha_1-1}]\\
&-\frac{d_1(ln\xi_1)^{\alpha_1-1}(lnt)^{\alpha_1-2}}{s\Delta_1}[\frac{a_2}{\Gamma(\alpha_1)}(ln\tfrac{T}{s})^{\alpha_1-1} + \frac{b_2}{\Gamma(\alpha_1-1)}(ln\tfrac{T}{s})^{\alpha_1-2}\\
&-\frac{d_2}{\Gamma(\alpha_1)}(ln\tfrac{\xi_2}{s})^{\alpha_1-1}];\ \xi_1\le s\le\xi_2,\ t\le s,
\end{aligned}\\[4pt]
\begin{aligned}
&\frac{1}{s\Gamma(\alpha_1)}(ln\tfrac{t}{s})^{\alpha_1-1}\\
&+\frac{d_1(ln\xi_1)^{\alpha_1-2}(lnt)^{\alpha_1-1}}{s\Delta_1}[\frac{a_2}{\Gamma(\alpha_1)}(ln\tfrac{T}{s})^{\alpha_1-1} + \frac{b_2}{\Gamma(\alpha_1-1)}(ln\tfrac{T}{s})^{\alpha_1-2}]\\
&-\frac{d_1(ln\xi_1)^{\alpha_1-1}(lnt)^{\alpha_1-2}}{s\Delta_1}[\frac{a_2}{\Gamma(\alpha_1)}(ln\tfrac{T}{s})^{\alpha_1-1} + \frac{b_2}{\Gamma(\alpha_1-1)}(ln\tfrac{T}{s})^{\alpha_1-2}];\ \xi_2\le s,\ s\le t,
\end{aligned}\\[4pt]
\begin{aligned}
&\frac{d_1(ln\xi_1)^{\alpha_1-2}(lnt)^{\alpha_1-1}}{s\Delta_1}[\frac{a_2}{\Gamma(\alpha_1)}(ln\tfrac{T}{s})^{\alpha_1-1} + \frac{b_2}{\Gamma(\alpha_1-1)}(ln\tfrac{T}{s})^{\alpha_1-2}]\\
&-\frac{d_1(ln\xi_1)^{\alpha_1-1}(lnt)^{\alpha_1-2}}{s\Delta_1}[\frac{a_2}{\Gamma(\alpha_1)}(ln\tfrac{T}{s})^{\alpha_1-1} + \frac{b_2}{\Gamma(\alpha_1-1)}(ln\tfrac{T}{s})^{\alpha_1-2}];\ \xi_2\le s,\ t\le s,
\end{aligned}
\end{cases}$$

with

$$\Delta_1 = d_1(ln\xi_1)^{\alpha_1-1}[a_2(lnT)^{\alpha_1-2} + \frac{b_2}{T}(\alpha_1-2)(lnT)^{\alpha_1-3} - d_2(ln\xi_2)^{\alpha_1-2}]$$

$$- d_1(ln\xi_1)^{\alpha_1-2}[a_2(lnT)^{\alpha_1-1} + \frac{b_2}{T}(\alpha_1-1)(lnT)^{\alpha_1-2} - d_2(ln\xi_2)^{\alpha_1-1}] \ne 0,$$

and

$G_2(t,s)$

$$= \begin{cases}
\frac{1}{s\Gamma(\alpha_2)}(\ln\frac{t}{s})^{\alpha_2-1} + \frac{d_3(\ln\xi_3)^{\alpha_2-2}(\ln t)^{\alpha_2-1}}{s\Delta_2}[\frac{a_4}{\Gamma(\alpha_2)}(\ln\frac{T}{s})^{\alpha_2-1} + \frac{b_4}{\Gamma(\alpha_2-1)}(\ln\frac{T}{s})^{\alpha_2-2} \\
\quad - \frac{d_4}{\Gamma(\alpha_2)}(\ln\frac{\xi_4}{s})^{\alpha_2-1}] \\
\quad - \frac{d_3(\ln t)^{\alpha_2-1}}{s\Delta_2\Gamma(\alpha_2)}(\ln\frac{\xi_3}{s})^{\alpha_2-1}[a_4(\ln T)^{\alpha_2-2} + \frac{b_4}{T}(\alpha_2-2)(\ln T)^{\alpha_2-3} - d_4(\ln\xi_4)^{\alpha_2-2}] \\
\quad + \frac{d_3(\ln t)^{\alpha_2-2}}{s\Delta_2\Gamma(\alpha_2)}(\ln\frac{\xi_3}{s})^{\alpha_2-1}[a_4(\ln T)^{\alpha_2-1} + \frac{b_4}{T}(\alpha_2-1)(\ln T)^{\alpha_2-2} - d_4(\ln\xi_4)^{\alpha_2-1}] \\
\quad - \frac{d_3(\ln\xi_3)^{\alpha_2-1}(\ln t)^{\alpha_2-2}}{s\Delta_2}[\frac{a_4}{\Gamma(\alpha_2)}(\ln\frac{T}{s})^{\alpha_2-1} + \frac{b_4}{\Gamma(\alpha_2-1)}(\ln\frac{T}{s})^{\alpha_2-2} \\
\quad - \frac{d_4}{\Gamma(\alpha_2)}(\ln\frac{\xi_4}{s})^{\alpha_2-1}]; \quad s \leq \xi_3,\ s \leq t, \\[2ex]
\frac{d_3(\ln\xi_3)^{\alpha_2-2}(\ln t)^{\alpha_2-1}}{s\Delta_2}[\frac{a_4}{\Gamma(\alpha_2)}(\ln\frac{T}{s})^{\alpha_2-1} + \frac{b_4}{\Gamma(\alpha_2-1)}(\ln\frac{T}{s})^{\alpha_2-2} - \frac{d_4}{\Gamma(\alpha_2)}(\ln\frac{\xi_4}{s})^{\alpha_2-1}] \\
\quad - \frac{d_3(\ln t)^{\alpha_2-1}}{s\Delta_2\Gamma(\alpha_2)}(\ln\frac{\xi_3}{s})^{\alpha_2-1}[a_4(\ln T)^{\alpha_2-2} + \frac{b_4}{T}(\alpha_2-2)(\ln T)^{\alpha_2-3} - d_4(\ln\xi_4)^{\alpha_2-2}] \\
\quad + \frac{d_3(\ln t)^{\alpha_2-2}}{s\Delta_2\Gamma(\alpha_2)}(\ln\frac{\xi_3}{s})^{\alpha_2-1}[a_4(\ln T)^{\alpha_2-1} + \frac{b_4}{T}(\alpha_2-1)(\ln T)^{\alpha_2-2} - d_4(\ln\xi_4)^{\alpha_2-1}] \\
\quad - \frac{d_3(\ln\xi_3)^{\alpha_2-1}(\ln t)^{\alpha_2-2}}{s\Delta_2}[\frac{a_4}{\Gamma(\alpha_2)}(\ln\frac{T}{s})^{\alpha_2-1} + \frac{b_4}{\Gamma(\alpha_2-1)}(\ln\frac{T}{s})^{\alpha_2-2} \\
\quad - \frac{d_4}{\Gamma(\alpha_2)}(\ln\frac{\xi_4}{s})^{\alpha_2-1}]; \quad s \leq \xi_3,\ t \leq s, \\[2ex]
\frac{1}{s\Gamma(\alpha_2)}(\ln\frac{t}{s})^{\alpha_2-1} \\
\quad + \frac{d_3(\ln\xi_3)^{\alpha_2-2}(\ln t)^{\alpha_2-1}}{s\Delta_2}[\frac{a_4}{\Gamma(\alpha_2)}(\ln\frac{T}{s})^{\alpha_2-1} + \frac{b_4}{\Gamma(\alpha_2-1)}(\ln\frac{T}{s})^{\alpha_2-2} - \frac{d_4}{\Gamma(\alpha_2)}(\ln\frac{\xi_4}{s})^{\alpha_2-1}] \\
\quad - \frac{d_3(\ln\xi_3)^{\alpha_2-1}(\ln t)^{\alpha_2-2}}{s\Delta_2}[\frac{a_4}{\Gamma(\alpha_2)}(\ln\frac{T}{s})^{\alpha_2-1} + \frac{b_4}{\Gamma(\alpha_2-1)}(\ln\frac{T}{s})^{\alpha_2-2} \\
\quad - \frac{d_4}{\Gamma(\alpha_2)}(\ln\frac{\xi_4}{s})^{\alpha_2-1}]; \quad \xi_3 \leq s \leq \xi_4,\ s \leq t, \\[2ex]
\frac{d_3(\ln\xi_3)^{\alpha_2-2}(\ln t)^{\alpha_2-1}}{s\Delta_2}[\frac{a_4}{\Gamma(\alpha_2)}(\ln\frac{T}{s})^{\alpha_2-1} + \frac{b_4}{\Gamma(\alpha_2-1)}(\ln\frac{T}{s})^{\alpha_2-2} - \frac{d_4}{\Gamma(\alpha_2)}(\ln\frac{\xi_4}{s})^{\alpha_2-1}] \\
\quad - \frac{d_3(\ln\xi_3)^{\alpha_2-1}(\ln t)^{\alpha_2-2}}{s\Delta_2}[\frac{a_4}{\Gamma(\alpha_2)}(\ln\frac{T}{s})^{\alpha_2-1} + \frac{b_4}{\Gamma(\alpha_2-1)}(\ln\frac{T}{s})^{\alpha_2-2} \\
\quad - \frac{d_4}{\Gamma(\alpha_2)}(\ln\frac{\xi_4}{s})^{\alpha_2-1}]; \quad \xi_3 \leq s \leq \xi_4,\ t \leq s, \\[2ex]
\frac{1}{s\Gamma(\alpha_2)}(\ln\frac{t}{s})^{\alpha_2-1} \\
\quad + \frac{d_3(\ln\xi_3)^{\alpha_2-2}(\ln t)^{\alpha_2-1}}{s\Delta_2}[\frac{a_4}{\Gamma(\alpha_2)}(\ln\frac{T}{s})^{\alpha_2-1} + \frac{b_4}{\Gamma(\alpha_2-1)}(\ln\frac{T}{s})^{\alpha_2-2}] \\
\quad - \frac{d_3(\ln\xi_3)^{\alpha_2-1}(\ln t)^{\alpha_2-2}}{s\Delta_2}[\frac{a_4}{\Gamma(\alpha_2)}(\ln\frac{T}{s})^{\alpha_2-1} + \frac{b_4}{\Gamma(\alpha_2-1)}(\ln\frac{T}{s})^{\alpha_2-2}]; \quad \xi_4 \leq s,\ s \leq t, \\[2ex]
\frac{d_3(\ln\xi_3)^{\alpha_2-2}(\ln t)^{\alpha_2-1}}{s\Delta_2}[\frac{a_4}{\Gamma(\alpha_2)}(\ln\frac{T}{s})^{\alpha_2-1} + \frac{b_4}{\Gamma(\alpha_2-1)}(\ln\frac{T}{s})^{\alpha_2-2}] \\
\quad - \frac{d_3(\ln\xi_3)^{\alpha_2-1}(\ln t)^{\alpha_2-2}}{s\Delta_2}[\frac{a_4}{\Gamma(\alpha_2)}(\ln\frac{T}{s})^{\alpha_2-1} + \frac{b_4}{\Gamma(\alpha_2-1)}(\ln\frac{T}{s})^{\alpha_2-2}]; \quad \xi_4 \leq s,\ t \leq s,
\end{cases}$$

with

$$\Delta_2 = d_3(\ln\xi_3)^{\alpha_2-1}[a_4(\ln T)^{\alpha_2-2} + \frac{b_4}{T}(\alpha_2-2)(\ln T)^{\alpha_2-3} - d_4(\ln\xi_4)^{\alpha_2-2}]$$

$$- d_3(\ln\xi_3)^{\alpha_2-2}[a_4(\ln T)^{\alpha_2-1} + \frac{b_4}{T}(\alpha_2-1)(\ln T)^{\alpha_2-2} - d_4(\ln\xi_4)^{\alpha_2-1}] \neq 0.$$

Definition 8.8. By a random solution of the problem (8.15)–(8.16) we mean a coupled measurable functions $(u, v) \in C(I) \times C(I)$ satisfying the boundary conditions (8.16) and Eqs. (8.15) on I.

The following hypotheses will be used in the sequel.

(8.8.1) The functions f_i; $i = 1, 2$ are Carathéodory.
(8.8.2) There exist continuous functions $p_i, q_i, r_i : I \to L^\infty(\Omega, \mathbb{R}_+)$; $i = 1, 2$ such that

$$\|f_i(t, u_1, v_1, x_1, w) - f_i(t, u_2, v_2, x_2, w)\| \le p_i(t, w)\|u_1 - u_2\| + q_i(t, w)\|v_1 - v_2\|$$
$$+ r_i(t, w)\|x_1 - x_2\|; \text{ for a.e. } t \in I, \text{ and each } u_i, v_i, x_i \in \mathbb{R}^m, \ i = 1, 2.$$

(8.8.3) There exist continuous functions $k_i, l_i, \bar{k}_i, \bar{l}_i : I \to L^\infty(\Omega, \mathbb{R}_+)$; $i = 1, 2$ such that

$$\|f_i(t, u, v, x, w)\| \le k_i(t, w) + l_i(t, w)\|u\| + \bar{k}_i(t, w)\|v\| + \bar{l}_i(t, w)\|x\|;$$

for a.e. $t \in I$, and each $u, v, x \in \mathbb{R}^m$.

First, we prove an existence and uniqueness result for the coupled system (8.15)–(8.16) by using Banach random fixed-point theorem type in generalized Banach spaces. We define the operators $N_1, N_2 : \mathcal{C} \times \Omega \to C(I)$ by

$$(N_i(u, v))(t, w) = \int_1^T G_i(t, s)g_i(s, w)ds; \ i = 1, 2, \tag{8.17}$$

where $g_i(\cdot, w) \in C(I)$ for $w \in \Omega$, such that

$$g_i(t, w) = f_i\left(t, \int_1^T G_1(t, s)g_1(s, w)ds, \int_1^T G_2(t, s)g_2(s, w)ds, g_i(t, w), w\right),$$

for $i = 1, 2$. Set

$$G_i^* = \sup_{t \in [1, T]} \int_1^t |G_i(t, s)|ds; \ i = 1, 2.$$

Theorem 8.9. *Assume that the hypotheses (8.8.1) and (8.8.2) hold. If, for every $w \in \Omega$, the matrix*

$$M(w) := \begin{pmatrix} \frac{G_1^* \|p_1(\cdot, w)\|_\infty}{1 - \|r_1(\cdot, w)\|_\infty} & \frac{G_1^* \|q_1(\cdot, w)\|_\infty}{1 - \|r_1(\cdot, w)\|_\infty} \\ \frac{G_2^* \|p_2(\cdot, w)\|_\infty}{1 - \|r_2(\cdot, w)\|_\infty} & \frac{G_2^* \|q_2(\cdot, w)\|_\infty}{1 - \|r_2(\cdot, w)\|_\infty} \end{pmatrix}$$

converges to 0, then the coupled system (8.15)–(8.16) has a unique random solution.

Proof. Consider the operator $N : \mathcal{C} \times \Omega \to \mathcal{C}$ defined by

$$(N(u, v))(t, w) = ((N_1(u, v))(t, w), (N_2(u, v))(t, w)). \tag{8.18}$$

Clearly, the fixed points of the operator N are random solutions of the coupled system (8.15)–(8.16). Let us show that N is a random operator on $\mathcal{C}$. Since f_i; $i = 1, 2$ are

Carathéodory functions, then $w \to f_i(t, u, v, x, w)$ are measurable maps, in view of Proposition 4.1.2, we conclude that the maps

$$w \to (N_1(u, v))(t, w) \text{ and } w \to (N_2(u, v))(t, w),$$

are measurable. As a result, N is a random operator on $\mathcal{C} \times \Omega$ into $\mathcal{C}$. We show that N satisfies all conditions of Theorem 2.81.

For any $w \in \Omega$ and each $(u_1, v_1), (u_2, v_2) \in \mathcal{C}$ and $t \in I$, we have

$$\|(N_i(u_1, v_1))(t, w) - (N_i(u_2, v_2))(t, w)\|$$

$$\leq \int_1^T |G_i(t, s)| \|g_i(t, w) - \bar{g}_i(t, w)\| ds,$$

where, $g_i(\cdot, w), \bar{g}_i(\cdot, w) \in C(I)$ for $w \in \Omega$, such that

$$g_i(t, w) = f_i \left(t, \int_1^T G_1(t, s) g_1(s, w) ds, \int_1^T G_2(t, s) g_2(s, w) ds, g_i(t, w), w \right);$$

$$\bar{g}_i(t, w) = f_i \left(t, \int_1^T G_1(t, s) \bar{g}_1(s, w) ds, \int_1^T G_2(t, s) \bar{g}_2(s, w) ds, \bar{g}_i(t, w), w \right),$$

for $i = 1, 2$. Then, from (7.7.2), we get

$$\|g_i(t, w) - \bar{g}_i(t, w)\|$$
$$\leq p_i(t, w) \|u_1(s, w) - v_1(s, w)\| + q_i(t, w) \|u_2(s, w) - v_2(s, w)\|$$
$$+ r_i(t, w) \|g_i(t, w) - \bar{g}_i(t, w)\|; \ i = 1, 2.$$

Thus

$$\|g_i(t, w) - \bar{g}_i(t, w)\|$$
$$\leq \frac{\|p_i(\cdot, w)\|_\infty}{1 - \|r_i(\cdot, w)\|_\infty} \|u_1(\cdot, w) - v_1(\cdot, w)\|_\infty$$
$$+ \frac{\|q_i(\cdot, w)\|_\infty}{1 - \|r_i(\cdot, w)\|_\infty} \|u_2(\cdot, w) - v_2(\cdot, w)\|_\infty; \ i = 1, 2.$$

Hence

$$\|(N_i(u_1, v_1))(\cdot, w) - (N_i(u_2, v_2))(\cdot, w)\|_\infty$$
$$\leq \frac{G_i^* \|p_i(\cdot, w)\|_\infty}{1 - \|r_i(\cdot, w)\|_\infty} \|u_1(\cdot, w) - v_1(\cdot, w)\|_\infty$$
$$+ \frac{G_i^* \|q_i(\cdot, w)\|_\infty}{1 - \|r_i(\cdot, w)\|_\infty} \|u_2(\cdot, w) - v_2(\cdot, w)\|_\infty; \ i = 1, 2.$$

Consequently

$$d((N(u_1, v_1))(\cdot, w), (N(u_2, v_2))(\cdot, w))$$
$$\leq M(w) d((u_1(\cdot, w), v_1(\cdot, w)), (u_2(\cdot, w), v_2(\cdot, w))),$$

where

$$d((u_1(\cdot, w), v_1(\cdot, w)), (u_2(\cdot, w), v_2(\cdot, w))) = \begin{pmatrix} \|u_1(\cdot, w) - v_1(\cdot, w)\|_\infty \\ \|u_2(\cdot, w) - v_2(\cdot, w)\|_\infty \end{pmatrix}.$$

Since for every $w \in \Omega$, the matrix $M(w)$ converges to zero, then Theorem 2.81 implies that coupled system (8.15)–(8.16) has a unique random solution. $\qquad\square$

Now, we prove an existence result for the coupled system (8.15)–(8.16) by using Leray–Schauder random fixed-point theorem type in generalized Banach space. Set

$$C(w) := \max \left\{ \frac{G_1^* \|l_1(\cdot, w)\|_\infty}{1 - \|\bar{l}_1(\cdot, w)\|_\infty} + \frac{G_2^* \|l_2(\cdot, w)\|_\infty}{1 - \|\bar{l}_2(\cdot, w)\|_\infty}, \frac{G_1^* \|\bar{k}_1(\cdot, w)\|_\infty}{1 - \|\bar{l}_1(\cdot, w)\|_\infty} \right.$$
$$\left. + \frac{G_2^* \|\bar{k}_2(\cdot, w)\|_\infty}{1 - \|\bar{l}_2(\cdot, w)\|_\infty} \right\}.$$

Theorem 8.10. *Assume that the hypotheses* (8.8.1) *and* (8.8.3) *hold. If* $\|\bar{l}_i(\cdot, w)\|_\infty < 1$; $i = 1, 2$, *and* $C(w) < 1$, *then the coupled system* (8.15)–(8.16) *has at least a random solution.*

Proof. Let $N : C \times \Omega \to C$ be the operator defined in (8.18). We show that N satisfies all conditions of Theorem 2.82. The proof will be given in several steps.

Step 1. $N(\cdot, \cdot, w)$ is continuous.
Let $(u_n, v_n)_n$ be a sequence such that $(u_n, v_n) \to (u, v) \in C$ as $n \to \infty$. For any $w \in \Omega$ and each $t \in I$, we have

$$\|(N_1(u_n, v_n))(t, w) - (N_1(u, v))(t, w)\|$$
$$\leq \int_1^T |G_1(t, s)| \| f_1(s, u_n(s, w), v_n(s, w), (^{Hc}D_1^{\alpha_1} u_n)(s, w), w)$$
$$- f_1(s, u(s, w), v(s, w), (^{Hc}D_1^{\alpha_1} u)(s, w), w)) \| ds$$
$$\leq G_1^* \| f_1(\cdot, u_n(\cdot, w), v_n(\cdot, w), (^{Hc}D_1^{\alpha_1} u_n)(\cdot, w), w)$$
$$- f_1(\cdot, u(\cdot, w), v(\cdot, w), (^{Hc}D_1^{\alpha_1} u)(\cdot, w), w)) \|_\infty.$$

Since f_1 is Carathéodory, we have

$$\|(N_1(u_n, v_n))(\cdot, w) - (N_1(u, v))(\cdot, w)\|_\infty \to 0 \text{ as } n \to \infty.$$

On the other hand, for any $w \in \Omega$ and each $t \in I$, we obtain

$$\|(N_2(u_n, v_n))(t, w) - (N_2(u, v))(t, w)\|$$
$$\leq \int_1^T |G_2(t, s)| \| f_2(\cdot, u_n(\cdot, w), v_n(\cdot, w), (^{Hc}D_1^{\alpha_2} v_n)(\cdot, w), w)$$
$$- f_2(\cdot, u(\cdot, w), v(\cdot, w), (^{Hc}D_1^{\alpha_2} v)(\cdot, w), w) \|_\infty ds.$$

Also, the fact that f_2 is Carathéodory, we get

$$\|(N_2(u_n, v_n))(\cdot, w) - (N_2(u, v))(\cdot, w)\|_\infty \to 0 \text{ as } n \to \infty.$$

Hence $N(\cdot, \cdot, w)$ is continuous.

Step 2. $N(\cdot, \cdot, w)$ maps bounded sets into bounded sets in $\mathcal{C}$.

Let $R > 0$ and set

$$B_R := \{(\mu, v) \in \mathcal{C} : \|\mu\|_\infty \le R, \|v\|_\infty \le R\}.$$

For any $w \in \Omega$ and each $(u, v) \in B_R$ and $t \in I$, we have

$$\|(N_1(u, v))(t, w)\| \le \int_1^T |G_1(t, s)| \|g_1(s, w)\| ds,$$

where, $g_1(\cdot, w) \in C(I)$ for $w \in \Omega$, such that

$$g_1(t, w) = f_1\left(t, \int_1^T G_1(t, s) g_1(s, w) ds, \int_1^T G_2(t, s) g_2(s, w) ds, g_1(t, w), w\right).$$

From (7.7.3) we have

$$\|g_1(t, w)\| \le k_1(t, w) + l_1(t, w)\|u(s, w)\| + \bar{k}_1(t, w)\|v(s, w)\|$$
$$+ \bar{l}_1(t, w)\|g_1(t, w)\|.$$

This gives

$$\|g_1(\cdot, w)\| \le \frac{\|k_1(\cdot, w)\|_\infty}{1 - \|\bar{l}_1(\cdot, w)\|_\infty}$$
$$+ \frac{\|l_1(\cdot, w)\|_\infty}{1 - \|\bar{l}_1(\cdot, w)\|_\infty}\|u(\cdot, w)\|_\infty + \frac{\|\bar{k}_1(\cdot, w)\|_\infty}{1 - \|\bar{l}_1(\cdot, w)\|_\infty}\|v(\cdot, w)\|_\infty$$
$$\le \frac{\|k_1(\cdot, w)\|_\infty}{1 - \|\bar{l}_1(\cdot, w)\|_\infty}$$
$$+ \frac{R\|l_1(\cdot, w)\|_\infty}{1 - \|\bar{l}_1(\cdot, w)\|_\infty} + \frac{R\|\bar{k}_1(\cdot, w)\|_\infty}{1 - \|\bar{l}_1(\cdot, w)\|_\infty}.$$

Thus we get

$$\|(N_1(u, v))(t, w)\|$$
$$\le \int_1^T |G_1(t, s)| \|g_1(s, w)\| ds$$
$$\le G_1^*\left(\frac{\|k_1(\cdot, w)\|_\infty}{1 - \|\bar{l}_1(\cdot, w)\|_\infty} + \frac{R\|l_1(\cdot, w)\|_\infty}{1 - \|\bar{l}_1(\cdot, w)\|_\infty} + \frac{R\|\bar{k}_1(\cdot, w)\|_\infty}{1 - \|\bar{l}_1(\cdot, w)\|_\infty}\right)$$
$$:= \ell_1(w).$$

Hence

$$\|(N_1(u, v))(\cdot, w)\|_\infty \leq \ell_1(w).$$

Also, for any $w \in \Omega$ and each $(u, v) \in B_R$ and $t \in I$, we get

$$\|(N_2(u, v))(\cdot, w)\|_\infty \leq G_2^* \left(\frac{\|k_2(\cdot, w)\|_\infty}{1 - \|\bar{l}_2(\cdot, w)\|_\infty} + \frac{R\|l_2(\cdot, w)\|_\infty}{1 - \|\bar{l}_2(\cdot, w)\|_\infty} + \frac{R\|\bar{k}_2(\cdot, w)\|_\infty}{1 - \|\bar{l}_2(\cdot, w)\|_\infty} \right)$$

$$:= \ell_2(w).$$

Consequently

$$\|(N(u, v))(\cdot, w)\|_\mathcal{C} \leq \max\{\ell_1(w), \ell_2(w)\} := \ell(w).$$

Step 3. $N(\cdot, \cdot, w)$ maps bounded sets into equicontinuous sets in $\mathcal{C}$.
Let B_R be the ball defined in Step 2. For each $t_1, t_2 \in I$ with $t_1 \leq t_2$ and any $(u, v) \in B_R$ and $w \in \Omega$, we have

$$\|(N_1(u, v))(t_1, w) - (N_1(u, v))(t_2, w)\| \leq \int_1^T |G_1(t_1, s) - G_1(t_2, s)| \|g_1(s, w)\| ds,$$

where, $g_1(\cdot, w) \in C(I)$ for $w \in \Omega$, such that

$$g_1(t, w) = f_1 \left(t, \int_1^T G_1(t, s) g_1(s, w) ds, \int_1^T G_2(t, s) g_2(s, w) ds, g_1(t, w), w \right).$$

Thus

$$\|(N_1(u, v))(t_1, w) - (N_1(u, v))(t_2, w)\|$$
$$\leq \left(\frac{\|k_1(\cdot, w)\|_\infty}{1 - \|\bar{l}_1(\cdot, w)\|_\infty} + \frac{R\|l_1(\cdot, w)\|_\infty}{1 - \|\bar{l}_1(\cdot, w)\|_\infty} + \frac{R\|\bar{k}_1(\cdot, w)\|_\infty}{1 - \|\bar{l}_1(\cdot, w)\|_\infty} \right) \int_1^T |G_1(t_1, s) - G_1(t_2, s)| ds$$
$$\to 0 \text{ as } t_1 \to t_2.$$

Also we get

$$\|(N_2(u, v))(t_1, w) - (N_2(u, v))(t_2, w)\|$$
$$\leq \left(\frac{\|k_2(\cdot, w)\|_\infty}{1 - \|\bar{l}_2(\cdot, w)\|_\infty} + \frac{R\|l_2(\cdot, w)\|_\infty}{1 - \|\bar{l}_2(\cdot, w)\|_\infty} + \frac{R\|\bar{k}_2(\cdot, w)\|_\infty}{1 - \|\bar{l}_2(\cdot, w)\|_\infty} \right) \int_1^T |G_2(t_1, s) - G_1(t_2, s)| ds$$
$$\to 0 \text{ as } t_1 \to t_2.$$

As a consequence of Steps 1 to 3, with the Arzelà–Ascoli theorem, we conclude that $N(\cdot, \cdot, w)$ maps B_R into a precompact set in $\mathcal{C}$.

Step 4. The set $E(w)$ consisting of $(u(\cdot, w), v(\cdot, w)) \in \mathcal{C}$ such that $(u(\cdot, w), v(\cdot, w)) = \lambda(w)(N((u, v))(\cdot, w)$ for some measurable function $\lambda : \Omega \to (0, 1)$ is bounded in $\mathcal{C}$.

Let $(u(\cdot, w), v(\cdot, w)) \in C$ such that $(u(\cdot, w), v(\cdot, w)) = \lambda(w)(N((u, v))(\cdot, w)$. Then $u(\cdot, w) = \lambda(w)(N_1((u, v))(\cdot, w)$ and $v(\cdot, w) = \lambda(w)(N_2((u, v))(\cdot, w)$. Thus, for any $w \in \Omega$ and each $t \in I$, we have

$$\|u(t, w)\| \leq \int_1^T |G_1(t, s)| \|g_1(s, w)\| ds,$$

where, $g_1(\cdot, w) \in C(I)$ for $w \in \Omega$, such that

$$g_1(t, w) = f_1\left(t, \int_1^T G_1(t, s)g_1(s, w)ds, \int_1^T G_2(t, s)g_2(s, w)ds, g_1(t, w), w\right).$$

Hence

$$\begin{aligned}
\|u(t, w)\| \\
&\leq \int_1^T |G_1(t, s)| \left(\frac{\|k_1(\cdot, w)\|_\infty}{1 - \|\bar{\bar{l}}_1(\cdot, w)\|_\infty}\right. \\
&\quad + \frac{\|l_1(\cdot, w)\|_\infty}{1 - \|\bar{\bar{l}}_1(\cdot, w)\|_\infty} \|u(s, w\| + \left.\frac{\|\bar{k}_1(\cdot, w)\|_\infty}{1 - \|\bar{\bar{l}}_1(\cdot, w)\|_\infty} \|v(s, w\|\right) ds \\
&\leq \frac{G_1^* \|k_1(\cdot, w)\|_\infty}{1 - \|\bar{\bar{l}}_1(\cdot, w)\|_\infty} \\
&\quad + \int_1^T |G_1(t, s)| \left(\frac{\|l_1(\cdot, w)\|_\infty}{1 - \|\bar{\bar{l}}_1(\cdot, w)\|_\infty} \|u(s, w\| + \frac{\|\bar{k}_1(\cdot, w)\|_\infty}{1 - \|\bar{\bar{l}}_1(\cdot, w)\|_\infty} \|v(s, w\|\right) ds \\
&\leq \frac{G_1^* \|k_1(\cdot, w)\|_\infty}{1 - \|\bar{\bar{l}}_1(\cdot, w)\|_\infty} \\
&\quad + \frac{G_1^* \|l_1(\cdot, w)\|_\infty}{1 - \|\bar{\bar{l}}_1(\cdot, w)\|_\infty} \|u(\cdot, w)\|_\infty + \frac{G_1^* \|\bar{k}_1(\cdot, w)\|_\infty}{1 - \|\bar{\bar{l}}_1(\cdot, w)\|_\infty} \|v(\cdot, w)\|_\infty.
\end{aligned}$$

Also we get

$$\begin{aligned}
\|v(t, w)\| &\leq \frac{G_2^* \|k_2(\cdot, w)\|_\infty}{1 - \|\bar{\bar{l}}_2(\cdot, w)\|_\infty} \\
&\quad + \frac{G_2^* \|l_2(\cdot, w)\|_\infty}{1 - \|\bar{\bar{l}}_2(\cdot, w)\|_\infty} \|u(\cdot, w)\|_\infty + \frac{G_2^* \|\bar{k}_2(\cdot, w)\|_\infty}{1 - \|\bar{\bar{l}}_2(\cdot, w)\|_\infty} \|v(\cdot, w)\|_\infty.
\end{aligned}$$

Set

$$A(w) := \frac{G_1^* \|k_1(\cdot, w)\|_\infty}{1 - \|\bar{\bar{l}}_1(\cdot, w)\|_\infty} + \frac{G_2^* \|k_2(\cdot, w)\|_\infty}{1 - \|\bar{\bar{l}}_2(\cdot, w)\|_\infty}.$$

We obtain

$$\|u(\cdot, w)\|_\infty + \|v(\cdot, w)\|_\infty \leq A(w) + C(w)(\|u(\cdot, w)\|_\infty + \|v(\cdot, w)\|_\infty).$$

It follows that

$$\|u(\cdot, w)\|_\infty + \|v(\cdot, w)\|_\infty \le \frac{A(w)}{1 - C(w)}$$

$$:= L(w).$$

Hence we get

$$\|(u(\cdot, w), v(\cdot, w))\|_\mathcal{C} \le L(w).$$

This shows that the set $E(w)$ is bounded.

As a consequence of steps 1 to 4 together with Theorem 2.82, we can conclude that N has at least one random fixed point in B_R, which is a random solution of the coupled system (8.15)–(8.16). □

8.3.3 An example

Let $\Omega = (-\infty, 0)$ be equipped with the usual σ-algebra consisting of Lebesgue measurable subsets of $(-\infty, 0)$. Consider the following random coupled system of Caputo–Hadamard fractional differential equations

$$\begin{cases} (^{Hc}D_1^{\frac{3}{2}}u)(t, w) = f\left(t, u(t, w), v(t, w), (^{Hc}D_1^{\frac{3}{2}}u)(t, w), w\right); \\ (^{Hc}D_1^{\frac{3}{2}}v)(t, w) = g\left(t, u(t, w), v(t, w), (^{Hc}D_1^{\frac{3}{2}}v)(t, w), w\right); \end{cases} \tag{8.19}$$

where $w \in \Omega, t \in [1, e]$, with the four-point boundary conditions

$$\begin{cases} u(1, w) - u'(1, w) = u(2, w), \\ 2u(T, w) + u'(T, w) = 2u(\frac{3}{2}, w), \\ 3v(1, w) - v'(1, w) = 3v(\frac{5}{4}, w), \\ v(T, w) + 2v'(T, w) = v(2, w), \end{cases} \quad ; \quad w \in \Omega, \tag{8.20}$$

where

$$f(t, u, v, y, w) = \frac{ct^{\frac{-1}{4}}w^2 u(t) \sin t}{64(1 + w^2 + \sqrt{t})(1 + |u| + |v| + |y|)}; \quad t \in [1, e],$$

$$g(t, u, v, y, w) = \frac{cw^2 v(t) \cos t}{64(1 + w^2 + |u| + |v| + |y|)}; \quad w \in \Omega, \ t \in [1, e],$$

and $c < \frac{1}{\max\{G_1^*, G_2^*\}}$. The hypothesis (8.8.2) is satisfied with

$$p_2(t, w) = q_1(t, w) = r_1(t, w) = r_2(t, w) = 0,$$

$$p_1(t, w) = \frac{cw^2 \sin t}{64(1 + w^2)}, \quad q_2(t, w) = \frac{cw^2 \cos t}{64(1 + w^2)}.$$

Also, if for every $w \in \Omega$, the matrix

$$\frac{cw^2}{64(1+w^2)} \begin{pmatrix} G_1^* & 0 \\ 0 & G_2^* \end{pmatrix},$$

converges to 0. Hence, Theorem 8.9 implies that the system (8.19)–(8.20) has a unique random solution defined on $[1, e]$.

8.4 Notes and remarks

This chapter contains the studies from Abbas et al. [4,6,35]. The monographs [385,438] and the papers [4,53,232,234] provide more important conclusions and analysis about the subject.

Coupled Hilfer–Hadamard fractional differential systems in generalized Banach spaces

The outcome of our study in this chapter can be considered as a partial continuation of the problems raised recently in the monographs of Abbas et al. [48,69,70], Samko et al. [385], Kilbas et al. [281], Zhou et al. [440], and the papers [67,169,239,240,266,277,395,412].

9.1 Coupled Hilfer and Hadamard fractional differential systems

9.1.1 Introduction

In this section we discuss the existence and uniqueness of solutions for the following coupled system of Hilfer fractional differential equations

$$
\begin{cases}
(D_0^{\alpha_1,\beta_1} u)(t) = f_1(t, u(t), v(t)) \\
(D_0^{\alpha_2,\beta_2} v)(t) = f_2(t, u(t), v(t))
\end{cases} \quad ; \ t \in I := [0, T],
\tag{9.1}
$$

with the following initial conditions

$$
\begin{cases}
(I_0^{1-\gamma_1} u)(0) = \phi_1 \\
(I_0^{1-\gamma_2} v)(0) = \phi_2,
\end{cases}
\tag{9.2}
$$

where $T > 0$, $\alpha_i \in (0, 1)$, $\beta_i \in [0, 1]$, $\gamma_i = \alpha_i + \beta_i - \alpha_i \beta_i$, $\phi_i \in \mathbb{R}^m$, $f_i : I \times \mathbb{R}^m \times \mathbb{R}^m \to \mathbb{R}^m$; $i = 1, 2$, are given functions, $I_0^{1-\gamma_i}$ is the left-sided mixed Riemann–Liouville integral of order $1 - \gamma_i$, $\mathbb{R}^m$; $m \in \mathbb{N}^*$ is the Euclidean Banach space with a suitable norm $\| \cdot \|$, and $D_0^{\alpha_i,\beta_i}$ is the generalized Riemann–Liouville derivative (Hilfer) operator of order α_i and type $\beta_i : i = 1, 2$.

Next, we consider the following coupled system of Hilfer–Hadamard fractional differential equations

$$
\begin{cases}
({}^H D_1^{\alpha_1,\beta_1} u)(t) = g_1(t, u(t), v(t)) \\
({}^H D_1^{\alpha_2,\beta_2} v)(t) = g_2(t, u(t,), v(t))
\end{cases} \quad ; \ t \in [1, T],
\tag{9.3}
$$

Fractional Difference, Differential Equations, and Inclusions. https://doi.org/10.1016/B978-0-44-323601-3.00016-2
Copyright © 2024 Elsevier Inc. All rights reserved, including those for text and data mining, AI training, and similar technologies.

with the following initial conditions

$$\begin{cases} (^H I_1^{1-\gamma_1} u)(1) = \psi_1 \\ (^H I_1^{1-\gamma_2} v)(1) = \psi_2, \end{cases} \tag{9.4}$$

where $T > 1$, $\alpha_i \in (0, 1)$, $\beta_i \in [0, 1]$, $\gamma_i = \alpha_i + \beta_i - \alpha_i \beta_i$, $\psi_i \in \mathbb{R}^m$, $g_i : [1, T] \times \mathbb{R}^m \times \mathbb{R}^m \to \mathbb{R}^m$; $i = 1, 2$ are given functions, $^H I_1^{1-\gamma_i}$ is the left-sided mixed Hadamard integral of order $1 - \gamma_i$, and $^H D_1^{\alpha_i, \beta_i}$ is the Hilfer–Hadamard fractional derivative of order α_i and type β_i; $i = 1, 2$.

9.1.2 Coupled Hilfer fractional differential systems

In this section we are concerned with the existence and uniqueness results of the system (9.1)–(9.2).

Definition 9.1. By a solution of the problem (9.1)–(9.2) we mean a coupled continuous functions $(u, v) \in C_{\gamma_1} \times C_{\gamma_2}$ those satisfy Eq. (9.1) on I, and the conditions $(I_0^{1-\gamma_1} u)(0^+) = \phi_1$, and $(I_0^{1-\gamma_2} v)(0^+) = \phi_2$.

The following hypotheses will be used in the sequel.

(9.1.1) There exist continuous functions $p_i, q_i : I \to (0, \infty)$; $i = 1, 2$ such that

$$\|f_i(t, u_1, v_1) - f_i(t, u_2, v_2)\| \le p_i(t)\|u_1 - u_2\| + q_i(t)\|v_1 - v_2\|;$$

for a.e. $t \in I$, and each $u_i, v_i \in \mathbb{R}^m$, $i = 1, 2$.

(9.1.2) There exist continuous functions $a_i, b_i : I \to (0, \infty)$; $i = 1, 2$ such that

$$\|f_i(t, u, v)\| \le a_i(t)\|u\| + b_i(t)\|v\|; \text{ for a.e. } t \in I, \text{ and each } u, v \in \mathbb{R}^m.$$

First, we prove an existence and uniqueness result for the coupled system (9.1)–(9.2) by using Banach fixed-point theorem type in generalized Banach spaces. Set

$$p_i^* := \sup_{t \in I} p(t), \quad q_i^* := \sup_{t \in I} q(t); \ i = 1, 2.$$

Theorem 9.2. *Assume that the hypothesis* (9.1.1) *holds. If the matrix*

$$M := \begin{pmatrix} \frac{T^{\alpha_1}}{\Gamma(1+\alpha_1)} p_1^* & \frac{T^{\alpha_1}}{\Gamma(1+\alpha_1)} q_1^* \\ \frac{T^{\alpha_2}}{\Gamma(1+\alpha_2)} p_2^* & \frac{T^{\alpha_2}}{\Gamma(1+\alpha_2)} q_2^* \end{pmatrix}$$

converges to 0, *then the coupled system* (9.1)–(9.2) *has a unique solution.*

Proof. Define the operators $N_i : C \to C_{\gamma_i}$; $i = 1, 2$ by

$$(N_1(u, v))(t) = \frac{\phi_1}{\Gamma(\gamma_1)} t^{\gamma_1 - 1} + \int_0^t (t - s)^{\alpha_1 - 1} \frac{f_1(s, u(s), v(s))}{\Gamma(\alpha_1)} ds, \tag{9.5}$$

and

$$(N_2(u, v))(t) = \frac{\phi_2}{\Gamma(\gamma_2)} t^{\gamma_2-1} + \int_0^t (t - s)^{\alpha_2-1} \frac{f_2(s, u(s), v(s))}{\Gamma(\alpha_2)} ds. \tag{9.6}$$

Consider the operator $N : \mathcal{C} \to \mathcal{C}$ defined by

$$(N(u, v))(t) = ((N_1(u, v))(t), (N_2(u, v))(t)). \tag{9.7}$$

Clearly, the fixed points of the operator N are solutions of the system (9.1)–(9.2).

For any $i \in \{1, 2\}$ and each $(u_1, v_1), (u_2, v_2) \in \mathcal{C}$ and $t \in I$, we have

$$\|t^{1-\gamma_1}(N_1(u_1, v_1))(t) - t^{1-\gamma_1}(N_1(u_2, v_2))(t)\|$$

$$\leq \frac{t^{1-\gamma_1}}{\Gamma(\alpha_1)} \int_0^t (t - s)^{\alpha_1-1} \|f_1(s, u_1(s), v_1(s)) - f_1(s, u_2(s), v_2(s))\| ds$$

$$\leq \frac{t^{1-\gamma_1}}{\Gamma(\alpha_1)} \int_0^t (t - s)^{\alpha_1-1} (p_1(t)\|u_1(s) - v_1(s)\|$$
$$+ q_1(t)\|u_2(s) - v_2(s)\|) ds$$

$$\leq \frac{1}{\Gamma(\alpha_1)} \int_0^t (t - s)^{\alpha_1-1} (p_1(t) s^{1-\gamma_1} \|u_1(s) - v_1(s)\|$$
$$+ q_1(t) s^{1-\gamma_1} \|u_2(s) - v_2(s)\|) ds$$

$$\leq \frac{p_1(t)\|u_1 - v_1\|_{C_{\gamma_1}} + q_1(t)\|u_2 - v_2\|_{C_{\gamma_2}}}{\Gamma(\alpha_1)}$$

$$\times \int_0^t (t - s)^{\alpha_1-1} ds$$

$$\leq \frac{T^{\alpha_1}}{\Gamma(1 + \alpha_1)} (p_1(t)\|u_1 - v_1\|_{C_{\gamma_1}} + q_1(t)\|u_2 - v_2\|_{C_{\gamma_2}}).$$

Then

$$\|N_1(u_1, v_1) - N_1(u_2, v_2)\|_{C_{\gamma_1}}$$

$$\leq \frac{T^{\alpha_1}}{\Gamma(1 + \alpha_1)} (p_1^*\|u_1 - v_1\|_{C_{\gamma_1}} + q_1(^*\|u_2 - v_2\|_{C_{\gamma_2}}).$$

Also, for each $(u_1, v_1), (u_2, v_2) \in \mathcal{C}$ and $t \in I$, we get

$$\|N_2(u_1, v_1) - N_2(u_2, v_2)\|_{C_{\gamma_2}}$$

$$\leq \frac{T^{\alpha_2}}{\Gamma(1 + \alpha_2)} (p_2^*\|u_1 - v_1\|_{C_{\gamma_1}} + q_2^*\|u_2 - v_2\|_{C_{\gamma_2}}).$$

Thus

$$d(N(u_1, v_1), N(u_2, v_2)) \leq M d((u_1, v_1), (u_2, v_2)),$$

where

$$d((u_1, v_1), (u_2, v_2)) = \begin{pmatrix} \|u_1 - v_1\|_{C_{\gamma_1}} \\ \|u_2 - v_2\|_{C_{\gamma_2}} \end{pmatrix}.$$

Since the matrix M converges to zero, then Theorem 2.90 implies that system (9.1)–(9.2) has a unique solution. □

Now, we prove an existence result for the coupled system (9.1)–(9.2) by using the nonlinear alternative of Leray–Schauder type in generalized Banach space.

Theorem 9.3. *Assume that the hypothesis (9.1.2) holds. Then the coupled system (9.1)–(9.2) has at least one solution.*

Proof. We show that the operator $N : C \to C$ defined in (9.7) satisfies all conditions of Theorem 2.77. The proof will be given in four steps.

Step 1. *N is continuous.*
Let $(u_n, v_n)_n$ be a sequence such that $(u_n, v_n) \to (u, v) \in C$ as $n \to \infty$. For any $i \in \{1, 2\}$ and each $t \in I$, we have

$$\|t^{1-\gamma_i}(N_i(u_n, v_n))(t) - t^{1-\gamma_i}(N_i(u, v))(t)\|$$

$$\leq \frac{t^{1-\gamma_i}}{\Gamma(\alpha_i)} \int_0^t (t-s)^{\alpha_i-1} \|f_i(s, u_n(s), v_n(s)) - f_i(s, u(s), v(s))\| ds$$

$$\leq \frac{1}{\Gamma(\alpha_i)} \int_0^t (t-s)^{\alpha_i-1} s^{1-\gamma_i} \|f_i(s, u_n(s), v_n(s)) - f_i(s, u(s), v(s))\| ds$$

$$\leq \frac{T^{\alpha_i}}{\Gamma(1+\alpha_i)} \|f_i(\cdot, u_n(\cdot), v_n(\cdot)) - f_i(\cdot, u(\cdot), v(\cdot))\|_{C_{\gamma_1}}.$$

Since f_i is continuous, then by the Lebesgue dominated convergence theorem, we get

$$\|N_i(u_n, v_n) - N_i(u, v)\|_{C_{\gamma_1}} \to 0 \text{ as } n \to \infty.$$

Hence N is continuous.

Step 2. *N maps bounded sets into bounded sets in C.*
Set

$$a_i^* := \sup_{t \in I} a(t), \quad b_i^* := \sup_{t \in I} b(t) : \ i = 1, 2.$$

Let $R > 0$ and set

$$B_R := \{(u, v) \in C : \|u\|_{C_{\gamma_1}} \leq R, \|v\|_{C_{\gamma_2}} \leq R\}.$$

For each $(u, v) \in B_R$ and $t \in I$, we have

$$\|t^{1-\gamma_1}(N_1(u, v))(t)\|$$

$$\leq \frac{\|\phi_1\|}{\Gamma(\gamma_1)} + \frac{t^{1-\gamma_1}}{\Gamma(\alpha_1)} \int_0^t (t-s)^{\alpha_1-1} \|f_1(s, u(s), v(s))\| ds$$

$$\leq \frac{\|\phi_1\|}{\Gamma(\gamma_1)}$$

$$+ \frac{1}{\Gamma(\alpha_1)} \int_0^t (t-s)^{\alpha_1-1} s^{1-\gamma_1} (a_1(s)\|u(s)\| + b_1(s)\|v(s)\|) ds$$

$$\leq \frac{\|\phi_1\|}{\Gamma(\gamma_1)} + \frac{R}{\Gamma(\alpha_1)} \int_0^t (t-s)^{\alpha_1-1} s^{1-\gamma_1} (a_1(s) + b_1(s)) ds$$

$$\leq \frac{\|\phi_1\|}{\Gamma(\gamma_1)} + \frac{(a_1^* + b_1^* T^{\alpha_1}}{\Gamma(1+\alpha_1)}$$

$$:= \ell_1.$$

Thus

$$\|N_1(u,v)\|_{C_{\gamma_1}} \leq \ell_1.$$

Also, for each $(u,v) \in B_R$ and $t \in I$, we get

$$\|N_2(u,v)\|_{C_{\gamma_2}} \leq \frac{\|\phi_2\|}{\Gamma(\gamma_2)} + \frac{(a_2^* + b_2^* T^{\alpha_2}}{\Gamma(1+\alpha)}$$

$$:= \ell_2.$$

Hence

$$\|N(u,v)\|_C \leq \max\{\ell_1, \ell_2\} := \ell.$$

Step 3. *N maps bounded sets into equicontinuous sets in* C.
Let B_R be the ball defined in Step 2. For each $t_1, t_2 \in I$ with $t_1 \leq t_2$ and $(u,v) \in B_R$, we have

$$\|t_1^{1-\gamma_1}(N_1(u,v))(t_1) - t_2^{1-\gamma_1}(N_1(u,v))(t_2)\|$$

$$\leq \frac{t_2^{1-\gamma_1}}{\Gamma(\alpha_1)} \int_{t-1}^{t_2} (t_2-s)^{\alpha_1-1} \|f_1(s,u(s),v(s))\| ds$$

$$\leq \frac{T^{\alpha_1}}{\Gamma(1+\alpha_1)} (t_2-t_1)^{\alpha_1} (a_1^* \|u\|_{C_{\gamma_1}} + b_1(^* \|v\|_{C_{\gamma_2}})$$

$$\leq \frac{RT^{\alpha_1}(a_1^* + b_1^*)}{\Gamma(1+\alpha_1)} (t_2-t_1)^{\alpha_1}$$

$$\to 0 \text{ as } t_1 \to t_2.$$

Also we get

$$\|t_1^{1-\gamma_2}(N_2(u,v))(t_1) - t_2^{1-\gamma_2}(N_2(u,v))(t_2)\|$$

$$\leq \frac{RT^{\alpha_1 2}(a_2^* + b_2^*))}{\Gamma(1+\alpha_2)} (t_2-t_1)^{\alpha_2}$$

$$\to 0 \text{ as } t_1 \to t_2.$$

As a consequence of Steps 1 to 3, with the Arzelà–Ascoli theorem, we conclude that N maps B_R into a precompact set in C.

Step 4. The set E consisting of $(u, v) \in \mathcal{C}$ such that $(u, v) = \lambda N(u, v)$ for some $\lambda \in (0, 1)$ is bounded in $\mathcal{C}$.

Let $(u, v) \in \mathcal{C}$ such that $(u, v) = \lambda N(u, v)$. Then $u = \lambda N_1(u, v)$ and $v = \lambda N_2(u, v)$. Thus, for each $t \in I$, we have

$$\|t^{1-\gamma_1}u(t)\| \leq \frac{\|\phi_1\|}{\Gamma(\gamma_1)} + \frac{t^{1-\gamma_1}}{\Gamma(\alpha_1)} \int_0^t (t-s)^{\alpha_1-1} \|f_1(s, u(s), v(s))\| ds$$

$$\leq \frac{\|\phi_1\|}{\Gamma(\gamma_1)}$$

$$+ \frac{1}{\Gamma(\alpha_1)} \int_0^t (t-s)^{\alpha_1-1} s^{1-\gamma_1} (a_1^* \|u(s\| + b_1^* \|v(s)\|) ds.$$

Also we get

$$\|t^{1-\gamma_2}v(t)\| \leq \frac{\|\phi_2\|}{\Gamma(\gamma_2)}$$

$$+ \frac{1}{\Gamma(\alpha_2)} \int_0^t (t-s)^{\alpha_2-1} s^{1-\gamma_2} (a_2^* \|u(s)\| + b_2^* \|v(s)\|) ds.$$

Hence we obtain

$$\|t^{1-\gamma_1}u(t)\| + \|t^{1-\gamma_2}v(t)\| \leq a + bc \int_0^t (t-s)^{\alpha-1} (\|s^{1-\gamma_1}u(s)\| + \|s^{1-\gamma_2}v(s)\|) ds,$$

where

$$a := \frac{\|\phi_1\|}{\Gamma(\gamma_1)} + \frac{\|\phi_2\|}{\Gamma(\gamma_2)}, \quad b := \frac{1}{\Gamma(\alpha_1)} + \frac{1}{\Gamma(\alpha_2)},$$

$$c := \max\{a_1^* + a_2^*, b_1^* + b_2^*\}, \quad \alpha := \max\{\alpha_1, \alpha_2\}.$$

Lemma 2.100 implies that there exists $\rho := \rho(\alpha) > 0$ such that

$$\|t^{1-\gamma_1}u(t)\| + \|t^{1-\gamma_2}v(t)\| \leq a + abc\rho \int_0^t (t-s)^{\alpha-1} ds$$

$$\leq \frac{a + abc\rho T^\alpha}{\alpha}$$

$$= L.$$

This gives

$$\|u\|_{C_{\gamma_1}} + \|v\|_{C_{\gamma_2}} \leq L.$$

Hence

$$\|(u, v)\|_{\mathcal{C}} \leq L.$$

This shows that the set E is bounded.

As a consequence of steps 1 to 4 together with Theorem 2.77, we can conclude that N has at least one fixed point in B_R, which is a solution of the system (9.1)–(9.2). $\square$

9.1.3 Coupled Hilfer–Hadamard fractional differential systems

Now, we are concerned with the coupled system (9.3)–(9.4).

Set $C := C([1, T])$ and denote the weighted space of continuous functions defined by

$$C_{\gamma,\ln}([1, T]) = \{w(t) : (\ln t)^{1-\gamma} w(t) \in C\},$$

with the norm

$$\|w\|_{C_{\gamma,\ln}} := \sup_{t \in [1,T]} |(\ln t)^{1-r} w(t)|.$$

Also, by $C_{\gamma_1,\gamma_2,\ln}([1, T]) := C_{\gamma_1,\ln}([1, T]) \times C_{\gamma_2,\ln}([1, T])$ we denote the product weighted space with the norm

$$\|(u, v)\|_{C_{\gamma_1,\gamma_2,\ln}([1,T])} = \|u\|_{C_{\gamma_1,\ln}} + \|v\|_{C_{\gamma_2,\ln}}.$$

Let us recall some definitions and properties of Hadamard fractional integration and differentiation. We refer to [281] for a more detailed analysis.

Definition 9.4. [281] (Hadamard fractional integral). The Hadamard fractional integral of order $q > 0$ for a function $g \in L^1([1, T])$, is defined as

$$(^H I_1^q g)(x) = \frac{1}{\Gamma(q)} \int_1^x \left(\ln \frac{x}{s}\right)^{q-1} \frac{g(s)}{s} ds,$$

provided the integral exists.

Example 9.5. Let $0 < q < 1$. Let $g(x) = \ln x$, $x \in [1, e]$. Then

$$(^H I_1^q g)(x) = \frac{1}{\Gamma(2 + q)} (\ln x)^{1+q}; \text{ for a.e. } x \in [1, e].$$

Set

$$\delta = x \frac{d}{dx}, \ q > 0, \ n = [q] + 1,$$

and

$$AC_\delta^n := \{u : [1, T] \to E : \delta^{n-1}[u(x)] \in AC(I)\}.$$

Analogous to the Riemann–Liouville fractional derivative, the Hadamard fractional derivative is defined in terms of the Hadamard fractional integral in the following way:

Definition 9.6. [281] (Hadamard fractional derivative). The Hadamard fractional derivative of order $q > 0$ applied to the function $w \in AC_\delta^n$ is defined as

$$(^H D_1^q w)(x) = \delta^n (^H I_1^{n-q} w)(x).$$

In particular, if $q \in (0, 1]$, then

$$(^H D_1^q w)(x) = \delta(^H I_1^{1-q} w)(x).$$

Example 9.7. Let $0 < q < 1$. Let $w(x) = \ln x$, $x \in [1, e]$. Then

$$({}^{H}D_1^q w)(x) = \frac{1}{\Gamma(2-q)}(\ln x)^{1-q}, \text{ for a.e. } x \in [1, e].$$

It has been proved (see, e.g., Kilbas [280, Theorem 4.8]) that in the space $L^1(I)$, the Hadamard fractional derivative is the left-inverse operator to the Hadamard fractional integral, i.e.,

$$({}^{H}D_1^q)({}^{H}I_1^q w)(x) = w(x).$$

From Theorem 2.3 of [281], we have

$$({}^{H}I_1^q)({}^{H}D_1^q w)(x) = w(x) - \frac{({}^{H}I_1^{1-q}w)(1)}{\Gamma(q)}(\ln x)^{q-1}.$$

Analogous to the Hadamard fractional calculus, the Caputo–Hadamard fractional derivative is defined in the following way:

Definition 9.8. (Caputo–Hadamard fractional derivative) The Caputo–Hadamard fractional derivative of order $q > 0$ applied to the function $w \in AC_\delta^n$ is defined as

$$({}^{Hc}D_1^q w)(x) = ({}^{H}I_1^{n-q} \delta^n w)(x).$$

In particular, if $q \in (0, 1]$, then

$$({}^{Hc}D_1^q w)(x) = ({}^{H}I_1^{1-q} \delta w)(x).$$

From the Hadamard fractional integral, the Hilfer–Hadamard fractional derivative (introduced for the first time in [337]) is defined in the following way:

Definition 9.9. (Hilfer–Hadamard fractional derivative). Let $\alpha \in (0, 1)$, $\beta \in [0, 1]$, $\gamma = \alpha + \beta - \alpha\beta$, $w \in L^1(I)$, and ${}^{H}I_1^{(1-\alpha)(1-\beta)}w \in AC(I)$. The Hilfer–Hadamard fractional derivative of order α and type β applied to the function w is defined as

$$\begin{aligned}
({}^{H}D_1^{\alpha,\beta} w)(t) &= \left({}^{H}I_1^{\beta(1-\alpha)}({}^{H}D_1^\gamma w)\right)(t) \\
&= \left({}^{H}I_1^{\beta(1-\alpha)} \delta({}^{H}I_1^{1-\gamma} w)\right)(t); \text{ for a.e. } t \in [1, T].
\end{aligned} \tag{9.8}$$

This new fractional derivative (9.8) may be viewed as interpolating the Hadamard fractional derivative and the Caputo–Hadamard fractional derivative. Indeed, for $\beta = 0$, this derivative reduces to the Hadamard fractional derivative and when $\beta = 1$, we recover the Caputo–Hadamard fractional derivative.

$$^{H}D_1^{\alpha,0} = {}^{H}D_1^\alpha, \text{ and } {}^{H}D_1^{\alpha,1} = {}^{Hc}D_1^\alpha.$$

From Theorem 21 in [338], we concluded the following lemma.

Lemma 9.10. *Let $g : [1, T] \times \mathbb{R}^m \to \mathbb{R}^m$ be such that $g(\cdot, u(\cdot)) \in C_{\gamma, \ln}([1, T])$ for any $u \in C_{\gamma, \ln}([1, T])$. Then problem (9.3) is equivalent to the following Volterra integral equation*

$$u(t) = \frac{\phi_0}{\Gamma(\gamma)} (\ln t)^{\gamma - 1} + ({}^H I_1^\alpha g(\cdot, u(\cdot)))(t).$$

Definition 9.11. By a solution of the coupled system (9.3)–(9.4) we mean a coupled continuous functions $(u, v) \in C_{\gamma_1, \ln} \times C_{\gamma_2, \ln}$ those satisfy the conditions (9.4) and Eqs. (9.3) on $[1, T]$.

Now we give (without proof) similar existence and uniqueness results for the system (9.3)–(9.4). Let us introduce the following hypotheses:

(9.11.1) There exist continuous functions $p_i, q_i : [1, T] \to (0, \infty)$; $i = 1, 2$ such that

$$\|g_i(t, u_1, v_1) - g_i(t, u_2, v_2)\| \leq p_i(t)\|u_1 - u_2\| + q_i(t)\|v_1 - v_2\|;$$

for a.e. $t \in [1, T]$, and each $u_i, v_i \in \mathbb{R}^m$, $i = 1, 2$.
(9.11.2) There exist continuous functions $a_i, b_i : [1, T] \to (0, \infty)$; $i = 1, 2$ such that

$$\|g_i(t, u, v)\| \leq a_i(t)\|u\| + b_i(t)\|v\|; \text{ for a.e. } t \in [1, T], \text{ and each } u, v \in \mathbb{R}^m.$$

Theorem 9.12. *Assume that the hypothesis (9.11.1) holds. If the matrix*

$$\begin{pmatrix} \frac{(\ln T)^{\alpha_1}}{\Gamma(1+\alpha_1)} p_1^* & \frac{(\ln T)^{\alpha_1}}{\Gamma(1+\alpha_1)} q_1^* \\ \frac{(\ln T)^{\alpha_2}}{\Gamma(1+\alpha_2)} p_2^* & \frac{(\ln T)^{\alpha_2}}{\Gamma(1+\alpha_2)} q_2^* \end{pmatrix}$$

converges to 0, then the coupled system (9.3)–(9.4) has a unique solution defined on $[1, T]$.

Theorem 9.13. *Assume that the hypothesis (9.11.2) holds. Then the coupled system (9.3)–(9.4) has at least one solution defined on $[1, T]$.*

9.1.4 An example

Consider the following coupled system of Hilfer fractional differential equations

$$\begin{cases} (D_0^{\frac{1}{2}, \frac{1}{2}} u)(t) = f(t, u(t), v(t)); \\ (D_0^{\frac{1}{2}, \frac{1}{2}} v)(t) = g(t, u(t), v(t)); \\ (I_0^{\frac{1}{4}} u)(0) = 1, \\ (I_0^{\frac{1}{4}} v_n)(0) = 0, \end{cases} : t \in [0, 1], \tag{9.9}$$

where

$$f(t, u, v) = \frac{t^{\frac{-1}{4}} (u(t) + v(t)) \sin t}{64(1 + \sqrt{t})(1 + |u| + |v|)}; \ t \in [0, 1],$$

$$g(t, u, v) = \frac{(u(t) + v(t)) \cos t}{64(1 + |u| + |v|)}; \ t \in [0, 1].$$

Set $\alpha_i = \beta_i = \frac{1}{2}$; $i = 1, 2$, then $\gamma_i = \frac{3}{4}$; $i = 1, 2$. The hypothesis (9.11.1) is satisfied with

$$p_1(t) = p_2(t) = q_1(t) = q_2(t) = \frac{1}{64}.$$

Also the matrix

$$\frac{1}{64\sqrt{\pi}} \begin{pmatrix} 1 & 1 \\ 1 & 1 \end{pmatrix}$$

converges to 0. Hence Theorem 9.2 implies that the system (9.9) has a unique solution defined on $[0, 1]$.

9.2 Coupled Hilfer and Hadamard random fractional differential systems with finite delay

9.2.1 Introduction

In this section we discuss the existence of random solutions for the following system of Hilfer fractional differential equations

$$\begin{cases} (u(t, w), v(t, w)) = (\phi_1(t, w), \phi_2(t, w)); \ t \in [-h, 0], \\ (D_0^{\alpha_1, \beta_1} u)(t, w) = f_1(t, u_t(w), v_t(w), w); \ t \in I, \\ (D_0^{\alpha_2, \beta_2} v)(t, w) = f_2(t, u_t(w), v_t(w), w); \ t \in I, \\ ((I_0^{1-\gamma_1} u)(0, w), (I_0^{1-\gamma_2} v)(0, w)) = (\Phi_1(w), \Phi_2(w)) \end{cases} \quad ; \ w \in \Omega, \quad (9.10)$$

where $T > 0$, $I := [0, T]$, $\alpha_i \in (0, 1)$, $\beta_i \in [0, 1]$, $\gamma_i = \alpha_i + \beta_i - \alpha_i \beta_i$, $\Phi_i : \Omega \to \mathbb{R}^m$, $f_i : I \times C \times C \times \Omega \to \mathbb{R}^m$; $i = 1, 2$, are given functions, $C := C[-h, 0]$, $h > 0$, $\phi_i(w) \in C$ such that $((I_0^{1-\gamma_i} \phi_i)(0, w) = \Phi_i(w)$; $i = 1, 2$, $\mathbb{R}^m$; $m \in \mathbb{N}^*$ is the Euclidean Banach space with a suitable norm $\| \cdot \|$. Furthermore, $u_t : \Omega \times [-h, 0] \to \mathbb{R}^m$ such that $(u_t(w))(s) := u_t(s, w) = u(t + s, w)$; $s \in [-h, 0]$, $w \in \Omega$, $I_0^{1-\gamma_i}$ is the left-sided mixed Riemann–Liouville integral of order $1 - \gamma_i$, and $D_0^{\alpha_i, \beta_i}$ is the generalized Riemann–Liouville derivative (Hilfer) operator of order α_i and type $\beta_i : i = 1, 2$.

Next, we consider the following coupled system of Hilfer–Hadamard fractional differential equations

$$\begin{cases} (u(t, w), v(t, w)) = (\psi_1(t, w), \psi_2(t, w)); \ t \in [1 - h, 1], \\ (^H D_1^{\alpha_1, \beta_1} u)(t, w) = g_1(t, u_t(w), v_t(w), w); \ t \in [1, T], \\ (^H D_1^{\alpha_2, \beta_2} v)(t, w) = g_2(t, u_t(w), v_t(w), w); \ t \in [1, T], \\ ((^H I_0^{1-\gamma_1} u)(1, w), (^H I_0^{1-\gamma_2} v)(1, w)) = (\Psi_1(w), \Psi_2(w)) \end{cases} \quad ; \ w \in \Omega, \quad (9.11)$$

where $T > 1$, $\alpha_i \in (0, 1)$, $\beta_i \in [0, 1]$, $\gamma_i = \alpha_i + \beta_i - \alpha_i \beta_i$, $\Psi_i : \Omega \to \mathbb{R}^m$, $g_i : [1, T] \times C \times C \times \Omega \to \mathbb{R}^m$; $i = 1, 2$ are given functions, $\psi_i \in C[1 - h, 1]$, such that $((^H I_0^{1-\gamma_i} \psi_i)(1, w) = \Psi_i(w)$; $i = 1, 2$, $^H I_1^{1-\gamma_i}$ is the left-sided mixed Hadamard integral of order $1 - \gamma_i$, and $^H D_1^{\alpha_i, \beta_i}$ is the Hilfer–Hadamard fractional derivative of order α_i and type β_i; $i = 1, 2$.

9.2.2 Coupled Hilfer random fractional differential systems

In this section we are concerned with the existence and uniqueness of random solutions of the problem (9.10).

Definition 9.14. By a solution of the problem (9.10) we mean a coupled measurable functions $(u, v) \in C_{\gamma_1} \times C_{\gamma_2}$ those satisfy Eqs. (9.10) on I.

The following hypotheses will be used in the sequel.

(9.14.1) The functions f_i; $i = 1, 2$ are random Carathéodory.
(9.14.2) There exist measurable functions $p_i, q_i : \Omega \to (0, \infty)$; $i = 1, 2$ such that

$$\|f_i(t, u_1, v_1, w) - f_i(t, u_2, v_2, w)\| \le p_i(w)\|u_1 - u_2\|_C + q_i(w)\|v_1 - v_2\|_C;$$

for a.e. $t \in I$, $w \in \Omega$, and each $u_i, v_i \in C$, $i = 1, 2$.
(9.14.3) There exist measurable functions $a_i, b_i : \Omega \to (0, \infty)$; $i = 1, 2$ such that

$$\|f_i(t, u, v, w)\| \le a_i(w)\|u\|_C + b_i(w)\|v\|_C; \text{ for a.e. } t \in I, \ w \in \Omega, \ u, v \in C.$$

First, we prove an existence and uniqueness result for the problem (9.10) by using Banach random fixed-point theorem type in generalized Banach spaces.

Theorem 9.15. *Assume that the hypotheses (9.14.1) and (9.14.2) hold. If for every $w \in \Omega$, the matrix*

$$M(w) := \begin{pmatrix} \frac{T^{\alpha_1}}{\Gamma(1+\alpha_1)} p_1(w) & \frac{T^{\alpha_1}}{\Gamma(1+\alpha_1)} q_1(w) \\ \frac{T^{\alpha_2}}{\Gamma(1+\alpha_2)} p_2(w) & \frac{T^{\alpha_2}}{\Gamma(1+\alpha_2)} q_2(w) \end{pmatrix}$$

converges to 0, then the system (9.10) has a unique random solution.

Proof. Define the operators $N_i : C \times \Omega \to C_{\gamma_i}$; $i = 1, 2$ by

$$(N_i(u, v))(t, w) = \begin{cases} \phi_i(t, w); \ t \in [-h, 0], \\ \frac{\Phi_i(w)}{\Gamma(\gamma_i)} t^{\gamma_i - 1} + \int_0^t (t - s)^{\alpha_i - 1} \frac{f_i(s, u_s(w), v_s(w), w)}{\Gamma(\alpha_i)} ds; \ t \in I. \end{cases} \quad (9.12)$$

Consider the operator $N : C \times \Omega \to C$ defined by

$$(N(u, v))(t, w) = ((N_1(u, v))(t, w), (N_2(u, v))(t, w)). \quad (9.13)$$

Clearly, the fixed points of the operator N are random solutions of the system (9.10).

Let us show that N is a random operator on $\mathcal{C}$. Since f_i; $i = 1, 2$ are random Carathéodory functions, then $w \to f_i(t, u_t(w), v_t(w), w)$ are measurable maps. Thus, we conclude that the maps

$$w \to (N_i(u, v))(t, w); \quad i = 1, 2,$$

are measurable. As a result, N is a random operator on $\mathcal{C} \times \Omega$ into $\mathcal{C}$. We show that N satisfies all conditions of Theorem 2.81.

For any $w \in \Omega$ and each $(u_1, v_1), (u_2, v_2) \in \mathcal{C}$ and $t \in I$, we have

$$\|t^{1-\gamma_1}(N_1(u_1, v_1))(t, w) - t^{1-\gamma_1}(N_1(u_2, v_2))(t, w)\|$$

$$\leq \frac{t^{1-\gamma_1}}{\Gamma(\alpha_1)} \int_0^t (t-s)^{\alpha_1-1} \|f_1(s, u_{1_s}(w), v_{1_s}(w), w) - f_1(s, u_{2_s}(w), v_{2_s}(w), w)\| ds$$

$$\leq \frac{t^{1-\gamma_1}}{\Gamma(\alpha_1)} \int_0^t (t-s)^{\alpha_1-1} (p_1(w) \|u_{1_s}(w) - v_{1_s}(w)\|_C$$

$$+ q_1(w) \|u_{2_s}(w) - v_{2_s}(w)\|_C) ds$$

$$\leq \frac{1}{\Gamma(\alpha_1)} \int_0^t (t-s)^{\alpha_1-1} (p_1(w) s^{1-\gamma_1} \|u_{1_s}(w) - v_{1_s}(w)\|_C$$

$$+ q_1(w) s^{1-\gamma_1} \|u_{2_s}(w) - v_{2_s}(w)\|_C) ds$$

$$\leq \frac{p_1(w) \|u_1(\cdot, w) - v_1(\cdot, w)\|_{C_{\gamma_1}} + q_1(w) \|u_2(\cdot, w) - v_2(\cdot, w)\|_{C_{\gamma_2}}}{\Gamma(\alpha_1)} \int_0^t (t-s)^{\alpha_1-1} ds$$

$$\leq \frac{T^{\alpha_1}}{\Gamma(1+\alpha_1)} (p_1(w) \|u_1(\cdot, w) - v_1(\cdot, w)\|_{C_{\gamma_1}} + q_1(w) \|u_2(\cdot, w) - v_2(\cdot, w)\|_{C_{\gamma_2}}).$$

Then

$$\|(N_1(u_1, v_1))(\cdot, w) - (N_1(u_2, v_2))(\cdot, w)\|_{C_{\gamma_1}}$$

$$\leq \frac{T^{\alpha_1}}{\Gamma(1+\alpha_1)} (p_1(w) \|u_1(\cdot, w) - v_1(\cdot, w)\|_{C_{\gamma_1}} + q_1(w) \|u_2(\cdot, w) - v_2(\cdot, w)\|_{C_{\gamma_2}}).$$

Also, for any $w \in \Omega$ and each $(u_1, v_1), (u_2, v_2) \in \mathcal{C}$ and $t \in I$, we get

$$\|(N_2(u_1, v_1))(\cdot, w) - (N_2(u_2, v_2))(\cdot, w)\|_{C_{\gamma_2}}$$

$$\leq \frac{T^{\alpha_2}}{\Gamma(1+\alpha_2)} (p_2(w) \|u_1(\cdot, w) - v_1(\cdot, w)\|_{C_{\gamma_1}} + q_2(w) \|u_2(\cdot, w) - v_2(\cdot, w)\|_{C_{\gamma_2}}).$$

Thus

$$d((N(u_1, v_1))(\cdot, w), (N(u_2, v_2))(\cdot, w))$$

$$\leq M(w) d((u_1(\cdot, w), v_1(\cdot, w)), (u_2(\cdot, w), v_2(\cdot, w))),$$

where

$$d((u_1(\cdot, w), v_1(\cdot, w)), (u_2(\cdot, w), v_2(\cdot, w))) = \begin{pmatrix} \|u_1(\cdot, w) - v_1(\cdot, w)\|_{\mathcal{C}_{\gamma_1}} \\ \|u_2(\cdot, w) - v_2(\cdot, w)\|_{\mathcal{C}_{\gamma_2}} \end{pmatrix}.$$

Since for every $w \in \Omega$, the matrix $M(w)$ converges to zero, then Theorem 2.81 implies that system (9.10) has a unique random solution. $\qquad\square$

Now, we prove an existence result for the system (9.10) by using the random nonlinear alternative of Leray–Schauder type in generalized Banach space.

Theorem 9.16. *Assume that the hypotheses (9.14.1) and (9.14.3) hold. Then the system (9.10) has at least one random solution.*

Proof. We show that the operator $N : \mathcal{C} \times \Omega \to \mathcal{C}$ defined in (9.13) satisfies all conditions of Theorem 2.82. The proof will be given in four steps.

Step 1. $N(\cdot, \cdot, w)$ *is continuous.*
Let $(u_n, v_n)_n$ be a sequence such that $(u_n, v_n) \to (u, v) \in \mathcal{C}$ as $n \to \infty$. For any $w \in \Omega$ and each $t \in I$, we have

$$\|t^{1-\gamma_1}(N_1(u_n, v_n))(t, w) - t^{1-\gamma_1}(N_1(u, v))(t, w)\|$$
$$\leq \frac{t^{1-\gamma_1}}{\Gamma(\alpha_1)} \int_0^t (t-s)^{\alpha_1-1} \|f_1(s, u_{ns}(w), v_{ns}(w), w) - f_1(s, u_s(w), v_s(w), w)\| ds$$
$$\leq \frac{1}{\Gamma(\alpha_1)} \int_0^t (t-s)^{\alpha_1-1} s^{1-\gamma_1} \|f_1(s, u_{ns}(w), v_{ns}(w), w) - f_1(s, u_s(w), v_s(w), w)\| ds$$
$$\leq \frac{T^{\alpha_1}}{\Gamma(1+\alpha_1)} \|f_1(\cdot, u_n(\cdot, w), v_n(\cdot, w), w) - f_1(\cdot, u(\cdot, w), v(\cdot, w), w)\|_{\mathcal{C}_{\gamma_1}}.$$

Since f_1 is Carathéodory, we have

$$\|(N_1(u_n, v_n))(\cdot, w) - (N_1(u, v))(\cdot, w)\|_{\mathcal{C}_{\gamma_1}} \to 0 \text{ as } n \to \infty.$$

Also, for any $w \in \Omega$ and each $t \in I$, we get

$$\|t^{1-\gamma_2}(N_2(u_n, v_n))(t, w) - t^{1-\gamma_2}(N_2(u, v))(t, w)\|$$
$$\leq \frac{T^{\alpha_2}}{\Gamma(1+\alpha_2)} \|f_2(\cdot, u_n(\cdot, w), v_n(\cdot, w), w) - f_2(\cdot, u(\cdot, w), v(\cdot, w), w)\|_{\mathcal{C}_{\gamma_2}},$$

and since f_2 is Carathéodory, we obtain

$$\|(N_2(u_n, v_n))(\cdot, w) - (N_2(u, v))(\cdot, w)\|_{\mathcal{C}_{\gamma_2}} \to 0 \text{ as } n \to \infty.$$

Hence $N(\cdot, \cdot, w)$ is continuous.

Step 2. $N(\cdot, \cdot, w)$ maps bounded sets into bounded sets in $\mathcal{C}$.
For any $w \in \Omega$, we set $R > \|\phi_i(w)\|_\mathcal{C}$; $i = 1, 2$ and define the ball

$$B_{R(w)} := \{(\mu, v) \in \mathcal{C} : \|\mu\|_{\mathcal{C}_{\gamma_1}} \le R(w), \|v\|_{\mathcal{C}_{\gamma_2}} \le R(w)\}.$$

For any $w \in \Omega$ and each $(u, v) \in B_{R(w)}$ and $t \in I$, we have

$$\|t^{1-\gamma_i}(N_i(u, v))(t, w)\|$$
$$\le \frac{\|\phi_i(w)\|}{\Gamma(\gamma_i)} + \frac{t^{1-\gamma_i}}{\Gamma(\alpha_i)} \int_0^t (t-s)^{\alpha_i-1}\|f_i(s, u_s(w), v_s(w), w)\|ds$$
$$\le \frac{\|\phi_i(w)\|}{\Gamma(\gamma_i)}$$
$$+ \frac{1}{\Gamma(\alpha_i)} \int_0^t (t-s)^{\alpha_i-1}s^{1-\gamma_i}(a_i(w)\|u_s(w)\|_\mathcal{C} + b_i(w)\|v_s(w)\|_\mathcal{C})ds$$
$$\le \frac{\|\phi_i(w)\|}{\Gamma(\gamma_i)} + \frac{R(w)}{\Gamma(\alpha_i)} \int_0^t (t-s)^{\alpha_i-1}s^{1-\gamma_i}(a_i(w) + b_i(w))ds$$
$$\le \frac{\|\phi_i(w)\|}{\Gamma(\gamma_i)} + \frac{(a_i(w) + b_i(w))R(w)T^{\alpha_i}}{\Gamma(1+\alpha_i)}$$
$$:= \ell_i(w); \ i = 1, 2.$$

Thus

$$\|(N_i(u, v))(\cdot, w)\|_{\mathcal{C}_{\gamma_i}} \le \ell_i(w).$$

Hence

$$\|(N(u, v))(\cdot, w)\|_\mathcal{C} \le \max\{\ell_1(w), \ell_2(w)\} := \ell(w).$$

Step 3. $N(\cdot, \cdot, w)$ maps bounded sets into equicontinuous sets in $\mathcal{C}$.
Let B_R be the ball defined in Step 2. For each $t_1, t_2 \in I$ with $t_1 \le t_2$ and any $(u, v) \in B_{R(w)}$ and $w \in \Omega$, we have

$$\|t_1^{1-\gamma_1}(N_1(u, v))(t_1, w) - t_2^{1-\gamma_1}(N_1(u, v))(t_2, w)\|$$
$$\le \frac{t_2^{1-\gamma_1}}{\Gamma(\alpha_1)} \int_{t-1}^{t_2} (t_2-s)^{\alpha_1-1}\|f_1(s, u_s(w), v_s(w), w)\|ds$$
$$\le \frac{T^{\alpha_1}}{\Gamma(1+\alpha_1)}(t_2-t_1)^{\alpha_1}(a_1(w)\|u(\cdot, w)\|_{\mathcal{C}_{\gamma_1}} + b_1(w)\|v(\cdot, w)\|_{\mathcal{C}_{\gamma_2}})$$
$$\le \frac{R(w)T^{\alpha_1}(a_1(w) + b_1(w))}{\Gamma(1+\alpha_1)}(t_2-t_1)^{\alpha_1}$$
$$\to 0 \text{ as } t_1 \to t_2.$$

Also we get

$$\|t_1^{1-\gamma_2}(N_2(u, v))(t_1, w) - t_2^{1-\gamma_2}(N_2(u, v))(t_2, w)\|$$

$$\leq \frac{R(w)T^{\alpha_{12}}(a_2(w) + b_2(w))}{\Gamma(1+\alpha_2)}(t_2 - t_1)^{\alpha_2}$$

$$\to 0 \text{ as } t_1 \to t_2.$$

As a consequence of Steps 1 to 3, with the Arzelà–Ascoli theorem, we conclude that $N(\cdot, \cdot, w)$ maps B_R into a precompact set in $\mathcal{C}$.

Step 4. The set $E(w)$ consisting of $(u(\cdot, w), v(\cdot, w)) \in \mathcal{C}$ such that $(u(\cdot, w), v(\cdot, w)) = \lambda(w)(N((u, v))(\cdot, w)$ for some measurable function $\lambda : \Omega \to (0, 1)$ is bounded in $\mathcal{C}$.
Let $(u(\cdot, w), v(\cdot, w)) \in \mathcal{C}$ such that $(u(\cdot, w), v(\cdot, w)) = \lambda(w)(N((u, v))(\cdot, w)$. Then $u(\cdot, w) = \lambda(w)(N_1((u, v))(\cdot, w)$ and $v(\cdot, w) = \lambda(w)(N_2((u, v))(\cdot, w)$. Thus, for any $w \in \Omega$ and each $t \in I$, we have

$$\|t^{1-\gamma_1}u(t, w)\|$$

$$\leq \frac{\|\phi_1(w)\|}{\Gamma(\gamma_1)} + \frac{t^{1-\gamma_1}}{\Gamma(\alpha_1)}\int_0^t (t-s)^{\alpha_1-1}\|f_1(s, u_s(w), v_s(w), w)\|ds$$

$$\leq \frac{\|\phi_1(w)\|}{\Gamma(\gamma_1)}$$

$$+ \frac{1}{\Gamma(\alpha_1)}\int_0^t (t-s)^{\alpha_1-1}s^{1-\gamma_1}(a_1(w)\|u_s(w)\|_C + b_1(w)\|v_s(w)\|_C)ds.$$

Also

$$\|t^{1-\gamma_2}v(t, w)\|$$

$$\leq \frac{\|\phi_2(w)\|}{\Gamma(\gamma_2)} + \frac{1}{\Gamma(\alpha_2)}\int_0^t (t-s)^{\alpha_2-1}s^{1-\gamma_2}(a_2(w)\|u_s(w)\|_C + b_2(w)\|v_s(w)\|_C)ds.$$

Hence we get

$$\|t^{1-\gamma_1}u(t, w)\| + \|t^{1-\gamma_2}v(t, w)\|$$

$$\leq a(w) + b(w)c(w)\int_0^t (t-s)^{\alpha-1}(s^{1-\gamma_1}\|u_s(w\|_C + s^{1-\gamma_2}\|v_s(w\|_C)ds,$$

where

$$a(w) := \frac{\|\phi_1(w)\|}{\Gamma(\gamma_1)} + \frac{\|\phi_2(w)\|}{\Gamma(\gamma_2)}, \ b(w) := \frac{1}{\Gamma(\alpha_1)} + \frac{1}{\Gamma(\alpha_2)},$$

$$c(w) := \max\{a_1(w) + a_2(w), b_1(w) + b_2(w)\}, \ \alpha := \max\{\alpha_1, \alpha_2\}.$$

We consider the function τ defined by

$$\tau(t, w) = \sup\{z(s, w) : -h \leq s \leq t\}; \ t \in I, \ w \in \Omega,$$

where $z(t, w) = \|t^{1-\gamma_1}u(t, w)\| + \|t^{1-\gamma_2}v(t, w)\|$. Let $t^* \in [-h, t]$ be such that $\tau(t) = z(t^*)$.

If $t^* \in I$, then by the previous inequality, for any $w \in \Omega$ and each $t \in I$, we have

$$\tau(t, w) \le a(w) + b(w)c(w) \int_0^t (t-s)^{\alpha-1} \tau(s, w) ds,$$

and if $t^* \in [-h, 0]$, then $\tau(t, w) \le \|T^{1-\gamma_1} \phi_1(t, w)\| + \|T^{1-\gamma_2} \phi_2(t, w)\|$ and the previous inequality holds.

Now, Lemma 2.100 implies that there exists $\rho := \rho(\alpha) > 0$ such that

$$\tau(t, w) \le a(w) + a(w)b(w)c(w)\rho \int_0^t (t-s)^{\alpha-1} ds$$
$$\le \frac{a(w) + a(w)b(w)c(w)\rho T^\alpha}{\alpha}$$
$$= L(w).$$

Since for every $t \in I$, and any $w \in \Omega$; we have $\|u_t(w)\|_C \le \tau(t, w)$, then

$$\|(u(\cdot, w), v(\cdot, w))\|_C \le \max\{L(w), T^{1-\gamma_1}\|\phi_1(w)\|_C + T^{1-\gamma_2}\|\phi_2(w)\|_C\} := M(w).$$

This shows that the set $E(w)$ is bounded.

As a consequence of steps 1 to 4 together with Theorem 2.82, we can conclude that N has at least one fixed point in $B_{R(w)}$, which is a solution for the system (9.10). □

9.2.3 Coupled Hilfer–Hadamard random fractional differential systems

Now, we are concerned with the coupled system (9.11).

Set $C := C([1, T])$, and denote the weighted space of continuous functions defined by

$$C_{\gamma, \ln}([1, T]) = \{w(t) : (\ln t)^{1-\gamma} w(t) \in C\},$$

with the norm

$$\|w\|_{C_{\gamma, \ln}} := \sup_{t \in [1, T]} |(\ln t)^{1-r} w(t)|.$$

Also, by $C_{\gamma_1, \gamma_2, \ln}([1, T]) := C_{\gamma_1, \ln}([1, T]) \times C_{\gamma_2, \ln}([1, T])$ we denote the product weighted space with the norm

$$\|(u, v)\|_{C_{\gamma_1, \gamma_2, \ln}([1-h, T])} = \|u\|_{C_{\gamma_1, \ln}} + \|v\|_{C_{\gamma_2, \ln}}.$$

Let us recall some definitions and properties of Hadamard fractional integration and differentiation. We refer to [281] for a more detailed analysis.

Definition 9.17. [281] (Hadamard fractional integral). The Hadamard fractional integral of order $q > 0$ for a function $g \in L^1([1, T])$, is defined as

$$(^H I_1^q g)(x) = \frac{1}{\Gamma(q)} \int_1^x \left(\ln \frac{x}{s}\right)^{q-1} \frac{g(s)}{s} ds,$$

provided the integral exists.

Example 9.18. Let $0 < q < 1$. Let $g(x) = \ln x$, $x \in [0, e]$. Then

$$(^H I_1^q g)(x) = \frac{1}{\Gamma(2+q)}(\ln x)^{1+q}; \text{ for a.e. } x \in [0, e].$$

Set

$$\delta = x\frac{d}{dx}, \; q > 0, \; n = [q] + 1,$$

and

$$AC_\delta^n := \{u : [1, T] \to E : \delta^{n-1}[u(x)] \in AC(I)\}.$$

Analogous to the Riemann–Liouville fractional derivative, the Hadamard fractional derivative is defined in terms of the Hadamard fractional integral in the following way:

Definition 9.19. [281] (Hadamard fractional derivative). The Hadamard fractional derivative of order $q > 0$ applied to the function $w \in AC_\delta^n$ is defined as

$$(^H D_1^q w)(x) = \delta^n (^H I_1^{n-q} w)(x).$$

In particular, if $q \in (0, 1]$, then

$$(^H D_1^q w)(x) = \delta(^H I_1^{1-q} w)(x).$$

Example 9.20. Let $0 < q < 1$. Let $w(x) = \ln x$, $x \in [0, e]$. Then

$$(^H D_1^q w)(x) = \frac{1}{\Gamma(2-q)}(\ln x)^{1-q}, \text{ for a.e. } x \in [0, e].$$

It has been proved (see, e.g., Kilbas [280, Theorem 4.8]) that in the space $L^1(I)$, the Hadamard fractional derivative is the left-inverse operator to the Hadamard fractional integral, i.e.

$$(^H D_1^q)(^H I_1^q w)(x) = w(x).$$

From Theorem 2.3 of [281], we have

$$(^H I_1^q)(^H D_1^q w)(x) = w(x) - \frac{(^H I_1^{1-q} w)(1)}{\Gamma(q)}(\ln x)^{q-1}.$$

Analogous to the Hadamard fractional calculus, the Caputo–Hadamard fractional derivative is defined in the following way:

Definition 9.21. (Caputo–Hadamard fractional derivative) The Caputo–Hadamard fractional derivative of order $q > 0$ applied to the function $w \in AC_\delta^n$ is defined as

$$(^{Hc} D_1^q w)(x) = (^H I_1^{n-q} \delta^n w)(x).$$

In particular, if $q \in (0, 1]$, then

$$({}^{Hc}D_1^q w)(x) = ({}^H I_1^{1-q} \delta w)(x).$$

From the Hadamard fractional integral, the Hilfer–Hadamard fractional derivative (introduced for the first time in [337]) is defined in the following way:

Definition 9.22. (Hilfer–Hadamard fractional derivative). Let $\alpha \in (0, 1)$, $\beta \in [0, 1]$, $\gamma = \alpha + \beta - \alpha\beta$, $w \in L^1(I)$, and ${}^H I_1^{(1-\alpha)(1-\beta)} w \in AC(I)$. The Hilfer–Hadamard fractional derivative of order α and type β applied to the function w is defined as

$$
\begin{aligned}
({}^H D_1^{\alpha,\beta} w)(t) &= \left({}^H I_1^{\beta(1-\alpha)} ({}^H D_1^\gamma w)\right)(t) \\
&= \left({}^H I_1^{\beta(1-\alpha)} \delta ({}^H I_1^{1-\gamma} w)\right)(t); \quad \text{for a.e. } t \in [1, T].
\end{aligned}
\tag{9.14}
$$

This new fractional derivative (9.14) may be viewed as interpolating the Hadamard fractional derivative and the Caputo–Hadamard fractional derivative. Indeed, for $\beta = 0$, this derivative reduces to the Hadamard fractional derivative and when $\beta = 1$, we recover the Caputo–Hadamard fractional derivative.

$$
{}^H D_1^{\alpha,0} = {}^H D_1^\alpha, \text{ and } {}^H D_1^{\alpha,1} = {}^{Hc} D_1^\alpha.
$$

From Theorem 21 in [338], we concluded the following lemma.

Lemma 9.23. *Let $g : [1, T] \times E \to E$ be such that $g(\cdot, u(\cdot)) \in C_{\gamma,\ln}([1, T])$ for any $u \in C_{\gamma,\ln}([1, T])$. Then problem (9.11) is equivalent to the following Volterra integral equation*

$$
u(t) = \frac{\phi_0}{\Gamma(\gamma)}(\ln t)^{\gamma-1} + ({}^H I_1^\alpha g(\cdot, u(\cdot)))(t).
$$

Definition 9.24. By a random solution of the system (9.11) we mean a coupled measurable functions $(u, v) \in C_{\gamma_1,\ln} \times C_{\gamma_2,\ln}$ those satisfy Eqs. (9.11) on $[1, T]$.

Now we give (without proof) similar existence and uniqueness results for the system (9.11). Let us introduce the following hypotheses:

(9.24.1) The functions g_i; $i = 1, 2$ are random Carathéodory.
(9.24.2) There exist measurable functions $p_i, q_i : \Omega \to (0, \infty)$; $i = 1, 2$ such that

$$
\|g_i(t, u_1, v_1) - g_i(t, u_2, v_2)\| \le p_i(w)\|u_1 - u_2\|_C + q_i(w)\|v_1 - v_2\|_C;
$$

for a.e. $t \in [1, T]$, and each $u_i, v_i \in C, i = 1, 2$.
(9.24.3) There exist measurable functions $a_i, b_i : \Omega \to (0, \infty)$; $i = 1, 2$ such that

$$
\|g_i(t, u, v)\| \le a_i(w)\|u\|_C + b_i(w)\|v\|_C; \text{ for a.e. } t \in [1, T], \ u, v \in C.
$$

Theorem 9.25. *Assume that the hypotheses (9.24.1) and (9.24.2) hold. If for every $w \in \Omega$, the matrix*

$$\begin{pmatrix} \frac{(\ln T)^{\alpha_1}}{\Gamma(1+\alpha_1)} p_1(w) & \frac{(\ln T)^{\alpha_1}}{\Gamma(1+\alpha_1)} q_1(w) \\ \frac{(\ln T)^{\alpha_2}}{\Gamma(1+\alpha_2)} p_2(w) & \frac{(\ln T)^{\alpha_2}}{\Gamma(1+\alpha_2)} q_2(w) \end{pmatrix}$$

converges to 0, then the problem (9.11) has a unique random solution.

Theorem 9.26. *Assume that the hypotheses (9.24.1) and (9.24.3) hold. Then the system (9.11) has at least one random solution.*

9.2.4 An example

Let $\Omega = (-\infty, 0)$ be equipped with the usual σ-algebra consisting of Lebesgue measurable subsets of $(-\infty, 0)$. Consider the following random coupled system of Hilfer fractional differential equations

$$\begin{cases} (u(t, w), v(t, w)) = (t \cos w, t \sin w); \ t \in [-2, 0], \\ (D_0^{\frac{1}{2}, \frac{1}{2}} u)(t, w) = f(t, u_t(w), v_t(w), w); \ t \in [0, 1], \\ (D_0^{\frac{1}{2}, \frac{1}{2}} v)(t, w) = g(t, u_t(w), v_t(w), w); \ t \in [0, 1], \\ ((I_0^{\frac{1}{4}} u)(0, w), (I_0^{\frac{1}{4}} v)(0, w)) = (0, 0), \end{cases} ; \ w \in \Omega, \qquad (9.15)$$

where

$$f(t, u_t(w), v_t(w), w) = \frac{1 + t^{\frac{-1}{4}} w^2 (u(t) + v(t)) \sin t}{(1 + w^2 + \sqrt{t})(1 + \|u_t(w)\|_{C([-2,0])} + \|v_t(w)\|_{C([-2,0])})},$$

and

$$g(t, u_t(w), v_t(w), w) = \frac{2 + w^2 (u_t(w) + v_t(w)) \cos t}{(1 + \|u_t(w)\|_{C([-2,0])} + \|v_t(w)\|_{C([-2,0])})},$$

for $t \in [0, 1]$ and $w \in \Omega$.

Set $\alpha_i = \beta_i = \frac{1}{2}$; $i = 1, 2$, then $\gamma_i = \frac{3}{4}$; $i = 1, 2$. The hypothesis (9.24.2) is satisfied with

$$p_1(w) = p_2(w) = q_1(w) = q_2(w) = \frac{w^2}{1 + w^2}.$$

Furthermore, for every $w \in \Omega$, the matrix

$$\frac{w^2}{\sqrt{\pi}(1 + w^2)} \begin{pmatrix} 1 & 1 \\ 1 & 1 \end{pmatrix}$$

converges to 0. Hence, Theorem 9.15 implies that the system (9.15) has a unique random solution defined on $[-2, 1]$.

9.3 Random coupled Hilfer and Hadamard fractional differential systems

9.3.1 Introduction

In this section we discuss the existence of random solutions for the following coupled system of Hilfer fractional differential equations

$$\begin{cases} (D_0^{\alpha_1,\beta_1} u)(t, w) = f_1(t, u(t, w), v(t, w), w) \\ (D_0^{\alpha_2,\beta_2} v)(t, w) = f_2(t, u(t, w), v(t, w), w) \end{cases} \quad ; t \in I := [0, T], \ w \in \Omega, \tag{9.16}$$

with the following initial conditions

$$\begin{cases} (I_0^{1-\gamma_1} u)(0, w) = \phi_1(w) \\ (I_0^{1-\gamma_2} v)(0, w) = \phi_2(w) \end{cases} \quad ; w \in \Omega, \tag{9.17}$$

where $T > 0$, $\alpha_i \in (0, 1)$, $\beta_i \in [0, 1]$, $\gamma_i = \alpha_i + \beta_i - \alpha_i \beta_i$, $\phi_i : \Omega \to \mathbb{R}^m$, $f_i : I \times \mathbb{R}^m \times \mathbb{R}^m \times \Omega \to \mathbb{R}^m$; $i = 1, 2$, are given functions, $I_0^{1-\gamma_i}$ is the left-sided mixed Riemann–Liouville integral of order $1 - \gamma_i$, and $D_0^{\alpha_i,\beta_i}$ is the generalized Riemann–Liouville derivative (Hilfer) operator of order α_i and type $\beta_i : i = 1, 2$.

Next, we consider the following coupled system of Hilfer–Hadamard fractional differential equations

$$\begin{cases} (^H D_1^{\alpha_1,\beta_1} u)(t, w) = g_1(t, u(t, w), v(t, w), w) \\ (^H D_1^{\alpha_2,\beta_2} v)(t, w) = g_2(t, u(t, w), v(t, w), w) \end{cases} \quad ; t \in [1, T], \ w \in \Omega, \tag{9.18}$$

with the following initial conditions

$$\begin{cases} (^H I_1^{1-\gamma_1} u)(1, w) = \psi_1(w) \\ (^H I_1^{1-\gamma_2} v)(1, w) = \psi_2(w) \end{cases} \quad ; w \in \Omega, \tag{9.19}$$

where $T > 1$, $\alpha_i \in (0, 1)$, $\beta_i \in [0, 1]$, $\gamma_i = \alpha_i + \beta_i - \alpha_i \beta_i$, $\psi_i : \Omega \to \mathbb{R}^m$, $g_i : [1, T] \times \mathbb{R}^m \times \mathbb{R}^m \times \Omega \to \mathbb{R}^m$; $i = 1, 2$ are given functions, $\mathbb{R}^m$; $m \in \mathbb{N}^*$ is the Euclidean Banach space with a suitable norm $\| \cdot \|$, $^H I_1^{1-\gamma_i}$ is the left-sided mixed Hadamard integral of order $1 - \gamma_i$, and $^H D_1^{\alpha_i,\beta_i}$ is the Hilfer–Hadamard fractional derivative of order α_i and type β_i; $i = 1, 2$.

9.3.2 Coupled Hilfer fractional differential systems

In this section we are concerned with the existence and uniqueness results of the system (9.16)–(9.17).

Definition 9.27. By a solution of the problem (9.16)–(9.17) we mean a coupled measurable functions $(u, v) \in C_{\gamma_1} \times C_{\gamma_2}$ those satisfy Eq. (9.16) on I, and the conditions $(I_0^{1-\gamma_1} u)(0^+) = \phi_1$, and $(I_0^{1-\gamma_2} v)(0^+) = \phi_2$.

The following hypotheses will be used in the sequel.

(9.27.1) The functions f_i; $i = 1, 2$ are Carathéodory.
(9.27.2) There exist measurable functions $p_i, q_i : \Omega \to (0, \infty)$; $i = 1, 2$ such that

$$\|f_i(t, u_1, v_1) - f_i(t, u_2, v_2)\| \le p_i(w)\|u_1 - u_2\| + q_i(w)\|v_1 - v_2\|;$$

for a.e. $t \in I$, and each $u_i, v_i \in \mathbb{R}^m, i = 1, 2$.
(9.27.3) There exist measurable functions $a_i, b_i : \Omega \to (0, \infty)$; $i = 1, 2$ such that

$$\|f_i(t, u, v)\| \le a_i(w)\|u\| + b_i(w)\|v\|; \text{ for a.e. } t \in I, \text{ and each } u, v \in \mathbb{R}^m.$$

First, we prove an existence and uniqueness result for the coupled system (9.16)–(9.17) by using Banach random fixed-point theorem type in generalized Banach spaces.

Theorem 9.28. *Assume that the hypotheses (9.27.1) and (9.27.2) hold. If for every $w \in \Omega$, the matrix*

$$M(w) := \begin{pmatrix} \frac{T^{\alpha_1}}{\Gamma(1+\alpha_1)} p_1(w) & \frac{T^{\alpha_1}}{\Gamma(1+\alpha_1)} q_1(w) \\ \frac{T^{\alpha_2}}{\Gamma(1+\alpha_2)} p_2(w) & \frac{T^{\alpha_2}}{\Gamma(1+\alpha_2)} q_2(w) \end{pmatrix}$$

converges to 0, then the coupled system (9.16)–(9.17) has a unique random solution.

Proof. Define the operators $N_1 : C \times \Omega \to C_{\gamma_1}$ and $N_2 : C \times \Omega \to C_{\gamma_2}$ by

$$(N_1(u, v))(t, w) = \frac{\phi_1(w)}{\Gamma(\gamma_1)} t^{\gamma_1 - 1} + \int_0^t (t - s)^{\alpha_1 - 1} \frac{f(s, u(s, w), v(s, w), w)}{\Gamma(\alpha_1)} ds, \quad (9.20)$$

and

$$(N_2(u, v))(t, w) = \frac{\phi_2(w)}{\Gamma(\gamma_2)} t^{\gamma_2 - 1} + \int_0^t (t - s)^{\alpha_2 - 1} \frac{f(s, u(s, w), v(s, w), w)}{\Gamma(\alpha_2)} ds. \quad (9.21)$$

Consider the operator $N : C \times \Omega \to C$ defined by

$$(N(u, v))(t, w) = ((N_1(u, v))(t, w), (N_2(u, v))(t, w)). \quad (9.22)$$

Clearly, the fixed points of the operator N are random solutions of the system (9.16)–(9.17).
Let us show that N is a random operator on C. Since f_i; $i = 1, 2$ are Carathéodory functions, then $w \to f_i(t, u, v, w)$ are measurable maps. Thus, we conclude that the maps

$$w \to (N_1(u, v))(t, w) \text{ and } w \to (N_2(u, v))(t, w)$$

are measurable. As a result, N is a random operator on $C \times \Omega$ into C. We show that N satisfies all conditions of Theorem 2.81.

For any $w \in \Omega$ and each $(u_1, v_1), (u_2, v_2) \in \mathcal{C}$ and $t \in I$, we have

$$\|t^{1-\gamma_1}(N_1(u_1, v_1))(t, w) - t^{1-\gamma_1}(N_1(u_2, v_2))(t, w)\|$$

$$\leq \frac{t^{1-\gamma_1}}{\Gamma(\alpha_1)} \int_0^t (t-s)^{\alpha_1-1} \|f_1(s, u_1(s, w), v_1(s, w), w) - f_1(s, u_2(s, w), v_2(s, w), w)\| ds$$

$$\leq \frac{t^{1-\gamma_1}}{\Gamma(\alpha_1)} \int_0^t (t-s)^{\alpha_1-1} (p_1(w)\|u_1(s, w) - v_1(s, w)\|$$

$$+ q_1(w)\|u_2(s, w) - v_2(s, w)\|) ds$$

$$\leq \frac{1}{\Gamma(\alpha_1)} \int_0^t (t-s)^{\alpha_1-1} (p_1(w)s^{1-\gamma_1}\|u_1(s, w) - v_1(s, w)\|$$

$$+ q_1(w)s^{1-\gamma_1}\|u_2(s, w) - v_2(s, w)\|) ds$$

$$\leq \frac{p_1(w)\|u_1(\cdot, w) - v_1(\cdot, w)\|_{C_{\gamma_1}} + q_1(w)\|u_2(\cdot, w) - v_2(\cdot, w)\|_{C_{\gamma_2}}}{\Gamma(\alpha_1)}$$

$$\times \int_0^t (t-s)^{\alpha_1-1} ds$$

$$\leq \frac{T^{\alpha_1}}{\Gamma(1+\alpha_1)} (p_1(w)\|u_1(\cdot, w) - v_1(\cdot, w)\|_{C_{\gamma_1}} + q_1(w)\|u_2(\cdot, w) - v_2(\cdot, w)\|_{C_{\gamma_2}}).$$

Then

$$\|(N_1(u_1, v_1))(\cdot, w) - (N_1(u_2, v_2))(\cdot, w)\|_{C_{\gamma_1}}$$

$$\leq \frac{T^{\alpha_1}}{\Gamma(1+\alpha_1)} (p_1(w)\|u_1(\cdot, w) - v_1(\cdot, w)\|_{C_{\gamma_1}} + q_1(w)\|u_2(\cdot, w) - v_2(\cdot, w)\|_{C_{\gamma_2}}).$$

Also, for any $w \in \Omega$ and each $(u_1, v_1), (u_2, v_2) \in \mathcal{C}$ and $t \in I$, we get

$$\|(N_2(u_1, v_1))(\cdot, w) - (N_2(u_2, v_2))(\cdot, w)\|_{C_{\gamma_2}}$$

$$\leq \frac{T^{\alpha_2}}{\Gamma(1+\alpha_2)} (p_2(w)\|u_1(\cdot, w) - v_1(\cdot, w)\|_{C_{\gamma_1}} + q_2(w)\|u_2(\cdot, w) - v_2(\cdot, w)\|_{C_{\gamma_2}}).$$

Thus

$$d((N(u_1, v_1))(\cdot, w), (N(u_2, v_2))(\cdot, w))$$

$$\leq M(w)d((u_1(\cdot, w), v_1(\cdot, w)), (u_2(\cdot, w), v_2(\cdot, w))),$$

where

$$d((u_1(\cdot, w), v_1(\cdot, w)), (u_2(\cdot, w), v_2(\cdot, w))) = \begin{pmatrix} \|u_1(\cdot, w) - v_1(\cdot, w)\|_{C_{\gamma_1}} \\ \|u_2(\cdot, w) - v_2(\cdot, w)\|_{C_{\gamma_2}} \end{pmatrix}.$$

Since for every $w \in \Omega$, the matrix $M(w)$ converges to zero, then Theorem 2.81 implies that system (9.16)–(9.17) has a unique random solution. $\square$

Now, we prove an existence result for the coupled system (9.16)–(9.17) by using the random nonlinear alternative of Leray–Schauder type in generalized Banach space.

Theorem 9.29. *Assume that the hypotheses* (9.27.1) *and* (9.27.3) *hold. Then the coupled system* (9.16)–(9.17) *has at least one random solution.*

Proof. We show that the operator $N : C \times \Omega \to C$ defined in (9.22) satisfies all conditions of Theorem 2.82. The proof will be given in four steps.

Step 1. $N(\cdot, \cdot, w)$ *is continuous.*
Let $(u_n, v_n)_n$ be a sequence such that $(u_n, v_n) \to (u, v) \in C$ as $n \to \infty$. For any $w \in \Omega$ and each $t \in I$, we have

$$\|t^{1-\gamma_1}(N_1(u_n, v_n))(t, w) - t^{1-\gamma_1}(N_1(u, v))(t, w)\|$$
$$\leq \frac{t^{1-\gamma_1}}{\Gamma(\alpha_1)} \int_0^t (t - s)^{\alpha_1 - 1} \|f_1(s, u_n(s, w), v_n(s, w), w$$
$$- f_1(s, u(s, w), v(s, w), w)\| ds$$
$$\leq \frac{1}{\Gamma(\alpha_1)} \int_0^t (t - s)^{\alpha_1 - 1} s^{1-\gamma_1} \|f_1(s, u_n(s, w), v_n(s, w), w)$$
$$- f_1(s, u(s, w), v(s, w), w)\| ds$$
$$\leq \frac{T^{\alpha_1}}{\Gamma(1 + \alpha_1)} \|f_1(\cdot, u_n(\cdot, w), v_n(\cdot, w), w) - f_1(\cdot, u(\cdot, w), v(\cdot, w), w)\|_{C_{\gamma_1}}.$$

Since f_1 is Carathéodory, we have

$$\|(N_1(u_n, v_n))(\cdot, w) - (N_1(u, v))(\cdot, w)\|_{C_{\gamma_1}} \to 0 \text{ as } n \to \infty.$$

On the other hand, for any $w \in \Omega$ and each $t \in I$, we obtain

$$\|t^{1-\gamma_2}(N_2(u_n, v_n))(t, w) - t^{1-\gamma_2}(N_2(u, v))(t, w)\|$$
$$\leq \frac{T^{\alpha_2}}{\Gamma(1 + \alpha_2)} \|f_2(\cdot, u_n(\cdot, w), v_n(\cdot, w), w) - f_2(\cdot, u(\cdot, w), v(\cdot, w), w)\|_{C_{\gamma_2}}.$$

Also the fact that f_2 is Carathéodory, we get

$$\|(N_2(u_n, v_n))(\cdot, w) - (N_2(u, v))(\cdot, w)\|_{C_{\gamma_2}} \to 0 \text{ as } n \to \infty.$$

Hence $N(\cdot, \cdot, w)$ is continuous.

Step 2. $N(\cdot, \cdot, w)$ *maps bounded sets into bounded sets in* C.
Let $R > 0$ and set

$$B_R := \{(\mu, v) \in C : \|\mu\|_{C_{\gamma_1}} \leq R, \|v\|_{C_{\gamma_2}} \leq R\}.$$

For any $w \in \Omega$ and each $(u, v) \in B_R$ and $t \in I$, we have

$$\|t^{1-\gamma_1}(N_1(u, v))(t, w)\|$$

$$\leq \frac{\|\phi_1(w)\|}{\Gamma(\gamma_1)} + \frac{t^{1-\gamma_1}}{\Gamma(\alpha_1)} \int_0^t (t-s)^{\alpha_1 - 1} \|f_1(s, u(s, w), v(s, w), w)\| ds$$

$$\leq \frac{\|\phi_1(w)\|}{\Gamma(\gamma_1)}$$

$$+ \frac{1}{\Gamma(\alpha_1)} \int_0^t (t-s)^{\alpha_1 - 1} s^{1-\gamma_1} (a_1(w) \|u(s, w\| + b_1(w) \|v(s, w\|) ds$$

$$\leq \frac{\|\phi_1(w)\|}{\Gamma(\gamma_1)} + \frac{R}{\Gamma(\alpha_1)} \int_0^t (t-s)^{\alpha_1 - 1} s^{1-\gamma_1} (a_1(w) + b_1(w)) ds$$

$$\leq \frac{\|\phi_1(w)\|}{\Gamma(\gamma_1)} + \frac{(a_1(w) + b_1(w) T^{\alpha_1}}{\Gamma(1 + \alpha_1)}$$

$$:= \ell_1.$$

Thus

$$\|(N_1(u, v))(\cdot, w)\|_{C_{\gamma_1}} \leq \ell_1.$$

Also, for any $w \in \Omega$ and each $(u, v) \in B_R$ and $t \in I$, we get

$$\|(N_2(u, v))(\cdot, w)\|_{C_{\gamma_2}} \leq \frac{\|\phi_2(w)\|}{\Gamma(\gamma_2)} + \frac{(a_2(w) + b_2(w) T^{\alpha_2}}{\Gamma(1 + \alpha)}$$

$$:= \ell_2.$$

Hence

$$\|(N(u, v))(\cdot, w)\|_C \leq \max\{\ell_1, \ell_2\} := \ell.$$

Step 3. $N(\cdot, \cdot, w)$ maps bounded sets into equicontinuous sets in C.
Let B_R be the ball defined in Step 2. For each $t_1, t_2 \in I$ with $t_1 \leq t_2$ and any $(u, v) \in B_R$ and $w \in \Omega$, we have

$$\|t_1^{1-\gamma_1}(N_1(u, v))(t_1, w) - t_2^{1-\gamma_1}(N_1(u, v))(t_2, w)\|$$

$$\leq \frac{t_2^{1-\gamma_1}}{\Gamma(\alpha_1)} \int_{t-1}^{t_2} (t_2 - s)^{\alpha_1 - 1} \|f_1(s, u(s, w), v(s, w), w)\| ds$$

$$\leq \frac{T^{\alpha_1}}{\Gamma(1 + \alpha_1)} (t_2 - t_1)^{\alpha_1} (a_1(w) \|u(\cdot, w)\|_{C_{\gamma_1}} + b_1(w) \|v(\cdot, w)\|_{C_{\gamma_2}})$$

$$\leq \frac{RT^{\alpha_1}(a_1(w) + b_1(w))}{\Gamma(1 + \alpha_1)} (t_2 - t_1)^{\alpha_1}$$

$$\to 0 \text{ as } t_1 \to t_2.$$

Also we get

$$\|t_1^{1-\gamma_2}(N_2(u,v))(t_1,w) - t_2^{1-\gamma_2}(N_2(u,v))(t_2,w)\|$$

$$\leq \frac{RT^{\alpha_{12}}(a_2(w) + b_2(w))}{\Gamma(1+\alpha_2)}(t_2 - t_1)^{\alpha_2}$$

$$\to 0 \text{ as } t_1 \to t_2.$$

As a consequence of Steps 1 to 3, with the Arzelà–Ascoli theorem, we conclude that $N(\cdot, \cdot, w)$ maps B_R into a precompact set in $\mathcal{C}$.

Step 4. The set $E(w)$ consisting of $(u(\cdot, w), v(\cdot, w)) \in \mathcal{C}$ such that

$$(u(\cdot, w), v(\cdot, w)) = \lambda(w)(N((u,v))(\cdot, w)$$

for some measurable function $\lambda : \Omega \to (0,1)$ is bounded in $\mathcal{C}$.

Let $(u(\cdot, w), v(\cdot, w)) \in \mathcal{C}$ such that $(u(\cdot, w), v(\cdot, w)) = \lambda(w)(N((u,v))(\cdot, w)$. Then $u(\cdot, w) = \lambda(w)(N_1((u,v))(\cdot, w)$ and $v(\cdot, w) = \lambda(w)(N_2((u,v))(\cdot, w)$. Thus, for any $w \in \Omega$ and each $t \in I$, we have

$$\|t^{1-\gamma_1}u(t,w)\| \leq \frac{\|\phi_1(w)\|}{\Gamma(\gamma_1)} + \frac{t^{1-\gamma_1}}{\Gamma(\alpha_1)}\int_0^t (t-s)^{\alpha_1-1}\|f_1(s,u(s,w),v(s,w),w)\|ds$$

$$\leq \frac{\|\phi_1(w)\|}{\Gamma(\gamma_1)}$$

$$+ \frac{1}{\Gamma(\alpha_1)}\int_0^t (t-s)^{\alpha_1-1}s^{1-\gamma_1}(a_1(w)\|u(s,w\| + b_1(w)\|v(s,w\|)ds.$$

Also we get

$$\|t^{1-\gamma_2}v(t,w)\| \leq \frac{\|\phi_2(w)\|}{\Gamma(\gamma_2)}$$

$$+ \frac{1}{\Gamma(\alpha_2)}\int_0^t (t-s)^{\alpha_2-1}s^{1-\gamma_2}(a_2(w)\|u(s,w\| + b_2(w)\|v(s,w\|)ds.$$

Hence we obtain

$$\|t^{1-\gamma_1}u(t,w)\| + \|t^{1-\gamma_2}v(t,w)\|$$

$$\leq a + bc\int_0^t (t-s)^{\alpha-1}(\|s^{1-\gamma_1}u(s,w\| + \|s^{1-\gamma_2}v(s,w\|)ds,$$

where

$$a := \frac{\|\phi_1(w)\|}{\Gamma(\gamma_1)} + \frac{\|\phi_2(w)\|}{\Gamma(\gamma_2)}, \quad b := \frac{1}{\Gamma(\alpha_1)} + \frac{1}{\Gamma(\alpha_2)},$$

$$c := \max\{a_1(w) + a_2(w), b_1(w) + b_2(w)\}, \quad \alpha := \max\{\alpha_1, \alpha_2\}.$$

Lemma 2.100 implies that there exists $\rho := \rho(\alpha) > 0$ such that

$$\|t^{1-\gamma_1}u(t,w)\| + \|t^{1-\gamma_2}v(t,w)\| \le a + abc\rho \int_0^t (t-s)^{\alpha-1}ds$$

$$\le \frac{a + abc\rho T^\alpha}{\alpha}$$

$$= L.$$

This gives

$$\|u(\cdot,w)\|_{C_{\gamma_1}} + \|v(\cdot,w)\|_{C_{\gamma_2}} \le L.$$

Hence

$$\|(u(\cdot,w),v(\cdot,w))\|_C \le L.$$

This shows that the set $E(w)$ is bounded.

As a consequence of steps 1 to 4 together with Theorem 2.82, we can conclude that N has at least one fixed point in B_R, which is a solution for system (9.16)–(9.17). □

9.3.3 Coupled Hilfer–Hadamard fractional differential systems

Now, we are concerned with the coupled system (9.18)–(9.19).

Set $C := C([1,T])$, and denote the weighted space of continuous functions defined by

$$C_{\gamma,\ln}([1,T]) = \{w(t) : (\ln t)^{1-\gamma}w(t) \in C\},$$

with the norm

$$\|w\|_{C_{\gamma,\ln}} := \sup_{t \in [1,T]} |(\ln t)^{1-r}w(t)|.$$

Also, by $C_{\gamma_1,\gamma_2,\ln}([1,T]) := C_{\gamma_1,\ln}([1,T]) \times C_{\gamma_2,\ln}([1,T])$ we denote the product weighted space with the norm

$$\|(u,v)\|_{C_{\gamma_1,\gamma_2,\ln}([1,T])} = \|u\|_{C_{\gamma_1,\ln}} + \|v\|_{C_{\gamma_2,\ln}}.$$

Let us recall some definitions and properties of Hadamard fractional integration and differentiation. We refer to [281] for a more detailed analysis.

Definition 9.30. [281] (Hadamard fractional integral). The Hadamard fractional integral of order $q > 0$ for a function $g \in L^1([1,T])$, is defined as

$$(^H I_1^q g)(x) = \frac{1}{\Gamma(q)} \int_1^x \left(\ln \frac{x}{s}\right)^{q-1} \frac{g(s)}{s}ds,$$

provided the integral exists.

Example 9.31. Let $0 < q < 1$. Let $g(x) = \ln x$, $x \in [0,e]$. Then

$$(^H I_1^q g)(x) = \frac{1}{\Gamma(2+q)}(\ln x)^{1+q}; \text{ for a.e. } x \in [0,e].$$

Set

$$\delta = x\frac{d}{dx}, \; q > 0, \; n = [q] + 1,$$

and

$$AC^n_\delta := \{u : [1, T] \to E : \delta^{n-1}[u(x)] \in AC(I)\}.$$

Analogous to the Riemann–Liouville fractional derivative, the Hadamard fractional derivative is defined in terms of the Hadamard fractional integral in the following way:

Definition 9.32. [281] (Hadamard fractional derivative). The Hadamard fractional derivative of order $q > 0$ applied to the function $w \in AC^n_\delta$ is defined as

$$({}^H D^q_1 w)(x) = \delta^n ({}^H I^{n-q}_1 w)(x).$$

In particular, if $q \in (0, 1]$, then

$$({}^H D^q_1 w)(x) = \delta({}^H I^{1-q}_1 w)(x).$$

Example 9.33. Let $0 < q < 1$. Let $w(x) = \ln x$, $x \in [0, e]$. Then

$$({}^H D^q_1 w)(x) = \frac{1}{\Gamma(2-q)}(\ln x)^{1-q}, \; \text{for a.e. } x \in [0, e].$$

It has been proved (see, e.g., Kilbas [280, Theorem 4.8]) that in the space $L^1(I)$, the Hadamard fractional derivative is the left-inverse operator to the Hadamard fractional integral, i.e.,

$$({}^H D^q_1)({}^H I^q_1 w)(x) = w(x).$$

From Theorem 2.3 of [281], we have

$$({}^H I^q_1)({}^H D^q_1 w)(x) = w(x) - \frac{({}^H I^{1-q}_1 w)(1)}{\Gamma(q)}(\ln x)^{q-1}.$$

Analogous to the Hadamard fractional calculus, the Caputo–Hadamard fractional derivative is defined in the following way:

Definition 9.34. (Caputo–Hadamard fractional derivative) The Caputo–Hadamard fractional derivative of order $q > 0$ applied to the function $w \in AC^n_\delta$ is defined as

$$({}^{Hc} D^q_1 w)(x) = ({}^H I^{n-q}_1 \delta^n w)(x).$$

In particular, if $q \in (0, 1]$, then

$$({}^{Hc} D^q_1 w)(x) = ({}^H I^{1-q}_1 \delta w)(x).$$

From the Hadamard fractional integral, the Hilfer–Hadamard fractional derivative (introduced for the first time in [337]) is defined in the following way:

Definition 9.35. (Hilfer–Hadamard fractional derivative). Let $\alpha \in (0,1)$, $\beta \in [0,1]$, $\gamma = \alpha + \beta - \alpha\beta$, $w \in L^1(I)$, and $^H I_1^{(1-\alpha)(1-\beta)} w \in AC(I)$. The Hilfer–Hadamard fractional derivative of order α and type β applied to the function w is defined as

$$
\begin{aligned}
(^H D_1^{\alpha,\beta} w)(t) &= \left(^H I_1^{\beta(1-\alpha)} (^H D_1^{\gamma} w)\right)(t) \\
&= \left(^H I_1^{\beta(1-\alpha)} \delta (^H I_1^{1-\gamma} w)\right)(t); \quad \text{for a.e. } t \in [1, T].
\end{aligned}
\tag{9.23}
$$

This new fractional derivative (9.23) may be viewed as interpolating the Hadamard fractional derivative and the Caputo–Hadamard fractional derivative. Indeed, for $\beta = 0$, this derivative reduces to the Hadamard fractional derivative and when $\beta = 1$, we recover the Caputo–Hadamard fractional derivative.

$$
^H D_1^{\alpha,0} = {}^H D_1^{\alpha}, \text{ and } {}^H D_1^{\alpha,1} = {}^{Hc} D_1^{\alpha}.
$$

From Theorem 21 in [338], we concluded the following lemma.

Lemma 9.36. *Let $g : [1,T] \times E \to E$ be such that $g(\cdot, u(\cdot)) \in C_{\gamma, \ln}([1,T])$ for any $u \in C_{\gamma, \ln}([1,T])$. Then problem (9.18) is equivalent to the following Volterra integral equation*

$$
u(t) = \frac{\phi_0}{\Gamma(\gamma)} (\ln t)^{\gamma-1} + (^H I_1^{\alpha} g(\cdot, u(\cdot)))(t).
$$

Definition 9.37. By a random solution of the coupled system (9.18)–(9.19) we mean a coupled measurable functions $(u,v) \in C_{\gamma_1, \ln} \times C_{\gamma_2, \ln}$ those satisfy the conditions (9.19) and Eqs. (9.18) on $[1, T]$.

Now we give (without proof) similar existence and uniqueness results for the system (9.18)–(9.19). Let us introduce the following hypotheses:

(9.37.1) The functions g_i; $i = 1, 2$ are Carathéodory.
(9.37.2) There exist measurable functions $p_i, q_i : \Omega \to (0, \infty)$; $i = 1, 2$ such that

$$
\|g_i(t, u_1, v_1) - g_i(t, u_2, v_2)\| \le p_i(w)\|u_1 - u_2\| + q_i(w)\|v_1 - v_2\|;
$$

for a.e. $t \in [1, T]$, and each $u_i, v_i \in \mathbb{R}^m$, $i = 1, 2$.
(9.37.3) There exist measurable functions $a_i, b_i : \Omega \to (0, \infty)$; $i = 1, 2$ such that

$$
\|g_i(t, u, v)\| \le a_i(w)\|u\| + b_i(w)\|v\|; \text{ for a.e. } t \in [1, T], \text{ and each } u, v \in \mathbb{R}^m.
$$

Theorem 9.38. *Assume that the hypotheses (9.37.1) and (9.37.2) hold. If for every $w \in \Omega$, the matrix*

$$
\begin{pmatrix}
\frac{(\ln T)^{\alpha_1}}{\Gamma(1+\alpha_1)} p_1(w) & \frac{(\ln T)^{\alpha_1}}{\Gamma(1+\alpha_1)} q_1(w) \\
\frac{(\ln T)^{\alpha_2}}{\Gamma(1+\alpha_2)} p_2(w) & \frac{(\ln T)^{\alpha_2}}{\Gamma(1+\alpha_2)} q_2(w)
\end{pmatrix}
$$

converges to 0, then the coupled system (9.18)–(9.19) has a unique random solution.

Theorem 9.39. *Assume that the hypotheses (9.37.1) and (9.37.3) hold. Then the coupled system (9.18)–(9.19) has at least a random solution.*

9.3.4 An example

Let $\Omega = (-\infty, 0)$ be equipped with the usual σ-algebra consisting of Lebesgue measurable subsets of $(-\infty, 0)$. Consider the following random coupled system of Hilfer fractional differential equations

$$
\begin{cases}
(D_0^{\frac{1}{2},\frac{1}{2}} u)(t, w) = f(t, u(t, w), v(t, w), w); \\
(D_0^{\frac{1}{2},\frac{1}{2}} v)(t) = g(t, u(t, w), v(t, w), w); \\
(I_0^{\frac{1}{4}} u)(0, w) = \cos w, \\
(I_0^{\frac{1}{4}} v_n)(0, w) = \sin w,
\end{cases}
\quad ; \ w \in \Omega, \ t \in [0, 1], \tag{9.24}
$$

where

$$
f(t, u, v, w) = \frac{t^{\frac{-1}{4}} w^2 (u(t) + v(t)) \sin t}{64(1 + w^2 + \sqrt{t})(1 + |u| + |v|)}; \ t \in [0, 1],
$$

$$
g(t, u, v) = \frac{w^2 (u(t) + v(t)) \cos t}{64(1 + |u| + |v|)}; \ w \in \Omega, \ t \in [0, 1].
$$

Set $\alpha_i = \beta_i = \frac{1}{2}$; $i = 1, 2$, then $\gamma_i = \frac{3}{4}$; $i = 1, 2$. The hypothesis (9.37.2) is satisfied with

$$
p_1(w) = p_2(w) = q_1(w) = q_2(w) = \frac{w^2}{64(1 + w^2)}.
$$

Also if, for every $w \in \Omega$, the matrix

$$
\frac{w^2}{64(1 + w^2)\Gamma(\frac{1}{2})} \begin{pmatrix} 1 & 1 \\ 1 & 1 \end{pmatrix}
$$

converges to 0. Hence, Theorem 9.28 implies that the system (9.24) has a unique random solution defined on $[0, 1]$.

9.4 Notes and remarks

The results of Chapter 9 are taken from [5,74]. For further detail on the topics covered in this chapter, we direct the reader to the monographs [69,70,440] and the publications [18, 252,331,378,388].

10

Oscillation and nonoscillation results for fractional q-difference equations and inclusions

The results obtained in this chapter are studied and presented as a consequence of the monographs [48,69,70,281,385,438,439] and the papers [1–4,7,13,14,20,21,53,76,233,234]. In recent years there has been much research activity concerning the oscillation and nonoscillation of solutions of different types of dynamic equations and inclusions. We refer the reader to the papers [167,259] and the references cited therein.

10.1 Oscillation and nonoscillation results for Caputo fractional q-difference equations and inclusions

10.1.1 Introduction

This section deals with some existence, oscillatory and nonoscillatory of solutions for some classes of Caputo fractional q-difference equations and inclusions by using set-valued analysis, fixed-point theory, and the method of upper and lower solutions.

First, we discuss the existence and the oscillatory and nonoscillatory of solutions to the fractional q-difference equation

$$({}^{c}D_{q}^{\alpha}u)(t) = f(t, u(t)); \ t \in I := [0, T], \tag{10.1}$$

with the initial condition

$$u(0) = u_0 \in \mathbb{R}, \tag{10.2}$$

where $q \in (0, 1)$, $\alpha \in (0, 1]$, $T > 0$, $f : I \times \mathbb{R} \to \mathbb{R}$ is a given continuous function, ${}^{c}D_{q}^{\alpha}$ is the Caputo fractional q-difference derivative of order α.

Next, we investigate the existence and the oscillatory and nonoscillatory of solutions to the fractional q-difference inclusion

$$({}^{c}D_{q}^{\alpha}u)(t) \in F(t, u(t)); \ t \in I, \tag{10.3}$$

with the initial condition (10.2), where $F : I \times \mathbb{R} \to \mathcal{P}(\mathbb{R})$ is a multi-valued map, $\mathcal{P}(\mathbb{R})$ is the family of all nonempty subsets of $\mathbb{R}$.

Fractional Difference, Differential Equations, and Inclusions. https://doi.org/10.1016/B978-0-44-323601-3.00017-4

233

Copyright © 2024 Elsevier Inc. All rights are reserved, including those for text and data mining, AI training, and similar technologies.

10.1.2 Caputo fractional q-difference equations

We begin by defining what we mean by a solution, an upper solution, and a lower solution to problem (10.1)–(10.2).

Definition 10.1. A function $u \in C(I)$ is said to be a solution of problem (10.1)–(10.2), if $u(0) = u_0$, and $^C D_q^\alpha u(t) = f(t, u(t))$ on I.

Definition 10.2. A function $w \in C(I)$ is said to be an upper solution of (10.1)–(10.2) if $w(0) \geq u_0$, and $^C D_q^\alpha w(t) \geq f(t, w(t))$ on I. Similarly, a function $v \in C(I)$ is said to be a lower solution of (10.1)–(10.2) if $v(0) \leq u_0$ and $^C D_q^\alpha v(t) \leq f(t, v(t))$ on I.

In the sequel we need the following fixed-point theorem:

Theorem 10.3. *(Schauder fixed-point theorem) [253] Let B be a closed, convex, and nonempty subset of a Banach space X. Let $N : B \to B$ be a continuous mapping such that $N(B)$ is a relatively compact subset of X. Then N has at least one fixed point in B.*

We present an existence result for the problem (10.1)–(10.2).

Theorem 10.4. *Assume that the following hypothesis holds.*

(10.4.1) *There exist v and $w \in C$, lower and upper solutions for the problem (10.1)–(10.2) such that $v \leq w$,*

Then the problem (10.1)–(10.2) has at least one solution u such that

$$v(t) \leq u(t) \leq w(t); \quad \text{for all } t \in I.$$

Proof. Consider the following modified problem

$$\begin{cases} (^C D_q^\alpha u)(t) = g(t, u(t)); \ t \in I := [0, T], \\ u(0) = u_0, \end{cases} \tag{10.4}$$

where

$$g(t, u(t)) = f(t, h(t, u(t))),$$
$$h(t, u(t)) = \max\{v(t), \min\{u(t), w(t)\}\},$$

for each $t \in I$.

A solution of problem (10.4) is a fixed point of the operator $N : C \to C$ defined by,

$$(Nu)(t) = u_0 + \int_0^t \frac{(t - qs)^{(\alpha-1)}}{\Gamma_q(\alpha)} f(s, u(s)) d_q s.$$

Notice that g is a continuous function, and from (10.4.1) there exists $M > 0$ such that

$$|g(t, u)| \leq M, \text{ for each } t \in I, \text{ and } u \in \mathbb{R}. \tag{10.5}$$

Set

$$\eta = |u_0| + \frac{MT^{(\alpha)}}{\Gamma_q(1 + \alpha)},$$

and

$$D = \{u \in C : \|u\|_C \leq \eta\}.$$

Clearly, D is a closed, bounded convex subset of C and that N maps D into itself. We shall show that N satisfies the assumptions of Theorem 10.3. The proof will be given in several steps.

Step 1. *N is continuous and $N(D)$ is bounded.*
It is clear that $N(D)$ is bounded, since $N(D) \subset D$ and D is bounded.
   Next, let $\{u_n\}$ be a sequence such that $u_n \to u$ in D. Then

$$
\begin{aligned}
&|(Nu_n)(t) - (Nu)(t)| \\
&\leq \int_0^t \frac{(t - qs)^{(\alpha-1)}}{\Gamma_q(\alpha)} |g(s, u_n(s)) - g(s, u(s))| d_q s \\
&\leq \int_0^t \frac{(t - qs)^{(\alpha-1)}}{\Gamma_q(\alpha)} \sup_{s \in I} |g(s, u_n(s)) - g(s, u(s))| d_q s.
\end{aligned}
$$

For each $t \in I$, set $(g \circ u)(t) := g(t, u(t))$. Thus we get

$$
\begin{aligned}
&(Nu_n)(t) - (Nu)(t)| \\
&\leq \int_0^t \frac{(t - qs)^{(\alpha-1)}}{\Gamma_q(\alpha)} \sup_{s \in I} |(g \circ u_n)(s) - (g \circ u)(s)| d_q s \\
&\leq \int_0^t \frac{(t - qs)^{(\alpha-1)}}{\Gamma_q(\alpha)} \|g \circ u_n - g \circ u\|_C d_q s \\
&\leq \frac{T^{(\alpha)}}{\Gamma_q(1 + \alpha)} \|g \circ u_n - g \circ u\|_C.
\end{aligned}
$$

From Lebesgue's dominated convergence theorem and the continuity of the function g, we get

$$|(Nu_n)(t) - (Nu)(t)| \to 0 \text{ as } n \to \infty.$$

Step 2. *N(D) is equicontinuous.*

Let $t_1, t_2 \in I$, $t_1 < t_2$, and let $u \in D$. Then

$$\|(Nu)(t_2) - (Nu)(t_1)\|$$

$$\leq \int_{t_1}^{t_2} \frac{(t_2 - qs)^{(\alpha-1)}}{\Gamma_q(\alpha)} |g(s, u(s))| d_q s$$

$$+ \int_0^{t_1} \left| \frac{(t_2 - qs)^{(\alpha-1)}}{\Gamma_q(\alpha)} - \frac{(t_1 - qs)^{(\alpha-1)}}{\Gamma_q(\alpha)} \right| |g(s, u(s))| d_q s$$

$$\leq M \int_{t_1}^{t_2} \frac{(t_2 - qs)^{(\alpha-1)}}{\Gamma_q(\alpha)} d_q s$$

$$+ M \int_0^{t_1} \left| \frac{(t_2 - qs)^{(\alpha-1)}}{\Gamma_q(\alpha)} - \frac{(t_1 - qs)^{(\alpha-1)}}{\Gamma_q(\alpha)} \right| d_q s.$$

As $t_1 \to t_2$, the right-hand side of the above inequality tends to zero.

As a consequence of the above two steps and the Arzelà–Ascoli theorem, we can conclude that N is a continuous and completely continuous operator. From an application of Theorem 10.3, we deduce that N has a fixed point u, which is a solution of the problem (10.4).

Step 3. *The solution u of (10.4) satisfies*

$$v(t) \leq u(t) \leq w(t) \quad \text{for all } t \in I.$$

Let u be the above solution to (10.4). We prove that

$$u(t) \leq w(t) \quad \text{for all } t \in I.$$

Assume that $u - w$ attains a positive maximum on I at $\bar{t} \in I$, that is,

$$(u - w)(\bar{t}) = \max\{u(t) - w(t) : t \in I\} > 0.$$

We distinguish the following cases.

Case 1. If $\bar{t} \in (0, T)$ then, there exists $t^* \in [0, \bar{t})$ such that

$$0 < u(t) - w(t) \leq u(\bar{t}) - w(\bar{t}); \quad \text{for all } t \in [t^*, \bar{t}]. \tag{10.6}$$

By the definition of h one has

$$({}^c D_q^\alpha u)(t) = g(t, u(t)), \tag{10.7}$$

for all $t \in [t^*, \bar{t}]$, where

$$g(t, u(t)) = f(t, w(t)); \quad t \in [t^*, \bar{t}].$$

An integration of Eq. (10.7), for each $t \in [t^*, \bar{t}]$ yields

$$u(t) = u_0 + \int_{t^*}^{t} \frac{(t - qs)^{(\alpha-1)}}{\Gamma_q(\alpha)} f(s, w(s)) d_q s. \qquad (10.8)$$

Using the fact that w is an upper solution to (10.1)–(10.2), we get

$$u(t) - u(t^*) \le w(t) - w(t^*) - \int_{t^*}^{t} \frac{(t - qs)^{(\alpha-1)}}{\Gamma_q(\alpha)} f(s, w(s)) d_q s$$

$$< w(t) - w(t^*) \qquad (10.9)$$

Thus, from (10.6) and (10.9), we obtain the contradiction

$$u(\bar{t}) - w(\bar{t}) < u(t^*) - w(t^*); \text{ for all } t \in [t^*, \bar{t}].$$

Case 2. If $\bar{t} = 0$, then

$$w(1) < u(1) \le w(1)$$

which is a contradiction. Thus

$$u(t) \le w(t) \text{ for all } t \in I.$$

Analogously, we can prove that

$$u(t) \ge v(t), \text{ for all } t \in I.$$

This shows that the integral problem (10.4) has a solution u satisfying $v \le u \le w$, which is solution of problem (10.1)–(10.2). $\qquad \square$

The following theorem gives sufficient conditions to ensure the nonoscillation of solutions of problem (10.1)–(10.2).

Definition 10.5. A solution $u \in C$ of problem (10.1)–(10.2) is said to be oscillatory if it is neither eventually positive nor eventually negative. Otherwise u is called nonoscillatory.

Theorem 10.6. *Assume that* (10.4.1) *and the following hypothesis hold.*

(10.6.1) *v is an eventually positive nondecreasing or w is an eventually negative nonincreasing.*

Then every solution u of (10.1)–(10.2) such that $u \in [v, w]$ is nonoscillatory.

Proof. Assume that v is an eventually positive. Thus there exists $T_v > 0$ such that

$$v(t) > 0; \text{ for all } t > T_v.$$

Hence $u(t) > 0$ for all $t > T_v$. This means that y is nonoscillatory.

Analogously, if w is an eventually negative, then there exists $T_w > 0$ such that $u(t) < 0$; for all $t > T_w$, which means that u is nonoscillatory. $\qquad \square$

The following theorem presents the oscillatory of the solutions of problem (10.1)–(10.2).

Theorem 10.7. *Assume that (10.4.1) and the following hypothesis hold.*

(10.7.1) *v and w are oscillatory.*

Then every solution u of (10.1)–(10.2) such that $u \in [v, w]$ is oscillatory.

Proof. Assume that problem (10.1)–(10.2) has a nonoscillatory solution u on I. Then there exists $T_u > 0$, such that $u(t) > 0$, for all $t > T_u$, or $u(t) < 0$, for all $t > T_u$. In the case when $u(t) > 0$, for all $t > T_u$, we have $w(t) > 0$, for all $t > T_u$, which is a contradiction since w is an oscillatory upper solution. Analogously in the case when $u(t) < 0$, for all $t > T_u$. We have $v(t) < 0$, for all $t > T_u$, which is a contradiction since v is an oscillatory lower solution. $\square$

10.1.3 Caputo fractional q-difference inclusions

Definition 10.8. A function $u \in AC(I)$ is said to be a solution of problem (10.3)–(10.2), if $u(0) = u_0$, and there exists a function $f \in S_{F \circ u}$ such that $^C D_q^\alpha u(t) = f(t)$ a.e. $t \in I$.

Definition 10.9. A function $w \in AC(I)$ is said to be an upper solution of (10.3)–(10.2) if $w(0) \geq u_0$, and there exists a function $v_1 \in S_{F \circ w}$ such that $^C D_q^\alpha w(t) \geq v_1(t)$ a.e. $t \in I$. Similarly, a function $v \in AC(I)$ is said to be a lower solution of (10.3)–(10.2) if $v(0) \geq u_0$, and there exists a function $v_2 \in S_{F \circ v}$ such that $^C D^\alpha v(t) \leq v_2(t)$ a.e. $t \in I$.

Next, we need the following fixed-point theorem:

Theorem 10.10. *(Martelli's fixed-point theorem) [305] Let X be a Banach space and $N : X \to \mathcal{P}_{cl,cv}(X)$ be an upper semicontinuous and condensing multi-valued operator. If the set $\Omega := \{u \in X : \lambda u \in N(u)$ for some $\lambda > 1\}$ is bounded, then N has a fixed point.*

Theorem 10.11. *Assume that the following hypotheses hold:*

(10.11.1) *$F : I \times \mathbb{R} \to \mathcal{P}_{cp,cv}(\mathbb{R})$ is Carathéodory;*
(10.11.2) *There exist $v, w \in AC(I)$, which are the lower and upper solutions, respectively, for problem (10.3)–(10.2) such that $v \leq w$;*
(10.11.3) *There exists $l \in L^1(I, \mathbb{R}_+)$ such that*

$$H_d(F(t, u), F(t, \bar{u})) \leq l(t)|u - \bar{u}| \text{ for every } u, \bar{u} \in \mathbb{R},$$

and

$$d(0, F(t, 0)) \leq l(t) \text{ a.e. } t \in I.$$

Then the problem (10.3)–(10.2) has at least one solution u defined on I such that

$$v \leq u \leq w.$$

Proof. Consider the multi-valued operator $N : C(I) \to \mathcal{P}(C(I))$ defined by:

$$N(u) = \left\{ h \in C(I) : h(t) = u_0 + \int_0^t \frac{(t - qs)^{(\alpha - 1)}}{\Gamma_q(\alpha)} v(s) d_q s; \ v \in S_{F \circ u} \right\}. \tag{10.10}$$

Clearly, the fixed points of N are solutions of the problem (10.3)–(10.2).
Consider the following modified problem

$$^C D_q^\alpha u(t) \in F(t, \tau(u(t))), \ \text{for a.e. } t \in I, \tag{10.11}$$

$$u(0) = u_0 \tag{10.12}$$

where

$$\tau(u(t)) = \max\{v(t), \min\{u(t), w(t)\}\},$$

and

$$\bar{u}(t) = \tau(u(t)).$$

A solution to (10.11)–(10.12) is a fixed point of the operator $N : C(I) \to \mathcal{P}(C(I))$ defined by

$$N(u) = \{ h \in C(I) \ : \ h(t) = u(0) + (I_q^\alpha v)(t) \},$$

where

$$v \in \{ x \in \widetilde{S}^1_{F \circ \tau(u)} : x(t) \geq v_1(t) \text{ on } A_1 \text{ and } x(t) \leq v_2(t) \text{ on } A_2 \},$$

$$S^1_{F \circ \tau(y)} = \{ x \in L^1(I) : x(t) \in F(t, (\tau u)(t)), \ \text{a.e. } t \in I \},$$

$$A_1 = \{ t \in I : u(t) < v(t) \leq w(t) \}, \quad A_2 = \{ t \in I : v(t) \leq w(t) < u(t) \}.$$

Remark 10.12. **(1)** For each $u \in C(I)$, the set $\widetilde{S}^1_{F \circ \tau(u)}$ is nonempty. In fact, (10.11.1) implies that there exists $v_3 \in S^1_{F \circ \tau(u)}$, so we set

$$v = v_1 \chi_{A_1} + v_2 \chi_{A_2} + v_3 \chi_{A_3},$$

where

$$A_3 = \{ t \in I : v(t) \leq u(t) \leq w(t) \}.$$

Then by decomposability, $x \in \widetilde{S}^1_{F \circ \tau(u)}$.

(2) From the definition of τ, it is clear that $F(\cdot, \tau u(\cdot))$ is an L^1-Carathéodory multi-valued map with compact convex values and there exists $\phi_1 \in C(I, \mathbb{R}_+)$ such that

$$\| F(t, \tau u(t)) \|_{\mathcal{P}} \leq \phi_1(t) \text{ for each } u \in \mathbb{R}.$$

Now set

$$R := |u_0| + \frac{\| \phi_1 \|_\infty T^{(\alpha)}}{\Gamma_q(1 + \alpha)},$$

and consider the closed and convex subset of $C(I)$ given by

$$B = \{u \in C(I) : \|u\|_\infty \le R\}.$$

We shall show that the operator $N : B \to \mathcal{P}_{cl,cv}(B)$ satisfies all the assumptions of Theorem 10.10. The proof will be given in steps.

Step 1. *$N(u)$ is convex for each $u \in B$.*
Let h_1, h_2 belong to $N(u)$; then there exist $v_1, v_2 \in \widetilde{S}^1_{F \circ \tau(u)}$ such that, for each $t \in I$ and any $i = 1, 2$, we have

$$h_i(t) = u(0) + (I_q^\alpha v_i)(t).$$

Let $0 \le d \le 1$. Then, for each $t \in I$, we have

$$(dh_1 + (1-d)h_2)(t) = u(0) + \int_0^t \frac{(t-qs)^{(\alpha-1)}}{\Gamma_q(\alpha)}[dv_1(s) + (1-d)v_2(s)]d_q s.$$

Since $S_{F \circ \tau(u)}$ is convex (because F has convex values), we have

$$dh_1 + (1-d)h_2 \in N(u).$$

Step 2. *N maps bounded sets into bounded sets in B.*
For each $h \in N(u)$, there exists $v \in \widetilde{S}^1_{F \circ \tau(u)}$ such that

$$h(t) = u(0) + \int_0^t \frac{(t-qs)^{(\alpha-1)}}{\Gamma_q(\alpha)} v(s)d_q s.$$

From conditions (10.11.1)–(10.11.3), for each $t \in I$, we have

$$|h(t)| \le |u(0)| + \left| \int_0^t \frac{(t-qs)^{(\alpha-1)}}{\Gamma_q(\alpha)} |v(s)|d_q s \right|$$

$$\le |u_0| + \frac{\|\phi_1\|_\infty T^{(\alpha)}}{\Gamma_q(1+\alpha)}.$$

Thus,

$$\|h\|_\infty \le R.$$

Step 3. *N maps bounded sets into equicontinuous sets of B.*
Let $t_1, t_2 \in I$ with $t_1 < t_2$, and let $u \in B$ and $h \in N(u)$. Then

$$|h(t_2) - h(t_1)| = \left| \int_0^{t_1} \frac{|(t_2-qs)^{(\alpha-1)} - (t_1-qs)^{(\alpha-1)}|}{\Gamma_q(\alpha)} v(s)d_q s \right.$$

$$\left. + \int_{t_1}^{t_2} \frac{(t_2-qs)^{(\alpha-1)}}{\Gamma_q(\alpha)} v(s)d_q s \right|$$

$$\leq \int_0^{t_1} \frac{|(t_2 - qs)^{(\alpha-1)} - (t_1 - qs)^{(\alpha-1)}|}{\Gamma_q(\alpha)} |v(s)| d_q s$$

$$+ \int_{t_1}^{t_2} \frac{|(t_2 - qs)^{(\alpha-1)}|}{\Gamma_q(\alpha)} |v(s)| d_q s$$

$$\leq \|\phi_1\|_\infty \int_0^{t_1} \frac{|(t_2 - qs)^{(\alpha-1)} - (t_1 - qs)^{(\alpha-1)}|}{\Gamma_q(\alpha)} d_q s$$

$$+ \|\phi_1\|_\infty \int_{t_1}^{t_2} \frac{|(t_2 - qs)^{(\alpha-1)}|}{\Gamma_q(\alpha)} d_q s$$

$$\to 0 \text{ as } t_1 \to t_2.$$

As a consequence of the three steps above, we can conclude from the Arzelà–Ascoli theorem that $N : C(I) \to \mathcal{P}(C(I))$ is continuous and completely continuous.

Step 4. *N has a closed graph.*

Let $u_n \to u_*$, $h_n \in N(u_n)$, and $h_n \to h_*$. We need to show that $h_* \in N(u_*)$. Now $h_n \in N(u_n)$ implies there exists $v_n \in \widetilde{S}^1_{F \circ \tau(u_n)}$ such that, for each $t \in I$,

$$h_n(t) = u(0) + \int_0^t \frac{(t - qs)^{(\alpha-1)}}{\Gamma_q(\alpha)} v_n(s) d_q s.$$

We must show that there exists $v_* \in \widetilde{S}^1_{F \circ \tau(u_*)}$ such that, for each $t \in I$,

$$h_*(t) = u(0) + \int_0^t \frac{(t - qs)^{(\alpha-1)}}{\Gamma_q(\alpha)} v_*(s) d_q s.$$

Since $F(t, \cdot)$ is upper semi-continuous, for every $\epsilon > 0$, there exists a natural number $n_0(\epsilon)$ such that, for every $n \geq n_0(\epsilon)$, we have

$$v_n(t) \in F(t, \tau u_n(t)) \subset F(t, u_*(t)) + \epsilon B(0, 1) \quad \text{a.e. } t \in I.$$

Since $F(\cdot, \cdot)$ has compact values, there exists a subsequence $v_{n_m}(\cdot)$ such that

$$v_{n_m}(\cdot) \to v_*(\cdot) \quad \text{as} \quad m \to \infty,$$

and

$$v_*(t) \in F(t, \tau u_*(t)) \quad \text{a.e. } t \in I.$$

For every $w \in F(t, \tau u_*(t))$, we have

$$|v_{n_m}(t) - v_*(t)| \leq |v_{n_m}(t) - w| + |w - v_*(t)|.$$

Hence

$$|v_{n_m}(t) - v_*(t)| \leq d(v_{n_m}(t), F(t, \tau u_*(t))).$$

We obtain an analogous relation by interchanging the roles of v_{n_m} and v_* to obtain

$$|v_{n_m}(t) - v_*(t)| \leq H_d(F(t, \tau u_n(t)), F(t, \tau u_*(t))) \leq l(t)\|y_n - y_*\|_\infty.$$

Thus

$$|h_{n_m}(t) - h_*(t)| \leq \int_0^t \frac{|(t-qs)^{(\alpha-1)}|}{\Gamma_q(\alpha)}|v_{n_m}(s) - v_*(s)|d_q s$$

$$\leq \|u_{n_m} - u_*\|_\infty \int_0^t \frac{|(t-qs)^{(\alpha-1)}|}{\Gamma_q(\alpha)}l(s)d_q s.$$

Therefore

$$\|h_{n_m} - h_*\|_\infty \leq \|u_{n_m} - u_*\|_\infty \int_0^t \frac{|(t_1-qs)^{(\alpha-1)}|}{\Gamma_q(\alpha)}l(s)d_q s \to 0 \quad \text{as } m \to \infty,$$

so Lemma 2.28 implies that N is upper semicontinuous.

Step 5. *The set $\Omega = \{u \in C : \lambda u \in N(u) \text{ for some } \lambda > 1\}$ in bounded.*
Let $u \in \Omega$. Then there exists $f \in \Lambda(\tilde{S}_{Fog(u)})$ such that

$$\lambda u(t) = |u_0| + \int_0^t \frac{(t-qs)^{(\alpha-1)}}{\Gamma_q(\alpha)}|f(s)|d_q s.$$

As in Step 2, this implies that for each $t \in I$, we have

$$\|u\|_C \leq \frac{R}{\lambda} < \ell.$$

This shows that Ω is bounded. As a consequence of Theorem 2.78, we deduce that N has a fixed point, which is a solution of (10.13)–(10.14) on I.

Step 6. *Every solution u of (10.11)–(10.12) satisfies $v(t) \leq u(t) \leq w(t)$ for all $t \in I$.*
Let u be a solution of (10.11)–(10.12). To prove that $v(t) \leq u(t)$ for all $t \in I$, suppose this is not the case. Then there exist t_1, t_2, with $t_1 < t_2$, such that $v(t_1) = u(t_1)$ and $v(t) > u(t)$ for all $t \in (t_1, t_2)$. In view of the definition of τ,

$$^C D_q^\alpha u(t) \in F(t, v(t)) \text{ for all } t \in (t_1, t_2).$$

Thus there exists $y \in S_{Fo\tau(v)}$ with $y(t) \geq v_1(t)$ a.e. on (t_1, t_2) such that

$$^C D_q^\alpha u(t) = y(t) \text{ for all } t \in (t_1, t_2).$$

An integration on $(t_1, t]$, with $t \in (t_1, t_2)$, yields

$$u(t) - y(t_1) = \int_{t_1}^t \frac{(t-qs)^{(\alpha-1)}}{\Gamma_q(\alpha)}v(s)d_q.$$

Since v is a lower solution of (10.13)–(10.14),

$$v(t) - v(t_1) \leq \int_{t_1}^t \frac{(t - qs)^{(\alpha-1)}}{\Gamma_q(\alpha)} v_1(s)d_q, \ t \in (t_1, t_2).$$

From the facts that $u(t_0) = v(t_0)$ and $v(t) \geq v_1(t)$, it follows that

$$v(t) \leq u(t) \text{ for all } t \in (t_1, t_2).$$

This is a contradiction, since $v(t) > u(t)$ for all $t \in (t_1, t_2)$. Consequently,

$$v(t) \leq u(t) \text{ for all } t \in I.$$

Similarly, we can prove that

$$u(t) \leq w(t) \text{ for all } t \in I.$$

This shows that

$$v(t) \leq u(t) \leq w(t) \text{ for all } t \in I.$$

Therefore the problem (10.11)–(10.12) has a solution u, which is a solution of (10.13)–(10.14) satisfying $v \leq u \leq w$.   □

   As in Theorems 10.6 and 10.7, the following results ensure the nonoscillation and oscillation of solutions of problem (10.13)–(10.2).

Theorem 10.13. *Assume that (10.11.1)–(10.11.3) and the following hypothesis hold.*

(10.13.1) *v is an eventually positive nondecreasing, or w is an eventually negative nonincreasing.*

Then every solution u of (10.13)–(10.2) such that $u \in [v, w]$ is nonoscillatory.

Theorem 10.14. *Assume that (10.11.1)–(10.11.3) and the following hypothesis hold.*

(10.14.1) *v and w are oscillatory.*

Then every solution u of (10.13)–(10.14) such that $u \in [v, w]$ is oscillatory.

10.2 Existence and oscillation for coupled fractional *q*-difference systems

In this section we discuss the existence of solutions and their oscillation for the following coupled fractional q-difference system

$$\begin{cases} (^C D_q^{\alpha_1} u_1)(t) = f_1(t, u_2(t)), \\ (^C D_q^{\alpha_2} u_2)(t) = f_2(t, u_1(t)), \\ (u_1(0), u_2(0)) = (u_{01}, u_{02}), \ u \text{ and } v \text{ are bounded on } \mathbb{R}_+, \end{cases} \quad ; t \in \mathbb{R}_+, \quad (10.13)$$

where $q \in (0,1)$, $\alpha_i \in (0,1]$, $\mathbb{R}_+ := [0,+\infty)$, $f_i : \mathbb{R}_+ \times \mathbb{R} \to \mathbb{R}$; $i = 1,2$, are given functions, and $^C D_q^{\alpha_i}$ is the Caputo fractional q-difference derivative of order α_i; $i = 1,2$.

10.2.1 Existence of bounded solutions

In this section we are concerned with the existence of solutions of the coupled system (10.13).

Definition 10.15. By a solution of the coupled system (10.13) we mean a pair of bounded coupled functions $(u_1, u_2) \in C(I) \times C(I)$ that satisfies the system

$$\begin{cases} (^C D_q^{\alpha_1} u_1)(t) = f_1(t, u_2(t)), \\ (^C D_q^{\alpha_2} u_2)(t) = f_2(t, u_1(t)), \end{cases}$$

on $\mathbb{R}_+ \times \mathbb{R}_+$ and the initial conditions $(u_1(0), u_2(0)) = (u_{01}, u_{02})$.

For $n \in \mathbb{N}$, let $I_n := [0,n]$. We denote by $X_n := C(I_n) \times C(I_n)$ the Banach space with the norm

$$\|(u,v)\|_{X_n} = \|u\|_\infty + \|v\|_\infty.$$

The following hypotheses will be used in the sequel.

(10.15.1) The functions $t \mapsto f_1(t,v)$ and $t \mapsto f_2(t,u)$ are measurable on $I_n := [0,n]$, $n \in \mathbb{N}$, for each $u, v \in \mathbb{R}$, and the functions $u \mapsto f_1(t,v)$ and $v \mapsto f_2(t,u)$ is continuous for a.e. $t \in I_n$.

(10.15.2) There exist continuous functions $p_{in} : I_n \to \mathbb{R}_+$, $n = 1,2$, such that

$$|f_i(t, u_i)| \le p_{in}(t), \text{ for a.e. } t \in I_n, \text{ and each } u_i \in \mathbb{R}.$$

Set

$$p_{in}^* = \sup_{t \in I_n} p_{in}(t), \ i = 1,2.$$

Theorem 10.16. *Assume that hypotheses* (10.15.1) *and* (10.15.2) *hold. Then the problem* (10.13) *has at least one bounded solution defined on* $\mathbb{R}_+$.

Proof. The proof will be given in two parts. Fix $n \in \mathbb{N}$ and consider the problem

$$\begin{cases} (^C D_q^{\alpha_1} u_1)(t) = f_1(t, u_2(t)) \\ (^C D_q^{\alpha_2} u_2)(t) = f_2(t, u_1(t)) \quad ; t \in I_n := [0,n]. \\ (u_1(0), u_2(0)) = (u_{01}, u_{02}) \end{cases} \tag{10.14}$$

Part 1. We begin by showing that (10.14) has a solution $(u_{1n}, u_{2n}) \in X_n$ with

$$\|(u_{1n}, u_{2n})\|_{X_n} \le R_n := \|u_{01}\| + \|u_{02}\| + \sum_{i=1}^{2} \frac{n^\alpha p_{in}^*}{\Gamma_q(1 + \alpha_i)}.$$

Consider the operators $N_i : C(I_n) \to C(I_n)$, $i = 1, 2$, and $N : X_n \to X_n$ defined by

$$(N_1 u_1)(t) = u_{01} + (I_q^{\alpha_1} f_1(\cdot, u_2(\cdot)))(t), \ t \in I, \tag{10.15}$$

$$(N_2) u_2(t) = u_{02} + (I_q^{\alpha_2} f_2(\cdot, u_1(\cdot)))(t), \ t \in I, \tag{10.16}$$

and

$$(N(u_1, u_2))(t) = ((N_1 u_1)(t), (N_2 u_2)(t)). \tag{10.17}$$

Clearly, the fixed points of the operator N are solutions of our coupled system (10.14).

For any $n \in \mathbb{N}^* := \mathbb{N} \setminus \{0\}$, we consider the ball

$$B_{R_n} := B(0, R_n) = \{w = (w_1, w_2) \in X_n : \|w\|_{X_n} \le R_n\}.$$

We shall show that the operator $N : B_{R_n} \to B_{R_n}$ satisfies all the assumptions of Theorem 2.68. The proof will be given in several steps.

Step 1. $N : B_{R_n} \to B_{R_n}$ *is continuous.*

Let $\{u_k\}_{k \in \mathbb{N}}$ be a sequence such that $u_k := (u_{1k}, u_{2k}) \to u := (u_1, u_2)$ in B_{R_n}. Then, for each $t \in I_n$, we have

$$\|(N_1 u_{1k})(t) - (N_1 u_1)(t)\| \le \int_0^t \frac{(tq - s)^{(\alpha_1 - 1)}}{\Gamma_q(\alpha_1)} \|f_1(s, u_{2k}(s)) - f_1(s, u_2(s))\| d_q s,$$

and

$$\|(N_2 u_{2k})(t) - (N_2 u_2)(t)\| \le \int_0^t \frac{(tq - s)^{(\alpha_2 - 1)}}{\Gamma_q(\alpha_2)} \|f_2(s, u_{1k}(s)) - f_1(s, u_1(s))\| d_q s.$$

Since $u_{ik} \to u_i$ as $k \to \infty$, the Lebesgue dominated convergence theorem implies that

$$\|N_i(u_{ik}) - N_i(u_i)\|_\infty \to 0 \quad \text{as } k \to \infty.$$

Hence

$$\|N(u_k) - N(u)\|_{X_n} \to 0 \quad \text{as } k \to \infty.$$

Step 2. $N(B_{R_n})$ *is uniformly bounded.*

For any $u := (u_1, u_2) \in X_n$, and each $t \in I_n$, we have

$$|(N_1(u_1)(t)| \le |u_{01}| + \int_0^t \frac{(t - qs)^{(\alpha_1 - 1)}}{\Gamma_q(\alpha_1)} |f_1(s, u_2(s))| d_q s$$

$$\le |u_{01}| + \int_0^t \frac{(t - qs)^{(\alpha_1 - 1)}}{\Gamma_q(\alpha)} p_{1n}(s) d_q s$$

$$\le |u_{01}| + p_{1n}^* \int_0^t \frac{(t - qs)^{(\alpha_1 - 1)}}{\Gamma_q(\alpha_1)} d_q s$$

$$\le |u_{01}| + \frac{n^\alpha p_{1n}^*}{\Gamma_q(1 + \alpha_1)}.$$

Also

$$|(N_2(u_2)(t)| \le |u_{02}| + \int_0^t \frac{(t-qs)^{(\alpha_2-1)}}{\Gamma_q(\alpha_2)}|f_2(s,u_1(s))|d_q s$$

$$\le |u_{02}| + \frac{n^\alpha p_{2n}^*}{\Gamma_q(1+\alpha_2)}.$$

Thus we get

$$\|(Nu)(t)\| = \|(N_1 u_1)(t)\| + \|(N_2 u_2)(t)\|$$

$$\le \|u_{01}\| + \frac{n^{\alpha_1} p_{1n}^*}{\Gamma_q(1+\alpha_1)} + \|u_{02}\| + \frac{n^{\alpha_2} p_{2n}^*}{\Gamma_q(1+\alpha_2)}$$

$$= R_n.$$

Hence

$$\|N(u)\|_{X_n} \le R_n. \tag{10.18}$$

This proves that $N(B_{R_n}) \subset B_{R_n}$.

Step 3. $N(B_{R_n})$ *is equicontinuous.*
Let $t_1, t_2 \in I_n$, $t_1 < t_2$ and let $u := (u_1, u_2) \in B_{R_n}$. Thus we have

$$|(N_1 u_1)(t_2) - (N_1 u_1)(t_1)|$$

$$\le \int_0^{t_1} \frac{|(t_2-qs)^{(\alpha-1)} - (t_1-qs)^{(\alpha_1-1)}|}{\Gamma_q(\alpha_1)}|f+1(s,u_2(s))|d_q s$$

$$+ \int_{t_1}^{t_2} \frac{|(t_2-qs)^{(\alpha_1-1)}|}{\Gamma_q(\alpha_1)}|f_1(s,u_2(s))|d_q s$$

$$\le p_{1n}^* \int_0^{t_1} \frac{|(t_2-qs)^{(\alpha_1-1)} - (t_1-qs)^{(\alpha_1-1)}|}{\Gamma_q(\alpha_1)}d_q s$$

$$+ p_{1n}^* \int_{t_1}^{t_2} \frac{|(t_2-qs)^{(\alpha_1-1)}|}{\Gamma_q(\alpha_1)}d_q s.$$

Also we get

$$|(N_2 u_2)(t_2) - (N_2 u_2)(t_1)|$$

$$\le p_{2n}^* \int_0^{t_1} \frac{|(t_2-qs)^{(\alpha_2-1)} - (t_1-qs)^{(\alpha_2-1)}|}{\Gamma_q(\alpha_2)}d_q s$$

$$+ p_{2n}^* \int_{t_1}^{t_2} \frac{|(t_2-qs)^{(\alpha_1-1)}|}{\Gamma_q(\alpha_2)}d_q s.$$

Thus

$$|(Nu)(t_2) - (Nu)(t_1)|$$

$$= |(N_1 u_1)(t_2) - (N_1 u_1)(t_1)| + |(N_2 u_2)(t_2) - (N_2 u_2)(t_1)|$$

$$\leq p_{1n}^* \int_0^{t_1} \frac{|(t_2 - qs)^{(\alpha_1 - 1)} - (t_1 - qs)^{(\alpha_1 - 1)}|}{\Gamma_q(\alpha_1)} d_q s$$

$$+ p_{2n}^* \int_0^{t_1} \frac{|(t_2 - qs)^{(\alpha_2 - 1)} - (t_1 - qs)^{(\alpha_2 - 1)}|}{\Gamma_q(\alpha_2)} d_q s$$

$$+ p_{1n}^* \int_{t_1}^{t_2} \frac{|(t_2 - qs)^{(\alpha_1 - 1)}|}{\Gamma_q(\alpha_1)} d_q s + p_{2n}^* \int_{t_1}^{t_2} \frac{|(t_2 - qs)^{(\alpha_1 - 1)}|}{\Gamma_q(\alpha_2)} d_q s.$$

As $t_1 \longrightarrow t_2$, the right-hand side of the above inequality tends to zero.

As a consequence of Steps 1 to 3, together with the Arzelà–Ascoli theorem, we can conclude that N is continuous and compact. From an application of Theorem 2.68, we deduce that N has a fixed point (u_1, u_2), which is a solution of the problem (10.14).

Part 2. The diagonalization process.

Now, we use the following diagonalization process. For $k \in \mathbb{N}$ and $i = 1, 2$, we let $w_k = (w_{1k}, w_{2k})$, such that

$$\begin{cases} w_{ik}(t) = u_{in_k}(t); \ t \in [0, n_k], \\ w_{ik}(t) = u_{in_k}(n_k); \ t \in [n_k, \infty) \end{cases} ; \ i = 1, 2.$$

Here $\{n_k\}_{k \in \mathbb{N}^*}$ is a sequence of numbers satisfying

$$0 < n_1 < n_2 < \cdots n_k < \cdots \uparrow \infty.$$

Let $S = \{w_k\}_{k=1}^\infty$. Notice that

$$|w_{n_k}(t)| = \sum_{i=1}^2 |w_{in_k}(t)| \leq R_n, \text{ for } t \in [0, n_1], \ k \in \mathbb{N}.$$

Also, if $k \in \mathbb{N}$ and $t \in [0, n_1]$, we have

$$w_{1n_k}(t) = u_0 + \int_0^t \frac{(t - qs)^{(\alpha_1 - 1)}}{\Gamma_q(\alpha_1)} f(s, w_{2n_k}(s)) d_q s,$$

and

$$w_{2n_k}(t) = u_0 + \int_0^t \frac{(t - qs)^{(\alpha_2 - 1)}}{\Gamma_q(\alpha_2)} f(s, w_{1n_k}(s)) d_q s.$$

Thus, for $k \in \mathbb{N}$ and $t, x \in [0, n_1]$, we have

$$|w_{1n_k}(t) - w_{1n_k}(x)| \leq \int_0^{n_1} \frac{|(t - qs)^{(\alpha_1 - 1)} - (x - qs)^{(\alpha_1 - 1)}|}{\Gamma_q(\alpha_1)} |f_1(s, w_{2n_k}(s))| d_q s,$$

and

$$|w_{2n_k}(t) - w_{2n_k}(x)| \leq \int_0^{n_1} \frac{|(t - qs)^{(\alpha_2 - 1)} - (x - qs)^{(\alpha_2 - 1)}|}{\Gamma_q(\alpha_2)} |f_2(s, w_{1n_k}(s))| d_q s.$$

Hence

$$|w_{1n_k}(t) - w_{1n_k}(x)| \leq p_{1n_1}^* \int_0^{n_1} \frac{|(t-qs)^{(\alpha_1-1)} - (x-qs)^{(\alpha_1-1)}|}{\Gamma_q(\alpha_1)} d_q s,$$

and

$$|w_{2n_k}(t) - w_{2n_k}(x)| \leq p_{2n-1}^* \int_0^{n_1} \frac{|(t-qs)^{(\alpha_2-1)} - (x-qs)^{(\alpha_2-1)}|}{\Gamma_q(\alpha_2)} d_q s.$$

The Arzelà–Ascoli theorem guarantees that there is a subsequence $\mathbb{N}_1^*$ of $\mathbb{N}$ and a coupled function $z_1 := (z_{11}, z_{21}) \in X_{n_1}$ with $u_{n_k} \to z_1$ as $k \to \infty$ in X_{n_1} through $\mathbb{N}_1^*$. Let $\mathbb{N}_1 = \mathbb{N}_1^* \backslash \{1\}$.

Notice that

$$|w_{n_k}(t)| \leq R_n, \text{ for } t \in [0, n_2], \ k \in \mathbb{N}.$$

Also, if $k \in \mathbb{N}$ and $t, x \in [0, n_2]$, we have

$$|w_{1n_k}(t) - w_{1n_k}(x)| \leq p_{1n_2}^* \int_0^{n_2} \frac{|(t-qs)^{(\alpha_1-1)} - (x-qs)^{(\alpha-1-1)}|}{\Gamma_q(\alpha_1)} d_q s,$$

and

$$|w_{2n_k}(t) - w_{2n_k}(x)| \leq p_{2n_2}^* \int_0^{n_2} \frac{|(t-qs)^{(\alpha_2-1)} - (x-qs)^{(\alpha-2-1)}|}{\Gamma_q(\alpha_2)} d_q s.$$

The Arzelà–Ascoli theorem guarantees that there is a subsequence $\mathbb{N}_2^*$ of $\mathbb{N}_1$ and a function $z_2 := (z_{12}, z_{22}) \in X_{n_2}$ with $u_{n_k} \to z_2$ as $k \to \infty$ in X_{n_2} through $\mathbb{N}_2^*$. Note that $z_1 = z_2$ on $[0, n_1]$ since $\mathbb{N}_2^* \subset \mathbb{N}_1$. Let $\mathbb{N}_2 = \mathbb{N}_2^* \backslash \{2\}$. Proceed inductively to obtain for $m = 3, 4, \dots$ a subsequence $\mathbb{N}_m^*$ of $\mathbb{N}_{m-1}$ and a function $z_m := (z_{1m}, z_{2m}) \in X_{n_m}$ with $u_{n_k} \to z_m$ as $k \to \infty$ in X_{n_m} through $\mathbb{N}_m^*$. Let $\mathbb{N}_m = \mathbb{N}_m^* \backslash \{m\}$.

Define a function y as follows. Fix $t \in (0, \infty)$ and let $m \in \mathbb{N}$ with $t \leq n_m$. Then define $u(t) := z_m(t)$. Thus $u = (u_1, u_2) \in X_\infty = C[0, \infty) \times C[0, \infty)$, $u(0) = (u_{01}, u_{02})$ and $|u(t)| \leq R_n$, for $t \in [0, \infty)$.

Again fix $t \in (0, \infty)$ and let $m \in \mathbb{N}$ with $t \leq n_m$. Then, for $n \in \mathbb{N}_m$, we have

$$u_{1n_k}(t) = u_{01} + \int_0^{n_m} \frac{(t-qs)^{(\alpha_1-1)}}{\Gamma_q(\alpha_1)} f_1(s, w_{2n_k}(s)) d_q s,$$

and

$$u_{2n_k}(t) = u_{02} + \int_0^{n_m} \frac{(t-qs)^{(\alpha_2-1)}}{\Gamma_q(\alpha_2)} f_2(s, w_{1n_k}(s)) d_q s.$$

Let $n_k \to \infty$ through $\mathbb{N}_m$ to obtain

$$z_{1m}(t) = u_{01} + \int_0^{n_m} \frac{(t-qs)^{(\alpha_1-1)}}{\Gamma_q(\alpha_1)} f_1(s, z_{2m}(s)) d_q s,$$

and

$$z_{2m}(t) = u_{02} + \int_0^{n_m} \frac{(t - qs)^{(\alpha_2 - 1)}}{\Gamma_q(\alpha_2)} f_2(s, z_{1m}(s)) d_q s.$$

Thus for $t \in [0, n_m]$,

$$(^C D_q^{\alpha_1} u_1)(t) = f_1(t, u_2(t)),$$

and

$$(^C D_q^{\alpha_2} u_2)(t) = f_2(t, u_1(t)).$$

Hence the constructed function u is a solution of the coupled system (10.13). This completes the proof. $\qquad\square$

10.2.2 Oscillation and nonoscillation results

Definition 10.17. [91] A solution u of problem (10.13) is said to be oscillatory if it is neither eventually positive nor eventually negative. Otherwise u is called nonoscillatory.

Definition 10.18. [91] A solution $u = (u_1, u_2)$ of the coupled system (10.13) is said to be strongly (weakly) oscillatory if each (at least one) of its components is oscillatory. Otherwise, it is said to be strongly (weakly) nonoscillatory if each (at least one) of its nontrivial components is nonoscillatory.

The following hypothesis will be used in the sequel.

(10.18.1) There exist continuous functions $q_{in} : I_n \to \mathbb{R}_+$, $n = 1, 2$, and continuous, bounded and increasing real functions g_i; $i = 1, 2$, such that, for a.e. $t \in I_n$, and each $u_i \in \mathbb{R}$, and $v \in \mathbb{R}^* := \mathbb{R} \backslash \{0\}$,

$$|f_1(t, u_2)| = q_{1n}(t) g_1(u_2), \quad |f_2(t, u_1)| = q_{2n}(t) g_2(u_1), \quad \text{and } v g_i(v) > 0.$$

Remark 10.19. We can see that (10.18.1) implies (10.15.2) with $p_{in}(t) = M q_{in}(t)$, where

$$M = \sup_{v \in \mathbb{R}} |g_i(v)|.$$

The following theorem gives sufficient conditions to ensure the nonoscillation of solutions of the coupled system (10.13).

Theorem 10.20. *Assume that (10.15.1) and (10.18.1) hold. If $u = (u_1, u_2)$ is a weakly nonoscillatory solution of (10.13), such that u_1 and u_2 have the same sign, then the first component u_1 is also nonoscillatory.*

Proof. Assume to the contrary that u_1 is oscillatory but u_2 is eventually positive. Then, in view of (10.18.1), there exists $n_m > 0$, such that $q_{1n}(t) g_1(u_2(t)) \geq 0$ for t larger than n_m. Thus, for all $t > n_m$,

$$u_1(t) = u_1(n_m) + \int_{n_m}^{n_m+1} \frac{(t-qs)^{(\alpha_1-1)}}{\Gamma_q(\alpha_1)} q_{1n}(s)g_1(u_2(s))d_qs > 0.$$

Hence $u_1(t) > 0$ for all large t. This is a contradiction.

Analogously, the case when u_2 is an eventually negative is proved similarly. Indeed, if u_2 is an eventually negative, then from (10.18.1), there exists $n_m > 0$, such that for all $t > n_m$,

$$u_1(t) = u_1(n_m) + \int_{n_m}^{n_m+1} \frac{(t-qs)^{(\alpha_1-1)}}{\Gamma_q(\alpha_1)} q_{1n}(s)g_1(u_2(s))d_qs < 0.$$

Thus $u_1(t) < 0$ for all large t. This is again a contradiction. This means that u_1 is nonoscillatory. □

Corollary 10.21. *Assume that (10.15.1) and (10.18.1) hold. If $u = (u_1, u_2)$ is a weakly nonoscillatory solution of (10.13), such that u_1 and u_2 have the same sign, then the second component u_2 is also nonoscillatory.*

Corollary 10.22. *Assume that (10.15.1) and (10.18.1) hold. If $u = (u_1, u_2)$ is a weakly nonoscillatory solution of (10.13) such that u_1 and u_2 have the same sign, then u is a strongly nonoscillatory solution of (10.13).*

The following theorem presents the oscillatory result for the coupled system (10.13).

Theorem 10.23. *Assume that (10.15.1) and (10.18.1) hold. If $u = (u_1, u_2)$ is a weakly oscillatory solution of (10.13), such that u_1 and u_2 have the same sign, then the first component u_1 is also oscillatory.*

Proof. Assume to the contrary that u_1 is nonoscillatory. If u_1 is an eventually positive solution, then in view of (10.18.1), there exists $n_m > 0$, such that $q_{2n}(t)g_2(u_1(t)) \geq 0$ for t larger than n_m. Thus, for all $t > n_m$,

$$u_2(t) = u_2(n_m) + \int_{n_m}^{n_m+1} \frac{(t-qs)^{(\alpha_2-1)}}{\Gamma_q(\alpha_2)} q_{2n}(s)g_2(u_1(s))d_qs > 0.$$

Hence $u_2(t) > 0$ for all large t. This is a contradiction since u_2 is an oscillatory solution.

Analogously, the case when u_1 is an eventually negative is proved similarly. Indeed, if u_1 is an eventually negative, then from (10.18.1), there exists $n_m > 0$, such that for all $t > n_m$,

$$u_2(t) = u_2(n_m) + \int_{n_m}^{n_m+1} \frac{(t-qs)^{(\alpha_2-1)}}{\Gamma_q(\alpha_2)} q_{2n}(s)g_2(u_1(s))d_qs < 0.$$

Thus $u_2(t) < 0$ for all large t. This is again a contradiction since u_2 is an oscillatory solution. This means that u_1 is oscillatory. □

Corollary 10.24. *Assume that* (10.15.1) *and* (10.18.1) *hold. If* $u = (u_1, u_2)$ *is a weakly oscillatory solution of (10.13) such that* u_1 *and* u_2 *have the same sign, then* u_2 *is an oscillatory solution.*

Corollary 10.25. *Assume that* (10.15.1) *and* (10.18.1) *hold. If* $u = (u_1, u_2)$ *is a weakly oscillatory solution of (10.13), such that* u_1 *and* u_2 *have the same sign, then* u *is an oscillatory solution of (10.13).*

10.3 Notes and remarks

The results of Chapter 10 are taken from the papers [46,57]. For even further information and outcomes on the subject, consult monographs [438,439] and articles [52,77,96,103, 179,232].

11

A Filippov's theorem and topological structure of solution sets for fractional q-difference inclusions

We take as motivation for this chapter the paper [96,103,232,234,441] and references therein.

In this chapter we shall be concerned with a Filippov theorem and the existence of solutions and the topological structure of solution sets for the following fractional q-difference inclusion

$$(^{c}D_{q}^{\alpha}u)(t) \in F(t, u(t)), \ t \in I := [0, T], \tag{11.1}$$

with the initial condition

$$u(0) = u_0 \in E, \tag{11.2}$$

where $(E, \| \cdot \|)$ is a real or complex Banach space, $q \in (0, 1)$, $\alpha \in (0, 1]$, $T > 0$, $F : I \times E \to \mathcal{P}(E)$ is a multi-valued map, $\mathcal{P}(E)$ is the family of all nonempty subsets of E, $^{c}D_{q}^{\alpha}$ is the Caputo fractional q-difference derivative of order α.

11.1 Existence and topological structure of solution sets

First, we state the definition of a solution of the problem (11.1)–(11.2).

Definition 11.1. By a solution of the problem (11.1)–(11.2) we mean a function $u \in C(I)$ that satisfies the initial condition (11.2) and the equation $(^{C}D_{q}^{\alpha}u)(t) = v(t)$ on I, where $v \in S_{F \circ u}$.

Now, we recall the set-valued version of the Mönch fixed-point theorem.

Theorem 11.2. *(Mönch fixed-point theorem) [325] Let X be Banach space and $K \subset X$ be a closed and convex set. Also let U be a relatively open subset of K and $N : \overline{U} \to \mathcal{P}_{c}(K)$. Suppose that N maps compact sets into relatively compact sets, $\mathrm{graph}(N)$ is closed and for some $x_0 \in U$, we have*

$$\mathrm{conv}(x_0 \cup N(M)) \supset M \subset \overline{U} \text{ and } \overline{M} = \overline{U} \ (C \subset M \text{ countable}) \text{ imply } \overline{M} \text{ is compact} \tag{11.3}$$

and

Fractional Difference, Differential Equations, and Inclusions. https://doi.org/10.1016/B978-0-44-323601-3.00018-6
Copyright © 2024 Elsevier Inc. All rights are reserved, including those for text and data mining, AI training, and similar technologies.

$$x \notin (1-\lambda)x_0 + \lambda N(x) \quad \forall x \in \overline{U} \setminus U, \ \lambda \in (0,1).　(11.4)$$

Then there exists $x \in \overline{U}$ with $x \in N(x)$.

Also we recall the Schauder–Tikhonov fixed-point theorem:

Theorem 11.3. *(Schauder–Tikhonov fixed-point theorem) [132] Let X be a locally convex space, C a convex closed subset of X and $N : C \to C$ is a continuous, compact map. Then N has at least one fixed point in C.*

11.1.1 The upper semi-continuous case

In this section we present a global existence result and prove the compactness of solution set for our problem by using Mönch's fixed-point theorem for multi-valued maps combined with the measure of noncompactness.

Theorem 11.4. *If the following hypotheses hold.*

(11.4.1) *The multi-valued map $F : I \times E \to \mathcal{P}_{cp,c}(E)$ is Carathéodory.*

(11.4.2) *There exists a function $p \in L^\infty(I, \mathbb{R}_+)$ such that*

$$\|F(t,u)\|_{\mathcal{P}} = \sup\{\|v\|_C : v(t) \in F(t,u)\} \le p(t);$$

for a.e. $t \in I$ and each $u \in E$.

(11.4.3) *For each bounded set $B \subset E$ and for each $t \in I$, we have*

$$\mu(F(t,B)) \le p(t)\mu(B).$$

(11.4.4) *The function $\phi \equiv 0$ is the unique solution in $C(I)$ of the inequality*

$$\Phi(t) \le 2p^*(I_q^\alpha \Phi)(t),$$

where p is the function defined in (H_3) and

$$p^* = esssup_{t \in I} p(t).$$

Then the problem (11.1)–(11.2) has at least one solution defined on I. Moreover, the solution set

$$S_F(u_0) = \{u \in C(I) : u \text{ is a solution of problem (11.1)–(11.2)}\}$$

is compact and the multi-valued map $S_F : u_0 \to (S_F)(u_0)$ is u.s.c.

Proof. Consider the multi-valued operator $N : C(I) \to \mathcal{P}(C(I))$ defined by:

$$N(u) = \left\{ h \in C(I) : h(t) = u_0 + \int_0^t \frac{(t-qs)^{(\alpha-1)}}{\Gamma_q(\alpha)} v(s) d_q s; \ v \in S_{F \circ u} \right\}.　(11.5)$$

Clearly, the fixed points of N are solutions of the problem (11.1)–(11.2).

Step 1. Existence of solutions.

From Theorem 5 in [126], the multi-valued operator N satisfies all assumptions of Theorem 11.2, and we conclude that N has at least one fixed point $u \in C(I)$, which is a solution of problem (11.1)–(11.2).

Step 2. Compactness of the solution set.

For each a $u_0 \in E$, we consider the set $S_F(u_0)$. From Step 1, there exists $M > 0$ such that for every $u \in S_F(u_0) : \|u\|_\infty \le M$. Since N is completely continuous, $N(S_F(u_0))$ is relatively compact in $C(I)$. Let $u \in S_f(u_0)$; then $u \in N(u)$. Hence $S_f(u_0) \subset N(S_f(u_0))$. It remains to prove that $S_F(u_0)$ is a closed subset in $C(I)$. Let $\{u_n : n \in \mathbb{N}\} \subset S_F(u_0)$ be such that the sequence $(u_n)_{n \in \mathbb{N}}$ converges to u. For every $n \in \mathbb{N}$, there exists v_n such that $v_n(t) \in F(t, u_n(t))$; a.e. $t \in I$ and

$$u_n(t) = u_0 + \int_0^t \frac{(t - qs)^{(\alpha-1)}}{\Gamma_q(\alpha)} v_n(s) d_q s.$$

Since $u_n \to u$, Lemma 2.29 implies that there exists v such that $v(t) \in F(t, u(t))$; a.e. $t \in I$ and

$$u(t) = u_0 + \int_0^t \frac{(t - qs)^{(\alpha-1)}}{\Gamma_q(\alpha)} v(s) d_q s. \tag{11.6}$$

Therefore $u \in S_F(u_0)$, which yields that $S_F(u_0)$ is closed, hence compact in $C(I)$.

Step 3. $S_F(\cdot)$ is u.s.c.

To do this, we prove that the graph Γ_{S_F} of S_F is closed. We have

$$\Gamma_{S_F} = \{(u_0, u) : u \in S_F(u_0)\}.$$

Let $(u_{0n}, u_n) \in \Gamma_{S_F}$ be such that $(u_{0n}, u_n) \to (u_0, u)$; as $n \to \infty$. Since $u_n \in S_F(u_{0n})$, there exists $v_n \in L^1(I)$ such that

$$u_n(t) = u_{0n} + \int_0^t \frac{(t - qs)^{(\alpha-1)}}{\Gamma_q(\alpha)} v_n(s) d_q s. \tag{11.7}$$

From Lemma 2.29 we can prove that there exists $v \in S_{Fou}$ such that u satisfies (11.6). Thus $u \in S_F(u_0)$. Now we show that S_F maps bounded sets into relatively compact sets of $C(I)$. Let B be a bounded set in E and let $\{u_n\} \subset S_F(B)$. Then there exists $\{u_{0n}\} \subset B$ and $v_n \in S_{Fou_n}$; $n \in \mathbb{N}$ such that (11.7) is satisfied. Since $\{u_{0n}\}$ bounded sequence, there exists a subsequence of $\{u_{0n}\}$ converging to u_0. As in the proof of Theorem 5 in [126], we can show that $\{u_n\}$ is compact on I. We conclude that there exists a subsequence of $\{u_n\}$ converging to u in $C(I)$. Also, from Lemma 2.29, we can prove that u satisfies (11.6) for some $v \in S_{Fou}$. Hence $S_F(u_0)$ is u.s.c. $\qquad\square$

11.1.2 The lower semi-continuous case

Our next existence result for problem (11.1)–(11.2) deals with the case where the nonlinearity is lower semi-continuous with respect to the second argument and does not have necessarily convex values. We will make use of the Mönch's fixed point theorem for multivalued maps combined with a selection theorem for lower semicontinuous multivalued maps with decomposable values.

Let us we state the celebrated selection theorem of Fryszkowski:

Lemma 11.5. *[238] Let X be a separable metric space and let E be a Banach space. Then every l.s.c. multi-valued operator $N : X \to \mathcal{P}_{cl}(L^1(I, E))$ with nonempty closed decomposable values has a continuous selection, i.e. there exists a continuous single-valued function $f : X \to L^1(I, E)$ such that $f(x) \in N(x)$ for every $x \in X$.*

In the sequel we need the following hypothesis.

(11.5.1) The multi-valued map F is nonempty compact valued such that:
   (a) the mapping $(t, u) \to F(t, u)$ is $\mathcal{L} \otimes \mathcal{B}$ measurable;
   (b) The mapping $u \to F(t, u)$ is lower semi-continuous for each $t \in I$.

Lemma 11.6. *[237] Let $N : I \times E \to \mathcal{P}_{cp}(L^1(I, E))$ be a locally integrable bounded multivalued map satisfying (11.5.1). Then N is of lower semi-continuous type.*

Theorem 11.7. *If (11.4.2) and (11.5.1) hold, then the problem (11.1)–(11.2) has at least one solution defined on I.*

Proof. By Lemma 11.6, F is of lower semi-continuous type. From Lemma 11.5, there exists a continuous selection $f : C(I) \to L^1(I)$ such that $f(u) \in S_F(u)$ for every $u \in C(I)$. Consider the problem

$$\begin{cases} (^cD_q^\alpha u)(t) = (fu)(t); \ t \in I, \\ u(0) = u_0 \in E, \end{cases} \tag{11.8}$$

and the operator $G : C(I) \to C(I)$ defined by

$$(Gu)(t) = u_0 + \int_0^t \frac{(t - qs)^{(\alpha-1)}}{\Gamma_q(\alpha)}(fu)(s)d_qs.$$

Clearly, the fixed points of the operator G are solutions of problem (11.1)–(11.2).

Let $u \in C(I)$. Then, for each $t \in I$, we have

$$(Gu)(t) = u_0 + \int_0^t \frac{(t - qs)^{(\alpha-1)}}{\Gamma_q(\alpha)}v(s)d_qs,$$

for some $v \in S_{Fou}$. On the other hand,

$$\|(Gu)(t)\| \leq \|u_0\| + \int_0^t \frac{(t - qs)^{(\alpha-1)}}{\Gamma_q(\alpha)}(fu)(s)d_qs$$

$$\leq \|u_0\| + \int_0^t \frac{(t - qs)^{(\alpha-1)}}{\Gamma_q(\alpha)}p(s)d_qs$$

$$\leq \|u_0\| + \frac{p^* T^{(\alpha)}}{\Gamma_q(1+\alpha)}$$

$$= R.$$

Hence $\|Gu\|_\infty \leq R$ and so $G(B_R) \subset B_R$, where $B_R := \{u \in C(I) : \|u\|_\infty \leq R\}$ is the bounded, closed and convex ball of $C(I)$. We will show that the multi-valued operator $G : B_R \to B_R$ satisfies all assumptions of Theorem 11.3. It remains for us to demonstrate that $G(B_R)$ is relatively compact.

Let (h_n) be any sequence in $G(B_R)$. By Arzelà–Ascoli compactness criterion in $C(I)$, we show (h_n) has a convergent subsequence. Since $h_n \in G(B_R)$ there are $u_n \in B_R$ and $v_n \in S_{F \circ u_n}$ such that

$$h_n(t) = \mu_0 + \int_0^t \frac{(t-qs)^{(\alpha-1)}}{\Gamma_q(\alpha)} v_n(s) d_q s.$$

We can prove that $\{h_n(t) : n \geq 1\}$ is relatively compact for each $t \in I$. In addition, for each t_1 and t_2 from I, with $t_1 < t_2$, we have

$$\|h_n(t_2) - h_n(t_1)\|$$

$$\leq \left\| \int_0^{t_2} \frac{(t_2-qs)^{(\alpha-1)}}{\Gamma_q(\alpha)} p(s) d_q s - \int_0^{t_1} \frac{(t_1-qs)^{(\alpha-1)}}{\Gamma_q(\alpha)} p(s) d_q s \right\|$$

$$\leq \int_{t_1}^{t_2} \frac{(t_2-qs)^{(\alpha-1)}}{\Gamma_q(\alpha)} p(s) d_q s$$

$$+ \int_0^{t_1} \frac{|(t_2-qs)^{(\alpha-1)} - (t_1-qs)^{(\alpha-1)}|}{\Gamma_q(\alpha)} p(s) d_q s \qquad (11.9)$$

$$\leq \frac{p^* T^\alpha}{\Gamma_q(1+\alpha)} (t_2-t_1)^\alpha$$

$$+ p^* \int_0^{t_1} \frac{|(t_2-qs)^{(\alpha-1)} - (t_1-qs)^{(\alpha-1)}|}{\Gamma_q(\alpha)} d_q s$$

$$\to 0 \text{ as } t_1 \longrightarrow t_2.$$

This shows that $\{h_n : n \geq 1\}$ is equicontinuous. Consequently, by the Arzelà–Ascoli theorem, $\{h_n : n \geq 1\}$ is relatively compact in B_R. By Theorem 11.3, we conclude that G has at least one fixed point, which is a solution of problem (11.1)–(11.2). $\qquad \square$

11.2 Filippov theorem

Lemma 11.8. *[326] Let $G : I \to \mathcal{P}_{cl}(E)$ be a measurable multifunction and $u : I \to E$ be a measurable function. Assume that there exists $p \in L^1(I, E)$ such that $G(t) \subset p(t) B(0, 1)$, where $B(0, 1)$ denotes the closed ball in E. Then there exists a measurable selection g of G such that for a.e. $t \in I$,*

$$\|u(t) - g(t)\| \leq d(u(t), G(t)).$$

Let $x_0 \in E$, $g \in L^1(I, E)$, and let $x \in C(I)$ be a solution of the fractional q-difference problem:

$$\begin{cases} (^cD_q^\alpha x)(t) = g(t), \ t \in I, \\ x(0) = x_0. \end{cases} \qquad (11.10)$$

In the sequel we need the following hypotheses.

(11.8.1) The multi-valued map $F : I \times E \to \mathcal{P}(E)$ satisfies:
    (11.8.1)$_a$ the map $t \mapsto F(t, u)$ is measurable; for all $u \in E$;
    (11.8.1)$_b$ the map $\lambda : t \mapsto d(f(t), F(t, x(t)))$ is integrable.
(11.8.2) There exists a function $p \in L^\infty(I, \mathbb{R}_+)$ such that

$$H_d(F(t, u), F(t, v)) \le p(t)\|u - v\|,$$

    for a.e. $t \in I$, and each $u, v \in E$.

Remark 11.9. From Assumptions $(11.8.1)_a$ and $(11.8.1)_b$, the multi-function $t \mapsto F(t, u(t))$ is measurable. It follows that $v(t) = d(f(t), F(t, x(t)))$ is measurable.

Set

$$p^* = esssup_{t \in I} \, p(t).$$

Theorem 11.10. *If the hypotheses (11.8.1) and (11.8.2) hold, then the problem (11.1)–(11.2) has at least one solution u defined on I. Moreover, for a.e. $t \in I$, u satisfies the estimates:*

$$\|u(t) - x(t)\| \le \|u_0 - x_0\| + \frac{p^* T^{(\alpha-1)}}{\Gamma_q(\alpha)} \int_0^t \sum_{i=2}^\infty \|x_i(s) - x_{i-1}(s)\| d_q s,$$

and

$$\|(^cD_q^\alpha u)(t) - g(t)\| \le p^* \sum_{i=2}^\infty \|x_i(t) - x_{i-1}(t)\|,$$

where

$$\|x_n(t) - x_{n-1}(t)\| \le \left(\frac{p^* T^{(\alpha-1)}}{\Gamma_q(\alpha)} \right)^{n-1} \int_0^t \int_0^{s_1} \int_0^{s_2} \cdots \int_0^{s_{n-2}} \left(\|u_0 - x_0\| \right.$$
$$+ \left. \frac{T^{(\alpha-1)}}{\Gamma_q(\alpha)} \int_0^{s_{n-1}} \lambda(\tau) d_q \tau \right) d_q s_{n-1} d_q s_{n-2} \cdots d_q s_1.$$

Proof. First, we construct a sequence of functions $(u_n)_{n \in \mathbb{N}}$, which will be shown to converge to a solution of problem (11.1)–(11.2) on I.

Let $f_0 = g$ on I. So we have

$$x(t) = x_0 + \int_0^t \frac{(t - qs)^{(\alpha-1)}}{\Gamma_q(\alpha)} f_0(s) d_q s.$$

Then define the multi-valued map $U_1 : I \to \mathcal{P}(E)$ by

$$U_1(t) + F(t, x(t)) \cap (f_0(t) + \lambda(t)B(0, 1)).$$

Since f_0 and λ are measurable, the ball $(f_0(t) + \lambda(t)B(0, 1))$ is measurable from Theorem III.4.1 in [201]. Moreover, $F(t, x(t))$ is measurable and U_1 is nonempty. It is clear that for a.e. $t \in I$,

$$d(0, F(t, 0)) \le d(0, f_0(t)) + d(f_0(t), F(t, x(t))) + H_d(F(t, x(t)), F(t, 0))$$
$$\le \|f_0(t)\| + \lambda(t) + p(t)\|x(t)\|.$$

Hence, for all $w \in F(t, x(t))$, we have

$$\|w\| \le d(0, F(t, 0)) + H_d(F(t, 0), F(t, x(t)))$$
$$\le \|f_0(t)\| + \lambda(t) + 2p(t)\|x(t)\| := M(t).$$

This implies that

$$F(t, x(t)) \subset M(t)B(0, 1); \ t \in I.$$

From Lemma 11.8, there exists a function u, which is a measurable selection of $F(t, x(t))$ such that

$$\|u(t) - f_0(t)\| \le d(f_0(t), F(t, x(t))) = \lambda(t).$$

Then $u \in U_1(t)$, proving our claim. We deduce that the intersection multi-valued operator $U_1(t)$ is measurable and that there exists a function $t \to f_1(t)$, which is a measurable selection for U_1 (see, [201,326]). Consider

$$x_1(t) = u_0 + \int_0^t \frac{(t - qs)^{(\alpha-1)}}{\Gamma_q(\alpha)} f_1(s)d_q s.$$

For each $t \in I$, we have

$$\|x_1(t) - x(t)\| \le \|u_0 - x_0\| + \int_0^t \frac{(t - qs)^{(\alpha-1)}}{\Gamma_q(\alpha)} \|f_1(s) - f_0(s)\|d_q s$$
$$\le \|u_0 - x_0\| + \frac{T^{(\alpha-1)}}{\Gamma_q(\alpha)} \int_0^t \lambda(s)d_q s. \tag{11.11}$$

Next, from Lemma 1.4 in [236], $F(t, x_1(t))$ is measurable.

The ball $(f_1(t) + p(t)\|x_1 t) - x(t)\|B(0, 1))$ is also measurable by Theorem III.4.1 in [201]. The set $U_2(t) = F(t, x_1(t)) \cap (f_1(t) + p(t)|x_1(t) - x(t)\|B(0, 1))$ is nonempty. Indeed, since f_1 is a measurable function, Lemma 11.8 yields a measurable selection u of $F(t, x_1(t))$ such that

$$\|u(t) - f_1(t)\| \le d(f_1(t), F(t, x_1(t))).$$

Then using (11.8.2), we get

$$\|u(t) - f_1(t)\| \le d(f_1(t), F(t, x_1(t)))$$
$$\le H_d(F(t, x(t)), F(t, x_1(t)))$$
$$\le p(t)\|x(t) - x_1(t)\|.$$

Thus $u \in U_2(t)$, proving our claim. Now, since the intersection multi-valued operator U_2 defined above is measurable, there exists a measurable selection $f_2(t) \in U_2(t)$. Hence

$$\|f_2(t) - f_1(t)\| \le p(t)\|x_1(t) - x(t)\|. \tag{11.12}$$

Define

$$x_2(t) = x_0 + \int_0^t \frac{(t - qs)^{(\alpha-1)}}{\Gamma_q(\alpha)} f_2(s) d_q s.$$

Using (11.11) and (11.12), for every $t \in I$,

$$\|x_2(t) - x_1(t)\| \le \frac{T^{(\alpha-1)}}{\Gamma_q(\alpha)} \int_0^t \|f_2(s) - f_1(s)\| d_q s$$
$$\le \frac{T^{(\alpha-1)}}{\Gamma_q(\alpha)} \int_0^t p(s)\|x_1(s) - x(s)\| d_q s$$
$$\le \frac{T^{(\alpha-1)}}{\Gamma_q(\alpha)} \int_0^t p(s) \left(\|u_0 - x_0\| + \frac{T^{(\alpha-1)}}{\Gamma_q(\alpha)} \int_0^s \lambda(\tau) d_q \tau \right) d_q s$$
$$\le \frac{p^* T^{(\alpha-1)}}{\Gamma_q(\alpha)} \int_0^t \left(\|u_0 - x_0\| + \frac{T^{(\alpha-1)}}{\Gamma_q(\alpha)} \int_0^s \lambda(\tau) d_q \tau \right) d_q s.$$

Let $U_3(t) = F(t, x_2(t)) \cap (f_2(t) + p(t)\|x_2(t) - x_1(t)\| B(0, 1))$. Arguing as for U_2, we can prove that U_3 is a measurable multi-valued map with nonempty values, so there exists a measurable selection $f_3(t) \in U_3(t)$. This allows us to define

$$x_3(t) = x_0 + \int_0^t \frac{(t - qs)^{(\alpha-1)}}{\Gamma_q(\alpha)} f_3(s) d_q s.$$

Then, for each $t \in I$,

$$\|x_3(t) - x_2(t)\|$$
$$\le \frac{T^{(\alpha-1)}}{\Gamma_q(\alpha)} \int_0^t \|f_3(s) - f_2(s)\| d_q s$$
$$\le \frac{T^{(\alpha-1)}}{\Gamma_q(\alpha)} \int_0^t p(s)\|x_2(s) - x_1(s)\| d_q s$$
$$\le \frac{p^* T^{(\alpha-1)}}{\Gamma_q(\alpha)} \int_0^t \left(\frac{p^* T^{(\alpha-1)}}{\Gamma_q(\alpha)} \int_0^{s_1} \left(\|u_0 - x_0\| + \frac{T^{(\alpha-1)}}{\Gamma_q(\alpha)} \int_0^{s_2} \lambda(\tau) d_q \tau \right) d_q s_2 \right) d_q s_1$$

$$\leq \left(\frac{p^* T^{(\alpha-1)}}{\Gamma_q(\alpha)} \right)^2 \int_0^t \int_0^{s_1} \left(\|u_0 - x_0\| + \frac{T^{(\alpha-1)}}{\Gamma_q(\alpha)} \int_0^{s_2} \lambda(\tau) d_q \tau \right) d_q s_2 d_q s_1.$$

Repeating the process for $n = 1, 2, \cdots$, for each $t \in I$,

$$\|x_n(t) - x_{n-1}(t)\| \leq \frac{T^{(\alpha-1)}}{\Gamma_q(\alpha)} \int_0^t \|f_n(s) - f_{n-1}(s)\| d_q s$$

$$\leq \frac{T^{(\alpha-1)}}{\Gamma_q(\alpha)} \int_0^t p(s)\|x_n(s) - x_{n-1}(s)\| d_q s.$$

Hence we get

$$\|x_n(t) - x_{n-1}(t)\| \leq \left(\frac{p^* T^{(\alpha-1)}}{\Gamma_q(\alpha)} \right)^{n-1} \int_0^t \int_0^{s_1} \int_0^{s_2} \cdots \int_0^{s_{n-2}} \left(\|u_0 - x_0\| \right.$$

$$\left. + \frac{T^{(\alpha-1)}}{\Gamma_q(\alpha)} \int_0^{s_{n-1}} \lambda(\tau) d_q \tau \right) d_q s_{n-1} d_q s_{n-2} \cdots d_q s_1. \tag{11.13}$$

By induction, suppose that (11.13) holds for some n and check (11.13) for $n+1$.

Let $U_{n+1}(t) = F(t, x_n(t)) \cap (f_n + p(t)\|x_n(t) - x_{n-1}(t)\| B(0,1))$. Since U_{n+1} is a nonempty measurable set, there exists a measurable selection $f_{n+1}(t) \in U_{n+1}(t)$, which allows us to define for $n \in \mathbb{N}$,

$$x_{n+1}(t) = x_0 + \int_0^t \frac{(t-qs)^{(\alpha-1)}}{\Gamma_q(\alpha)} f_{n+1}(s) d_q s. \tag{11.14}$$

Therefore, for a.e. $t \in I$, we have

$$\|x_{n+1}(t) - x_n(t)\| \leq \frac{T^{(\alpha-1)}}{\Gamma_q(\alpha)} \int_0^t \|f_{n+1}(s) - f_n(s)\| d_q s$$

$$\leq \frac{T^{(\alpha-1)}}{\Gamma_q(\alpha)} \int_0^t p(s)\|x_{n+1}(s) - x_n(s)\| d_q s$$

$$\leq \left(\frac{p^* T^{(\alpha-1)}}{\Gamma_q(\alpha)} \right)^{n-1} \int_0^t \int_0^{s_1} \int_0^{s_2} \cdots \int_0^{s_{n-1}} \left(\|u_0 - x_0\| \right.$$

$$\left. + \frac{T^{(\alpha-1)}}{\Gamma_q(\alpha)} \int_0^{s_n} \lambda(\tau) d_q \tau \right) d_q s_n d_q s_{n-1} \cdots d_q s_1.$$

Consequently, (11.13) is true for all $n \in \mathbb{N}$. We infer that $\{x_n\}_n$ is a Cauchy sequence in $C(I)$, converging uniformly to a limit function $u \in C(I)$.

Moreover, from the definition of $\{U_n\}_n$, we have

$$\|f_{n+1} - f_n\| \leq p(t)\|x_n - x_{n-1}\|; \quad \text{a.e. } t \in I.$$

Hence, for almost every $t \in I$, $\{f_n(t)\}_n$ is also a Cauchy sequence in E and then converges almost everywhere to some measurable function $f(\cdot)$ in E. In addition, since $f_0 = g$, we have, for a.e. $t \in I$,

$$\|f_n(t)\| \le \sum_{i=1}^{n} \|f_i(t) - f_{i-1}(t)\| + \|f_0(t)\|$$

$$\le p(t) \sum_{i=2}^{\infty} \|x_i(t) - x_{i-1}(t)\| + \|u_0 - x_0\| + \|f_0(t)\|.$$

From the Lebesgue dominated convergence theorem, we deduce that $\{f_n\}_n$ converges to $f \in L^1(I, E)$. Passing to the limit in (11.14), we find that the function

$$u(t) = u_0 + \int_0^t \frac{(t - qs)^{(\alpha-1)}}{\Gamma_q(\alpha)} f(s) d_q s,$$

is a solution of problem (11.1)–(11.2).

Moreover, for a.e. $t \in I$, we have

$$\|u(t) - x(t)\| \le \|u_0 - x_0\| + \frac{T^{(\alpha-1)}}{\Gamma_q(\alpha)} \int_0^t \|f(s) - f_0(s)\| d_q s$$

$$\le \|u_0 - x_0\| + \frac{T^{(\alpha-1)}}{\Gamma_q(\alpha)} \int_0^t \|f(s) - f_n(s)\| d_q s$$

$$+ \frac{T^{(\alpha-1)}}{\Gamma_q(\alpha)} \int_0^t \|f_n(s) - f_0(s)\| d_q s$$

$$\le \|u_0 - x_0\| + \frac{T^{(\alpha-1)}}{\Gamma_q(\alpha)} \int_0^t \|f(s) - f_n(s)\| d_q s$$

$$+ \frac{p^* T^{(\alpha-1)}}{\Gamma_q(\alpha)} \int_0^t \sum_{i=2}^{\infty} \|x_i(s) - x_{i-1}(s)\| d_q s.$$

As $n \to \infty$, we get

$$\|u(t) - x(t)\| \le \|u_0 - x_0\| + \frac{p^* T^{(\alpha-1)}}{\Gamma_q(\alpha)} \int_0^t \sum_{i=2}^{\infty} \|x_i(s) - x_{i-1}(s)\| d_q s.$$

Next, we give an estimate of $\|(^c D_q^\alpha u)(t) - g(t)\|$; for $t \in I$. We have

$$\|(^c D_q^\alpha u)(t) - g(t)\| = \|f(t) - f_0(t)\|$$

$$\le \|f_n(t) - f_0(t)\| + \|f_n(t) - f(t)\|$$

$$\le \|f_n(t) - f(t)\| + p^* \sum_{i=2}^{\infty} \|x_i(t) - x_{i-1}(t)\|.$$

As $n \to \infty$, we get

$$\|(^c D_q^\alpha u)(t) - g(t)\| \leq p^* \sum_{i=2}^{\infty} \|x_i(t) - x_{i-1}(t)\|. \qquad \square$$

11.2.1 An example

Let

$$E = l^1 = \left\{ u = (u_1, u_2, \ldots, u_n, \ldots), \sum_{n=1}^{\infty} |u_n| < \infty \right\}$$

be the Banach space with the norm

$$\|u\|_E = \sum_{n=1}^{\infty} |u_n|.$$

Consider now the following problem of fractional $\frac{1}{4}$-difference inclusion

$$\begin{cases} (^c D_{\frac{1}{4}}^{\frac{1}{2}} u_n)(t) \in F_n(t, u(t)); \ t \in [0, e], \\ u(0) = (1, 0, \ldots, 0, \ldots), \end{cases} \tag{11.15}$$

where

$$F_n(t, u(t)) = \frac{t^2 e^{-4-t}}{1 + \|u(t)\|_E} [u_n(t) - 1, u_n(t)]; \ t \in [0, e],$$

with $u = (u_1, u_2, \ldots, u_n, \ldots)$. Set $\alpha = \frac{1}{2}$, and $F = (F_1, F_2, \ldots, F_n, \ldots)$.

For each $u \in E$ and $t \in [0, e]$, we have

$$\|F(t, u)\|_{\mathcal{P}} \leq ct^2 e^{-t-4}.$$

Hence, the hypothesis (11.8.2) is satisfied with $p^* = ce^{-2}$. A simple computation shows that conditions of Theorem 11.4 are satisfied. Hence the problem (11.15) has at least one solution defined on $[0, e]$. Moreover, the solution set $S_F(u_0)$ is compact and the multi-valued map $S_F : u_0 \to (S_F)(u_0)$ is u.s.c.

11.3 Notes and remarks

This chapter contains the studies from Abbas et al. [27]. For some more and pertinent results, we recommend the reader to publications [202–205,263] and their references.

12

On ψ-Caputo fractional differential equations in Banach spaces

The findings of our research in this chapter can be viewed as a partial continuation of the problems highlighted recently in the papers [89,317,437].

As motivation, we used the paper of S. Zhang [432], where he studied the following boundary value problem with Caputo fractional derivative of the form

$$\begin{cases} {}^{c}D^{q}u(t) = f(t, u(t)), & t \in [0, 1], 1 < q \le 2, \\ u(0) = a \neq 0, & u(1) = b \neq 0, \end{cases}$$

where ${}^{c}D^{q}$ is the Caputo fractional derivative of order q and $f \in C(J \times \mathbb{R}, \mathbb{R})$. The author obtained the existence of solutions by means of Schauder fixed-point theorem.

In [89] Agarwal et al. discussed the following Caputo FDEs with boundary conditions:

$$\begin{cases} {}^{c}D^{q}u(t) = f(t, u(t)), & t \in [0, T], 1 < q \le 2, \\ u(0) = u_0, & u(T) = u_T, \end{cases}$$

where ${}^{c}D^{q}$ is the Caputo fractional derivative of order q, $f : [0, T] \times E \longrightarrow E$ is a given function and T is a fixed positive constant, E is a Banach space with norm $\| \cdot \|$, and $u_0, u_T \in E$. They obtained the existence of solutions by employing Mönch fixed-point theorem combined with the technique of measure of noncompactness.

In 2016, using the ideas of Meir–Keeler condensing operators together with the notion of measure of noncompactness Mursaleen and Rizvi [317] studied the solvability of infinite systems of second-order differential equations in the sequence spaces c_0 and ℓ_1.

$$\begin{cases} u_i''(t) = -f_i(t, u(t)), & t \in [0, T], \\ u_i(0) = u_i(T) = 0, & i = 1, 2, \dots. \end{cases}$$

12.1 Boundary value problem for fractional differential equations via densifiability techniques

12.1.1 Introduction

In this section we consider the boundary value problem containing ψ-Caputo fractional derivative in Banach spaces:

Fractional Difference, Differential Equations, and Inclusions. https://doi.org/10.1016/B978-0-44-323601-3.00019-8
Copyright © 2024 Elsevier Inc. All rights are reserved, including those for text and data mining, AI training, and similar technologies.

$$\begin{cases} {}^{c}D_{a^{+}}^{\alpha;\psi} u(t) = f(t, u(t)), & t \in J := [a, b], \\ u(a) = u(a) = \theta, \end{cases} \tag{12.1}$$

where $\alpha \in (1, 2]$ and ${}^{c}D_{a^{+}}^{\alpha;\psi}$ denotes the ψ-Caputo fractional derivative of order α and $f : J \times E \longrightarrow E$ is a given function that fulfills certain conditions that will be specified hereafter, E is a Banach space with norm $\| \cdot \|$, and θ refers to the null vector in the space E.

12.1.2 Existence results

In this section we prove the existence outcomes of the suggested system (12.1).

The following lemma is essential for the existence of solutions to the BVP (12.1).

Lemma 12.1 ([147]). *For* $f \in L^{1}(J, \mathbb{R})$, *the following BVP*

$$\begin{cases} {}^{c}D_{a^{+}}^{\alpha;\psi} u(t) = f(t), & 1 < \alpha \le 2, t \in J, \\ u(a) = u(b) = 0, \end{cases} \tag{12.2}$$

has unique solution given by

$$\begin{aligned} u(t) = & \int_{a}^{t} \frac{\psi'(\varrho)(\psi(t) - \psi(\varrho))^{\alpha-1}}{\Gamma(\alpha)} f(\varrho) d\varrho \\ & - \frac{(\psi(t) - \psi(a))}{\Gamma(\alpha)(\psi(b) - \psi(a))} \int_{a}^{b} \psi'(\varrho)(\psi(b) - \psi(\varrho))^{\alpha-1} f(\varrho) d\varrho. \end{aligned} \tag{12.3}$$

Theorem 12.2. *Let us assume that the following hypotheses are satisfied.*

(12.2.1) *The function* $f : J \times E \to E$ *satisfies Carathéodory conditions.*

(12.2.2) *There exist a function* $\mu \in L^{\infty}(J, \mathbb{R}_{+})$ *and a continuous nondecreasing function* $\Theta : \mathbb{R}_{+} \to \mathbb{R}_{+}$ *such that*

$$\| f(t, z) \| \le \mu(t) \Theta(\|z\|), \quad \text{for all } z \in E.$$

(12.2.3) *There exists a number* $\upsilon > 0$ *such that*

$$2 \frac{\|\mu\|_{L^{\infty}} (\psi(b) - \psi(a))^{\alpha}}{\Gamma(\alpha + 1)} \Theta(\upsilon) \le \upsilon. \tag{12.4}$$

(12.2.4) *There are two functions* $g \in L^{\infty}(J, \mathbb{R}_{+})$ *and* $\beta \in \mathcal{A}$ *such that for any non-empty, bounded and convex* $\mathbb{P} \subset E$ *the inequality*

$$\omega\big(f(t, \mathbb{P})\big) \le g(t) \beta \big(\omega(\mathbb{P})\big),$$

holds for a.e. $t \in J$, *where* ω *is the degree of nondensifiability.*

Then the BVP (12.1) admits at least one solution provided that

$$2 \frac{\|\mu\|_{L^{\infty}} (\psi(b) - \psi(a))^{\alpha}}{\Gamma(\alpha + 1)} \le 1.$$

Proof. Using Lemma 12.1 the BVP (12.1) can be switched into an equivalent integral equation, defined as follows:

$$u(t) = \int_a^t \frac{\psi'(\varrho)(\psi(t) - \psi(\varrho))^{\alpha-1}}{\Gamma(\alpha)} f(\varrho, u(\varrho)) d\varrho$$

$$- \frac{\psi(t) - \psi(a)}{\Gamma(\alpha)(\psi(b) - \psi(a))} \int_a^b \psi'(\varrho)(\psi(b) - \psi(\varrho))^{\alpha-1} f(\varrho, u(\varrho)) d\varrho. \qquad (12.5)$$

Thus to investigate the existence of a solution for the BVP (12.1), we turn it into a fixed-point problem (FPP) for the operator $\mathbb{O} : C(J, E) \longrightarrow C(J, E)$ defined by the following formula:

$$\mathbb{O}u(t) = \int_a^t \frac{\psi'(\varrho)(\psi(t) - \psi(\varrho))^{\alpha-1}}{\Gamma(\alpha)} f(\varrho, u(\varrho)) d\varrho$$

$$- \frac{\psi(t) - \psi(a)}{\Gamma(\alpha)(\psi(b) - \psi(a))} \int_a^b \psi'(\varrho)(\psi(b) - \psi(\varrho))^{\alpha-1} f(\varrho, u(\varrho)) d\varrho. \qquad (12.6)$$

To demonstrate the intended outcome, we first let

$$\mathbb{B}_\upsilon = \{u \in C(J, E) : \|u\|_\infty \leq \upsilon\},$$

where υ satisfy the inequality (12.4). We will prove that the operator $\mathbb{O}$ fulfills all the hypotheses of Theorem 2.94.

First, we show $\mathbb{O}\mathbb{B}_\upsilon \subset \mathbb{B}_\upsilon$. Indeed, for any $u \in \mathbb{B}_\upsilon$ and under hypothesis (12.2.2), we get

$$\|\mathbb{O}u(t)\| \leq \int_a^t \frac{\psi'(\varrho)(\psi(t) - \psi(\varrho))^{\alpha-1}}{\Gamma(\alpha)} \|f(\varrho, u(\varrho))\| d\varrho$$

$$+ \frac{\psi(t) - \psi(a)}{\Gamma(\alpha)(\psi(b) - \psi(a))} \int_a^b \psi'(\varrho)(\psi(b) - \psi(\varrho))^{\alpha-1} \|f(\varrho, u(\varrho))\| d\varrho$$

$$\leq \int_a^t \frac{\psi'(\varrho)(\psi(t) - \psi(\varrho))^{\alpha-1}}{\Gamma(\alpha)} \mu(\varrho)\Theta(\|u(\varrho)\|) d\varrho$$

$$+ \frac{\psi(t) - \psi(a)}{\Gamma(\alpha)(\psi(b) - \psi(a))} \int_a^b \psi'(\varrho)(\psi(b) - \psi(\varrho))^{\alpha-1} \mu(\varrho)\Theta(\|u(\varrho)\|) d\varrho$$

$$\leq \frac{2\|\mu\|_{L^\infty}\Theta(\upsilon)(\psi(b) - \psi(a))^\alpha}{\Gamma(\alpha+1)}$$

$$\leq \upsilon,$$

which implies that $\|\mathbb{O}u\| \leq \upsilon$, and so $\mathbb{O}\mathbb{B}_\upsilon \subset \mathbb{B}_\upsilon$. Furthermore, by combining the assumptions (12.2.1), (12.2.2), and the Lebesgue dominated convergence theorem, we can get easily that the operator $\mathbb{O}$ is continuous on $\mathbb{B}_\upsilon$.

Now, we prove that $\mathbb{O}$ satisfies the contractive condition appearing in Theorem 2.94. To do this, let V be any nonempty and convex subset of $\mathbb{B}_\upsilon$, for each $\varrho \in J$, let $\epsilon_\varrho := \omega(V(\varrho))$. Through the hypothesis (12.2.4), there is $g \in L^\infty(J, \mathbb{R}_+)$ and $\beta \in \mathcal{A}$ such that for a.e. $\varrho \in J$:

$$\omega(f(\varrho, V(\varrho))) \leq g(\varrho)\beta(\epsilon_\varrho).$$

Therefore, given any $\varepsilon > 0$, there is a continuous mapping $\sigma_\varrho \colon \Delta \to E$, with $\sigma_\varrho(\Delta) \subset f(\varrho, V(\varrho))$, such that for all $u \in V$ there is $\zeta \in \Delta$ with

$$\|f(\varrho, u(\varrho)) - \sigma_\varrho(\zeta)\| \le g(\varrho)\beta(\epsilon_\varrho) + \varepsilon, \text{ for a.e. } \varrho \in J. \tag{12.7}$$

Construct now the mapping $\tilde{\sigma} \colon \Delta \to (C(J, E), \|\cdot\|_\infty)$ as follows:

$$\zeta \in \Delta \longmapsto \tilde{\sigma}(\zeta, t)$$
$$:= \int_a^t \frac{\psi'(\varrho)(\psi(t) - \psi(\varrho))^{\alpha-1}}{\Gamma(\alpha)} \sigma_\varrho(\zeta) d\varrho$$
$$- \frac{\psi(t) - \psi(a)}{\Gamma(\alpha)(\psi(b) - \psi(a))} \int_a^b \psi'(\varrho)(\psi(b) - \psi(\varrho))^{\alpha-1}\sigma_\varrho(\zeta) d\varrho, \ t \in J.$$

It is clear that $\tilde{\sigma}$ is continuous and $\tilde{\sigma}(\Delta) \subset \mathbb{O}(V)$. Additionally, by (12.7), given $u \in V$, we can find $\zeta \in \Delta$ such that:

$$\|\mathbb{O}u(t) - \tilde{\sigma}(\zeta, t)\|$$
$$\le \int_a^t \frac{\psi'(\varrho)(\psi(t) - \psi(\varrho))^{\alpha-1}}{\Gamma(\alpha)} \|f(\varrho, u(\varrho)) - \sigma_\varrho(\zeta)\| d\varrho$$
$$+ \frac{(\psi(t) - \psi(a))}{\Gamma(\alpha)(\psi(b) - \psi(a))} \int_a^b \psi'(\varrho)(\psi(b) - \psi(\varrho))^{\alpha-1} \|f(\varrho, u(\varrho)) - \sigma_\varrho(\zeta)\| d\varrho$$
$$\le \int_a^t \frac{\psi'(\varrho)(\psi(t) - \psi(\varrho))^{\alpha-1}}{\Gamma(\alpha)} (g(\varrho)\beta(\epsilon_\varrho) + \varepsilon) d\varrho$$
$$+ \frac{(\psi(t) - \psi(a))}{\Gamma(\alpha)(\psi(b) - \psi(a))} \int_a^b \psi'(\varrho)(\psi(b) - \psi(\varrho))^{\alpha-1}(g(\varrho)\beta(\epsilon_\varrho) + \varepsilon) d\varrho.$$

Applying Lemma 2.55 and the features of β, setting $\epsilon := \omega(V)$, we can deduce that $\beta(\epsilon_\varrho) \le \beta(\epsilon)$ for a.e. $\varrho \in J$ and from the last estimate, we get:

$$\|\mathbb{O}u(t) - \tilde{\sigma}(\zeta, t)\|$$
$$\le \int_a^t \frac{\psi'(\varrho)(\psi(t) - \psi(\varrho))^{\alpha-1}}{\Gamma(\alpha)} (g(\varrho)\beta(\epsilon) + \varepsilon) d\varrho$$
$$+ \frac{\psi(t) - \psi(a)}{\Gamma(\alpha)(\psi(b) - \psi(a))} \int_a^b \psi'(\varrho)(\psi(b) - \psi(\varrho))^{\alpha-1}(g(\varrho)\beta(\epsilon) + \varepsilon) d\varrho.$$

Since the previous inequality is valid, for every $\varepsilon > 0$, we conclude

$$\|\mathbb{O}u(t) - \tilde{\sigma}(\zeta, t)\| \le 2\frac{\|g\|_{L^\infty}(\psi(b) - \psi(a))^\alpha}{\Gamma(\alpha + 1)}\beta(\epsilon)$$
$$\le \beta(\epsilon),$$

which means, from the arbitrariness of $t \in J$, that $\omega(\mathbb{O}V) \le \beta(\epsilon)$.

Since the assumptions in Theorem 2.94 are fulfilled, the intended result follows. $\square$

12.1.3 Some examples

Take $E = c_0 = \{u = (u_1, u_2, \cdots, u_i, \cdots) : u_i \to 0 \, (i \to \infty)\}$, the Banach space of real sequences converging to zero, with the standard norm

$$\|u\|_\infty = \sup_i |u_i|.$$

Let $a = 1$, $\psi(t) = \ln t$ and fixed $1 < \alpha < \frac{\ln 0.25}{\ln \ln (\sqrt{\ln 4 - 1} + 1)} \simeq 1.9$ and let $1 < b < \exp 0.25^{\frac{1}{\alpha}}$.

Example 12.3. Consider the BVP:

$$\begin{cases} {}^c_H D^{\alpha;\psi}_{1^+} u(t) = f(t, u(t)), & t \in J := [1, b], \\ u(1) = u(b) = \theta. \end{cases} \tag{12.8}$$

System (12.8) is a special case of BVP (12.1), with $f : J \times c_0 \to c_0$ defined as

$$f(t, u) = \left\{ \frac{2i}{(i + 1)((t - 1)^2 + 1)} \left(\frac{1}{i} + \ln(1 + |u_i|) \right) \right\}_{i \geq 1},$$

for $t \in J$, $u = \{u_i\}_{i \geq 1} \in c_0$. Obviously, assumption (12.2.1) of Theorem 12.2 is satisfied. Moreover, for each $t \in J$ and $u \in c_0$, we get

$$\|f(t, u)\|_\infty \leq \left\| \frac{2i}{(i + 1)((t - 1)^2 + 1)} \left(\frac{1}{i} + |u_i| \right) \right\|_\infty$$

$$\leq \frac{2}{(t - 1)^2 + 1} (\|u\|_\infty + 1)$$

$$= \mu(t) \Theta(\|u\|).$$

Hence (12.2.2) holds for $\mu(t) = \frac{2}{(t-1)^2+1}$, $t \in J$ and $\Theta(\ell) = 1 + \ell$, $\ell \in [0, \infty)$. The inequality appearing in (12.2.3) has the following expression

$$\frac{4 (\log b)^\alpha (\upsilon + 1)}{\Gamma(\alpha + 1)} \leq \upsilon.$$

Then υ can be chosen as

$$\upsilon \geq \frac{1}{\Gamma(\alpha + 1) - 1},$$

so (12.2.3) is satisfied. On the other hand, for any nonempty, bounded, and convex subset V of $C(J, c_0)$ and $t \in J$ fixed, let σ be an ϵ_t-dense curve in $V(t)$ for some $\epsilon_t > 0$. Then, for $u \in V$, there is $\zeta \in \Delta$ satisfying:

$$\|u(t) - \sigma(\zeta, t)\|_\infty \leq \epsilon_t.$$

Therefore we have

$$\|f(t, u(t)) - f(t, \sigma(\zeta, t))\|_\infty \leq \frac{2}{(t-1)^2 + 1} \left\| \ln\left(1 + \frac{\|u(t) - \sigma(\zeta, t)\|}{1 + \|\sigma(\zeta, t)\|}\right) \right\|$$

$$\leq \frac{2}{(t-1)^2 + 1} \ln(1 + \|u(t) - \sigma(\zeta, t)\|)$$

$$\leq \frac{2}{(t-1)^2 + 1} \ln(1 + \epsilon_t),$$

and $\beta(t) = \ln(1 + t)$. This function is continuous and it is easily seen that $\beta \in \mathcal{A}$ and so condition (12.2.4) is satisfied taking $g(t) := \frac{2}{(t-1)^2+1}$. Hence all the conditions of Theorem 12.2 are satisfied and consequently problem (12.8) has at least one solution $u \in C(J, c_0)$.

Remark 12.4. We would like to point out that in the aforementioned example, the Darbo fixed-point theorem (DFPT) for the Hausdorff MNC χ, can not be implemented. To begin, remember that the Hausdorff MNC χ in the space c_0 may be obtained by the following expression

$$\chi(V) = \lim_{i \to \infty} \left\{ \sup_{u \in V} \left(\sup_{k \geq i} |u_k| \right) \right\}, \tag{12.9}$$

where $V \in \mathfrak{M}_{c_0}$ (cf., [95]). Next, by taking the standard basis of c_0, $V = \{e_i : i \in \mathbb{N}\}$, given $t \in J$ from (12.9), we have

$$\chi(\{f(t, e_i) : e_i \in V\})$$

$$= \lim_{i \to \infty} \left\{ \sup_{e_i \in V} \left(\sup_{k \geq i} \frac{2k}{(k+1)((t-1)^2+1)} \left(\frac{1}{k} + \ln(1 + |e_k|) \right) \right) \right\}$$

$$= \frac{2}{(t-1)^2+1} \lim_{i \to \infty} \left\{ \sup_{k \geq i} \left(\frac{1}{k} + \ln 2 \right) \right\} \geq \frac{\ln 4}{(b-1)^2+1},$$

and

$$\frac{\ln 4}{(b-1)^2+1} \geq 1 \Longleftrightarrow b \leq \sqrt{\ln 4 - 1} + 1 \simeq 1.6215.$$

This shows that $\chi(\{f(t, e_i) : e_i \in V\})$ is strictly greater than $\chi(V) = 1$.

12.2 Application of Meir–Keeler condensing operators

12.2.1 Introduction

In this section we consider the existence of solutions for a new boundary value problems with ψ-Caputo differential equations in Banach spaces. More precisely, we will consider the following problem

$$\begin{cases} {}^cD_{a+}^{\alpha;\psi}u(t) = f(t,u(t)), & t \in J := [a,b], \\ u(a) = \theta_a, \quad u(b) = u_b, \end{cases} \tag{12.10}$$

where ${}^cD_{a+}^{\alpha;\psi}$ is the ψ-Caputo fractional derivative of order $\alpha \in (1,2]$, $f : [a,b] \times E \longrightarrow E$ is a given function satisfying some assumptions that will be specified later, E is a Banach space with norm $\|\cdot\|$ and $\theta_a, u_b \in E$.

12.2.2 Existence results

Let us recall the definition and lemma of a solution for problem (12.10). First of all, we define what we mean by a solution for the boundary value problem (12.10).

Definition 12.5. A function $u \in AC^1(J,E)$ is said to be a solution of Eq. (12.10) if u satisfies the equation ${}^cD_{a+}^{\alpha;\psi}u(t) = f(t,u(t))$, a.e. on J, and the conditions $u(a) = \theta_a$, $u(b) = u_b$.

For the existence of solutions for the problem (12.10) we need the following lemma:

Lemma 12.6. *For a given $h \in L^1(J,\mathbb{R})$, the unique solution of the linear fractional boundary value problem*

$$\begin{cases} {}^cD_{a+}^{\alpha;\psi} = h(t), & 1 < \alpha \le 2, \ t \in J := [a,b], \\ u(a) = \theta_a, \quad u(b) = u_b, \end{cases} \tag{12.11}$$

is given by

$$\begin{aligned} u(t) &= \theta_a + I_{a+}^{\alpha;\psi}h(t) + \frac{(u_b - \theta_a)(\psi(t) - \psi(a))}{(\psi(b) - \psi(a))} - \frac{(\psi(t) - \psi(a))}{(\psi(b) - \psi(a))}I_{a+}^{\alpha;\psi}h(b) \\ &= \frac{1}{\Gamma(\alpha)}\int_a^t \psi'(s)(\psi(t) - \psi(s))^{\alpha-1}h(s)\mathrm{d}s \\ &\quad - \frac{(\psi(t) - \psi(a))}{\Gamma(\alpha)(\psi(b) - \psi(a))}\int_a^b \psi'(s)(\psi(b) - \psi(s))^{\alpha-1}h(s)\mathrm{d}s \\ &\quad + \theta_a + \frac{(u_b - \theta_a)(\psi(t) - \psi(a))}{(\psi(b) - \psi(a))}. \end{aligned} \tag{12.12}$$

Proof. Taking the ψ-Riemann–Liouville fractional integral of order α to the first equation of (12.11), we get

$$u(t) = I_{a+}^{\alpha;\psi}h(t) + c_0 + c_1(\psi(t) - \psi(a)), \quad c_0, c_1 \in \mathbb{R}. \tag{12.13}$$

Substituting $t = a$ in (12.13) and applying the first boundary condition of (12.11), it follows that

$$c_0 = \theta_a.$$

For $t = b$ in (12.13) and using the second boundary condition of (12.11), it yields

$$u(b) = u_b = I_{a^+}^{\alpha;\psi} h(b) + \theta_a + c_1(\psi(b) - \psi(a)). \tag{12.14}$$

Solving (12.14), we find that

$$c_1 = \frac{u_b - \theta_a}{(\psi(b) - \psi(a))} - \frac{1}{(\psi(b) - \psi(a))} I_{a^+}^{\alpha;\psi} h(b). \tag{12.15}$$

Substituting the values of c_0 and c_1 into (12.13), we get the integral equation (12.12). The converse follows by direct computation, which completes the proof. □

Now, we will present our main result concerning the existence of solutions of problem (12.10).

Theorem 12.7. *Assume that the following hypotheses are satisfied.*

(12.7.1) *The function* $f : [a, b] \times E \longrightarrow E$ *satisfies Carathéodory conditions.*
(12.7.2) *There exist a continuous function* $p_f : J \to \mathbb{R}_+$ *and a continuous nondecreasing function* $\phi : \mathbb{R}_+ \longrightarrow \mathbb{R}_+$ *such that*

$$\|f(t, u)\| \le p_f(t)\phi(\|u\|), \quad \text{for a.e. } t \in J, \text{ and each } u \in E.$$

(12.7.3) *For each bounded set* $B \subset E$ *and each* $t \in J$, *the following inequality holds*

$$\chi(f(t, B)) \le p_f(t)\chi(B),$$

where χ *is the Hausdorff measure of noncompactness.*
(12.7.4) *There exists a constant* $r > 0$ *such that*

$$r \ge 2\|\theta_a\| + \|u_b\| + \phi(r)\mathcal{M}_\psi. \tag{12.16}$$

If

$$4\mathcal{M}_\psi < 1, \tag{12.17}$$

where $\mathcal{M}_\psi = \frac{2p_f^*(\psi(b) - \psi(a))^\alpha}{\Gamma(\alpha+1)}$ *and* $p_f^* := \sup_{t \in J} p_f(t)$, *then the problem (12.10) has at least one solution defined on* J.

Proof. Consider the operator $\mathcal{T} : C(J, E) \longrightarrow C(J, E)$ defined by:

$$\mathcal{T}u(t) = \frac{1}{\Gamma(\alpha)} \int_a^t \psi'(s)(\psi(t) - \psi(s))^{\alpha-1} f(s, u(s)) ds$$

$$- \frac{(\psi(t) - \psi(a))}{\Gamma(\alpha)(\psi(b) - \psi(a))} \int_a^b \psi'(s)(\psi(b) - \psi(s))^{\alpha-1} f(s, u(s)) ds$$

$$+ \theta_a + \frac{(u_b - \theta_a)(\psi(t) - \psi(a))}{(\psi(b) - \psi(a))}. \tag{12.18}$$

It is obvious that $\mathcal{T}$ is well defined due to (12.7.1) and (12.7.2). Then the fractional integral equation (12.12) can be written as the following operator equation

$$u = \mathcal{T}u. \tag{12.19}$$

Thus the existence of a solution for Eq. (12.10) is equivalent to the existence of a fixed point for operator $\mathcal{T}$, which satisfies operator equation (12.19).

Define the set

$$\Omega_r = \{w \in C(J, E) : \|w\|_\infty \le r\}.$$

Notice that Ω_r is closed, convex and bounded subset of the Banach space $C(J, E)$. We will show that the operator $\mathcal{T}$ satisfies all the assumptions of Theorem 2.98. We split the proof into four steps.

Step 1: The operator $\mathcal{T}$ maps the set Ω_r into itself. By the assumption (12.7.2), we have

$$
\begin{aligned}
\|\mathcal{T}u(t)\| &\le \frac{1}{\Gamma(\alpha)} \int_a^t \psi'(s)(\psi(t) - \psi(s))^{\alpha-1} \|f(s, u(s))\| ds \\
&+ \frac{(\psi(t) - \psi(a))}{\Gamma(\alpha)(\psi(b) - \psi(a))} \int_a^b \psi'(s)(\psi(b) - \psi(s))^{\alpha-1} \|f(s, u(s))\| ds \\
&+ \|\theta_a\| + \frac{(\|u_b\| + \|\theta_a\|)(\psi(t) - \psi(a))}{(\psi(b) - \psi(a))} \\
&\le \frac{1}{\Gamma(\alpha)} \int_a^t \psi'(s)(\psi(t) - \psi(s))^{\alpha-1} p_f(s)\phi(\|u(s)\|) ds \\
&+ \frac{(\psi(t) - \psi(a))}{\Gamma(\alpha)(\psi(b) - \psi(a))} \int_a^b \psi'(s)(\psi(b) - \psi(s))^{\alpha-1} p_f(s)\phi(\|u(s)\|) ds \\
&+ \|\theta_a\| + \frac{(\|u_b\| + \|\theta_a\|)(\psi(t) - \psi(a))}{(\psi(b) - \psi(a))} \\
&\le 2\|\theta_a\| + \|u_b\| + \phi(r)\mathcal{M}_\psi \\
&\le r.
\end{aligned}
$$

Thus

$$\|\mathcal{T}u\| \le r.$$

This proves that $\mathcal{T}$ transforms the ball Ω_r into itself.

Step 2: The operator $\mathcal{T}$ is continuous. Suppose that $\{u_n\}$ is a sequence such that $u_n \to u$ in Ω_r as $n \to \infty$. It is easy to see that $f(s, u_n(s)) \to f(s, u(s))$, as $n \to +\infty$, due to the Carathéodory continuity of f. On the other hand, taking (12.7.2) into consideration, we get the following relations:

$$
\begin{aligned}
&\psi'(s)(\psi(t) - \psi(s))^{\alpha-1} \|f(s, u_n(s)) - f(s, u(s))\| \\
&\le 2p_f(s)\phi(r)\psi'(s)(\psi(t) - \psi(s))^{\alpha-1},
\end{aligned}
$$

$$\psi'(s)(\psi(b) - \psi(s))^{\alpha-1} \| f(s, u_n(s)) - f(s, u(s)) \|$$
$$\leq 2p_f(s)\phi(r)\psi'(s)(\psi(b) - \psi(s))^{\alpha-1},$$

which implies that each term on the left is integrable. By the Lebesgue dominated convergent theorem, we obtain

$$\frac{1}{\Gamma(\alpha)} \int_a^t \psi'(s)(\psi(t) - \psi(s))^{\alpha-1} \| f(s, u_n(s)) - f(s, u(s)) \| ds \to 0 \quad \text{as } n \to +\infty$$

$$\frac{1}{\Gamma(\alpha)} \int_a^b \psi'(s)(\psi(b) - \psi(s))^{\alpha-1} \| f(s, u_n(s)) - f(s, u(s)) \| ds \to 0 \quad \text{as } n \to +\infty.$$

It follows that $\|\mathcal{T}u_n - \mathcal{T}u\| \to 0$ as $n \to +\infty$. Which implies the continuity of the operator $\mathcal{T}$.

Step 3: The set $\mathcal{T}(\Omega_r)$ is equicontinuous. For any $a < t_1 < t_2 < b$ and $u \in \Omega_r$, we get

$$\|\mathcal{T}(u)(t_2) - \mathcal{T}(u)(t_1)\|$$

$$\leq \frac{1}{\Gamma(\alpha)} \int_a^{t_1} \psi'(s) \left[(\psi(t_2) - \psi(s))^{\alpha-1} - (\psi(t_1) - \psi(s))^{\alpha-1} \right] \| f(s, u(s)) \| ds$$

$$+ \frac{1}{\Gamma(\alpha)} \int_{t_1}^{t_2} \psi'(s)(\psi(t_2) - \psi(s))^{\alpha-1} \| f(s, u(s)) \| ds$$

$$+ \frac{(\psi(t_2) - \psi(t_1))}{\Gamma(\alpha)(\psi(b) - \psi(a))} \int_a^b \psi'(s)(\psi(b) - \psi(s))^{\alpha-1} \| f(s, u(s)) \| ds$$

$$+ \frac{\|u_b - \theta_a\|(\psi(t_2) - \psi(t_1))}{(\psi(b) - \psi(a))}$$

$$\leq \frac{\|p_f\|_{L^\infty}\phi(r)}{\Gamma(\alpha + 1)} \Big[(\psi(t_2) - \psi(a))^{\alpha} - (\psi(t_1) - \psi(a))^{\alpha}$$

$$+ (\psi(t_2) - \psi(t_1))(\psi(b) - \psi(a))^{\alpha-1} \Big] + \frac{\|u_b - \theta_a\|(\psi(t_2) - \psi(t_1))}{(\psi(b) - \psi(a))}.$$

As $t_2 \to t_1$, the right-hand side of the above inequality tends to zero independently of $u \in \Omega_r$. Hence, we conclude that $\mathcal{T}(\Omega_r) \subseteq C(J, E)$ is bounded and equicontinuous.

Step 4: Now, we prove that $\mathcal{T} : \Omega_r \to \Omega_r$ is a Meir–Keeler condensing operator. To do this, suppose $\varepsilon > 0$ is given. We will prove that there exists $\delta > 0$ such that

$$\varepsilon \leq \chi_C(B) < \varepsilon + \delta \Rightarrow \chi_C(\mathcal{T}B) < \varepsilon, \quad \text{for any } B \subset \Omega_r.$$

For every bounded subset $B \subset \Omega_r$ and $\varepsilon' > 0$ using Lemma 2.40 and the properties of χ, there exists sequences $\{u_n\}_{n=1}^\infty \subset B$ such that

$$\chi(\mathcal{T}(B)(t)) \leq 2\chi \left\{ \frac{1}{\Gamma(\alpha)} \int_a^t \psi'(s)(\psi(t) - \psi(s))^{\alpha-1} f\left(s, \{u_n(s)\}_{n=1}^\infty\right) ds \right\}$$

$$+ 2\chi \left\{ \frac{(\psi(t) - \psi(a))}{\Gamma(\alpha)(\psi(b) - \psi(a))} \int_a^b \psi'(s)(\psi(t) - \psi(s))^{\alpha-1} f\left(s, \{u_n(s)\}_{n=1}^\infty\right) ds \right\} + \varepsilon'.$$

Next, by Lemma 2.41 and (12.7.3), we have

$$\chi\left(\mathcal{T}(B)(t)\right) \leq \frac{4}{\Gamma(\alpha)} \int_a^t \psi'(s)(\psi(t)-\psi(s))^{\alpha-1}\chi\left(f\left(s,\{u_n(s)\}_{n=1}^\infty\right)\right)ds$$

$$+\frac{4(\psi(t)-\psi(a))}{\Gamma(\alpha)(\psi(b)-\psi(a))}\int_a^b \psi'(s)(\psi(t)-\psi(s))^{\alpha-1}\chi\left(f\left(s,\{u_n(s)\}_{n=1}^\infty\right)\right)ds+\varepsilon'$$

$$\leq \frac{4}{\Gamma(\alpha)}\int_a^t \psi'(s)(\psi(t)-\psi(s))^{\alpha-1}p_f(s)\chi\left(\{u_n(s)\}_{n=1}^\infty\right)ds$$

$$+\frac{4(\psi(t)-\psi(a))}{\Gamma(\alpha)(\psi(b)-\psi(a))}\int_a^b \psi'(s)(\psi(t)-\psi(s))^{\alpha-1}p_f(s)\chi\left(\{u_n(s)\}_{n=1}^\infty\right)ds+\varepsilon'$$

$$\leq 4\frac{2p_f^*\left(\psi(b)-\psi(a)\right)^\alpha}{\Gamma(\alpha+1)}\chi_C(B)+\varepsilon'$$

$$=4\mathcal{M}_\psi\chi_C(B)+\varepsilon'.$$

As the last inequality is true, for every $\varepsilon' > 0$, we infer

$$\chi\left(\mathcal{T}(B)(t)\right)\leq 4\mathcal{M}_\psi\chi_C(B).$$

Since $\mathcal{T}(B)\subset\Omega_r$ is bounded and equicontinuous, we know from Lemma 2.40 that

$$\chi_C\left(\mathcal{T}(B)\right)=\max_{t\in J}\chi\left(\mathcal{T}(B)(t)\right).$$

Therefore we have

$$\chi_C\left(\mathcal{T}(B)\right)\leq 4\mathcal{M}_\psi\chi_C(B).$$

Observe that from the last estimates $\chi_C\left(\mathcal{T}(B)\right)\leq 4\mathcal{M}_\psi\chi_C(B)<\varepsilon\Rightarrow\chi_C(B)<\frac{1}{4\mathcal{M}_\psi}\varepsilon$.

Consequently, for given $\varepsilon>0$, and taking $\delta=\frac{1-4\mathcal{M}_\psi}{4\mathcal{M}_\psi}\varepsilon$, we get the following implication:

$$\varepsilon\leq\chi_C(B)<\varepsilon+\delta\Rightarrow\chi_C(\mathcal{T}B)<\varepsilon,\quad\text{for any }B\subset\Omega_r.$$

Which means that $\mathcal{T}:\Omega_r\to\Omega_r$ is a Meir–Keeler condensing operator. It follows from Theorem 2.98 that the operator $\mathcal{T}$ defined by (12.18) has at least one fixed point $u\in\Omega_r$, which is just the solution of the boundary value problem (12.10). This completes the proof of Theorem 12.7.   □

12.2.3 Some examples

In this section we give two examples to illustrate the usefulness of our main result. Let

$$E=c_0=\{u=(u_1,u_2,\ldots,u_n,\ldots):u_n\to 0\,(n\to\infty)\},$$

be the Banach space of real sequences converging to zero, endowed its usual norm

$$\|u\|_\infty=\sup_{n\geq 1}|u_n|.$$

Example 12.8. Consider the following boundary value problem of a fractional differential posed in c_0:

$$\begin{cases} {}^{CH}D_{1^+}^{1.8}u(t) = f(t, u(t)), \ t \in J := [1, e], \\ u(1) = (0, 0, \ldots, 0, \ldots), \ u(1) = (0, 0, \ldots, 0, \ldots). \end{cases} \qquad (12.20)$$

Note that this problem is a particular case of BVP (12.10), where

$$\alpha = 1.8, a = 1, b = e, \psi(t) = \ln(t),$$

and $f : J \times c_0 \longrightarrow c_0$ given by

$$f(t, u) = \left\{ \frac{1}{e^t + 3} \left(\frac{1}{n^2} + \arctan(|u_n|) \right) \right\}_{n \geq 1}, \quad \text{for } t \in J, u = \{u_n\}_{n \geq 1} \in c_0.$$

It is clear that condition (12.7.1) hold and

$$\|f(t, u)\| \leq \frac{1}{e^t + 3} (1 + \|u\|)$$

$$= p_f(t)\phi(\|u\|).$$

Therefore assumption (12.7.2) of the Theorem 12.2 is satisfied with $p_f(t) = \frac{1}{e^t + 3}$, $t \in J$ and $\phi(x) = 1 + x$, $x \in [0, \infty)$. On the other hand, for any bounded set $B \subset c_0$, we have

$$\chi(f(t, B)) \leq p_f(t)\chi(B), \text{ a.e. } t \in J.$$

Hence (12.7.3) is satisfied. Now, we will check if condition (12.17) is satisfied. Indeed,

$$4\mathcal{M}_\psi = 0.8345 < 1,$$

and

$$(1 + r)\mathcal{M}_\psi \leq r,$$

thus

$$r \geq \frac{\mathcal{M}_\psi}{1 - \mathcal{M}_\psi} = 5.0421.$$

Then r can be chosen as $r = 5.5$. Consequently, all the hypotheses of Theorem 12.2 are satisfied and we conclude that the BVP (12.20) has at least one solution $u \in C(J, c_0)$.

Let

$$E = \ell^1 = \left\{ x = (u_1, u_2, \ldots, u_n, \ldots), \sum_{n=1}^{\infty} |u_n| < \infty \right\}$$

be the Banach space with the norm

$$\|u\|_E = \sum_{n=1}^{\infty} |u_n|.$$

Example 12.9. Consider the following boundary value problem of a fractional differential posed in ℓ^1:

$$\begin{cases} {}^cD_{0+}^{\frac{3}{2}} u(t) = f(t, u(t)), \ t \in J := [0, 1], \\ u(0) = (0.5, 0, \ldots, 0, \ldots), \ \ u(1) = (1, 0, \ldots, 0, \ldots). \end{cases} \tag{12.21}$$

Note that, this problem is a particular case of BVP (12.10), where

$$\alpha = \frac{3}{2}, a = 0, b = 1, \psi(t) = t$$

and $f : J \times \ell^1 \longrightarrow \ell^1$ given by

$$f(t, u) = \left\{ \frac{1}{(t+3)^2} \left(\frac{t}{2^n} + \frac{u_n}{\|u\| + 1} \right) \right\}_{n \geq 1}, \quad \text{for } t \in J, u = \{u_n\}_{n \geq 1} \in \ell^1.$$

It is clear that condition (12.7.1) hold and

$$\|f(t, u)\| \leq \frac{1}{(t+3)^2} (1 + \|u\|)$$
$$= p_f(t)\phi(\|u\|).$$

Therefore assumption (12.7.2) of the Theorem 12.2 is satisfied with $p_f(t) = \frac{1}{(t+3)^2}, t \in J$ and $\phi(x) = 1 + x, x \in [0, \infty)$. On the other hand, for any bounded set $B \subset c_0$, we have

$$\chi(f(t, B)) \leq \frac{1}{(t+3)^2} \chi(B), \text{ a.e. } t \in J.$$

Hence (12.7.3) is satisfied. Now, we will check if condition (12.17) is satisfied. Indeed,

$$4\mathcal{M}_\psi = 0.6687 < 1,$$

and

$$2 + (1 + r)\mathcal{M}_\psi \leq r,$$

thus

$$r \geq \frac{2 + \mathcal{M}_\psi}{1 - \mathcal{M}_\psi} = 8.0544.$$

Then r can be chosen as $r = 8.5$. Consequently, all the hypotheses of Theorem 12.2 are satisfied and we conclude that the BVP (12.21) has at least one solution $u \in C(J, \ell^1)$.

12.3 Notes and remarks

The results of Chapter 12 are taken from the papers [148,221]. For more relevant results and studies, one can see the monographs [267,281,304,332,357,392,439] and the papers [199,209,210,216,243–246,314,315,317,356,363,364,389,398].

13

Ulam stability for ψ-Caputo fractional differential equations and systems

We took as motivation the following papers [88,112,149,190,211,231,260,276,285–288,298, 302,303,330,331,335,336,387,390,397,406,407,413,414,416–418,431,432].

13.1 Existence and Mittag–Leffler–Ulam stability of fractional partial differential equations

13.1.1 Introduction

In this section we obtain results on existence, uniqueness, and Mittag–Leffler–Ulam–Hyers–Rassias stability of solutions for a ψ-Caputo fractional partial fractional differential equation of the following type

$$
\begin{cases}
\left({}^{c}D_{\theta}^{\alpha;\psi}x\right)(t,\mathrm{u}) = f(t,\mathrm{u},x(t,\mathrm{u})), & (t,\mathrm{u}) \in \tilde{\mathrm{I}} := [a,b] \times [a,c] \\
x(t,a) = \eta(t), & t \in [a,b], \\
x(a,\mathrm{u}) = v(\mathrm{u}), & \mathrm{u} \in [a,c], \\
x(a,a) = \eta(a) = v(a),
\end{cases}
\tag{13.1}
$$

where a, b, and c are positive constants, $f : \tilde{\mathrm{I}} \times E \longrightarrow E$ is a given function satisfying some assumptions that will be specified later, E is a real Banach space with norm $\|\cdot\|$, and $\eta : [a,b] \longrightarrow E$, $v : [a,c] \longrightarrow E$ are given absolutely continuous functions. ${}^{c}D_{\theta}^{\alpha;\psi}$ is the ψ-Caputo fractional derivative of order $\alpha = (\alpha_1,\alpha_2) \in (0,1] \times (0,1]$ and $\theta = (a,a)$.

13.1.2 Existence, uniqueness, and stability results

Let us start by defining what we mean by a solution of problem (13.1).

Definition 13.1 ([407]). A function $x \in C(\tilde{\mathrm{I}}, E)$ is said to be a solution of (13.1) if x satisfies the conditions $x(t,a) = \eta(t)$, $x(a,\mathrm{u}) = v(\mathrm{u})$, with $x(a,a) = \eta(a) = v(a)$ and the equation $\left({}^{c}D_{\theta}^{\alpha;\psi}x\right)(t,\mathrm{u}) = f(t,\mathrm{u},x(t,\mathrm{u}))$ on $\tilde{\mathrm{I}}$.

To prove our main result, we need the following lemma:

Fractional Difference, Differential Equations, and Inclusions. https://doi.org/10.1016/B978-0-44-323601-3.00020-4
Copyright © 2024 Elsevier Inc. All rights reserved, including those for text and data mining, AI training, and similar technologies.

Lemma 13.2 ([407]). *A function $x \in C(\tilde{I}, E)$ is a solution of problem* (13.1) *if and only if x satisfies the following integral equation*

$$x(t, u) = \zeta(t, u) + \int_a^t \int_a^u \frac{\psi'(s)\psi'(t)(\psi(t) - \psi(s))^{\alpha_1-1}(\psi(u) - \psi(t))^{\alpha_2-1}}{\Gamma(\alpha_1)\Gamma(\alpha_2)} \qquad (13.2)$$
$$\times f(s, t, x(s, t))\, dt ds, \quad (t, u) \in \tilde{I},$$

where

$$\zeta(t, u) = \eta(t) + v(u) - \eta(a).$$

Now, we consider the $\mathbb{E}_{(\alpha_1,\alpha_2)}$–Ulam–Hyers stability for problem (13.1). Let $\varepsilon > 0$, $\mathbb{L} \geq 0$ and $\Phi : \tilde{I} \to \mathbb{R}^+$, be a continuous function. We consider the following inequalities:

$$\left\| (^cD_\theta^{\alpha;\psi} y)(t, u) - f(t, u, y(t, u)) \right\| \leq \varepsilon, \quad (t, u) \in \tilde{I}; \qquad (13.3)$$

$$\left\| (^cD_\theta^{\alpha;\psi} y)(t, u) - f(t, u, y(t, u)) \right\| \leq \Phi(t, u), \quad (t, u) \in \tilde{I}; \qquad (13.4)$$

$$\left\| (^cD_\theta^{\alpha;\psi} y)(t, u) - f(t, u, y(t, u)) \right\| \leq \varepsilon\Phi(t, u), \quad (t, u) \in \tilde{I}. \qquad (13.5)$$

Definition 13.3 ([417,418]). Eq. (13.1) is $\mathbb{E}_{(\alpha_1,\alpha_2)}$–Ulam–Hyers stable if there exists a real number $c > 0$ such that, for each $\varepsilon > 0$ and for each solution $y \in C(\tilde{I}, E)$ of inequality (13.3), there exists a solution $x \in C(\tilde{I}, E)$ of (13.1) with

$$\|y(t, u) - x(t, u)\| \leq c\varepsilon\mathbb{E}_{(\alpha_1,\alpha_2)}\left(\mathbb{L}(\psi(t) - \psi(a))^{\alpha_1}(\psi(u) - \psi(a))^{\alpha_2}\right).$$

Definition 13.4 ([417,418]). Eq. (13.1) is generalized $\mathbb{E}_{(\alpha_1,\alpha_2)}$–Ulam–Hyers stable if there exists $\omega : C(\mathbb{R}_+, \mathbb{R}_+)$ with $\omega(0) = 0$ such that, for each $\varepsilon > 0$ and for each solution $y \in C(\tilde{I}, \mathbb{R})$ of inequality (13.3), there exists a solution $x \in C(\tilde{I}, E)$ of (13.1) with

$$\|y(t, u) - x(t, u)\| \leq \omega(\varepsilon)\mathbb{E}_{(\alpha_1,\alpha_2)}\left(\mathbb{L}(\psi(t) - \psi(a))^{\alpha_1}(\psi(u) - \psi(a))^{\alpha_2}\right), \quad (t, u) \in \tilde{I}.$$

Definition 13.5 ([417,418]). Eq. (13.1) is $\mathbb{E}_{(\alpha_1,\alpha_2)}$–Ulam–Hyers–Rassias stable with respect to Φ if there exists a real number $c_\Phi > 0$ such that, for each $\varepsilon > 0$ and for each solution $y \in C(\tilde{I}, E)$ of inequality (13.5), there exists a solution $x \in C(\tilde{I}, E)$ of (13.1) with

$$\|y(t, u) - x(t, u)\|$$
$$\leq c_\Phi\varepsilon\Phi(t, u)\mathbb{E}_{(\alpha_1,\alpha_2)}\left(\mathbb{L}(\psi(t) - \psi(a))^{\alpha_1}(\psi(u) - \psi(a))^{\alpha_2}\right), \quad (t, u) \in \tilde{I}.$$

Definition 13.6 ([417,418]). Eq. (13.1) is generalized $\mathbb{E}_{(\alpha_1,\alpha_2)}$–Ulam–Hyers–Rassias stable with respect to Φ if there exists a real number $c_\Phi > 0$ such that, for each solution $y \in C(\tilde{I}, E)$ of inequality (13.4), there exists a solution $x \in C(\tilde{I}, E)$ of (13.1) with

$$\|y(t, u) - x(t, u)\|$$
$$\leq c_\Phi\Phi(t, u)\mathbb{E}_{(\alpha_1,\alpha_2)}\left(\mathbb{L}(\psi(t) - \psi(a))^{\alpha_1}(\psi(u) - \psi(a))^{\alpha_2}\right), \quad (t, u) \in \tilde{I}.$$

Remark 13.7 ([417,418]). It is clear that

(i) Definition 13.3 $\Rightarrow$ Definition 13.4,
(ii) Definition 13.5 $\Rightarrow$ Definition 13.6,
(iii) Definition 13.5 for $\Phi(\cdot,\cdot) = 1 \Rightarrow$ Definition 13.3.

Remark 13.8 ([417,418]). A function $y \in C(\tilde{I}, E)$ is a solution of inequality (13.5) if and only if there exists a function $g \in C(\tilde{I}, E)$ (which depends on solution y) such that

(i) $\|g(t, \mathbf{u})\| \leq \varepsilon \Phi(t, \mathbf{u}), \quad (t, \mathbf{u}) \in \tilde{I}.$
(ii) $(^{c}D_{\theta}^{\alpha;\psi} y)(t, \mathbf{u}) = f(t, \mathbf{u}, y(t, \mathbf{u})) + g(t, \mathbf{u}), \quad (t, \mathbf{u}) \in \tilde{I}.$

Now, we are ready to present our main results. The first existence result is based on Weissinger fixed-point theorem.

Theorem 13.9. *Assume that the following hypotheses hold.*

(13.9.1) *The function $f : \tilde{I} \times E \longrightarrow E$ is continuous.*
(13.9.2) *There exists $\mathbb{L} > 0$ such that*

$$\|f(t, \mathbf{u}, x) - f(t, \mathbf{u}, y)\| \leq \mathbb{L}\|x - y\|, \quad (t, \mathbf{u}) \in \tilde{I}, x, y \in E.$$

Then there exists a unique solution of problem (13.1) on $\tilde{I}$. Furthermore, if the following hypothesis holds:

(13.9.3) *There exists $\lambda_{\Phi} > 0$ such that for each $(t, \mathbf{u}) \in \tilde{I}$, we have*

$$\left(I_{\theta}^{\alpha;\psi} \Phi\right)(t, \mathbf{u}) \leq \lambda_{\Phi} \Phi(t, \mathbf{u}),$$

then the problem (13.2) is $\mathbb{E}_{(\alpha_1, \alpha_2)}$ Ulam–Hyers–Rassias stable and consequently generalized $\mathbb{E}_{(\alpha_1, \alpha_2)}$–Ulam–Hyers–Rassias stable.

Proof. In view of Lemma 13.2, we transform the integral representation (13.2) of the initial value problem (13.1) into

$$x = \mathbb{T}x, \quad x \in C(\tilde{I}, E),$$

where $\mathbb{T} : C(\tilde{I}, E) \longrightarrow C(\tilde{I}, E)$ is defined by

$$\mathbb{T}x(t, \mathbf{u}) = \zeta(t, \mathbf{u}) + \int_{a}^{t} \int_{a}^{\mathbf{u}} \frac{\psi'(s)\psi'(t)(\psi(t) - \psi(s))^{\alpha_1 - 1}(\psi(\mathbf{u}) - \psi(t))^{\alpha_2 - 1}}{\Gamma(\alpha_1)\Gamma(\alpha_2)} \quad (13.6)$$
$$\times f(s, t, x(s, t))\mathrm{dtds}.$$

Clearly the operator $\mathbb{T}$ is well defined. Now, we apply Weissinger fixed-point theorem to prove that $\mathbb{T}$ has a unique fixed point. Indeed, it enough to show that $\mathbb{T}^n$ is a contraction operator for sufficiently large n. By the induction method, for any $x, y \in C(\tilde{I}, E)$ and $(t, \mathbf{u}) \in \tilde{I}$, we will verify that

$$\|\mathbb{T}^n x - \mathbb{T}^n y\|_\infty \le \frac{\mathbb{L}^n(\psi(t)-\psi(a))^{n\alpha_1}(\psi(u)-\psi(a))^{n\alpha_2}}{\Gamma(n\alpha_1+1)\Gamma(n\alpha_2+1)}\|x-y\|_\infty, \quad n\in\mathbb{N}. \tag{13.7}$$

For $n=0$, the above inequality is trivially true. We assume that (13.7) is true for $n=k$ and prove it for $n=k+1$. From the definition of the operator $\mathbb{T}$ and assumption (13.9.2), we can get

$$\begin{aligned}
\left\|\mathbb{T}^{k+1}x(t,u)-\mathbb{T}^{k+1}y(t,u)\right\| &= \left\|\mathbb{T}(\mathbb{T}^k x(t,u))-\mathbb{T}(\mathbb{T}^k y(t,u))\right\| \\
&\le \int_a^t\int_a^u \frac{\psi'(s)\psi'(t)(\psi(t)-\psi(s))^{\alpha_1-1}(\psi(u)-\psi(t))^{\alpha_2-1}}{\Gamma(\alpha_1)\Gamma(\alpha_2)} \\
&\quad \times \left\|f(s,t,\mathbb{T}^k x(s,t))-f(s,t,\mathbb{T}^k y(s,t))\right\| dtds \\
&\le \frac{\mathbb{L}^{k+1}\|x-y\|_\infty}{\Gamma(k\alpha_1+1)\Gamma(k\alpha_2+1)}\int_a^t\int_a^u \frac{\psi'(s)\psi'(t)(\psi(t)-\psi(s))^{\alpha_1-1}(\psi(u)-\psi(t))^{\alpha_2-1}}{\Gamma(\alpha_1)\Gamma(\alpha_2)} \\
&\quad \times (\psi(s)-\psi(a))^{k\alpha_1}(\psi(t)-\psi(a))^{k\alpha_2} dtds \\
&\le \frac{\mathbb{L}^{k+1}\|x-y\|_\infty}{\Gamma(k\alpha_1+1)\Gamma(k\alpha_2+1)}\int_a^t \frac{\psi'(s)(\psi(t)-\psi(s))^{\alpha_1-1}}{\Gamma(\alpha_1)}(\psi(s)-\psi(a))^{k\alpha_1} ds \\
&\quad \times \int_a^u \frac{\psi'(t)(\psi(u)-\psi(t))^{\alpha_2-1}}{\Gamma(\alpha_2)}(\psi(t)-\psi(a))^{k\alpha_2} dt.
\end{aligned}$$

Note that

$$\begin{aligned}
\int_a^z &\frac{\psi'(w)(\psi(z)-\psi(w))^{v-1}}{\Gamma(v)}(\psi(w)-\psi(a))^{kv} dw \\
&= \frac{(\psi(z)-\psi(a))^{(k+1)v}}{\Gamma(v)}\int_0^1 (1-\theta)^{v-1}\theta^{kv} d\theta \\
&= \frac{(\psi(z)-\psi(a))^{(k+1)v}}{\Gamma(v)} B(v,kv+1) \\
&= \frac{\Gamma(kv+1)(\psi(z)-\psi(a))^{(k+1)v}}{\Gamma((k+1)v+1)},
\end{aligned}$$

where we have used the variable substitution $y=\frac{\psi(w)-\psi(a)}{\psi(z)-\psi(a)}$, and the relationship between the Beta and the Gamma functions.

Using the above arguments, we get

$$\|\mathbb{T}^{k+1}x-\mathbb{T}^{k+1}y\|_\infty \le \frac{\mathbb{L}^{k+1}(\psi(t)-\psi(a))^{(k+1)\alpha_1}(\psi(u)-\psi(a))^{(k+1)\alpha_2}}{\Gamma((k+1)\alpha_1+1)\Gamma((k+1)\alpha_2+1)}\|x-y\|_\infty.$$

Therefore, by the method of mathematical induction, we know that the inequality (13.7) is true for any $n\in\mathbb{N}$ and $(t,u)\in\tilde{I}$. Hence we have

$$\|\mathbb{T}^n x-\mathbb{T}^n y\|_\infty \le \frac{\mathbb{L}^n(\psi(b)-\psi(a))^{n\alpha_1}(\psi(c)-\psi(a))^{n\alpha_2}}{\Gamma(n\alpha_1+1)\Gamma(n\alpha_2+1)}\|x-y\|_\infty, \quad n\in\mathbb{N}.$$

Setting

$$\vartheta_n = \frac{\left(\mathbb{L}(\psi(b) - \psi(a))^{\alpha_1} (\psi(c) - \psi(a))^{\alpha_2}\right)^n}{\Gamma(n\alpha_1 + 1)\Gamma(n\alpha_2 + 1)}.$$

We observe that

$$\sum_{n=0}^{\infty} \vartheta_n = \sum_{n=0}^{\infty} \frac{\left(\mathbb{L}(\psi(b) - \psi(a))^{\alpha_1} (\psi(c) - \psi(a))^{\alpha_2}\right)^n}{\Gamma(n\alpha_1 + 1)\Gamma(n\alpha_2 + 1)}$$

$$= \mathbb{E}_{(\alpha_1, \alpha_2)}\left(\mathbb{L}(\psi(b) - \psi(a))^{\alpha_1} (\psi(c) - \psi(a))^{\alpha_2}\right).$$

Thus $\mathbb{T}^n$ is a contraction mapping. Therefore by Weissinger fixed-point theorem, $\mathbb{T}$ has a unique fixed point. That is (13.1) has a unique solution.

Now we complete the proof by studying the Mittag–Leffler–Ulam–Hyers–Rassias stability of the proposed problem (13.1).

Let $\varepsilon > 0$, let $y \in C(\tilde{I}, E)$ be a function which satisfies the inequality (13.5), and let $x \in C(\tilde{I}, E)$ be the unique solution of the following problem,

$$\begin{cases} (^c D_\theta^{\alpha;\psi} x)(t, u) = f(t, u, x(t, u)), & (t, u) \in \tilde{I} := [a, b] \times [a, c], \\ x(t, a) = \eta(t), & t \in [a, b], \\ x(a, u) = v(x), & u \in [a, c], \\ x(a, a) = \eta(a) = v(a). \end{cases}$$

By Lemma 13.2, we have

$$x(t, u) = \zeta(t, u) + \int_a^t \int_a^u \frac{\psi'(s)\psi'(t)(\psi(t) - \psi(s))^{\alpha_1 - 1}(\psi(u) - \psi(t))^{\alpha_2 - 1}}{\Gamma(\alpha_1)\Gamma(\alpha_2)}$$
$$\times f(s, t, x(s, t)) \, dtds.$$

Since we have assumed that y is a solution of (13.5), hence we have, by Remark 13.8,

$$\begin{cases} (^c D_\theta^{\alpha;\psi} y)(t, u) = f(t, u, y(t, u)) + g(t, u), & (t, u) \in \tilde{I}, \\ y(t, a) = \eta(t), & t \in [a, b], \\ y(a, u) = v(u), & u \in [a, c], \\ y(a, a) = \eta(a) = v(a). \end{cases}$$

Again by Lemma 13.2, we have

$$y(t, u) = \zeta(t, u) + \int_a^t \int_a^u \frac{\psi'(s)\psi'(t)(\psi(t) - \psi(s))^{\alpha_1 - 1}(\psi(u) - \psi(t))^{\alpha_2 - 1}}{\Gamma(\alpha_1)\Gamma(\alpha_2)}$$
$$\times \left(f(s, t, y(s, t)) + g(s, t)\right) dtds.$$

On the other hand, we have, for each $(t, u) \in \tilde{I}$,

$$
\begin{aligned}
&\|x(t, u) - y(t, u)\| \\
&\leq \int_a^t \int_a^u \frac{\psi'(s)\psi'(t)(\psi(t) - \psi(s))^{\alpha_1 - 1}(\psi(u) - \psi(t))^{\alpha_2 - 1}}{\Gamma(\alpha_1)\Gamma(\alpha_2)} \|g(s, t)\| dtds \\
&+ \int_a^t \int_a^u \frac{\psi'(s)\psi'(t)(\psi(t) - \psi(s))^{\alpha_1 - 1}(\psi(u) - \psi(t))^{\alpha_2 - 1}}{\Gamma(\alpha_1)\Gamma(\alpha_2)} \\
&\times \|f(s, t, x(s, t)) - f(s, t, y(s, t))\| dtds.
\end{aligned}
$$

Hence using part (i) of Remark 13.8, (13.9.2), and (13.9.3), we obtain

$$
\begin{aligned}
&\|x(t, u) - y(t, u)\| \\
&\leq \varepsilon \int_a^t \int_a^u \frac{\psi'(s)\psi'(t)(\psi(t) - \psi(s))^{\alpha_1 - 1}(\psi(u) - \psi(t))^{\alpha_2 - 1}}{\Gamma(\alpha_1)\Gamma(\alpha_2)} \Phi(s, t) dtds \\
&+ \mathbb{L} \int_a^t \int_a^u \frac{\psi'(s)\psi'(t)(\psi(t) - \psi(s))^{\alpha_1 - 1}(\psi(u) - \psi(t))^{\alpha_2 - 1}}{\Gamma(\alpha_1)\Gamma(\alpha_2)} \\
&\times \|x(s, t) - y(s, t)\| dtds \\
&\leq \varepsilon \lambda_\Phi \Phi(t, u) + \mathbb{L} \int_a^t \int_a^u \frac{\psi'(s)\psi'(t)(\psi(t) - \psi(s))^{\alpha_1 - 1}(\psi(u) - \psi(t))^{\alpha_2 - 1}}{\Gamma(\alpha_1)\Gamma(\alpha_2)} \\
&\times \|x(s, t) - y(s, t)\| dtds.
\end{aligned}
$$

Applying the Gronwall inequality Eq. (2.18) to the above inequality with

$$
v(t, u) = \|y(t, u) - x(t, u)\|, \quad w(t, u) = \varepsilon \lambda_\Phi \Phi(t, u), \quad c = \frac{\mathbb{L}}{\Gamma(\alpha_1)\Gamma(\alpha_2)},
$$

we have

$$
\|y(t, u) - x(t, u)\| \leq w(t, u) \mathbb{E}_{(\alpha_1, \alpha_2)} \left(\mathbb{L}(\psi(t) - \psi(a))^{\alpha_1} (\psi(u) - \psi(a))^{\alpha_2} \right),
$$

which yields that

$$
\|y(t, u) - x(t, u)\| \leq \varepsilon \lambda_\Phi \Phi(t, u) \mathbb{E}_{(\alpha_1, \alpha_2)} \left(\mathbb{L}(\psi(t) - \psi(a))^{\alpha_1} (\psi(u) - \psi(a))^{\alpha_2} \right). \qquad (13.8)
$$

Taking for simplicity

$$
c_\Phi = \lambda_\Phi,
$$

then (13.8) becomes

$$
\|y(t, u) - x(t, u)\| \leq c_\Phi \varepsilon \Phi(t, u) \mathbb{E}_{(\alpha_1, \alpha_2)} \left(\mathbb{L}(\psi(t) - \psi(a))^{\alpha_1} (\psi(u) - \psi(a))^{\alpha_2} \right).
$$

Thus problem (13.1) is $\mathbb{E}_{(\alpha_1, \alpha_2)}$–Ulam–Hyers–Rassias stable. Furthermore, if we set $\varepsilon = 1$, then problem (13.1) is generalized $\mathbb{E}_{(\alpha_1, \alpha_2)}$–Ulam–Hyers–Rassias stable. This completes the proof. $\qquad \square$

The second result is based on Mönch fixed-point theorem.

Set

$$\mathbb{M}_\psi = \frac{(\psi(b) - \psi(a))^{\alpha_1}(\psi(c) - \psi(a))^{\alpha_2}}{\Gamma(\alpha_1 + 1)\Gamma(\alpha_2 + 1)}, \quad f^* = \sup_{(t,u)\in\tilde{I}} \|f(t, u, 0)\|.$$

Theorem 13.10. *Assume that the hypotheses* (13.9.1) *and* (13.9.2) *hold. If*

$$\mathbb{L}\mathbb{M}_\psi < 1, \tag{13.9}$$

then problem (13.1) *has at least one solution defined on* $\tilde{I}$.

Proof. In order to use Mönch fixed-point theorem, we should define a subset $\mathfrak{B}_r$ of $C(\tilde{I}, E)$ by

$$\mathfrak{B}_r = \{w \in C(\tilde{I}, E) : \|w\|_\infty \leq r\},$$

with $r > 0$, such that

$$r \geq \frac{\|\zeta\| + f^*\mathbb{M}_\psi}{1 - \mathbb{L}\mathbb{M}_\psi}.$$

Notice that $\mathfrak{B}_r$ is a closed, convex, and bounded subset of the Banach space $C(\tilde{I}, E)$. We will prove that $\mathbb{T}$ satisfies all conditions of Mönch fixed point theorem.

Step 1: The operator $\mathbb{T}$ maps the set $\mathfrak{B}_r$ into itself. Indeed, for any $x \in \mathfrak{B}_r$ and for each $(t, u) \in \tilde{I}$, from the definition of the operator $\mathbb{T}$ and assumption (13.9.2), we can get

$$\begin{aligned}
&\|\mathbb{T}x(t, u)\| \\
&\leq \|\zeta(t, u)\| + \int_a^t \int_a^u \frac{\psi'(s)\psi'(t)(\psi(t) - \psi(s))^{\alpha_1-1}(\psi(u) - \psi(t))^{\alpha_2-1}}{\Gamma(\alpha_1)\Gamma(\alpha_2)} \\
&\quad \times \left(\|f(s, t, x(s, t)) - f(s, t, 0)\| + \|f(s, t, 0)\|\right) dtds \\
&\leq \|\zeta\| + (\mathbb{L}r + f^*) \int_a^t \int_a^u \frac{\psi'(s)\psi'(t)(\psi(t) - \psi(s))^{\alpha_1-1}(\psi(u) - \psi(t))^{\alpha_2-1}}{\Gamma(\alpha_1)\Gamma(\alpha_2)} dtds \\
&= \|\zeta\| + \mathbb{M}_\psi(\mathbb{L}r + f^*).
\end{aligned}$$

Thus

$$\|\mathbb{T}x\| \leq r.$$

This proves that $\mathbb{T}$ transforms the ball $\mathfrak{B}_r$ into itself.

Step 2: The operator $\mathbb{T}$ is continuous. Suppose that $\{x_n\}$ is a sequence such that $x_n \to x$ in $\mathfrak{B}_r$ as $n \to \infty$. By (13.9.2), we get

$$\|\mathbb{T}x_n(t, \mathrm{u}) - \mathbb{T}x(t, \mathrm{u})\|$$

$$\leq \mathbb{L} \int_a^t \int_a^{\mathrm{u}} \frac{\psi'(s)\psi'(t)(\psi(t) - \psi(s))^{\alpha_1 - 1}(\psi(\mathrm{u}) - \psi(t))^{\alpha_2 - 1}}{\Gamma(\alpha_1)\Gamma(\alpha_2)}$$

$$\times \|x_n(s, t) - y_n(s, t)\| \mathrm{dt} \mathrm{ds}.$$

Hence

$$\|\mathbb{T}x_n - \mathbb{T}x\| \leq \mathbb{L}M_\psi \|x_n - x\|. \tag{13.10}$$

Since $x_n \to x$ as $n \to +\infty$, then Eq. (13.10) implies

$$\|\mathbb{T}x_n - \mathbb{T}x\| \to 0 \text{ as } n \to +\infty,$$

which implies the continuity of the operator $\mathbb{T}$.

Step 3: $\mathbb{T}(\mathfrak{B}_r)$ is equicontinuous.

Let $(t_1, \mathrm{u}_1), (t_2, \mathrm{u}_2) \in \tilde{\mathrm{I}}$, with $t_1 < t_2$ and $\mathrm{u}_1 < \mathrm{u}_2$, and let $x \in \mathfrak{B}_r$. Taking (13.9.2) into consideration, we get

$$\|\mathbb{T}x(t_2, \mathrm{u}_2) - \mathbb{T}x(t_1, \mathrm{u}_1)\| \leq |\zeta(t_2, \mathrm{u}_2) - \zeta(t_1, \mathrm{u}_1)|$$

$$+ \frac{\mathbb{L}r + f^*}{\Gamma(\alpha_1)\Gamma(\alpha_2)} \int_a^{t_1} \int_a^{\mathrm{u}_1} \psi'(s)\psi'(t) \big[(\psi(t_1) - \psi(s))^{\alpha_1 - 1}(\psi(\mathrm{u}_1) - \psi(t))^{\alpha_2 - 1}$$

$$- (\psi(t_2) - \psi(s))^{\alpha_1 - 1}(\psi(\mathrm{u}_2) - \psi(t))^{\alpha_2 - 1} \big] \mathrm{dt} \mathrm{ds}$$

$$+ \frac{\mathbb{L}r + f^*}{\Gamma(\alpha_1)\Gamma(\alpha_2)} \int_{t_1}^{t_2} \int_{\mathrm{u}_1}^{\mathrm{u}_2} \psi'(s)\psi'(t)(\psi(t_2) - \psi(s))^{\alpha_1 - 1}(\psi(\mathrm{u}_2) - \psi(t))^{\alpha_2 - 1} \mathrm{dt} \mathrm{ds}$$

$$+ \frac{\mathbb{L}r + f^*}{\Gamma(\alpha_1)\Gamma(\alpha_2)} \int_{t_1}^{t_2} \int_a^{\mathrm{u}_1} \psi'(s)\psi'(t)(\psi(t_2) - \psi(s))^{\alpha_1 - 1}(\psi(\mathrm{u}_2) - \psi(t))^{\alpha_2 - 1} \mathrm{dt} \mathrm{ds}$$

$$+ \frac{\mathbb{L}r + f^*}{\Gamma(\alpha_1)\Gamma(\alpha_2)} \int_a^{t_1} \int_{\mathrm{u}_1}^{\mathrm{u}_2} \psi'(s)\psi'(t)(\psi(t_2) - \psi(s))^{\alpha_1 - 1}(\psi(\mathrm{u}_2) - \psi(t))^{\alpha_2 - 1} \mathrm{dt} \mathrm{ds}$$

$$\leq |\zeta(t_2, \mathrm{u}_2) - \zeta(t_1, \mathrm{u}_1)|$$

$$+ \frac{2(\mathbb{L}r + f^*)(\psi(\mathrm{u}_2) - \psi(\mathrm{u}_1))^{\alpha_2} \big[(\psi(t_2) - \psi(a))^{\alpha_1} - (\psi(t_2) - \psi(t_1))^{\alpha_1} \big]}{\Gamma(\alpha_1 + 1)\Gamma(\alpha_2 + 1)}$$

$$+ \big[(\psi(\mathrm{u}_2) - \psi(a))^{\alpha_2}(\psi(t_2) - \psi(t_1))^{\alpha_1} - (\psi(\mathrm{u}_1) - \psi(a))^{\alpha_2}(\psi(t_1) - \psi(a))^{\alpha_1} \big]$$

$$\times \frac{(\mathbb{L}r + f^*)}{\Gamma(\alpha_1 + 1)\Gamma(\alpha_2 + 1)}.$$

As $t_1 \to t_2$ and $\mathrm{u}_1 \to \mathrm{u}_2$, the right-hand side of the above inequality tends to zero, independently of $x \in \mathfrak{B}_r$. Hence, we conclude that $\mathbb{T}(\mathfrak{B}_r) \subseteq C(\tilde{\mathrm{I}}, E)$ is bounded and equicontinuous.

Step 4: Our aim in this step is to show that the operator $\mathbb{T}$ satisfies the Mönch condition on $\mathfrak{B}_r$. Let $\mathcal{V}$ be a subset of $\mathfrak{B}_r$ such that $\mathcal{V} \subset \overline{\text{conv}}(\mathbb{T}(\mathcal{V}) \cup \{0\})$. Then, $\mathcal{V}$ is bounded and equicontinuous and therefore the function $(t, \mathfrak{u}) \to \gamma(t, \mathfrak{u}) = \mu(\mathcal{V}(t, \mathfrak{u}))$ is continuous on $\tilde{\mathbb{I}}$. From (13.9.2) and the properties of the measure of noncompactness μ and Lemma 2.42, we find

$$
\begin{aligned}
&\gamma(t, \mathfrak{u}) \\
&\leq \mu(\overline{\text{conv}}(\mathbb{T}(\mathcal{V})(t, \mathfrak{u}) \cup \{0\}) \leq \mu(\mathbb{T}(\mathcal{V})(t, \mathfrak{u})) \\
&\leq \mu\left\{ \int_a^t \int_a^{\mathfrak{u}} \frac{\psi'(s)\psi'(t)(\psi(t) - \psi(s))^{\alpha_1 - 1}(\psi(\mathfrak{u}) - \psi(t))^{\alpha_2 - 1}}{\Gamma(\alpha_1)\Gamma(\alpha_2)} f(s, t, x(s, t)) dt ds \right\} \\
&\leq \int_a^t \int_a^{\mathfrak{u}} \frac{\psi'(s)\psi'(t)(\psi(t) - \psi(s))^{\alpha_1 - 1}(\psi(\mathfrak{u}) - \psi(t))^{\alpha_2 - 1}}{\Gamma(\alpha_1)\Gamma(\alpha_2)} \\
&\quad \times \mu\{ f(s, t, x(s, t)), x \in \mathcal{V}\} dt ds \\
&\leq \mathbb{L} \int_a^t \int_a^{\mathfrak{u}} \frac{\psi'(s)\psi'(t)(\psi(t) - \psi(s))^{\alpha_1 - 1}(\psi(\mathfrak{u}) - \psi(t))^{\alpha_2 - 1}}{\Gamma(\alpha_1)\Gamma(\alpha_2)} \gamma(s, t) dt ds.
\end{aligned}
$$

Hence thanks to the Gronwall inequality (Lemma 2.102), we obtain $\gamma(t, \mathfrak{u}) = \mu(\mathcal{V}(t, \mathfrak{u})) = 0$, for each $(t, \mathfrak{u}) \in \tilde{\mathbb{I}}$, and then $\mathcal{V}(t, \mathfrak{u})$ is relatively compact in E. In view of the Ascoli–Arzelà theorem, $\mathcal{V}$ is relatively compact in $\mathfrak{B}_r$. By Theorem 2.75, we conclude that $\mathbb{T}$ has a fixed point, which is a solution to problem (13.1). This completes the proof. $\qquad \square$

Our last result is based on the concept of Meir–Keeler condensing operators.

Theorem 13.11. *Assume that the hypotheses* (13.9.1) *and* (13.9.2) *hold. If*

$$
4\mathbb{L}\mathbb{M}_\psi < 1, \tag{13.11}
$$

then the IVP (13.1) *has at least one solution defined on* $\tilde{\mathbb{I}}$.

Proof. Consider the operator $\mathbb{T}$ defined in (13.6). We will show that $\mathbb{T}$ satisfies all the assumptions of Theorem 2.98. We know that $\mathbb{T} : \mathfrak{B}_r \to \mathfrak{B}_r$ is bounded and continuous. It is enough to prove that $\mathbb{T}$ is a Meir–Keeler condensing operator. To do this, suppose $\varepsilon > 0$ is given, and we will prove that there exists $\delta > 0$ such that

$$
\varepsilon \leq \mu_C(B) < \varepsilon + \delta \Rightarrow \mu_C(\mathbb{T}B) < \varepsilon, \quad \text{for any } B \subset \mathfrak{B}_r.
$$

For every bounded subset $B \subset \mathfrak{B}_r$ and $\varepsilon' > 0$, using Lemma 2.40 and the properties of μ, there exists a sequence $\{x_n\}_{n=1}^{\infty} \subset B$ such that

$$
\begin{aligned}
&\mu(\mathbb{T}(B)(t, \mathfrak{u})) \\
&\leq 2\mu\left\{ \int_a^t \int_a^{\mathfrak{u}} \frac{\psi'(s)\psi'(t)(\psi(t) - \psi(s))^{\alpha_1 - 1}(\psi(\mathfrak{u}) - \psi(t))^{\alpha_2 - 1}}{\Gamma(\alpha_1)\Gamma(\alpha_2)} f(s, t, x_n(s, t)) dt ds \right\} + \varepsilon'.
\end{aligned}
$$

Next, by Lemma 2.40 and (13.9.2), we have

$$\mu\left(\mathbb{T}(B)(t, \mathrm{u})\right)$$

$$\leq 4 \int_a^t \int_a^{\mathrm{u}} \frac{\psi'(s)\psi'(t)(\psi(t)-\psi(s))^{\alpha_1-1}(\psi(\mathrm{u})-\psi(t))^{\alpha_2-1}}{\Gamma(\alpha_1)\Gamma(\alpha_2)} \mu\left(f\left(s, t, \{x_n(s, t)\}_{n=1}^{\infty}\right)\right) \mathrm{dt ds} + \varepsilon'$$

$$\leq 4 \mathbb{L} M_\psi \mu_C(B) + \varepsilon'.$$

As the last inequality is true, for every $\varepsilon' > 0$, we infer

$$\mu\left(\mathbb{T}(B)(t, \mathrm{u})\right) \leq 4\mathbb{L} M_\psi \mu_C(B).$$

Since $\mathbb{T}(B) \subset \mathfrak{B}_r$ is bounded and equicontinuous, we know from Lemma 2.40 that

$$\mu_C\left(\mathbb{T}(B)\right) = \max_{t \in J} \mu\left(\mathbb{T}(B)(t, \mathrm{u})\right).$$

Therefore we have

$$\mu_C\left(\mathbb{T}(B)\right) \leq 4\mathbb{L} M_\psi \mu_C(B).$$

Observe that from the last estimates

$$\mu_C\left(\mathbb{T}(B)\right) \leq 4\mathbb{L} M_\psi \mu_C(B) < \varepsilon \Rightarrow \mu_C(B) < \frac{1}{4\mathbb{L} M_\psi}\varepsilon.$$

Let us now take

$$\delta = \frac{1 - 4\mathbb{L} M_\psi}{4\mathbb{L} M_\psi}\varepsilon,$$

we get

$$\varepsilon \leq \mu_C(B) < \varepsilon + \delta,$$

which means that $\mathbb{T} : \mathfrak{B}_r \to \mathfrak{B}_r$ is a Meir–Keeler condensing operator. It follows from Theorem 2.98 that the operator $\mathbb{T}$ defined by (13.6) has at least one fixed point $x \in \mathfrak{B}_r$, which is the desired solution of problem (13.1). This completes the proof of Theorem 13.9. $\square$

13.1.3 Some examples

In this section we give two examples to illustrate our above results.

Example 13.12. Let

$$E = \ell^1 = \left\{ x = (x_1, x_2, \ldots, x_k, \ldots), \quad \sum_{k=1}^{\infty} |x_k| < \infty \right\},$$

be the Banach space with the norm

$$\|x\|_{\ell^1} = \sum_{k=1}^{\infty} |x_k|.$$

Consider the following infinite system of partial hyperbolic fractional differential equations of the form

$$\begin{cases} (^C_H D^\alpha_\theta x_k)(t, u) = \dfrac{1}{t+u} \dfrac{x_k(t,u)}{1+|x_k(t,u)|}, & (t, u) \in \tilde{I} := [1, e] \times [1, e], \\ x(t, 1) = (t, 0, \ldots, 0, \ldots), & t \in [1, e], \\ x(1, u) = (u, 0, \ldots, 0, \ldots), & u \in [1, e], \end{cases} \tag{13.12}$$

where

$$\alpha = (\alpha_1, \alpha_2) = (0.5, 0.5), \quad a = 1, \quad b = c = e, \quad \psi(\cdot) = \ln(\cdot), \quad E = \ell^1.$$

Set

$$x = (x_1, x_2, \ldots, x_k, \ldots), \quad f = (f_1, f_2, \ldots, f_k, \ldots).$$

Clearly, the function $f : \tilde{I} \times \ell^1 \longrightarrow \ell^1$ is continuous. Moreover, for any $x, y \in \ell^1$ and $(t, u) \in \tilde{I}$, we have

$$\|f(t, u, x) - f(t, u, y)\|_{\ell^1} \leq \frac{1}{2}\|x - y\|_{\ell^1}.$$

Thus hypothesis (13.9.2) is satisfied with $\mathbb{L} = \frac{1}{2}$. An application of Theorem 13.9 shows that problem (13.12) has a unique solution in $C(\tilde{I}, \ell^1)$. Moreover, by letting $\Phi(t, u) = \ln t \ln u$, we get

$$^H I^\alpha_\theta \Phi(t, u) = \frac{(\ln t \ln u)^{1.5}}{\Gamma(2.5)\Gamma(2.5)} \leq \frac{16}{9\pi} \ln t \ln u = \lambda_\Phi \Phi(t).$$

So condition (13.9.3) is satisfied with $\Phi(t, u) = \ln t \ln u$ and $\lambda_\Phi = \frac{16}{9\pi}$. It follows from Theorem 13.9 that problem (13.12) is $\mathbb{E}_{(\alpha_1, \alpha_2)}$–Ulam–Hyers–Rassias stable and consequently it is generalized $\mathbb{E}_{(\alpha_1, \alpha_2)}$–Ulam–Hyers–Rassias stable.

Example 13.13. Now let

$$E = c_0 = \{x = (x_1, x_2, \ldots, x_k, \ldots) : x_k \to 0 \ (k \to \infty)\},$$

be the Banach space of real sequences converging to zero, endowed with its usual norm

$$\|x\|_{c_0} = \sup_{k \geq 1} |x_k|.$$

Consider the following infinite system of partial hyperbolic fractional differential equations of the form

$$\begin{cases} (^cD_\theta^{\alpha;\psi}x_k)(t,u) = \frac{1}{9+e^{t+u}}\left(\frac{1}{k^2} + \ln(1+|x_k(t,u)|)\right), (t,u) \in \tilde{I} := [0,1] \times [0,1], \\ x(t,0) = (0,0,\dots,0,\dots), \quad t \in [0,1], \\ x(0,u) = (0,0,\dots,0,\dots), \quad u \in [0,1], \end{cases} \tag{13.13}$$

where

$$\alpha = (\alpha_1,\alpha_2) = (0.5, 0.5), \quad a=0, \quad b=c=1, \quad E=c_0.$$

Taking also $\psi(\cdot) = \sigma(\cdot)$ where $\sigma(\cdot)$ is the Sigmoid function, which can be expressed in the following form

$$\sigma(z) = \frac{1}{1+e^{-z}},$$

a convenience of the Sigmoid function is its derivative

$$\sigma'(z) = \sigma(z)(1-\sigma(z)).$$

Set

$$x = (x_1, x_2, \dots, x_k, \dots), \quad f = (f_1, f_2, \dots, f_k, \dots).$$

Clearly, the function $f : \tilde{I} \times c_0 \longrightarrow c_0$ is continuous. Moreover, for any $x, y \in c_0$ and $(t,u) \in \tilde{I}$, we have

$$\|f(t,u,x) - f(t,u,y)\|_{c_0} \le \frac{1}{10}\|x-y\|_{c_0}.$$

Therefore, assumption (13.9.2) is satisfied with $\mathbb{L} = \frac{1}{10}$. We will check if condition (13.9) is satisfied. Indeed,

$$\mathbb{L}\mathbb{M}_\psi \simeq 0.03.$$

Consequently, all the hypothesis of Theorem 13.10 are satisfied. Hence the problem (13.13) has at least one solution $x \in C(\tilde{I}, c_0)$. Moreover, the condition (13.11) is satisfied. Then, by Theorem 13.11, problem (13.13) has at least one solution $x \in C(\tilde{I}, c_0)$.

13.2 Coupled system of fractional differential equations without and with delay in generalized Banach spaces

13.2.1 Introduction

In this section, we deal with the existence and uniqueness results as well as the Ulam–Hyers stability of solutions for the following system of differential equations involving the

ψ-Caputo derivative of fractional order:

$$\begin{cases} (^c D_{a+}^{\alpha;\psi} x)(t) = \mathbb{A}_1 x(t) + f_1(t, x(t), y(t)), \\ (^c D_{a+}^{\beta;\psi} y)(t) = \mathbb{A}_2 y(t) + f_2(t, x(t), y(t)), \end{cases} \quad t \in J := [a, b], \tag{13.14}$$

subject to the initial conditions

$$\begin{cases} x(a) = \phi_1, \\ y(a) = \phi_2, \end{cases} \tag{13.15}$$

where $^c D_{a+}^{\alpha;\psi}$, $^c D_{a+}^{\beta;\psi}$ are the ψ-Caputo fractional derivative of order $\alpha, \beta \in (0, 1]$, respectively, which was recently proposed by Almeida [119]. $f_1, f_2 : J \times \mathbb{R}^n \times \mathbb{R}^n \longrightarrow \mathbb{R}^n$ are a given continuous functions, a and b are positive constants such that $a < b, \phi_1, \phi_2 \in \mathbb{R}^n$, and $\mathbb{A}_1, \mathbb{A}_2 \in \mathbb{R}^{n \times n}$.

Next, we turn our attention to study the existence and uniqueness of solutions to the following delayed coupled system of the form:

$$\begin{cases} (^c D_{a+}^{\alpha;\psi} x)(t) = g_1(t, x_t, y_t), \\ (^c D_{a+}^{\beta;\psi} y)(t) = g_2(t, x_t, y_t), \end{cases} \quad t \in J, \tag{13.16}$$

along with the initial conditions

$$\begin{cases} x(t) = \vartheta_1(t), \\ y(t) = \vartheta_2(t), \end{cases} \quad t \in [a - \delta, a], \tag{13.17}$$

where $\delta > 0$ is a constant delay and $g_1, g_2 : J \times C([-\delta, 0], \mathbb{R}^n) \times C([-\delta, 0], \mathbb{R}^n) \longrightarrow \mathbb{R}^n$, are given continuous functions and $\vartheta_1, \vartheta_2 : [a - \delta, a] \longrightarrow \mathbb{R}^n$ are two continuous functions. For any function z defined on $[a - \delta, a]$ and any $t \in J$, we denote by z_t the element of $C([-\delta, 0], \mathbb{R}^n)$ defined by

$$z_t(\rho) = z(t + \rho), \quad \rho \in [-\delta, 0].$$

Hence $z_t(\cdot)$ represents the history of the state from times $t - \delta$ up to the present time t.

Before engaging the existence results of this section, we introduce the essential functional spaces that we will adopt in this work. We denote by $C([a, b], \mathbb{R}^n)$ the Banach space of all continuous functions z from $[a, b]$ into $\mathbb{R}^n$ with the supremum norm

$$\|z\|_{[a,b]} = \sup_{t \in [a,b]} \|z(t)\|.$$

Let $\mathfrak{C} = C([a - \delta, b], \mathbb{R}^n)$, denote the Banach space of functions from $[a - \delta, b]$ into $\mathbb{R}^n$ equipped with the supremum norm $\|z\|_{\mathfrak{C}}$. In addition, let us denote by $\mathfrak{C}_\delta := C([-\delta, 0], \mathbb{R}^n)$ the Banach space of functions w from $[-\delta, 0]$ into $\mathbb{R}^n$, endowed with the norm

$$\|w\|_{\mathfrak{C}_\delta} = \sup_{\rho \in [-\delta, 0]} \|w(\rho)\|.$$

The following lemma has an important role in proving our main results.

Lemma 13.14 ([147]). *Let $\omega, \theta > 0$. Then, for all $t \in [a, b]$, we have*

$$I_{a+}^{\omega;\psi} e^{\theta(\psi(t)-\psi(a))} \leq \frac{e^{\theta(\psi(t)-\psi(a))}}{\theta^\omega}.$$

Remark 13.15 ([399,400]). On the space $C(J, \mathbb{R}^n)$ we define a Bielecki type norm $\|\cdot\|_{\mathfrak{B}}$ as below

$$\|z\|_{\mathfrak{B}} := \sup_{t \in J} \frac{\|z(t)\|}{e^{\theta(\psi(t)-\psi(a))}}, \quad \theta > 0. \tag{13.18}$$

Consequently, we have the following proprieties

1. $(C(J, \mathbb{R}^n), \|\cdot\|_{\mathfrak{B}})$ is a Banach space.
2. The norms $\|\cdot\|_{\mathfrak{B}}$ and $\|\cdot\|_\infty$ are equivalent on $C(J, \mathbb{R}^n)$, where $\|\cdot\|_\infty$ denotes the Chebyshev norm on $C(J, \mathbb{R}^n)$, i.e.,

$$\iota_1 \|\cdot\|_{\mathfrak{B}} \leq \|\cdot\|_\infty \leq \iota_2 \|\cdot\|_{\mathfrak{B}},$$

where

$$\iota_1 = 1, \quad \iota_2 = e^{\theta(\psi(b)-\psi(a))}.$$

For more properties on Bielecki type norm, see [213,399,400].

Let $x, y \in \mathbb{R}^m$ with $x = (x_1, x_2, \ldots, x_m)$, $y = (y_1, y_2, \ldots, y_m)$. By $x \leq y$ we mean $x_i \leq y_i$, $i = 1, \ldots, m$. Also,

$$|x| = (|x_1|, |x_2|, \ldots, |x_m|),$$
$$\max(x, y) = (\max(x, y), \max(\bar{x}, \bar{y}), \ldots, \max(x_m, y_m)),$$

and

$$\mathbb{R}_+^m = \{x \in \mathbb{R}^m : x_i \in \mathbb{R}_+, i = 1, \ldots, m\}.$$

If $c \in \mathbb{R}$, then $x \leq c$ means $x_i \leq c, i = 1, \ldots, m$.

Definition 13.16 ([331]). Let X be a nonempty set. By a vector-valued metric on X we mean a map $d: X \times X \to \mathbb{R}^m$ with the following properties:

(i) $d(x, y) \geq 0$ for all $x, y \in X$, and if $d(x, y) = 0$, then $x = y$;
(ii) $d(x, y) = d(y, x)$ for all $x, y \in X$;
(iii) $d(x, z) \leq d(x, y) + d(y, z)$ for all $x, y, z \in X$.

We call the pair (X, d) a generalized metric space with

$$d(x, y) := \begin{pmatrix} d_1(x, y) \\ d_2(x, y) \\ \vdots \\ d_m(x, y) \end{pmatrix}.$$

Notice that d is a generalized metric space on X if and only if $d_i, i = 1, \ldots, m$, are metrics on X.

Definition 13.17 ([404]). A square matrix $\mathbb{A}$ of real numbers is said to be convergent to zero if and only if its spectral radius $\rho(\mathbb{A})$ is strictly less than 1. In other words, this means that all the eigenvalues of $\mathbb{A}$ are in the open unit disc, i.e., $|\lambda| < 1$ for every $\lambda \in \mathbb{C}$ with $\det(\mathbb{A} - \lambda I) = 0$, where I denotes the unit matrix of $\mathbb{A}_{m \times m}(\mathbb{R})$.

Theorem 13.18 ([404]). *For any nonnegative square matrix $\mathbb{A}$, the following properties are equivalent*

1. $\mathbb{A}$ *is convergent to zero;*
2. $\rho(\mathbb{A}) < 1$;
3. *the matrix $I - \mathbb{A}$ is nonsingular and*

$$(I - \mathbb{A})^{-1} = I + \mathbb{A} + \cdots + \mathbb{A}^n + \cdots;$$

4. $I - \mathbb{A}$ *is nonsingular and $(I - \mathbb{A})^{-1}$ is a nonnegative matrix.*

Example 13.19 ([335]). The matrix $\mathbb{A} \in \mathbb{A}_{2 \times 2}(\mathbb{R})$ defined by

$$\mathbb{A} = \begin{pmatrix} a & b \\ c & d \end{pmatrix}$$

converges to zero in the following cases:

(1) $b = c = 0, a, d > 0$, and $\max\{a, d\} < 1$.
(2) $c = 0, a, d > 0, a + d < 1$, and $-1 < b < 0$.
(3) $a + b = c + d = 0, a > 1, c > 0$, and $|a - c| < 1$.

Definition 13.20 ([336,345]). Let $(\mathbb{E}, d)$ be a generalized metric space. An operator $\mathbb{T}: \mathbb{E} \to \mathbb{E}$ is said to be contractive if there exists a matrix $\mathbb{A}$ convergent to zero such that

$$d(\mathbb{T}(x), \mathbb{T}(y)) \leq \mathbb{A}d(x, y), \quad \text{for all } x, y \in \mathbb{E}.$$

13.2.2 Existence, uniqueness, and stability results

In this section we prove the existence and uniqueness of solutions for the given problem (13.14)–(13.15). We also study the Ulam–Hyers stability of the mentioned system.

Before starting and proving our main result, let us define what we mean by a solution of the problem (13.14)–(13.15).

Definition 13.21. By a solution of problem (13.14)–(13.15) we mean a coupled function $(x, y) \in C(J, \mathbb{R}^n) \times C(J, \mathbb{R}^n)$ that satisfies the system

$$\begin{cases} (^c D_{a+}^{\alpha;\psi} x)(t) = \mathbb{A}_1 x(t) + f_1(t, x(t), y(t)), \\ (^c D_{a+}^{\beta;\psi} y)(t) = \mathbb{A}_2 y(t) + f_2(t, x(t), y(t)), \end{cases} \quad t \in J,$$

and the initial conditions

$$\begin{cases} x(a) = \phi_1, \\ y(a) = \phi_2. \end{cases}$$

For the existence of solutions for the problem (13.14)–(13.15), we need the following lemma:

Lemma 13.22. *Let $\omega \in (0, 1]$ be fixed, $\mathbb{A} \in \mathbb{R}^{n \times n}$ and $h \in C(J \times \mathbb{R}^n, \mathbb{R}^n)$. Then the Cauchy problem*

$$\begin{cases} (^c D_{a+}^{\omega;\psi} z)(t) = \mathbb{A}z(t) + f(t, z(t)), & t \in J, \\ z(a) = \phi \in \mathbb{R}^n, \end{cases} \tag{13.19}$$

is equivalent to the following integral equation,

$$z(t) = \phi + \int_a^t \frac{\psi'(s)(\psi(t) - \psi(s))^{\omega-1}}{\Gamma(\omega)} (\mathbb{A}z(s) + f(s, z(s))) ds. \tag{13.20}$$

Proof. Let $z(t)$ be a solution of the problem (13.19). Define $h(t) = \mathbb{A}z(t) + f(t, z(t))$. Then

$$(^c D_{a+}^{\omega;\psi} z)(t) = h(t), \ 0 < \omega \le 1,$$

that is

$$(^c D_{a+}^{\omega;\psi} z)(t) = I_{a+}^{1-\omega;\psi} \left(\frac{1}{\psi'(t)} \frac{d}{dt} z \right)(t) = h(t), \ 0 < \omega \le 1.$$

Taking the ψ–Riemann–Liouville fractional integral of order ω to the above equation, we get

$$I_{a+}^{1;\psi} \left(\frac{1}{\psi'(t)} \frac{d}{dt} z \right)(t) = \mathbb{I}_{a+}^{\omega;\psi} h(t), \ 0 < \omega \le 1.$$

Since

$$I_{a+}^{1;\psi} \left(\frac{1}{\psi'(t)} \frac{d}{dt} z \right)(t) = I_{a+}^1 \left(\frac{d}{dt} z \right)(t) = z(t) - z(a),$$

we get

$$z(t) = \phi + I_{a+}^{\omega;\psi} h(t).$$

Using the definition of $h(t)$, we obtain (13.20). Conversely, suppose that $z(t)$ is the solution of Eq. (13.20). Then it can be written as

$$z(t) = \phi + I_{a^+}^{\omega;\psi} h(t), \tag{13.21}$$

where $h(t) = \mathbb{A}z(t) + f(t, z(t))$. Since $h(t)$ is continuous and ϕ is a constant vector, operating the ψ-Caputo fractional differential operator $^cD_{a^+}^{\omega;\psi}$ on both sides of Eq. (13.21), we obtain

$$(^cD_{a^+}^{\omega;\psi} z)(t) = {}^cD_{a^+}^{\omega;\psi} \phi + (^cD_{a^+}^{\omega;\psi} z)\mathbb{I}_{a^+}^{\omega;\psi} h(t).$$

Using Lemma 2.5 yields

$$(^cD_{a^+}^{\omega;\psi} z)(t) = \mathbb{A}z(t) + f(t, z(t)).$$

From (13.21), we get $z(a) = \phi$. This proves that $z(t)$ is the solution of Cauchy problem (13.19), which completes the proof. $\qquad\square$

As a consequence of Lemma 13.22, we have the following result, which is useful in our main results.

Lemma 13.23. *Let $\alpha, \beta \in (0, 1]$ be fixed, $\mathbb{A}_1, \mathbb{A}_2 \in \mathbb{R}^{n \times n}$ and $f_1, f_2 \in C(J \times \mathbb{R}^n \times \mathbb{R}^n, \mathbb{R}^n)$. Then the coupled systems (13.14)–(13.15) is equivalent to the following integral equations*

$$\begin{cases} x(t) = \phi_1 + \int_a^t \frac{\psi'(s)(\psi(t)-\psi(s))^{\alpha-1}}{\Gamma(\alpha)} \big(\mathbb{A}_1 x(s) + f_1(s, x(s), y(s))\big)\mathrm{d}s, \\ y(t) = \phi_2 + \int_a^t \frac{\psi'(s)(\psi(t)-\psi(s))^{\beta-1}}{\Gamma(\beta)} \big(\mathbb{A}_2 y(s) + f_2(s, x(s), y(s))\big)\mathrm{d}s, \end{cases} \quad t \in J. \tag{13.22}$$

Our first result on the uniqueness is based on the Perov fixed-point theorem combined with the Bielecki norm.

Theorem 13.24. *Let the following assumptions hold:*

(13.24.1) *$f_1, f_2 : J \times \mathbb{R}^n \times \mathbb{R}^n \longrightarrow \mathbb{R}^n$ are continuous functions.*
(13.24.2) *There exist continuous functions $p_i, q_i : J \to \mathbb{R}_+$, $i = 1, 2$, such that*

$$\|f_i(t, x_1, y_1) - f_i(t, x_2, y_2)\| \leq p_i(t)\|x_1 - x_2\| + q_i(t)\|y_1 - y_2\|$$

for all $t \in J$ and each $x_1, y_1, x_2, y_2 \in \mathbb{R}^n$.

Then the coupled system (13.16)–(13.17) has a unique solution.

For computational convenience, we introduce the following notations:

$$p_i^* := \sup_{t \in J} p_i(t), \quad q_i^* := \sup_{t \in J} q_i(t), \quad \phi_i^* := \|\phi_i\|, \quad \mathbb{A}_i^* = \|\mathbb{A}_i\|,$$

$$f_i^* := \sup_{t \in J} \|f_i(t, 0, 0)\|, \quad \ell_\psi^\alpha := \frac{(\psi(b) - \psi(a))^\alpha}{\Gamma(\alpha + 1)}, \quad \ell_\psi^\beta := \frac{(\psi(b) - \psi(a))^\beta}{\Gamma(\beta + 1)}.$$

Proof. Consider the Banach space $C(J, \mathbb{R}^n)$ equipped with a Bielecki norm type $\|\cdot\|_{\mathcal{B}}$ defined in (13.18). Consequently, the product space $\mathbb{X} := C(J, \mathbb{R}^n) \times C(J, \mathbb{R}^n)$ is a generalized Banach space, endowed with the Bielecki vector-valued norm

$$\|(x, y)\|_{\mathbb{X}, \mathcal{B}} = \begin{pmatrix} \|x\|_{\mathcal{B}} \\ \|y\|_{\mathcal{B}} \end{pmatrix}.$$

We define an operator $\mathbb{T} = (\mathbb{T}_1, \mathbb{T}_2) : \mathbb{X} \to \mathbb{X}$ by:

$$\mathbb{T}(x, y) = \big(\mathbb{T}_1(x, y), \mathbb{T}_2(x, y)\big), \tag{13.23}$$

where

$$(\mathbb{T}_1(x, y))(t) = \phi_1 + \int_a^t \frac{\psi'(s)(\psi(t) - \psi(s))^{\alpha-1}}{\Gamma(\alpha)} \big(\mathbb{A}_1 x(s) + f_1(s, x(s), y(s))\big) ds, \tag{13.24}$$

and

$$(\mathbb{T}_2(x, y))(t) = \phi_2 + \int_a^t \frac{\psi'(s)(\psi(t) - \psi(s))^{\beta-1}}{\Gamma(\beta)} \big(\mathbb{A}_2 y(s) + f_2(s, x(s), y(s))\big) ds. \tag{13.25}$$

Now, we apply Perov fixed-point theorem to prove that $\mathbb{T}$ has a unique fixed point. Indeed, it enough to show that $\mathbb{T}$ is $\mathbb{A}_\theta$-contraction mapping on $\mathbb{X}$ via the Bielecki vector-valued norm. For this end, given $(x_1, y_1), (x_2, y_2) \in \mathbb{X}$ and $t \in J$, using (13.24.2), and Lemma 13.14, we can get

$$\big\|(\mathbb{T}_1(x_1, y_1))(t) - (\mathbb{T}_1(x_2, y_2))(t)\big\|$$

$$\leq \int_a^t \frac{\psi'(s)(\psi(t) - \psi(s))^{\alpha-1}}{\Gamma(\alpha)} \big(p_1(s)\|x_1(s) - x_2(s)\| + q_1(s)\|y_1(s) - y_2(s)\|\big) ds$$

$$+ \|\mathbb{A}_1\| \int_a^t \frac{\psi'(s)(\psi(t) - \psi(s))^{\alpha-1}}{\Gamma(\alpha)} \|x_1(s) - x_2(s)\| ds$$

$$\leq \int_a^t \frac{\psi'(s)(\psi(t) - \psi(s))^{\alpha-1}}{\Gamma(\alpha)}$$

$$\times \frac{p_1(s)\|x_1(s) - x_2(s)\| + q_1(s)\|y_1(s) - y_2(s)\|}{e^{\theta(\psi(s) - \psi(a))}} e^{\theta(\psi(s) - \psi(a))} ds$$

$$+ \|\mathbb{A}_1\| \int_a^t \frac{\psi'(s)(\psi(t) - \psi(s))^{\alpha-1}}{\Gamma(\alpha)} \frac{\|x_1(s) - x_2(s)\|}{e^{\theta(\psi(s) - \psi(a))}} e^{\theta(\psi(s) - \psi(a))} ds$$

$$\leq \big((p_1^* + \mathbb{A}_1^*)\|x_1 - x_2\|_{\mathcal{B}} + q_1^*\|y_1 - y_2\|_{\mathcal{B}}\big)$$

$$\times \int_a^t \frac{\psi'(s)(\psi(t) - \psi(s))^{\alpha-1}}{\Gamma(\alpha)} e^{\theta(\psi(s) - \psi(a))} ds$$

$$\leq \frac{e^{\theta(\psi(t) - \psi(a))}}{\theta^\alpha} \big((p_1^* + \mathbb{A}_1^*)\|x_1 - x_2\|_{\mathcal{B}} + q_1^*\|y_1 - y_2\|_{\mathcal{B}}\big).$$

Hence

$$\left\| \mathbb{T}_1(x_1, y_1) - \mathbb{T}_1(x_2, y_2) \right\|_{\mathfrak{B}} \leq \frac{p_1^* + \mathbb{A}_1^*}{\theta^\alpha} \|x_1 - x_2\|_{\mathfrak{B}} + \frac{q_1^*}{\theta^\alpha} \|y_1 - y_2\|_{\mathfrak{B}}.$$

By the same technique, we can also get

$$\left\| \mathbb{T}_2(x_1, y_1) - \mathbb{T}_2(x_2, y_2) \right\|_{\mathfrak{B}} \leq \frac{p_2^*}{\theta^\beta} \|x_1 - x_2\|_{\mathfrak{B}} + \frac{q_2^* + \mathbb{A}_2^*}{\theta^\beta} \|y_1 - y_2\|_{\mathfrak{B}}.$$

This implies that

$$\left\| \mathbb{T}(x_1, y_1) - \mathbb{T}(x_2, y_2) \right\|_{\mathbb{X},\mathfrak{B}} \leq \mathbb{A}_\theta \left\| (x_1, y_1) - (x_2, y_2) \right\|_{\mathbb{X},\mathfrak{B}},$$

where

$$\mathbb{A}_\theta = \begin{pmatrix} \frac{p_1^* + \mathbb{A}_1^*}{\theta^\alpha} & \frac{q_1^*}{\theta^\alpha} \\ \frac{p_2^*}{\theta^\beta} & \frac{q_2^* + \mathbb{A}_2^*}{\theta^\beta} \end{pmatrix}. \tag{13.26}$$

Taking θ large enough it follows that the matrix $\mathbb{A}$ is convergent to zero and thus, an application of Perov theorem shows that $\mathbb{T}$ has a unique fixed point. So the coupled system (13.14)–(13.15) has a unique solution in $\mathbb{X}$. $\qquad\square$

Now we give our existence result for problem (13.14)–(13.15). The arguments are based on the Krasnoselskii type fixed-point theorem in generalized Banach spaces.

Theorem 13.25. *Let the assumptions* (13.24.1) *and* (13.24.2) *are satisfied. Then the coupled system* (13.14)–(13.15) *has at least one solution.*

Proof. In order to use the Krasnoselskii fixed-point theorem to prove our main result, we define a subset $\mathbb{B}_\xi$ of $\mathbb{X}$ by

$$\mathbb{B}_\xi = \left\{ (x, y) \in \mathbb{X} : \|(x, y)\|_{\mathbb{X},\mathfrak{B}} \leq \xi \right\},$$

with $\xi := (\xi_1, \xi_2) \in \mathbb{R}_+^2$ such that

$$\begin{cases} \xi_1 \geq \gamma_1 \mathbb{M}_1 + \gamma_2 \mathbb{M}_2, \\ \xi_2 \geq \gamma_3 \mathbb{M}_1 + \gamma_4 \mathbb{M}_2, \end{cases}$$

where $\mathbb{M}_1$, $\mathbb{M}_2$ and γ_i, $i = \overline{1,4}$ are positive real numbers that will be specified later. Moreover, notice that $\mathbb{B}_\xi$ is closed, convex, and bounded subset of the generalized Banach space $\mathbb{X}$, and construct the operators $\mathbb{U} = (\mathbb{U}_1, \mathbb{U}_2)$ and $\mathbb{V} = (\mathbb{V}_1, \mathbb{V}_2)$ on $\mathbb{B}_\xi$ as

$$\begin{cases} \mathbb{U}_1(x, y)(t) = \int_a^t \frac{\psi'(s)(\psi(t) - \psi(s))^{\alpha-1}}{\Gamma(\alpha)} f_1(s, x(s), y(s)) ds, \\ \mathbb{U}_2(x, y)(t) = \int_a^t \frac{\psi'(s)(\psi(t) - \psi(s))^{\beta-1}}{\Gamma(\beta)} f_2(s, x(s), y(s)) ds, \end{cases}$$

and

$$\begin{cases} \mathbb{V}_1(x,y)(t) = \phi_1 + \int_a^t \frac{\psi'(s)(\psi(t)-\psi(s))^{\alpha-1}}{\Gamma(\alpha)} \mathbb{A}_1 x(s) ds, \\ \mathbb{V}_2(x,y)(t) = \phi_2 + \int_a^t \frac{\psi'(s)(\psi(t)-\psi(s))^{\beta-1}}{\Gamma(\beta)} \mathbb{A}_2 y(s) ds. \end{cases}$$

Obviously, both $\mathbb{U}$ and $\mathbb{V}$ are well defined due to (13.24.1) and (13.24.2). Furthermore, by Lemma 13.23 the operators form of system (13.22) may be written as

$$(x,y) = (\mathbb{U}_1(x,y), \mathbb{U}_2(x,y)) + (\mathbb{V}_1(x,y), \mathbb{V}_2(x,y)) := \mathbb{T}(x,y). \tag{13.27}$$

Thus, the fixed point of operator $\mathbb{T}$ coincides with the solution of the coupled system (13.14)–(13.15). We will prove that $\mathbb{U}$ and $\mathbb{V}$, satisfy all conditions of Theorem 2.93. For better readability, we break the proof into a sequence of steps.

Step 1: $\mathbb{U}(x,y) + \mathbb{V}(\bar{x},\bar{y}) \in \mathbb{B}_\xi$, for any $(x,y), (\bar{x},\bar{y}) \in \mathbb{B}_\xi$. Indeed, for $(x,y), (\bar{x},\bar{y}) \in X$ and for each $t \in J$, from the definition of the operator $\mathbb{U}_1$ and assumption (13.24.2), we can get

$$\|\mathbb{U}_1(x,y)(t)\|$$
$$\leq \int_a^t \frac{\psi'(s)(\psi(t)-\psi(s))^{\alpha-1}}{\Gamma(\alpha)} (\|f_1(s,x(s),y(s)) - f_1(s,0,0)\| + \|f_1(s,0,0)\|) ds$$
$$\leq \frac{e^{\theta(\psi(t)-\psi(a))}}{\theta^\alpha} (p_1^* \|x\|_{\mathfrak{B}} + q_1^* \|y\|_{\mathfrak{B}}) + f_1^* \int_a^t \frac{\psi'(s)(\psi(t)-\psi(s))^{\alpha-1}}{\Gamma(\alpha)} ds$$
$$= \frac{e^{\theta(\psi(t)-\psi(a))}}{\theta^\alpha} (p_1^* \|x\|_{\mathfrak{B}} + q_1^* \|y\|_{\mathfrak{B}}) + f_1^* \ell_\psi^\alpha.$$

Hence

$$\|\mathbb{U}_1(x,y)\|_{\mathfrak{B}} \leq \frac{p_1^*}{\theta^\alpha} \|x\|_{\mathfrak{B}} + \frac{q_1^*}{\theta^\alpha} \|y\|_{\mathfrak{B}} + f_1^* \ell_\psi^\alpha.$$

By similar procedure, we get

$$\|\mathbb{U}_2(x,y)\|_{\mathfrak{B}} \leq \frac{p_2^*}{\theta^\beta} \|x\|_{\mathfrak{B}} + \frac{q_2^*}{\theta^\beta} \|y\|_{\mathfrak{B}} + f_2^* \ell_\psi^\beta.$$

Thus the above inequalities can be written in the vectorial form as follows

$$\|\mathbb{U}(x,y)\|_{X,\mathfrak{B}} := \begin{pmatrix} \|\mathbb{U}_1(x,y)\|_{\mathfrak{B}} \\ \|\mathbb{U}_2(x,y)\|_{\mathfrak{B}} \end{pmatrix} \leq \mathbb{B}_\theta \begin{pmatrix} \|x\|_{\mathfrak{B}} \\ \|y\|_{\mathfrak{B}} \end{pmatrix} + \begin{pmatrix} f_1^* \ell_\psi^\alpha \\ f_2^* \ell_\psi^\beta \end{pmatrix}, \tag{13.28}$$

where

$$\mathbb{B}_\theta = \begin{pmatrix} \frac{p_1^*}{\theta^\alpha} & \frac{q_1^*}{\theta^\alpha} \\ \frac{p_2^*}{\theta^\beta} & \frac{q_2^*}{\theta^\beta} \end{pmatrix}.$$

In a similar way, we get

$$\left\| \mathbb{V}(\bar{x}, \bar{y}) \right\|_{X, \mathfrak{B}} := \begin{pmatrix} \left\| \mathbb{V}_1(\bar{x}, \bar{y}) \right\|_{\mathfrak{B}} \\ \left\| \mathbb{V}_2(\bar{x}, \bar{y}) \right\|_{\mathfrak{B}} \end{pmatrix} \leq D_\theta \begin{pmatrix} \|\bar{x}\|_{\mathfrak{B}} \\ \|\bar{y}\|_{\mathfrak{B}} \end{pmatrix} + \begin{pmatrix} \phi_1^* \\ \phi_2^* \end{pmatrix}, \tag{13.29}$$

where

$$D_\theta = \begin{pmatrix} \frac{A_1^*}{\theta^\alpha} & 0 \\ 0 & \frac{A_2^*}{\theta^\beta} \end{pmatrix}.$$

Combining (13.28) and (13.29), it follows that

$$\left\| \mathbb{U}(x, y) \right\|_{X, \mathfrak{B}} + \left\| \mathbb{V}(\bar{x}, \bar{y}) \right\|_{X, \mathfrak{B}} \leq B_\theta \begin{pmatrix} \|x\|_{\mathfrak{B}} \\ \|y\|_{\mathfrak{B}} \end{pmatrix} + D_\theta \begin{pmatrix} \|\bar{x}\|_{\mathfrak{B}} \\ \|\bar{y}\|_{\mathfrak{B}} \end{pmatrix} + \begin{pmatrix} \ell_\psi^\alpha f_1^* + \phi_1^* \\ \ell_\psi^\beta f_2^* + \phi_2^* \end{pmatrix}. \tag{13.30}$$

Now we look for $\xi = (\xi_1, \xi_2) \in \mathbb{R}_+^2$ such that $\mathbb{U}(x, y) + \mathbb{V}(\bar{x}, \bar{y}) \in B_\xi$ for any $(x, y), (\bar{x}, \bar{y}) \in B_\xi$. To this end, according to (13.30), it is sufficient to show

$$A_\theta \begin{pmatrix} \xi_1 \\ \xi_2 \end{pmatrix} + \begin{pmatrix} M_1 \\ M_2 \end{pmatrix} \leq \begin{pmatrix} \xi_1 \\ \xi_2 \end{pmatrix},$$

where

$$\begin{pmatrix} M_1 \\ M_2 \end{pmatrix} = \begin{pmatrix} \ell_\psi^\alpha f_1^* + \phi_1^* \\ \ell_\psi^\beta f_2^* + \phi_2^* \end{pmatrix}.$$

Equivalently

$$\begin{pmatrix} M_1 \\ M_2 \end{pmatrix} \leq (I - A_\theta) \begin{pmatrix} \xi_1 \\ \xi_2 \end{pmatrix}. \tag{13.31}$$

For a sufficiently large θ, matrix A_θ is convergent to zero. It yields, from Theorem 13.18, that the matrix $(I - A_\theta)$ is nonsingular and $(I - A_\theta)^{-1}$ has nonnegative elements. Therefore (13.31) is equivalent to

$$\begin{pmatrix} \xi_1 \\ \xi_2 \end{pmatrix} \geq (I - A_\theta)^{-1} \begin{pmatrix} M_1 \\ M_2 \end{pmatrix}.$$

Moreover, if we denote

$$(I - A_\theta)^{-1} = \begin{pmatrix} \gamma_1 & \gamma_2 \\ \gamma_3 & \gamma_4 \end{pmatrix},$$

then we obtain

$$\begin{cases} \xi_1 \geq \gamma_1 M_1 + \gamma_2 M_2, \\ \xi_2 \geq \gamma_3 M_1 + \gamma_4 M_2. \end{cases}$$

Which means that $f(x, y) + \mathbb{H}(\bar{x}, \bar{y}) \in \mathbb{B}_\xi$.

Step 2: $\mathbb{V}$ is $\mathbb{D}_\theta$-contraction mapping on $\mathbb{B}_\xi$. In fact for each $t \in J$ and for any $(x_1, y_1), (x_2, y_2) \in \mathbb{B}_\xi$. By the same way of the proof of Theorem 13.24, we can easily show that

$$\left\|\mathbb{V}(x_1, y_1) - \mathbb{V}(x_2, y_2)\right\|_{\mathbb{X}, \mathfrak{B}} \leq \mathbb{D}_\theta \left\|(x_1, y_1) - (x_2, y_2)\right\|_{\mathbb{X}, \mathfrak{B}}.$$

Taking θ large enough it follows that the matrix $\mathbb{D}_\theta$ is convergent to zero and thus, $\mathbb{V}$ is an $\mathbb{D}_\theta$-contraction mapping on $\mathbb{B}_\xi$ with respect to the Bielecki norm.

Step 3: $\mathbb{U}$ is compact and continuous. Firstly, the continuity of $\mathbb{U}$ follows from the continuity of f_1 and f_2. Next, we prove that $\mathbb{U}$ is uniformly bounded on $\mathbb{B}_\xi$. From (13.28), and for each $(x, y) \in \mathbb{B}_\xi$, we can get

$$\left\|\mathbb{U}(x, y)\right\|_{\mathbb{X}, \mathfrak{B}} := \begin{pmatrix} \left\|\mathbb{U}_1(x, y)\right\|_{\mathfrak{B}} \\ \left\|\mathbb{U}_2(x, y)\right\|_{\mathfrak{B}} \end{pmatrix} \leq \mathbb{B}_\theta \begin{pmatrix} \xi_1 \\ \xi_2 \end{pmatrix} + \begin{pmatrix} f_1^* \ell_\psi^\alpha \\ f_2^* \ell_\psi^\beta \end{pmatrix} < \infty.$$

This proves that $\mathbb{U}$ is uniformly bounded.

Finally, it remains to show that $\mathbb{U}(\mathbb{B}_\xi)$ is equicontinuous. Let $(x, y) \in \mathbb{B}_\xi$ and any $t_1, t_2 \in J$, with $t_1 \leq t_2$. Taking assumption (13.24.2), into consideration, together with Remark 13.15, we can find

$$\left\|\mathbb{U}_1(x, y)(t_2) - \mathbb{U}_1(x, y)(t_1)\right\|$$

$$\leq \int_a^{t_1} \frac{\psi'(s)\left[(\psi(t_2) - \psi(s))^{\alpha-1} - (\psi(t_1) - \psi(s))^{\alpha-1}\right]}{\Gamma(\alpha)} \|f_1(s, x(s), y(s))\| ds$$

$$+ \int_{t_1}^{t_2} \frac{\psi'(s)(\psi(t_2) - \psi(s))^{\alpha-1}}{\Gamma(\alpha)} \|f_1(s, x(s), y(s))\| ds$$

$$\leq (p_1^* \|x\|_\infty + q_1^* \|y\|_\infty + f_1^*)$$

$$\times \int_a^{t_1} \frac{\psi'(s)\left[(\psi(t_2) - \psi(s))^{\alpha-1} - (\psi(t_1) - \psi(s))^{\alpha-1}\right]}{\Gamma(\alpha)} ds$$

$$+ (p_1^* \|x\|_\infty + q_1^* \|y\|_\infty + f_1^*) \int_{t_1}^{t_2} \frac{\psi'(s)(\psi(t_2) - \psi(s))^{\alpha-1}}{\Gamma(\alpha)} ds$$

$$\leq \frac{p_1^* \iota_2 \|x\|_{\mathfrak{B}} + q_1^* \iota_2 \|y\|_{\mathfrak{B}} + f_1^*}{\Gamma(\alpha + 1)}$$

$$\times \left[(\psi(t_1) - \psi(a))^\alpha + 2(\psi(t_2) - \psi(t_1))^\alpha - (\psi(t_2) - \psi(a))^\alpha\right]$$

$$\leq 2\frac{p_1^* \iota_2 \xi_1 + q_1^* \iota_2 \xi_2 + f_1^*}{\Gamma(\alpha + 1)}(\psi(t_2) - \psi(t_1))^\alpha.$$

Similarly,

$$\left\|\mathbb{U}_2(x, y)(t_2) - \mathbb{U}_2(x, y)(t_1)\right\| \leq 2\frac{p_2^* \iota_2 \xi_1 + q_2^* \iota_2 \xi_2 + f_2^*}{\Gamma(\beta + 1)}(\psi(t_2) - \psi(t_1))^\beta.$$

Therefore

$$\|\mathbb{U}(x,y)(t_2) - \mathbb{U}(x,y)(t_1)\| := \begin{pmatrix} \|\mathbb{U}_1(x,y)(t_2) - \mathbb{U}_1(x,y)(t_1)\| \\ \|\mathbb{U}_2(x,y)(t_2) - \mathbb{U}_2(x,y)(t_1)\| \end{pmatrix}$$

$$\leq 2 \begin{pmatrix} \frac{p_1^* \iota_2 \xi_1 + q_1^* \iota_2 \xi_2 + f_1^*}{\Gamma(\alpha+1)} (\psi(t_2) - \psi(t_1))^\alpha \\ \frac{p_2^* \iota_2 \xi_1 + q_2^* \iota_2 \xi_2 + f_2^*}{\Gamma(\beta+1)} (\psi(t_2) - \psi(t_1))^\beta \end{pmatrix}.$$

As $t_1 \to t_2$, the right-hand side of the above inequalities tends to zero independently of $(x,y) \in \mathbb{B}_\xi$. Hence we conclude that $\mathbb{T}(\mathbb{B}_\xi)$ is equicontinuous. By Arzelà–Ascoli's theorem, we deduce that $\mathbb{U}$ is a compact operator. Thus all the assumptions of Theorem 2.93 are satisfied. As a consequence of Krasnoselskii fixed-point theorem, we conclude that the operator $\mathbb{T} = \mathbb{U} + \mathbb{V}$ defined by (13.27) has at least one fixed point $(x,y) \in \mathbb{B}_\xi$, which is just the solution of system (13.14)–(13.15). This completes the proof of Theorem 13.25. $\qquad\square$

Now, we close this section by studying the Ulam–Hyers stability for problem (13.14)–(13.15) by means of integral representation of its solution given by

$$x(t) = \mathbb{T}_1(x,y)(t), \quad y(t) = \mathbb{T}_2(x,y)(t),$$

where $\mathbb{T}_1$ and $\mathbb{T}_2$ are defined by (13.24) and (13.25).

Define the following nonlinear operators $\mathbb{S}_1, \mathbb{S}_2 : \mathbb{X} \to C(J, \mathbb{R})$:

$$\begin{cases} ({}^c D_{a^+}^{\alpha;\psi} \tilde{x})(t) - \mathbb{A}_1 \tilde{x}(t) - f_1(t, \tilde{x}(t), \tilde{y}(t)) = \mathbb{S}_1(\tilde{x}, \tilde{y})(t), \\ ({}^c D_{a^+}^{\beta;\psi} \tilde{y})(t) - \mathbb{A}_2 \tilde{y}(t) - f_2(t, \tilde{x}(t), \tilde{y}(t)) = \mathbb{S}_2(\tilde{x}, \tilde{y})(t), \end{cases} \quad t \in J.$$

For some $\varepsilon_1, \varepsilon_2 > 0$, we consider the following inequality:

$$\begin{cases} \|\mathbb{S}_1(\tilde{x}, \tilde{y})(t)\| \leq \varepsilon_1, \\ \|\mathbb{S}_2(\tilde{x}, \tilde{y})(t)\| \leq \varepsilon_2, \end{cases} \quad t \in J. \tag{13.32}$$

Definition 13.26 ([397,431]). The coupled system (13.14)–(13.15) is Ulam–Hyers stable if we can find a positive constants $\omega_i, i = \overline{1,4}$ such that for every $\varepsilon_1, \varepsilon_1 > 0$ and for each solution $(\tilde{x}, \tilde{y}) \in \mathbb{X}$ of inequality (13.32), there exists a solution $(x,y) \in \mathbb{X}$ of (13.14)–(13.15) with

$$\begin{cases} \|\tilde{x}(t) - x(t)\| \leq \omega_1 \varepsilon_1 + \omega_2 \varepsilon_2, \\ \|\tilde{y}(t) - y(t)\| \leq \omega_3 \varepsilon_1 + \omega_4 \varepsilon_2, \end{cases} \quad t \in J.$$

Theorem 13.27. *Let the assumptions of Theorem 13.24 hold. Then problem (13.14)–(13.15) is Ulam–Hyers stable with respect to the Bielecki's norm.*

Proof. Let $(x, y) \in \mathbb{X}$ be the solution of problem (13.14)–(13.15) satisfying (13.24) and (13.25). Let $(\tilde{x}, \tilde{y})$ be any solution satisfying (13.32):

$$\begin{cases} ({}^{C}D_{a^+}^{\alpha;\psi} \tilde{x})(t) = \mathbb{A}_1 \tilde{x}(t) + f_1(t, \tilde{x}(t), \tilde{y}(t)) + \mathbb{S}_1(\tilde{x}, \tilde{y})(t), \\ ({}^{C}D_{a^+}^{\beta;\psi} \tilde{y})(t) = \mathbb{A}_2 \tilde{y}(t) + f_2(t, \tilde{x}(t), \tilde{y}(t)) + \mathbb{S}_2(\tilde{x}, \tilde{y})(t), \end{cases} \quad t \in \mathrm{J}.$$

So

$$\tilde{x}(t) = \mathbb{T}_1(\tilde{x}, \tilde{y})(t) + \int_a^t \frac{\psi'(s)(\psi(t) - \psi(s))^{\alpha-1}}{\Gamma(\alpha)} \mathbb{S}_1(\tilde{x}, \tilde{y})(s) ds, \tag{13.33}$$

and

$$\tilde{y}(t) = \mathbb{T}_2(\tilde{x}, \tilde{y})(t) + \int_a^t \frac{\psi'(s)(\psi(t) - \psi(s))^{\beta-1}}{\Gamma(\beta)} \mathbb{S}_2(\tilde{x}, \tilde{y})(s) ds. \tag{13.34}$$

It follows from (13.33) and (13.34) that

$$\left\| \tilde{x}(t) - \mathbb{T}_1(\tilde{x}, \tilde{y})(t) \right\| \leq \int_a^t \frac{\psi'(s)(\psi(t) - \psi(s))^{\alpha-1}}{\Gamma(\alpha)} \left\| \mathbb{S}_1(\tilde{x}, \tilde{y})(s) \right\| ds \leq \ell_\psi^\alpha \varepsilon_1, \tag{13.35}$$

and

$$\left\| \tilde{y}(t) - \mathbb{T}_2(\tilde{x}, \tilde{y})(t) \right\| \leq \int_a^t \frac{\psi'(s)(\psi(t) - \psi(s))^{\beta-1}}{\Gamma(\beta)} \left\| \mathbb{S}_2(\tilde{x}, \tilde{y})(s) \right\| ds \leq \ell_\psi^\beta \varepsilon_2. \tag{13.36}$$

Thus, by (13.24.2), Lemma 13.14 and inequalities (13.35), (13.36), we get

$$\begin{aligned} \left\| \tilde{x}(t) - x(t) \right\| &= \left\| \tilde{x}(t) - \mathbb{T}_1(\tilde{x}, \tilde{y})(t) + \mathbb{T}_1(\tilde{x}, \tilde{y})(t) - x(t) \right\| \\ &\leq \left\| \tilde{x}(t) - \mathbb{T}_1(\tilde{x}, \tilde{y})(t) \right\| + \left\| \mathbb{T}_1(\tilde{x}, \tilde{y})(t) - \mathbb{T}_1(x, y)(t) \right\| \\ &\leq \ell_\psi^\alpha \varepsilon_1 + \left(\frac{p_1^* + \mathbb{A}_1^*}{\theta^\alpha} \| \tilde{x} - x \|_\mathfrak{B} + \frac{q_1^*}{\theta^\alpha} \| \tilde{y} - y \|_\mathfrak{B} \right) e^{\theta(\psi(t) - \psi(a))}. \end{aligned}$$

Hence we get

$$\| \tilde{x} - x \|_\mathfrak{B} \leq \ell_\psi^\alpha \varepsilon_1 + \frac{p_1^* + \mathbb{A}_1^*}{\theta^\alpha} \| \tilde{x} - x \|_\mathfrak{B} + \frac{q_1^*}{\theta^\alpha} \| \tilde{y} - y \|_\mathfrak{B}. \tag{13.37}$$

Similarly, we have

$$\| \tilde{y} - y \|_\mathfrak{B} \leq \ell_\psi^\beta \varepsilon_1 + \frac{p_2^*}{\theta^\beta} \| \tilde{x} - x \|_\mathfrak{B} + \frac{q_2^* + \mathbb{A}_2^*}{\theta^\alpha} \| \tilde{y} - y \|_\mathfrak{B}. \tag{13.38}$$

Inequalities (13.37) and (13.37) can be rewritten in matrix form as

$$(I - \mathbb{A}_\theta) \begin{pmatrix} \| \tilde{x} - x \|_\mathfrak{B} \\ \| \tilde{y} - y \|_\mathfrak{B} \end{pmatrix} \leq \begin{pmatrix} \ell_\psi^\alpha \varepsilon_1 \\ \ell_\psi^\beta \varepsilon_2 \end{pmatrix}, \tag{13.39}$$

where $\mathbb{A}_\theta$ is the matrix given by (13.26). Since the matrix $\mathbb{A}_\theta$ is convergent to zero for sufficiently large θ, it yields, from Theorem 13.16, that the matrix $(I - \mathbb{A}_\theta)$ is nonsingular and $(I - \mathbb{A}_\theta)^{-1}$ has nonnegative elements. Therefore (13.39) is equivalent to

$$\begin{pmatrix} \|\tilde{x} - x\|_{\mathfrak{B}} \\ \|\tilde{y} - y\|_{\mathfrak{B}} \end{pmatrix} \leq (I - \mathbb{A}_\theta)^{-1} \begin{pmatrix} \ell_\psi^\alpha \varepsilon_1 \\ \ell_\psi^\beta \varepsilon_2 \end{pmatrix},$$

which yields that

$$\begin{cases} \|\tilde{x} - x\|_{\mathfrak{B}} \leq \gamma_1 \ell_\psi^\alpha \varepsilon_1 + \gamma_2 \ell_\psi^\beta \varepsilon_2, \\ \|\tilde{y} - y\|_{\mathfrak{B}} \leq \gamma_3 \ell_\psi^\alpha \varepsilon_1 + \gamma_4 \ell_\psi^\beta \varepsilon_2, \end{cases}$$

where $\gamma_i, i = \overline{1,4}$ are the elements of the matrix $(I - \mathbb{A}_\theta)^{-1}$.

Hence the coupled system (13.14)–(13.15) is Ulam–Hyers stable with respect to Bielecki's norm $\| \cdot \|_{\mathfrak{B}}$. $\square$

13.2.3 Existence and uniqueness results

In this section we focus on the existence and uniqueness of solutions for the given problem (13.16)–(13.17). Before proceeding to the main results, we start by the following definition.

Definition 13.28. By a solution of problem (13.16)–(13.17) we mean a coupled function $(x, y) \in C([a - \delta, b], \mathbb{R}^n) \times C([a - \delta, b], \mathbb{R}^n)$ that satisfies the system

$$\begin{cases} ({}^cD_{a^+}^{\alpha;\psi} x)(t) = g_1(t, x_t, y_t), \\ ({}^cD_{a^+}^{\beta;\psi} y)(t) = g_2(t, x_t, y_t), \end{cases} \quad t \in J,$$

and the initial conditions

$$\begin{cases} x(t) = \vartheta_1(t), \\ y(t) = \vartheta_2(t), \end{cases} \quad t \in [a - \delta, a].$$

To prove the existence of solutions to (13.16)–(13.17), we need the following lemma that was proven in the recent work of Almeida [120].

Lemma 13.29 ([120]). *Let* $g : [a, b] \times C([-\delta, 0], \mathbb{R}^n) \longrightarrow \mathbb{R}^n$ *be a continuous function. Then* $z \in C([a - \delta, b], \mathbb{R}^n)$ *is the solution of*

$$\begin{cases} {}^cD_{a^+}^{\alpha;\psi} z(t) = g(t, z_t), & t \in [a, b], \\ z(t) = \vartheta_1(t), & t \in [a - \delta, a], \end{cases}$$

if and only if it is the solution of the integral equation

$$z(t) = \begin{cases} \vartheta_1(a) + \int_a^t \frac{\psi'(s)(\psi(t) - \psi(s))^{\alpha-1}}{\Gamma(\alpha)} g(s, z_s) ds, & t \in [a, b], \\ \vartheta_1(t), & t \in [a - \delta, a]. \end{cases} \tag{13.40}$$

As a consequence of Lemma 13.29 we have the following result, which will be used in the sequel in the proofs of the main results.

Lemma 13.30. *Let $\alpha, \beta \in (0, 1]$ be fixed and $g_1, g_2 : J \times C([-\delta, 0], \mathbb{R}^n) \times C([-\delta, 0], \mathbb{R}^n) \longrightarrow \mathbb{R}^n$ are a given continuous functions. Then the coupled systems* (13.16)–(13.17) *is equivalent to the following integral equations*

$$x(t) = \begin{cases} \vartheta_1(a) + \int_a^t \frac{\psi'(s)(\psi(t)-\psi(s))^{\alpha-1}}{\Gamma(\alpha)} g_1(s, x_s, y_s) \mathrm{d}s, & t \in [a, b], \\ \vartheta_1(t), & t \in [a-\delta, a], \end{cases} \tag{13.41}$$

and

$$y(t) = \begin{cases} \vartheta_2(a) + \int_a^t \frac{\psi'(s)(\psi(t)-\psi(s))^{\beta-1}}{\Gamma(\beta)} g_2(s, x_s, y_s) \mathrm{d}s, & t \in [a, b], \\ \vartheta_2(t), & t \in [a-\delta, a]. \end{cases} \tag{13.42}$$

Define a square matrix $\mathbb{A}_\psi$ as

$$\mathbb{A}_\psi = \begin{pmatrix} \ell_\psi^\alpha \mathbb{L}_1 & \ell_\psi^\alpha \mathbb{M}_1 \\ \ell_\psi^\beta \mathbb{L}_2 & \ell_\psi^\beta \mathbb{M}_2 \end{pmatrix}. \tag{13.43}$$

Firstly, we prove the uniqueness result by means of the Perov fixed-point theorem.

Theorem 13.31. *If the following assumptions are true and the matrix $\mathbb{A}_\psi$ defined in* (13.43) *converges to zero.*

(13.31.1) $g_1, g_2 : J \times C([-\delta, 0], \mathbb{R}^n) \times C([-\delta, 0], \mathbb{R}^n) \longrightarrow \mathbb{R}^n$ *are continuous functions.*
(13.31.2) *There exist positive constants $\mathbb{L}_i, \mathbb{M}_i, i = 1, 2$ such that*

$$\|g_i(t, x, y) - g_i(t, \bar{x}, \bar{y})\| \le \mathbb{L}_i \|x - \bar{x}\|_{\mathfrak{C}_\delta} + \mathbb{M}_i \|y - \bar{y}\|_{\mathfrak{C}_\delta}$$

for all $t \in J$ and each $(x, y), (\bar{x}, \bar{y}) \in \mathfrak{C}_\delta \times \mathfrak{C}_\delta$.

Then the coupled system (13.16)–(13.17) *possesses a unique solution in the space $C([a - \delta, b], \mathbb{R}^n) \times C([a - \delta, b], \mathbb{R}^n)$*

Proof. Consider the Banach space $\mathfrak{C} = C([a - \delta, b], \mathbb{R}^n)$ equipped with the norm

$$\|z\|_{\mathfrak{C}} := \sup_{t \in [a-\delta, b]} \|z(t)\|.$$

Consequently, the product space $\mathcal{X} := \mathfrak{C} \times \mathfrak{C}$ is a generalized Banach space, endowed with the vector-valued norm

$$\|(x, y)\|_{\mathcal{X}} = \begin{pmatrix} \|x\|_{\mathfrak{C}} \\ \|y\|_{\mathfrak{C}} \end{pmatrix}.$$

For the sake of brevity, we set

$$g_i^* := \sup_{t \in J} \|g_i(t, 0, 0)\|.$$

We define an operator $\mathbb{K} = (\mathbb{K}_1, \mathbb{K}_2) : \mathcal{X} \to \mathcal{X}$ by:

$$\mathbb{K}(x, y) = \big(\mathbb{K}_1(x, y), \mathbb{K}_2(x, y)\big), \tag{13.44}$$

where

$$\mathbb{K}_1(x, y)(t) = \begin{cases} \vartheta_1(a) + \int_a^t \frac{\psi'(s)(\psi(t) - \psi(s))^{\alpha-1}}{\Gamma(\alpha)} g_1(s, x_s, y_s)\mathrm{d}s, & t \in [a, b], \\ \vartheta_1(t), & t \in [a - \delta, a], \end{cases} \tag{13.45}$$

and

$$\mathbb{K}_2(x, y)(t) = \begin{cases} \vartheta_2(a) + \int_a^t \frac{\psi'(s)(\psi(t) - \psi(s))^{\beta-1}}{\Gamma(\beta)} g_2(s, x_s, y_s)\mathrm{d}s, & t \in [a, b], \\ \vartheta_2(t), & t \in [a - \delta, a]. \end{cases} \tag{13.46}$$

Now, we apply Perov fixed-point theorem to prove that $\mathbb{K}$ has a unique fixed point. To do this, it is enough to show that $\mathbb{K}$ is $\mathbb{A}_\psi$-contraction mapping on $\mathcal{X}$. In fact, for all $t \in [a - \delta, b]$, $(x, y), (\bar{x}, \bar{y}) \in \mathcal{X}$. When $t \in [a - \delta, a]$, we have

$$\|\mathbb{K}_1(x, y)(t) - \mathbb{K}_1(\bar{x}, \bar{y})(t)\| = 0.$$

On the other hand, keeping in mind the definition of the operator $\mathbb{K}_1$ on $[a, b]$ together with assumption (13.31.2), we can get

$$\begin{aligned}
&\big\|\mathbb{K}_1(x, y)(t) - \mathbb{K}_1(\bar{x}, \bar{y})(t)\big\| \\
&\leq \int_a^t \frac{\psi'(s)(\psi(t) - \psi(s))^{\alpha-1}}{\Gamma(\alpha)} \|g_1(s, x_s, y_s) - g_1(s, \bar{x}_s, \bar{y}_s)\|\mathrm{d}s \\
&\leq \int_a^t \frac{\psi'(s)(\psi(t) - \psi(s))^{\alpha-1}}{\Gamma(\alpha)} \big(\mathbb{L}_1\|x_s - \bar{x}_s\|_{\mathfrak{C}_\delta} + \mathbb{M}_1\|y_s - \bar{y}_s\|_{\mathfrak{C}_\delta}\big)\mathrm{d}s \\
&\leq \big(\mathbb{L}_1\|x - \bar{x}\|_{\mathfrak{C}} + \mathbb{M}_1\|y - \bar{y}\|_{\mathfrak{C}}\big) \int_a^t \frac{\psi'(s)(\psi(t) - \psi(s))^{\alpha-1}}{\Gamma(\alpha)}\mathrm{d}s \\
&\leq \frac{(\psi(b) - \psi(a))^\alpha}{\Gamma(\alpha + 1)}\big(\mathbb{L}_1\|x - \bar{x}\|_{\mathfrak{C}} + \mathbb{M}_1\|y - \bar{y}\|_{\mathfrak{C}}\big) \\
&= \ell_\psi^\alpha \big(\mathbb{L}_1\|x - \bar{x}\|_{\mathfrak{C}} + \mathbb{M}_1\|y - \bar{y}\|_{\mathfrak{C}}\big).
\end{aligned}$$

Hence

$$\big\|\mathbb{K}_1(x, y) - \mathbb{K}_1(\bar{x}, \bar{y})\big\|_{[a,b]} \leq \ell_\psi^\alpha \big(\mathbb{L}_1\|x - \bar{x}\|_{\mathfrak{C}} + \mathbb{M}_1\|y - \bar{y}\|_{\mathfrak{C}}\big).$$

By similar procedure, we get

$$
\begin{cases}
\|\mathbb{K}_2(x, y) - \mathbb{K}_2(\bar{x}, \bar{y})\|_{[a-\delta, a]} = 0 \\
\|\mathbb{K}_2(x, y) - \mathbb{K}_2(\bar{x}, \bar{y})\|_{[a,b]} \le \ell_\psi^\beta \left(\mathbb{L}_2 \|x - \bar{x}\|_{\mathfrak{C}} + \mathbb{M}_2 \|y - \bar{y}\|_{\mathfrak{C}}\right).
\end{cases}
$$

Consequently,

$$
\|\mathbb{K}(x, y) - \mathbb{K}(\bar{x}, \bar{y})\|_{\mathcal{X}} := \begin{pmatrix} \|\mathbb{K}_1(x, y)\|_{\mathfrak{C}} \\ \|\mathbb{K}_2(x, y)\|_{\mathfrak{C}} \end{pmatrix} \le \mathbb{A}_\psi \|(x, y) - (\bar{x}, \bar{y})\|_{\mathcal{X}},
$$

where $\mathbb{A}_\psi$ is the matrix given by (13.43). Since the matrix $\mathbb{A}_\psi$ converges to zero, then Theorem 2.92 implies that coupled system (13.16)–(13.17) has a unique solution in $\mathcal{X}$.   □

Next, the following result is based on Schauder type fixed-point theorem in generalized Banach spaces.

Theorem 13.32. *Let the assumptions* (13.31.1) *and* (13.31.2) *are satisfied. Then the coupled system* (13.16)–(13.17) *has at least one solution, provided that the spectral radius of the matrix* $\mathbb{A}_\psi$ *is less than one.*

Proof. In order to apply Schauder fixed-point theorem type in a generalized Banach space, we need to construct a nonempty, closed, bounded convex set $\mathbb{B}_r \subset \mathcal{X}$ such that

$$
\mathbb{K}(\mathbb{B}_r) \subseteq \mathbb{B}_r, \tag{13.47}
$$

where the operator $\mathbb{K} : \mathcal{X} \to \mathcal{X}$ defined in (13.44). Let us consider the set

$$
\mathbb{B}_r = \{(x, y) \in \mathcal{X} : \|(x, y)\|_{\mathcal{X}} \le r\},
$$

where $r := (r_1, r_2) \in \mathbb{R}_+^2$ will be specified later. Now we try to find $r_1, r_2 \ge 0$ such that (13.47) holds. Indeed, for all $t \in [a - \delta, b]$, $(x, y) \in \mathcal{X}$. When $t \in [a - \delta, a]$, we have

$$
\|\mathbb{K}_1(x, y)(t)\| \le \|\vartheta_1\|_{[a-\delta, a]},
$$

which yields

$$
\|\mathbb{K}_1(x, y)\|_{[a-\delta, a]} \le \|\vartheta_1\|_{[a-\delta, a]}, \tag{13.48}
$$

and if $t \in [a, b]$, we have

$$
\|\mathbb{K}_1(x, y)(t)\| \le \|\vartheta_1(a)\| + \int_a^t \frac{\psi'(s)(\psi(t) - \psi(s))^{\alpha-1}}{\Gamma(\alpha)} \left(\|g_1(s, x_s, y_s) - g_1(s, 0, 0)\| \right.
$$
$$
\left. + \|g_1(s, 0, 0)\|\right) ds
$$
$$
\le \|\vartheta_1(a)\| + \ell_\psi^\alpha \left(\mathbb{L}_1 \|x\|_{\mathfrak{C}} + \mathbb{M}_1 \|y\|_{\mathfrak{C}} + g_1^*\right).
$$

Hence we get

$$\|\mathbb{K}_1(x, y)\|_{[a,b]} \le \|\vartheta_1(a)\| + \ell_\psi^\alpha \left(L_1 \|x\|_{\mathcal{C}} + M_1 \|y\|_{\mathcal{C}} + g_1^* \right). \tag{13.49}$$

So from (13.48) and (13.49), we get

$$\begin{aligned}
\|\mathbb{K}_1(x, y)\|_{\mathcal{C}} &\le \|\mathbb{K}_1(x, y)\|_{[a-\delta,a]} + \|\mathbb{K}_1(x, y)\|_{[a,b]} \\
&\le \|\vartheta_1\|_{[a-\delta,a]} + \|\vartheta_1(a)\| + \ell_\psi^\alpha \left(L_1 \|x\|_{\mathcal{C}} + M_1 \|y\|_{\mathcal{C}} + g_1^* \right).
\end{aligned}$$

In a similar way, we get

$$\|\mathbb{K}_2(x, y)\|_{\mathcal{C}} \le \|\vartheta_2\|_{[a-\delta,a]} + \|\vartheta_2(a)\| + \ell_\psi^\alpha \left(L_2 \|x\|_{\mathcal{C}} + M_2 \|y\|_{\mathcal{C}} + g_2^* \right).$$

Thus the above inequalities can be written in the vectorial form as follows

$$\|\mathbb{K}(x, y)\|_{\mathcal{X}} := \begin{pmatrix} \|\mathbb{K}_1(x, y)\|_{\mathcal{C}} \\ \|\mathbb{K}_2(x, y)\|_{\mathcal{C}} \end{pmatrix} \le \mathbb{A}_\psi \begin{pmatrix} \|x\|_{\mathcal{C}} \\ \|y\|_{\mathcal{C}} \end{pmatrix} + \begin{pmatrix} \mathbb{P}_1 \\ \mathbb{P}_2 \end{pmatrix}, \tag{13.50}$$

where $\mathbb{A}_\psi$ is the matrix given by (13.43), and

$$\begin{pmatrix} \mathbb{P}_1 \\ \mathbb{P}_2 \end{pmatrix} = \begin{pmatrix} \ell_\psi^\alpha g_1^* + \|\vartheta_1(a)\| + \|\vartheta_1\|_{[a-\delta,a]} \\ \ell_\psi^\beta g_2^* + \|\vartheta_2(a)\| + \|\vartheta_2\|_{[a-\delta,a]} \end{pmatrix}.$$

Now we look for $r = (r_1, r_2) \in \mathbb{R}_+^2$ such that $\|\mathbb{K}(x, y)\|_{\mathcal{C}} \le r$, for any $(x, y) \in \mathbb{B}_r$. To this end, according to (13.50), it is sufficient to show

$$\mathbb{A}_\psi \begin{pmatrix} r_1 \\ r_2 \end{pmatrix} + \begin{pmatrix} \mathbb{P}_1 \\ \mathbb{P}_2 \end{pmatrix} \le \begin{pmatrix} r_1 \\ r_2 \end{pmatrix}.$$

Equivalently

$$\begin{pmatrix} \mathbb{P}_1 \\ \mathbb{P}_2 \end{pmatrix} \le (I - \mathbb{A}_\psi) \begin{pmatrix} r_1 \\ r_2 \end{pmatrix}. \tag{13.51}$$

Since the matrix $\mathbb{A}_\psi$ is convergent to zero. It yields, from Theorem 13.16, that the matrix $(I - \mathbb{A}_\psi)$ is nonsingular and $(I - \mathbb{A}_\psi)^{-1}$ has nonnegative elements. Therefore (13.51) is equivalent to

$$\begin{pmatrix} r_1 \\ r_2 \end{pmatrix} \ge (I - \mathbb{A}_\psi)^{-1} \begin{pmatrix} \mathbb{P}_1 \\ \mathbb{P}_2 \end{pmatrix}. \tag{13.52}$$

Furthermore, if we denote

$$(I - \mathbb{A}_\psi)^{-1} = \begin{pmatrix} \kappa_1 & \kappa_2 \\ \kappa_3 & \kappa_4 \end{pmatrix}.$$

Then (13.52) becomes

$$\begin{cases} r_1 \geq \kappa_1 \mathbb{P}_1 + \kappa_2 \mathbb{P}_2, \\ r_2 \geq \kappa_3 \mathbb{P}_1 + \kappa_4 \mathbb{P}_2. \end{cases}$$

Which means that $\mathbb{K}(\mathbb{B}_r) \subseteq \mathbb{B}_r$. Moreover, by a similar process used in [120], it is easy to show that the operator $\mathbb{K}$ is continuous and $\mathbb{K}(\mathbb{B}_r)$ is relatively compact. Combining this facts, with Arzelà–Ascoli's theorem, we conclude that $\mathbb{K}$ is a compact operator. Invoking Theorem 2.91, we get a fixed point of $\mathbb{K}$ in $\mathbb{B}_r$, which is just a solution of system (13.16)–(13.17). This completes the proof of Theorem 13.32. $\qquad\square$

13.2.4 Some examples

In this section we provide some examples to illustrate our results constructed in the previous two sections

Example 13.33. Consider the following fractional relaxation differential systems

$$\begin{cases} (^cD_{0^+}^{0.5}x)(t) = 0.5x(t) + f_1(t, x(t), y(t)), \\ (^cD_{0^+}^{0.5}y)(t) = 0.5y(t) + f_2(t, x(t), y(t)), \end{cases} \quad t \in J := [0, 1], \tag{13.53}$$

with initial conditions

$$\begin{cases} x(0) = 1, \\ y(0) = 1, \end{cases} \tag{13.54}$$

where

$$\alpha = \beta = 0.5, \mathbb{A}_1 = \mathbb{A}_2 = c = 0.5, a = 0, b = 1, \psi(t) = t, n = 1,$$

and

$$f_1(t, x(t), y(t)) = (t+1)\ln(1 + |x(t)|) + e^t \arctan y(t),$$

$$f_2(t, x(t), y(t)) = \frac{t^2}{1 + |x(t)| + |y(t)|}.$$

Clearly, the functions $f_1, f_2 : J \times \mathbb{R}^2 \longrightarrow \mathbb{R}$ are continuous. Moreover, for any $x_1, y_1, x_2, y_2 \in \mathbb{R}$ and $t \in J$ we have

$$|f_1(t, x_1, y_1) - f_1(t, x_2, y_2)| \leq p_1(t)|x_1 - x_2| + q_1(t)|y_1 - y_2|$$

$$|f_2(t, x_1, y_1) - f_2(t, x_2, y_2)| \leq p_2(t)|x_1 - x_2| + q_2(t)|y_1 - y_2|,$$

where

$$p_1(t) = t + 1, \quad q_1(t) = e^t, \quad p_2(t) = q_2(t) = t^2.$$

Obviously,

$$p_1^* := 2, \ q_1^* := e^2, \ p_2^* = q_2^* := 1.$$

Furthermore, the matrix $\mathbb{A}_\theta$ given by (13.26) has the following form

$$\mathbb{A}_\theta = \frac{1}{\sqrt{\theta}} \begin{pmatrix} 2.5 & e^2 \\ 1 & 1.5 \end{pmatrix}.$$

Taking θ large enough it follows that the matrix $\mathbb{A}_\theta$ is convergent to zero and thus an application of Theorem 13.31 shows that the coupled system (13.53)–(13.54) has a unique solution and is Ulam–Hyers stable.

Example 13.34. Let us consider problem (13.1)–(13.17) with specific data:

$$\alpha = 0.8, \beta = 0.9, a = 0, b = 1, n = 2,$$

$$\mathbb{A}_1 = \begin{pmatrix} 2.5 & e^2 \\ 1 & 1.5 \end{pmatrix}, \tag{13.55}$$

$$\mathbb{A}_2 = \begin{pmatrix} 2.5 & e^2 \\ 1 & 1.5 \end{pmatrix}.$$

To illustrate Theorem 13.25, we take $\psi(t) = \sigma(t)$, where $\sigma(t)$ is the Sigmoid function, which can be expressed as

$$\sigma(t) = \frac{1}{1 + e^{-t}}, \tag{13.56}$$

and a convenience of the Sigmoid function is its derivative

$$\sigma'(t) = \sigma(t)(1 - \sigma(t)).$$

Taking also $f_1, f_2 : J \times \mathbb{R}^2 \times \mathbb{R}^2 \longrightarrow \mathbb{R}^2$ such that, $x = (x_1, x_2), y = (y_1, y_2)$ with

$$f_1(t, x(t), y(t)) = \begin{pmatrix} (x_1(t) + x_2(t))e^t \\ t \ln(1 + |y_1(t)| + |y_2(t)|) \end{pmatrix}. \tag{13.57}$$

$$f_2(t, x(t), y(t)) = \begin{pmatrix} (1 + t)e^{-(y_1(t) + y_2(t))} \\ e^{2t} \sin(x_1(t) + x_2(t)) \end{pmatrix}.$$

Clearly, the functions f_1, f_2 are continuous. Moreover, for any $x, y, \bar{x}, \bar{y} \in \mathbb{R}^2$ and $t \in J$ we have

$$\|f_1(t, x, y) - f_1(t, \bar{x}, \bar{y})\|_1 \le p_1(t)\|x - \bar{x}\|_1 + q_1(t)\|y - \bar{y}\|_1$$
$$\|f_2(t, x, y) - f_2(t, \bar{x}, \bar{y})\|_1 \le p_2(t)\|x - \bar{x}\|_1 + q_2(t)\|y - \bar{y}\|_1,$$

where $\| \cdot \|_1$ is a norm in $\mathbb{R}^2$ defined as follows

$$\|x\|_1 = |x_1| + |x_2|, \quad x = (x_1, x_2).$$

Hence the hypothesis (13.24.2) is satisfied with

$$p_1(t) = e^t, \quad q_1(t) = t, \quad p_2(t) = t+1, \quad q_2(t) = e^{2t}.$$

It follows from Theorem 13.25 that the system (13.14)–(13.15) with the data (13.55), (13.56), and (13.57) has at least one solution.

Example 13.35. Consider the following fractional delayed coupled system of the form:

$$\begin{cases} (^{CH}D_{1^+}^{0.5}x)(t) = g_1(t, x_t, y_t), \\ (^{CH}D_{1^+}^{0.5}y)(t) = g_2(t, x_t, y_t), \end{cases} \quad t \in J := [1, e], \tag{13.58}$$

with initial conditions

$$\begin{cases} x(t) = \vartheta_1(t) = (\vartheta_{11}(t), \vartheta_{12}(t)), \\ y(t) = \vartheta_2(t) = (\vartheta_{21}(t), \vartheta_{22}(t)), \end{cases} \quad t \in [1 - \delta, 1], \tag{13.59}$$

where

$$\alpha = \beta = 0.5, \, \psi(t) = \ln t, \, a = 1, b = e, \, \ell_\psi^\alpha = \ell_\psi^\beta = \frac{2}{\sqrt{\pi}},$$

$$g_1(t, x_t, y_t) = \begin{pmatrix} \frac{|x_{1,t}| + |x_{2,t}|}{e^{t+1}} \\ \frac{\sin(|y_{1,t}| + |y_{2,t}|)}{t+9} \end{pmatrix},$$

and

$$g_2(t, x_t, y_t) = \frac{1}{(t+1)^2} \begin{pmatrix} \ln(1 + |y_{1,t}| + |y_{2,t}|) \\ |x_{1,t}| + |x_{2,t}| \end{pmatrix}.$$

Clearly, the functions g_1, g_2 are continuous. Moreover, for any $x, y, \bar{x}, \bar{y} \in \mathcal{C}_\delta$, and $t \in J$, we have

$$\|g_1(t, x, y) - g_1(t, \bar{x}, \bar{y})\|_1 \le \mathbb{L}_1 \|x - \bar{x}\|_{\mathcal{C}_\delta} + \mathbb{M}_1 \|y - \bar{y}\|_{\mathcal{C}_\delta}$$
$$\|g_2(t, x, y) - g_2(t, \bar{x}, \bar{y})\|_1 \le \mathbb{L}_2 \|x - \bar{x}\|_{\mathcal{C}_\delta} + \mathbb{M}_2 \|y - \bar{y}\|_{\mathcal{C}_\delta}.$$

Hence the hypothesis (13.31.2) holds with

$$\mathbb{L}_1 = e^{-2}, \quad \mathbb{M}_1 = 0.1, \quad \mathbb{L}_2 = \mathbb{M}_2 = 0.25.$$

Furthermore, the matrix $\mathbb{A}_\psi$ given by (13.43) has the following form

$$\mathbb{A}_\psi = \frac{2}{\sqrt{\pi}} \begin{pmatrix} e^{-2} & 0.1 \\ 0.25 & 0.25 \end{pmatrix}.$$

Using the Matlab® program, we can get the eigenvalues of $\mathbb{A}_\psi$ as follows $\sigma_1 = 0.0276$, $\sigma_2 = 0.4072$, which show that $\mathbb{A}_\psi$ is converging to zero. Therefore, by Theorem 13.32, the coupled system (13.58)–(13.59) has a unique solution.

13.3 Coupled system of nonlinear hyperbolic partial fractional differential equations in generalized Banach spaces

13.3.1 Introduction

In this section we study the existence, uniqueness, as well as the Ulam–Hyers stability of solutions for the following coupled system:

$$\begin{cases} (^cD_{\tilde{a}}^{\alpha;\psi}x)(t, u) = \lambda_1 x(t, u) + f_1(t, u, x(t, u), y(t, u)), \\ (^cD_{\tilde{a}}^{\beta;\psi}y)(t, u) = \lambda_2 y(t, u) + f_2(t, u, x(t, u), y(t, u)), \end{cases} \quad (t, u) \in \tilde{I}, \qquad (13.60)$$

under the initial conditions

$$\begin{cases} x(t, a) = \eta_1(t), & t \in [a, b], \\ x(a, u) = v_1(u), & u \in [a, c], \\ x(a, a) = \eta_1(a) = v_1(a), \\ y(t, a) = \eta_2(t), & t \in [a, b], \\ y(a, u) = v_2(u), & u \in [a, c], \\ y(a, a) = \eta_2(a) = v_2(a), \end{cases} \qquad (13.61)$$

where $\tilde{I} := [a, b] \times [a, c]$, a, b and c are positive constants, $f_1, f_2 : \tilde{I} \times \mathbb{R}^2 \longrightarrow \mathbb{R}$ are given continuous functions. $\eta_1, \eta_2 : [a, b] \longrightarrow \mathbb{R}$, $v_1, v_2 : [a, c] \longrightarrow \mathbb{R}$ are given absolutely continuous functions. $^cD_{a^+}^{\alpha;\psi}$, $^cD_{a^+}^{\beta;\psi}$ are the ψ-Caputo fractional derivative of order $\alpha = (\alpha_1, \alpha_2)$, $\beta = (\beta_1, \beta_2) \in (0, 1] \times (0, 1]$, respectively, and $\tilde{a} = (a, a)$.

13.3.2 Existence results

Definition 13.36. By a solution of problem (13.60)–(13.61) we mean a coupled function $(x, y) \in C(\tilde{I}, \mathbb{R}) \times C(\tilde{I}, \mathbb{R})$ that satisfies the system

$$\begin{cases} (^cD_{\tilde{a}}^{\alpha;\psi}x)(t, u) = \lambda_1 x(t, u) + f_1(t, u, x(t, u), y(t, u)), \\ (^cD_{\tilde{a}}^{\beta;\psi}y)(t, u) = \lambda_2 y(t, u) + f_2(t, u, x(t, u), y(t, u)), \end{cases} \quad (t, u) \in \tilde{I},$$

and the initial conditions

$$\begin{cases} x(t, a) = \eta_1(t), & t \in [a, b], \\ x(a, u) = v_1(u), & u \in [a, c], \\ x(a, a) = \eta_1(a) = v_1(a), \\ y(t, a) = \eta_2(t), & t \in [a, b], \\ y(a, u) = v_2(u), & u \in [a, c], \\ y(a, a) = \eta_2(a) = v_2(a). \end{cases}$$

Lemma 13.37. *Let $f_1, f_2 \in C(\tilde{I}, \mathbb{R})$. Then the integral solution for the linear system of fractional differential equations:*

$$\begin{cases} (^c D_{\bar{a}}^{\alpha;\psi} x)(t, u) = f_1(t, u), \\ (^c D_{\bar{a}}^{\beta;\psi} y)(t, u) = f_2(t, u), \end{cases} \quad (t, u) \in \tilde{I},$$

supplemented with the initial conditions (13.61) is equivalent to the following integral equations

$$\begin{cases} x(t, u) = \zeta_1(t, u) + \int_a^t \int_a^u \frac{\psi'(s)\psi'(t)(\psi(t)-\psi(s))^{\alpha_1-1}(\psi(u)-\psi(t))^{\alpha_2-1}}{\Gamma(\alpha_1)\Gamma(\alpha_2)} f_1(s, t)\, dt ds, \\ y(t, u) = \zeta_2(t, u) + \int_a^t \int_a^u \frac{\psi'(s)\psi'(t)(\psi(t)-\psi(s))^{\beta_1-1}(\psi(u)-\psi(t))^{\beta_2-1}}{\Gamma(\beta_1)\Gamma(\beta_2)} f_2(s, t)\, dt ds, \end{cases}$$

where

$$\zeta_1(t, u) = \eta_1(t) + v_1(u) - \eta_1(a),$$
$$\zeta_2(t, u) = \eta_2(t) + v_2(u) - \eta_2(a).$$

Proof. The proof is similar to the one given in [407].   □

As a consequence of Lemma 13.37 we have the following result, which is useful for our main results.

Lemma 13.38. *Let $\alpha = (\alpha_1, \alpha_2)$, $\beta = (\beta_1, \beta_2) \in (0, 1] \times (0, 1]$ be fixed, $\lambda_1, \lambda_2 \in \mathbb{R}_+$ and $f_1, f_2 \in C(\tilde{I} \times \mathbb{R}^2, \mathbb{R})$. Then the coupled systems (13.60)–(13.61) is equivalent to the following integral equations*

$$\begin{cases} x(t, u) = \zeta_1(t, u) + \int_a^t \int_a^u \frac{\psi'(s)\psi'(t)(\psi(t)-\psi(s))^{\alpha_1-1}(\psi(u)-\psi(t))^{\alpha_2-1}}{\Gamma(\alpha_1)\Gamma(\alpha_2)} \\ \qquad\qquad \times (\lambda_1 x(s, t) + f_1(s, t, x(s, t), y(s, t)))\, dt ds, \\ y(t, u) = \zeta_2(t, u) + \int_a^t \int_a^u \frac{\psi'(s)\psi'(t)(\psi(t)-\psi(s))^{\beta_1-1}(\psi(u)-\psi(t))^{\beta_2-1}}{\Gamma(\beta_1)\Gamma(\beta_2)} \\ \qquad\qquad \times (\lambda_2 y(s, t) + f_2(s, t, x(s, t), y(s, t)))\, dt ds. \end{cases} \quad (13.62)$$

Our first result on the uniqueness is based on the Perov fixed-point theorem combined with the Bielecki norm.

Theorem 13.39. *Let the following assumptions hold:*

(13.39.1) *$f_1, f_2 : \tilde{I} \times \mathbb{R}^2 \to \mathbb{R}$ are continuous functions.*
(13.39.2) *There exist continuous functions $p_i, q_i : \tilde{I} \to \mathbb{R}_+$, $i = 1, 2$, such that*

$$|f_i(t, u, x_1, y_1) - f_i(t, u, x_2, y_2)| \leq p_i(t, u)|x_1 - x_2| + q_i(t, u)|y_1 - y_2|$$

for all $(t, u) \in \tilde{I}$ and each $(x_1, y_1), (x_2, y_2) \in \mathbb{R}^2$.

Then the coupled system (13.60)–(13.61) has a unique solution.

For the sake of brevity, we set

$$p_i^* := \sup_{(t,u)\in\tilde{I}} p_i(t,u), \; q_i^* := \sup_{(t,u)\in\tilde{I}} q_i(t,u), \; \zeta_i^* := \sup_{(t,u)\in\tilde{I}} |\zeta_i(t,u)|,$$

$$f_i^* := \sup_{(t,u)\in\tilde{I}} |f_i(t,u,0,0)|,$$

$$\mathrm{M}_\psi^1 := \frac{(\psi(b)-\psi(a))^{\alpha_1}(\psi(c)-\psi(a))^{\alpha_2}}{\Gamma(\alpha_1+1)\Gamma(\alpha_2+1)},$$

$$\mathrm{M}_\psi^2 := \frac{(\psi(b)-\psi(a))^{\beta_1}(\psi(c)-\psi(a))^{\beta_2}}{\Gamma(\beta_1+1)\Gamma(\beta_2+1)}.$$

Proof. Consider the Banach space $C(\tilde{I}, \mathbb{R})$ equipped with a Bielecki norm type $\|\cdot\|_\mathfrak{B}$ defined as below

$$\|x\|_\mathfrak{B} := \sup_{(t,u)\in\tilde{I}} \frac{|x(t,u)|}{e^{\theta(\psi(t)+\psi(u))}}, \quad \theta > 0. \tag{13.63}$$

Consequently, the product space $\mathbb{X} := C(\tilde{I}, \mathbb{R}) \times C(\tilde{I}, \mathbb{R})$ is a generalized Banach space, endowed with the Bielecki vector-valued norm

$$\|(x,y)\|_{\mathbb{X},\mathfrak{B}} = \begin{pmatrix} \|x\|_\mathfrak{B} \\ \|y\|_\mathfrak{B} \end{pmatrix}.$$

We define an operator $\mathfrak{S} = (\mathfrak{S}_1, \mathfrak{S}_2) : \mathbb{X} \to \mathbb{X}$ by:

$$\mathfrak{S}(x,y) = \big(\mathfrak{S}_1(x,y), \mathfrak{S}_2(x,y)\big), \tag{13.64}$$

where

$$(\mathfrak{S}_1(x,y))(t,u)$$
$$= \zeta_1(t,u) + \int_a^t \int_a^u \frac{\psi'(s)\psi'(t)(\psi(t)-\psi(s))^{\alpha_1-1}(\psi(u)-\psi(t))^{\alpha_2-1}}{\Gamma(\alpha_1)\Gamma(\alpha_2)}$$
$$\times (\lambda_1 x(s,t) + f_1(s,t,x(s,t),y(s,t)))\,dtds, \tag{13.65}$$

and

$$(\mathfrak{S}_2(x,y))(t,u)$$
$$= \zeta_2(t,u) + \int_a^t \int_a^u \frac{\psi'(s)\psi'(t)(\psi(t)-\psi(s))^{\beta_1-1}(\psi(u)-\psi(t))^{\beta_2-1}}{\Gamma(\beta_1)\Gamma(\beta_2)}$$
$$\times (\lambda_2 y(s,t) + f_2(s,t,x(s,t),y(s,t)))\,dtds. \tag{13.66}$$

Now, we apply Perov fixed-point theorem to prove that $\mathfrak{S}$ has a unique fixed point. Indeed, it is enough to show that $\mathfrak{S}$ is $\mathbb{A}$-contraction mapping on $\mathbb{X}$ via the Bielecki's vector-valued norm. To this end, given $(x_1, y_1), (x_2, y_2) \in \mathbb{X}$, and $(t, u) \in \tilde{I}$, using (13.39.2) and

Lemma 13.14, we can get

$$
\begin{aligned}
&\left|(\mathfrak{S}_1(x_1, y_1))(t, u) - (\mathfrak{S}_1(x_2, y_2))(t, u)\right| \\
&\leq \int_a^t \int_a^u \frac{\psi'(s)\psi'(t)(\psi(t) - \psi(s))^{\alpha_1-1}(\psi(u) - \psi(t))^{\alpha_2-1}}{\Gamma(\alpha_1)\Gamma(\alpha_2)} \\
&\quad \times (p_1(s,t)|x_1(s,t) - x_2(s,t)| + q_1(s,t)|y_1(s,t) - y_2(s,t)|)\, dtds \\
&\quad + \lambda_1 \int_a^t \int_a^u \frac{\psi'(s)\psi'(t)(\psi(t) - \psi(s))^{\alpha_1-1}(\psi(u) - \psi(t))^{\alpha_2-1}}{\Gamma(\alpha_1)\Gamma(\alpha_2)} \\
&\quad \times |x_1(s,t) - x_2(s,t)|\, dtds \\
&\leq \left(p_1^*\|x_1 - x_2\|_{\mathfrak{B}} + q_1^*\|y_1 - y_2\|_{\mathfrak{B}} + \lambda_1\|x_1 - x_2\|_{\mathfrak{B}}\right) \\
&\quad \times \int_a^t \int_a^u \frac{\psi'(s)\psi'(t)(\psi(t) - \psi(s))^{\alpha_1-1}(\psi(u) - \psi(t))^{\alpha_2-1}}{\Gamma(\alpha_1)\Gamma(\alpha_2)} e^{\theta(\psi(s)+\psi(t))}\, dtds \\
&= \frac{e^{\theta(\psi(t)+\psi(u))}}{\theta^{\alpha_1+\alpha_2}}\left(p_1^*\|x_1 - x_2\|_{\mathfrak{B}} + q_1^*\|y_1 - y_2\|_{\mathfrak{B}} + \lambda_1\|x_1 - x_2\|_{\mathfrak{B}}\right).
\end{aligned}
$$

Hence

$$
\left\|\mathfrak{S}_1(x_1, y_1) - \mathfrak{S}_1(x_2, y_2)\right\|_{\mathfrak{B}} \leq \frac{p_1^* + \lambda_1}{\theta^{\alpha_1+\alpha_2}}\|x_1 - x_2\|_{\mathfrak{B}} + \frac{q_1^*}{\theta^{\alpha_1+\alpha_2}}\|y_1 - y_2\|_{\mathfrak{B}}.
$$

Furthermore, for each $(x_1, y_1), (x_2, y_2) \in \mathbb{X}$ and $(t, u) \in \tilde{I}$, we get:

$$
\left\|\mathfrak{S}_2(x_1, y_1) - \mathfrak{S}_2(x_2, y_2)\right\|_{\mathfrak{B}} \leq \frac{p_2^*}{\theta^{\beta_1+\beta_2}}\|x_1 - x_2\|_{\mathfrak{B}} + \frac{q_2^* + \lambda_2}{\theta^{\beta_1+\beta_2}}\|y_1 - y_2\|_{\mathfrak{B}}.
$$

This implies that

$$
\left\|\mathfrak{S}(x_1, y_1) - \mathfrak{S}(x_2, y_2)\right\|_{\mathbb{X},\mathfrak{B}} \leq \mathbb{A}\|(x_1, y_1) - (x_2, y_2)\|_{\mathbb{X},\mathfrak{B}},
$$

where

$$
\mathbb{A} = \begin{pmatrix} \dfrac{p_1^*+\lambda_1}{\theta^{\alpha_1+\alpha_2}} & \dfrac{q_1^*}{\theta^{\alpha_1+\alpha_2}} \\ \dfrac{p_2^*}{\theta^{\beta_1+\beta_2}} & \dfrac{q_2^*+\lambda_2}{\theta^{\beta_1+\beta_2}} \end{pmatrix}. \tag{13.67}
$$

Taking θ large enough, it follows that the matrix $\mathbb{A}$ is convergent to zero and thus an application of Perov theorem shows that $\mathfrak{S}$ has a unique fixed point. So the coupled system (13.60)–(13.61) has a unique solution in $\mathbb{X}$. □

Now we give our existence result for problem (13.60)–(13.61). The arguments are based on the Krasnoselskii type fixed-point theorem in generalized Banach spaces.

Theorem 13.40. *Let the assumptions* (13.39.1) *and* (13.39.2) *be satisfied. Then the coupled system* (13.60)–(13.61) *has at least one solution, provided that the spectral radius of the ma-*

trix $\mathbb{K}_\psi$ *is less than one, where the matrix* $\mathbb{K}_\psi$ *is defined as below*

$$\mathbb{K}_\psi = \begin{pmatrix} \mathrm{M}^1_\psi(p_1^* + \lambda_1) & \mathrm{M}^1_\psi q_1^* \\ \mathrm{M}^2_\psi p_2^* & \mathrm{M}^2_\psi(q_2^* + \lambda_2) \end{pmatrix}. \tag{13.68}$$

Proof. In order to use the Krasnoselskii fixed-point theorem to prove our main result, we define a subset $\mathbb{B}_\mu$ of $\mathbb{X}$ by

$$\mathbb{B}_\mu = \left\{ (x, y) \in \mathbb{X} : \|(x, y)\|_{\mathbb{X}, \infty} \leq \mu \right\},$$

where $\mu := (\mu_1, \mu_2) \in \mathbb{R}^2_+$ will be specified later and $\| \cdot \|_{\mathbb{X}, \infty}$ denotes the Chebyshev norm. Moreover, notice that $\mathbb{B}_\mu$ is closed, convex, and bounded subset of the generalized Banach space $\mathbb{X}$, and construct the operators $\mathbb{G} = (\mathbb{G}_1, \mathbb{G}_2)$ and $\mathbb{H} = (\mathbb{H}_1, \mathbb{H}_2)$ on $\mathbb{B}_\mu$ as

$$\begin{cases} \mathbb{G}_1(x, y)(t, u) = \int_a^t \int_a^u \frac{\psi'(s)\psi'(t)(\psi(t)-\psi(s))^{\alpha_1-1}(\psi(u)-\psi(t))^{\alpha_2-1}}{\Gamma(\alpha_1)\Gamma(\alpha_2)} \\ \qquad\qquad \times f_1(s, t, x(s, t), y(s, t))\,dt\,ds, \\ \mathbb{G}_2(x, y)(t, u) = \int_a^t \int_a^u \frac{\psi'(s)\psi'(t)(\psi(t)-\psi(s))^{\beta_1-1}(\psi(u)-\psi(t))^{\beta_2-1}}{\Gamma(\beta_1)\Gamma(\beta_2)} \\ \qquad\qquad \times f_2(s, t, x(s, t), y(s, t))\,dt\,ds, \end{cases}$$

and

$$\begin{cases} \mathbb{H}_1(x, y)(t, u) = \zeta_1(t, u) + \lambda_1 \int_a^t \int_a^u \frac{\psi'(s)\psi'(t)(\psi(t)-\psi(s))^{\alpha_1-1}(\psi(u)-\psi(t))^{\alpha_2-1}}{\Gamma(\alpha_1)\Gamma(\alpha_2)} x(s, t)\,dt\,ds, \\ \mathbb{H}_2(x, y)(t, u) = \zeta_2(t, u) + \lambda_2 \int_a^t \int_a^u \frac{\psi'(s)\psi'(t)(\psi(t)-\psi(s))^{\beta_1-1}(\psi(u)-\psi(t))^{\beta_2-1}}{\Gamma(\beta_1)\Gamma(\beta_2)} y(s, t)\,dt\,ds. \end{cases}$$

Obviously, both $\mathbb{G}$ and $\mathbb{H}$ are well defined due to (13.39.1) and (13.39.2). Furthermore, by Lemma 13.38, the operators form of system (13.62) may be written as

$$(x, y) = (\mathbb{G}_1(x, y), \mathbb{G}_2(x, y)) + (\mathbb{H}_1(x, y), \mathbb{H}_2(x, y)) := \mathfrak{G}(x, y). \tag{13.69}$$

Thus the fixed point of operator $\mathfrak{G}$ coincides with the solution of the coupled system (13.60)–(13.61). We will prove that $\mathbb{G}$ and $\mathbb{H}$, satisfy all conditions of Theorem 2.93. For clarity, we will divide the remain of the proof into several steps.

Step 1: $\mathbb{G}(x, y) + \mathbb{H}(\bar{x}, \bar{y}) \in \mathbb{B}_\mu$, for any $(x, y), (\bar{x}, \bar{y}) \in \mathbb{B}_\mu$. Indeed, for $(x, y), (\bar{x}, \bar{y}) \in \mathbb{X}$ and for each $(t, u) \in \tilde{\mathrm{I}}$, from the definition of the operator $\mathbb{G}_1$ and assumption (13.39.2), we can get

$$|\mathbb{G}_1(x, y)(t, u)|$$
$$\leq \int_a^t \int_a^u \frac{\psi'(s)\psi'(t)(\psi(t) - \psi(s))^{\alpha_1-1}(\psi(u) - \psi(t))^{\alpha_2-1}}{\Gamma(\alpha_1)\Gamma(\alpha_2)}$$
$$\qquad \times \left(|f_1(s, t, x(s, t), y(s, t)) - f_1(s, t, 0, 0)| + |f_1(s, t, 0, 0)| \right)\,dt\,ds$$
$$\leq (p_1^* \|x\|_\infty + q_1^* \|y\|_\infty + f_1^*)$$

$$\times \int_a^t \int_a^u \frac{\psi'(s)\psi'(t)(\psi(t) - \psi(s))^{\alpha_1 - 1}(\psi(u) - \psi(t))^{\alpha_2 - 1}}{\Gamma(\beta_1)\Gamma(\alpha_2)} dt ds$$
$$= M_\psi^1 (p_1^* \|x\|_\infty + q_1^* \|y\|_\infty + f_1^*).$$

Hence

$$\|\mathbb{G}_1(x, y)\|_\infty \le M_\psi^1 (p_1^* \|x\|_\infty + q_1^* \|y\|_\infty + f_1^*).$$

By similar procedure, we get

$$\|\mathbb{G}_2(x, y)\|_\infty \le M_\psi^2 (p_2^* \|x\|_\infty + q_2^* \|y\|_\infty + f_2^*).$$

Thus the above inequalities can be written in the vectorial form as follows

$$\|\mathbb{G}(x, y)\|_{X,\infty} := \begin{pmatrix} \|\mathbb{G}_1(x, y)\|_\infty \\ \|\mathbb{G}_2(x, y)\|_\infty \end{pmatrix} \le \mathbb{B}_\psi \begin{pmatrix} \|x\|_\infty \\ \|y\|_\infty \end{pmatrix} + \begin{pmatrix} M_\psi^1 f_1^* \\ M_\psi^2 f_2^* \end{pmatrix}, \tag{13.70}$$

where

$$\mathbb{B}_\psi = \begin{pmatrix} M_\psi^1 p_1^* & M_\psi^1 q_1^* \\ M_\psi^2 p_2^* & M_\psi^2 q_2^* \end{pmatrix}.$$

In a similar way, we get

$$\|\mathbb{H}(\bar{x}, \bar{y})\|_{X,\infty} := \begin{pmatrix} \|\mathbb{H}_1(\bar{x}, \bar{y})\|_\infty \\ \|\mathbb{H}_2(\bar{x}, \bar{y})\|_\infty \end{pmatrix} \le D_\psi \begin{pmatrix} \|\bar{x}\|_\infty \\ \|\bar{y}\|_\infty \end{pmatrix} + \begin{pmatrix} \zeta_1^* \\ \zeta_2^* \end{pmatrix}, \tag{13.71}$$

where

$$D_\psi = \begin{pmatrix} \lambda_1 M_\psi^1 & 0 \\ 0 & \lambda_2 M_\psi^2 \end{pmatrix}.$$

Combining (13.70) and (13.71), it follows that

$$\|\mathbb{H}(\bar{x}, \bar{y})\|_{X,\infty} + \|\mathbb{G}(x, y)\|_{X,\infty} \le \mathbb{B}_\psi \begin{pmatrix} \|x\|_\infty \\ \|y\|_\infty \end{pmatrix} + D_\psi \begin{pmatrix} \|\bar{x}\|_\infty \\ \|\bar{y}\|_\infty \end{pmatrix}$$
$$+ \begin{pmatrix} M_\psi^1 f_1^* + \zeta_1^* \\ M_\psi^2 f_2^* + \zeta_2^* \end{pmatrix}. \tag{13.72}$$

Now we look for $\mu = (\mu_1, \mu_2) \in \mathbb{R}_+^2$ such that $\mathbb{G}(x, y) + \mathbb{H}(\bar{x}, \bar{y}) \in \mathbb{B}_\mu$ for any $(x, y), (\bar{x}, \bar{y}) \in \mathbb{B}_\mu$. To this end, according to (13.72), it is sufficient to show

$$\mathbb{K}_\psi \begin{pmatrix} \mu_1 \\ \mu_2 \end{pmatrix} + \begin{pmatrix} M_\psi^1 f_1^* + \zeta_1^* \\ M_\psi^2 f_2^* + \zeta_2^* \end{pmatrix} \le \begin{pmatrix} \mu_1 \\ \mu_2 \end{pmatrix},$$

where

$$\mathbb{K}_\psi = \mathbb{B}_\psi + D_\psi.$$

Equivalently

$$\begin{pmatrix} \mathrm{M}_\psi^1 f_1^* + \zeta_1^* \\ \mathrm{M}_\psi^2 f_2^* + \zeta_2^* \end{pmatrix} \leq (I - \mathbb{K}_\psi) \begin{pmatrix} \mu_1 \\ \mu_2 \end{pmatrix}. \tag{13.73}$$

Since the matrix $\mathbb{K}_\psi$ is convergent to zero. It yields that the matrix $(I - \mathbb{K}_\psi)$ is nonsingular and $(I - \mathbb{K}_\psi)^{-1}$ has nonnegative elements. Therefore (13.73) is equivalent to

$$\begin{pmatrix} \mu_1 \\ \mu_2 \end{pmatrix} \geq (I - \mathbb{K}_\psi)^{-1} \begin{pmatrix} \mathrm{M}_\psi^1 f_1^* + \zeta_1^* \\ \mathrm{M}_\psi^2 f_2^* + \zeta_2^* \end{pmatrix},$$

which means that $\mathbb{G}(x, y) + \mathbb{H}(\bar{x}, \bar{y}) \in \mathbb{B}_\mu$.

Step 2: $\mathbb{H}$ is $\tilde{D}_\psi$-contraction mapping on $\mathbb{B}_\mu$. In fact, for each $(t, \mathrm{u}) \in \tilde{I}$ and for any $(x_1, y_1), (x_2, y_2) \in \mathbb{B}_\mu$. By the same way of the proof of Theorem 13.39, we can easily show that

$$\|\mathbb{H}(x_1, y_1) - \mathbb{H}(x_2, y_2)\|_{\mathrm{X}, \mathfrak{B}} \leq \tilde{D}_\psi \|(x_1, y_1) - (x_2, y_2)\|_{\mathrm{X}, \mathfrak{B}},$$

where

$$\tilde{D}_\psi = \begin{pmatrix} \frac{\lambda_1}{\theta^{\alpha_1 + \alpha_2}} & 0 \\ 0 & \frac{\lambda_2}{\theta^{\beta_1 + \beta_2}} \end{pmatrix}.$$

Taking θ large enough it follows that the matrix $\tilde{D}_\psi$ is convergent to zero and thus $\mathbb{H}$ is an $\tilde{D}_\psi$-contraction mapping on $\mathbb{B}_\mu$ with respect to the Bielecki norm.

Step 3: $\mathbb{G}$ is compact and continuous. Firstly, the continuity of $\mathbb{G}$ follows from the continuity of f_1 and f_2. Next we prove that $\mathbb{G}$ is uniformly bounded on $\mathbb{B}_\mu$. From (13.70), and for each $(x, y) \in \mathbb{B}_\mu$, we can get

$$\|\mathbb{G}(x, y)\|_{\mathrm{X}, \infty} := \begin{pmatrix} \|\mathbb{G}_1(x, y)\|_\infty \\ \|\mathbb{G}_2(x, y)\|_\infty \end{pmatrix} \leq \mathbb{B}_\psi \begin{pmatrix} \mu_1 \\ \mu_2 \end{pmatrix} + \begin{pmatrix} \mathrm{M}_\psi^1 f_1^* \\ \mathrm{M}_\psi^2 f_2^* \end{pmatrix} < \infty.$$

This proves that $\mathbb{G}$ is uniformly bounded.

Finally, we show that $\mathbb{G}(\mathbb{B}_\mu)$ is equicontinuous. Let $(x, y) \in \mathbb{B}_\mu$ and any $(t_1, \mathrm{u}_1), (t_2, \mathrm{u}_2) \in \tilde{I}$, with $t_1 < t_2$ and $\mathrm{u}_1 < \mathrm{u}_2$. Taking (13.39.2) into consideration, we can find

$$|\mathbb{G}_1(x, y)(t_2, \mathrm{u}_2) - \mathbb{G}_1(x, y)(t_1, \mathrm{u}_1)|$$

$$\leq \frac{p_1^* \mu_1 + q_1^* \mu_2 + f_1^*}{\Gamma \alpha_1) \Gamma(\alpha_2)} \int_a^{t_1} \int_a^{\mathrm{u}_1} \psi'(s) \psi'(t) [(\psi(t_1) - \psi(s))^{\alpha_1 - 1} (\psi(\mathrm{u}_1) - \psi(t))^{\alpha_2 - 1}}$$

$$- (\psi(t_2) - \psi(s))^{\alpha_1 - 1}(\psi(u_2) - \psi(t))^{\alpha_2 - 1}] dt ds + \frac{p_1^* \mu_1 + q_1^* \mu_2 + f_1^*}{\Gamma(\alpha_1)\Gamma(\alpha_2)}$$

$$\times \int_{t_1}^{t_2} \int_{u_1}^{u_2} \psi'(s)\psi'(t)(\psi(t_2) - \psi(s))^{\alpha_1 - 1}(\psi(u_2) - \psi(t))^{\alpha_2 - 1} dt ds$$

$$+ \frac{p_1^* \mu_1 + q_1^* \mu_2 + f_1^*}{\Gamma(\alpha_1)\Gamma(\alpha_2)}$$

$$\times \int_{t_1}^{t_2} \int_{a}^{u_1} \psi'(s)\psi'(t)(\psi(t_2) - \psi(s))^{\alpha_1 - 1}(\psi(u_2) - \psi(t))^{\alpha_2 - 1} dt ds$$

$$+ \frac{p_1^* \mu_1 + q_1^* \mu_2 + f_1^*}{\Gamma(\alpha_1)\Gamma(\alpha_2)}$$

$$\times \int_{a}^{t_1} \int_{u_1}^{u_2} \psi'(s)\psi'(t)(\psi(t_2) - \psi(s))^{\alpha_1 - 1}(\psi(u_2) - \psi(t))^{\alpha_2 - 1} dt ds$$

$$\leq \frac{p_1^* \mu_1 + q_1^* \mu_2 + f_1^*}{\Gamma(\alpha_1 + 1)\Gamma(\alpha_2 + 1)}$$

$$\times \Big[2(\psi(u_2) - \psi(u_1))^{\alpha_2} \big[(\psi(t_2) - \psi(a))^{\alpha_1} - (\psi(t_2) - \psi(t_1))^{\alpha_1} \big]$$

$$+ \big[(\psi(u_2) - \psi(a))^{\alpha_2} (\psi(t_2) - \psi(t_1))^{\alpha_1} - (\psi(u_1) - \psi(a))^{\alpha_2} (\psi(t_1) - \psi(a))^{\alpha_1} \big] \Big].$$

Similarly,

$$|\mathbb{G}_2(x, y)(t_2, u_2) - \mathbb{G}_2 x(t_1, u_1)|$$

$$\leq \frac{p_1^* \mu_1 + q_1^* \mu_2 + f_1^*}{\Gamma(\beta_1 + 1)\Gamma(\beta_2 + 1)}$$

$$\times \Big[2(\psi(u_2) - \psi(u_1))^{\beta_2} \big[(\psi(t_2) - \psi(a))^{\beta_1} - (\psi(t_2) - \psi(t_1))^{\beta_1} \big]$$

$$+ \big[(\psi(u_2) - \psi(a))^{\beta_2} (\psi(t_2) - \psi(t_1))^{\beta_1} - (\psi(u_1) - \psi(a))^{\beta_2} (\psi(t_1) - \psi(a))^{\beta_1} \big] \Big].$$

As $t_1 \to t_2$ and $u_1 \to u_2$, the right-hand side of the above inequalities tends to zero independently of $(x, y) \in \mathbb{B}_\mu$. Hence the operators $\mathbb{G}_1$ and $\mathbb{G}_2$ are equicontinuous and thus the operator $\mathbb{G}$ is equicontinuous. By Arzelà–Ascoli theorem, we deduce that $\mathbb{G}$ is a compact operator. Thus all the assumptions of Theorem 2.93 are satisfied. As a consequence of Krasnoselskii fixed-point theorem, we conclude that the operator $\mathfrak{S} = \mathbb{G} + \mathbb{H}$ defined by (13.69) has at least one fixed point $(x, y) \in \mathbb{B}_\mu$, which is the solution of system (13.60)–(13.61). This completes the proof of Theorem 13.40. $\square$

Now, we complete this section by studying the Ulam–Hyers stability for problem (13.60)–(13.61) by means of integral representation of its solution given by

$$x(t, u) = \mathfrak{S}_1(x, y)(t, u), \quad y(t) = \mathfrak{S}_2(x, y)(t, u),$$

where $\mathfrak{S}_1$ and $\mathfrak{S}_2$ are defined by (13.65) and (13.66).

Define the following nonlinear operators $\mathbb{S}_1, \mathbb{S}_2 : \mathbb{X} \to C(\tilde{I}, \mathbb{R})$:

$$\begin{cases} (^cD_{\tilde{a}}^{\alpha;\psi}\tilde{x})(t,u) - \lambda_1\tilde{x}(t,u) - f_1(t,u,\tilde{x}(t,u),\tilde{y}(t,u)) = \mathbb{S}_1(\tilde{x},\tilde{y})(t,u), \\ (^cD_{\tilde{a}}^{\beta;\psi}\tilde{y})(t,u) - \lambda_2\tilde{y}(t,u) - f_2(t,u,\tilde{x}(t,u),\tilde{y}(t,u)) = \mathbb{S}_2(\tilde{x},\tilde{y})(t,u), \end{cases}$$

where $(t,u) \in \tilde{I}$. For some $\varepsilon_1, \varepsilon_2 > 0$, we consider the following inequality:

$$\begin{cases} |\mathbb{S}_1(\tilde{x},\tilde{y})(t,u)| \le \varepsilon_1, \\ |\mathbb{S}_2(\tilde{x},\tilde{y})(t,u)| \le \varepsilon_2, \end{cases} \quad (t,u) \in \tilde{I}. \tag{13.74}$$

Definition 13.41 ([397,431]). The coupled system (13.60)–(13.61) is Ulam–Hyers stable if we can find a positive constants a, b, c, and d such that for every $\varepsilon_1, \varepsilon_1 > 0$ and for each solution $(\tilde{x}, \tilde{y}) \in \mathbb{X}$ of inequality (13.74), there exists a solution $(x, y) \in \mathbb{X}$ of (13.60)–(13.61) with

$$\begin{cases} |\tilde{x}(t,u) - x(t,u)| \le a\varepsilon_1 + b\varepsilon_2, \\ |\tilde{y}(t,u) - y(t,u)| \le c\varepsilon_1 + d\varepsilon_2, \end{cases} \quad (t,u) \in \tilde{I}.$$

Theorem 13.42. *Let the assumptions of Theorem 13.39 hold. Then problem* (13.60)–(13.61) *is Ulam–Hyers stable with respect to the Bielecki norm.*

Proof. Let $(x, y) \in \mathbb{X}$ be the solution of problem (13.60)–(13.61) satisfying (13.65) and (13.66). Let $(\tilde{x}, \tilde{y})$ be any solution satisfying (13.74):

$$\begin{cases} (^cD_{\tilde{a}}^{\alpha;\psi}\tilde{x})(t,u) = \lambda_1\tilde{x}(t,u) + f_1(t,u,\tilde{x}(t,u),\tilde{y}(t,u)) + \mathbb{S}_1(\tilde{x},\tilde{y})(t,u), \\ (^cD_{\tilde{a}}^{\beta;\psi}\tilde{y})(t,u) = \lambda_2\tilde{y}(t,u) + f_2(t,u,\tilde{x}(t,u),\tilde{y}(t,u)) + \mathbb{S}_2(\tilde{x},\tilde{y})(t,u), \end{cases}$$

where $(t,u) \in \tilde{I}$. So

$$\tilde{x}(t,u) = \mathbb{S}_1(\tilde{x},\tilde{y})(t,u) + \int_a^t \int_a^u \frac{\psi'(s)\psi'(t)(\psi(t) - \psi(s))^{\alpha_1-1}(\psi(u) - \psi(t))^{\alpha_2-1}}{\Gamma(\alpha_1)\Gamma(\alpha_2)}$$
$$\times \mathbb{S}_1(\tilde{x},\tilde{y})(s,t)\,dtds, \tag{13.75}$$

and

$$\tilde{y}(t,u) = \mathbb{S}_2(\tilde{x},\tilde{y})(t,u) + \int_a^u \frac{\psi'(s)\psi'(t)(\psi(t) - \psi(s))^{\beta_1-1}(\psi(u) - \psi(t))^{\beta_2-1}}{\Gamma(\beta_1)\Gamma(\beta_2)}$$
$$\times \mathbb{S}_2(\tilde{x},\tilde{y})(s,t)\,dtds. \tag{13.76}$$

It follows from (13.75) and (13.76) that

$$|\tilde{x}(t,u) - \mathbb{S}_1(\tilde{x},\tilde{y})(t,u)|$$
$$\le \int_a^t \int_a^u \frac{\psi'(s)\psi'(t)(\psi(t) - \psi(s))^{\alpha_1-1}(\psi(u) - \psi(t))^{\alpha_2-1}}{\Gamma(\alpha_1)\Gamma(\alpha_2)} |\mathbb{S}_1(\tilde{x},\tilde{y})(s,t)|\,dtds$$
$$\le \mathbb{M}_\psi^1 \varepsilon_1, \tag{13.77}$$

and

$$\left| \tilde{y}(t, u) - \mathfrak{S}_2(\tilde{x}, \tilde{y})(t, u) \right|$$

$$\leq \int_a^t \int_a^u \frac{\psi'(s)\psi'(t)(\psi(t) - \psi(s))^{\beta_1 - 1}(\psi(u) - \psi(t))^{\beta_2 - 1}}{\Gamma(\beta_1)\Gamma(\beta_2)} \left| \mathbb{S}_2(\tilde{x}, \tilde{y})(s, t) \right| dtds$$

$$\leq M_\psi^2 \varepsilon_2. \tag{13.78}$$

Thus, by (H2), Lemma 13.14, and inequalities (13.77), (13.78), we get

$$\left| \tilde{x}(t, u) - x(t, u) \right|$$

$$= \left| \tilde{x}(t, u) - \mathfrak{S}_1(\tilde{x}, \tilde{y})(t, u) + \mathfrak{S}_1(\tilde{x}, \tilde{y})(t, u) - x(t, u) \right|$$

$$\leq \left| \tilde{x}(t, u) - \mathfrak{S}_1(\tilde{x}, \tilde{y})(t, u) \right| + \left| \mathfrak{S}_1(\tilde{x}, \tilde{y})(t, u) - \mathfrak{S}_1 x(t, u) \right|$$

$$\leq M_\psi^1 \varepsilon_1 + \left(\frac{p_1^* + \lambda_1}{\theta^{\alpha_1 + \alpha_2}} \| \tilde{x} - x \|_\mathfrak{B} + \frac{q_1^*}{\theta^{\alpha_1 + \alpha_2}} \| \tilde{y} - y \|_\mathfrak{B} \right) e^{\theta(\psi(t) + \psi(u))}.$$

Hence we get

$$\| \tilde{x} - x \|_\mathfrak{B} \leq M_\psi^1 \varepsilon_1 + \frac{p_1^* + \lambda_1}{\theta^{\alpha_1 + \alpha_2}} \| \tilde{x} - x \|_\mathfrak{B} + \frac{q_1^*}{\theta^{\alpha_1 + \alpha_2}} \| \tilde{y} - y \|_\mathfrak{B}. \tag{13.79}$$

Similarly, we have

$$\| \tilde{y} - y \|_\mathfrak{B} \leq M_\psi^2 \varepsilon_2 + \frac{p_2^*}{\theta^{\beta_1 + \beta_2}} \| \tilde{x} - x \|_\mathfrak{B} + \frac{q_1^* + \lambda_2}{\theta^{\beta_1 + \beta_2}} \| \tilde{y} - y \|_\mathfrak{B}. \tag{13.80}$$

Inequalities (13.79) and (13.79) can be rewritten in matrix form as

$$(I - \mathbb{A}) \begin{pmatrix} \| \tilde{x} - x \|_\mathfrak{B} \\ \| \tilde{y} - y \|_\mathfrak{B} \end{pmatrix} \leq \begin{pmatrix} M_\psi^1 \varepsilon_1 \\ M_\psi^2 \varepsilon_2 \end{pmatrix}. \tag{13.81}$$

Where $\mathbb{A}$ is the matrix given by (13.67). Since the matrix $\mathbb{A}$ is convergent to zero for sufficiently large θ. It yields that the matrix $(I - \mathbb{A})$ is nonsingular and $(I - \mathbb{A})^{-1}$ has nonnegative elements. Thus (13.81) is equivalent to

$$\begin{pmatrix} \| \tilde{x} - x \|_\mathfrak{B} \\ \| \tilde{y} - y \|_\mathfrak{B} \end{pmatrix} \leq (I - \mathbb{A})^{-1} \begin{pmatrix} M_\psi^1 \varepsilon_1 \\ M_\psi^2 \varepsilon_2 \end{pmatrix}. \tag{13.82}$$

Furthermore, if we denote

$$(I - \mathbb{A})^{-1} = \begin{pmatrix} \kappa_1 & \kappa_2 \\ \kappa_3 & \kappa_4 \end{pmatrix}.$$

Then (13.82) becomes

$$\begin{cases} \| \tilde{x} - x \|_\mathfrak{B} \leq \kappa_1 M_\psi^1 \varepsilon_1 + \kappa_2 M_\psi^2 \varepsilon_2, \\ \| \tilde{y} - y \|_\mathfrak{B} \leq \kappa_3 M_\psi^1 \varepsilon_1 + \kappa_4 M_\psi^2 \varepsilon_2. \end{cases}$$

Which means that the coupled system (13.60)–(13.61) is Ulam–Hyers stable with respect to Bielecki's norm $\| \cdot \|_{X, \mathfrak{B}}$.   □

13.3.3 Some examples

In this section we give two examples to illustrate our above results.

Example 13.43. Consider the coupled system

$$\begin{cases} ({}^C D_{\tilde{a}}^{\alpha} x)(t, u) = x(t, u) + f_1(t, u, x(t, u), y(t, u)), \\ ({}^C D_{\tilde{a}}^{\beta} y)(t, u) = y(t, u) + f_2(t, u, x(t, u), y(t, u)), \end{cases} \quad (t, u) \in \tilde{I} := [0, 1] \times [0, 1], \quad (13.83)$$

with initial conditions

$$\begin{cases} x(t, 0) = t, & t \in [0, 1], \\ x(0, u) = u^2, & u \in [0, 1], \\ x(0, 0) = 0, \\ y(t, 0) = te^t, & t \in [0, 1], \\ y(0, u) = u, & u \in [0, 1], \\ y(0, 0) = 0, \end{cases} \quad (13.84)$$

where $\alpha = (\alpha_1, \alpha_2) = \beta = (\beta_1, \beta_2) = (0.75, 0.75)$, $\lambda_1 = \lambda_2 = 1$, $a = 0$, $b = c = 1$, $\psi(t) = t$ and

$$f_1(t, u, x(t, u), y(t, u)) = e^{t+u} \ln(1 + |x(t, u)|) + e^{tu} \arctan y(t, u),$$

$$f_2(t, u, x(t, u), y(t, u)) = \frac{t + u}{(1 + |x(t, u)| + |y(t, u)|)}.$$

Clearly, the functions f_1, f_2 are continuous. Moreover, for any $x, y \in \mathbb{R}$ and $(t, u) \in \tilde{I}$, we have

$$|f_1(t, u, x_1, y_1) - f_1(t, u, x_2, y_2)| \le p_1(t, u)|x_1 - x_2| + q_1(t, u)|y_1 - y_2|$$
$$|f_2(t, u, x_1, y_1) - f_2(t, u, x_2, y_2)| \le p_2(t, u)|x_1 - x_2| + q_2(t, u)|y_1 - y_2|,$$

where

$$p_1(t, u) = e^{t+u}, \quad q_1(t, u) = e^{tu}, \quad p_2(t, u) = q_2(t, u) = t + u.$$

$$p_1^* := e^2, \ q_1^* := e, \ p_2^* = q_2^* := 2.$$

Furthermore, the matrix $\mathbb{A}$ given by (13.67) has the following form

$$\mathbb{A} = \frac{1}{\theta\sqrt{\theta}} \begin{pmatrix} e^2 + 1 & e \\ 2 & 3 \end{pmatrix}.$$

Taking θ large enough it follows that the matrix $\mathbb{A}$ is convergent to zero and thus an application of Theorem 13.39 shows that the coupled system (13.83)–(13.84) has a unique solution and is Ulam–Hyers stable.

Example 13.44. Consider the coupled system

$$
\begin{cases}
(^{CH}D_{\tilde{a}}^{\alpha}x)(t, \mathbf{u}) = 0.1x(t, \mathbf{u}) + f_1(t, \mathbf{u}, x(t, \mathbf{u}), y(t, \mathbf{u})), \\
(^{CH}D_{\tilde{a}}^{\beta}y)(t, \mathbf{u}) = 0.25y(t, \mathbf{u}) + f_2(t, \mathbf{u}, x(t, \mathbf{u}), y(t, \mathbf{u})),
\end{cases}
\tag{13.85}
$$

where $(t, \mathbf{u}) \in \tilde{I} := [1, e] \times [1, e]$, with initial conditions

$$
\begin{cases}
x(t, 1) = t, & t \in [1, e], \\
x(1, \mathbf{u}) = \mathbf{u}, & \mathbf{u} \in [1, e], \\
x(1, 1) = 1, \\
y(t, 1) = t, & t \in [1, e], \\
y(1, \mathbf{u}) = \mathbf{u}, & \mathbf{u} \in [1, e], \\
y(1, 1) = 1,
\end{cases}
\tag{13.86}
$$

where

$$
\alpha = (\alpha_1, \alpha_2) = \beta = (\beta_1, \beta_2) = (0.5, 0.5), \ \psi(\cdot) = \ln(\cdot), a = 1, b = c = e,
$$

$$
\lambda_1 = 0.1, \lambda_2 = 0.25, \zeta_1^* =, \zeta_2^* = 2e - 1, \quad \mathbb{M}_{\psi}^1 = \mathbb{M}_{\psi}^2 = \frac{4}{\pi},
$$

and

$$
f_1(t, \mathbf{u}, x(t, \mathbf{u}), y(t, \mathbf{u})) = \frac{x(t, \mathbf{u})}{e^{t+\mathbf{u}}(1 + |x(t, \mathbf{u})|)} + \frac{y(t, \mathbf{u})}{t + \mathbf{u} + 2},
$$

$$
f_2(t, \mathbf{u}, x(t, \mathbf{u}), y(t, \mathbf{u})) = \frac{1}{(t+\mathbf{u})^2}\left(\frac{x(t, \mathbf{u}) + \sqrt{1 + x^2(t, \mathbf{u})}}{2} + \sin|y(t, \mathbf{u})|\right).
$$

Clearly, the functions f_1, f_2 are continuous. Moreover, for any $x, y \in \mathbb{R}$ and $(t, \mathbf{u}) \in \tilde{I}$, we have

$$
|f_1(t, \mathbf{u}, x_1, y_1) - f_1(t, \mathbf{u}, x_2, y_2)| \leq p_1(t, \mathbf{u})|x_2 - x_2| + q_1(t, \mathbf{u})|y_1 - y_2|
$$

$$
|f_2(t, \mathbf{u}, x_1, y_1) - f_2(t, \mathbf{u}, x_2, y_2)| \leq p_2(t, \mathbf{u})|x_1 - x_2| + q_2(t, \mathbf{u})|y_1 - y_2|,
$$

where

$$
p_1(t, \mathbf{u}) = \frac{1}{e^{t+\mathbf{u}}}, \quad q_1(t, \mathbf{u}) = \frac{1}{t+\mathbf{u}+2}, \quad p_2(t, \mathbf{u}) = q_2(t, \mathbf{u}) = \frac{1}{(t+\mathbf{u})^2}.
$$

$$
p_1^* = e^{-2}, \quad q_1^* = 0.25, \quad p_2^* = 0.25, \quad q_2^* = 0.25.
$$

Furthermore, the matrix $\mathbb{K}_{\psi}$ given by (13.68) has the following form

$$
\mathbb{K}_{\psi} = \frac{4}{\pi}\begin{pmatrix} e^{-2} + 0.1 & 0.25 \\ 0.25 & 0.5 \end{pmatrix}.
$$

Using the Matlab program, we can get the eigenvalues of $\mathbb{K}_\psi$ as follows $\sigma_1 = 0.1080$, $\sigma_2 = 0.8283$. Which show that $\mathbb{K}_\psi$ is converging to zero. Therefore, by Theorem 13.40, the coupled system (13.85)–(13.86) has at least one solution.

13.4 Notes and remarks

We have established all the results contained in this chapter by taking into account the papers [150,152,219]. Fore more information of the subject, one may see the monographs [312,332,438] and the papers [17,33,34,401,403].

14

Monotone iterative technique for ψ-Caputo fractional differential equations

The results of our analysis in this chapter can be viewed as a conditional extension of the problems discussed fairly recently in the following papers [51,75,84,95,98,100,121–124,127, 141,166,208,235,288,299,306,318,383,384,414,422,432].

14.1 Initial value problem for nonlinear ψ-Caputo fractional differential equations

14.1.1 Introduction

In this section we deal with the existence and uniqueness of extremal solutions for the following initial value problem of fractional differential equations involving the ψ-Caputo derivative:

$$\begin{cases} {}^{c}D_{a^{+}}^{\alpha;\psi}x(t) = f(t,x(t)), \ t \in J := [a,b], \\ x(a) = a^{*}, \end{cases} \tag{14.1}$$

where ${}^{c}D_{a^{+}}^{\alpha;\psi}$ is the ψ-Caputo fractional derivative of order $\alpha \in (0,1]$, $f : [a,b] \times \mathbb{R} \longrightarrow \mathbb{R}$ is a given continuous function and $a^{*} \in \mathbb{R}$.

14.1.2 Existence results

Let us recall the definition and lemma of a solution for problem (14.1). First of all, we define what we mean by a solution for the boundary value problem (14.1).

Definition 14.1. A function $x \in C(J, \mathbb{R})$ is said to be a solution of Eq. (14.1) if x satisfies the equation ${}^{c}D_{a^{+}}^{\alpha;\psi}x(t) = f(t, x(t))$, for each $t \in J$, and the condition

$$x(a) = a^{*}.$$

For the existence of solutions for the problem (14.1), we need the following lemma for a general linear equation of $\alpha > 0$ that generalizes expression (3.1.34) in [281].

Fractional Difference, Differential Equations, and Inclusions. https://doi.org/10.1016/B978-0-44-323601-3.00021-6
Copyright © 2024 Elsevier Inc. All rights are reserved, including those for text and data mining, AI training, and similar technologies.

Lemma 14.2. *For a given $h \in C(J, \mathbb{R})$ and $\alpha \in (n-1, n]$, with $n \in \mathbb{N}$, the linear fractional initial value problem*

$$\begin{cases} {}^{c}D_{a+}^{\alpha;\psi} x(t) + rx(t) = h(t), & t \in J := [a, b], \\ x_{\psi}^{[k]}(a) = a_k, & k = 0, \ldots, n-1, \end{cases} \tag{14.2}$$

has unique solution given by

$$x(t) = \sum_{k=0}^{n-1} \frac{a_k}{k!} [\psi(t) - \psi(a)]^k - r\mathcal{I}_{a+}^{\alpha;\psi} x(t) + \mathcal{I}_{a+}^{\alpha;\psi} h(t)$$

$$= \sum_{k=0}^{n-1} \frac{a_k}{k!} [\psi(t) - \psi(a)]^k - \frac{r}{\Gamma(\alpha)} \int_a^t \psi'(s)(\psi(t) - \psi(s))^{\alpha-1} x(s) ds \tag{14.3}$$

$$+ \frac{1}{\Gamma(\alpha)} \int_a^t \psi'(s)(\psi(t) - \psi(s))^{\alpha-1} h(s) ds.$$

Moreover, the solution of the Volterra integral equation (14.3) can be represented by

$$x(t) = \sum_{k=0}^{n-1} a_k [\psi(t) - \psi(a)]^k \mathbb{E}_{\alpha, k+1}\left(-r(\psi(t) - \psi(a))^\alpha\right)$$

$$+ \int_a^t \psi'(s)(\psi(t) - \psi(s))^{\alpha-1} \mathbb{E}_{\alpha, \alpha}\left(-r(\psi(t) - \psi(a))^\alpha\right) h(s) ds, \tag{14.4}$$

where $\mathbb{E}_{\alpha, \beta}(\cdot)$ is the two-parametric Mittag–Leffler function defined in (2.11).

Proof. Since $\alpha \in (n-1, n]$, we know that the Cauchy problem (14.2) is equivalent to the following Volterra integral equation

$$x(t) = \sum_{k=0}^{n-1} \frac{a_k}{k!} [\psi(t) - \psi(a)]^k - r\mathcal{I}_{a+}^{\alpha;\psi} x(t) + \mathcal{I}_{a+}^{\alpha;\psi} h(t)$$

$$= \sum_{k=0}^{n-1} \frac{a_k}{k!} [\psi(t) - \psi(a)]^k - \frac{r}{\Gamma(\alpha)} \int_a^t \psi'(s)(\psi(t) - \psi(s))^{\alpha-1} x(s) ds$$

$$+ \frac{1}{\Gamma(\alpha)} \int_a^t \psi'(s)(\psi(t) - \psi(s))^{\alpha-1} h(s) ds.$$

Note that the above equation can be written in the following form

$$x(t) = \mathcal{T}x(t),$$

where the operator $\mathcal{T}$ is defined by

$$\mathcal{T}x(t) = \sum_{k=0}^{n-1} \frac{a_k}{k!} [\psi(t) - \psi(a)]^k - r\mathcal{I}_{a+}^{\alpha;\psi} x(t) + \mathcal{I}_{a+}^{\alpha;\psi} h(t).$$

We apply Weissinger fixed-point theorem to prove that $\mathcal{T}$ has a unique fixed point. Let $n \in \mathbb{N}$ and $x, y \in C(J, \mathbb{R})$. Then, using the semigroup property of the fractional integral operator $\mathcal{I}_{a+}^{\alpha;\psi}$, we have

$$
\begin{aligned}
|\mathcal{T}^n(x)(t) - \mathcal{T}^n(y)(t)| &= \left| -r \mathcal{I}_{a+}^{\alpha;\psi} \left(\mathcal{T}^{n-1} x(t) - \mathcal{T}^{n-1} y(t) \right) \right| \\
&= \left| -r \mathcal{I}_{a+}^{\alpha;\psi} \left(-r \mathcal{I}_{a+}^{\alpha;\psi} \left(\mathcal{T}^{n-2} x(t) - \mathcal{T}^{n-2} y(t) \right) \right) \right| \\
&\vdots \\
&= \left| (-r)^n \mathcal{I}_{a+}^{n\alpha;\psi} (x(t) - y(t)) \right| \\
&\leq \frac{\left(r(\psi(b) - \psi(a))^\alpha \right)^n}{\Gamma(n\alpha + 1)} \|x - y\|.
\end{aligned}
$$

Hence we have

$$
\|\mathcal{T}^n(x) - \mathcal{T}^n(y)\| \leq \frac{r^n (\psi(b) - \psi(a))^{n\alpha}}{\Gamma(n\alpha + 1)} \|x - y\|.
$$

It is well known that

$$
\sum_{n=0}^{\infty} \frac{r^n (\psi(b) - \psi(a))^{n\alpha}}{\Gamma(n\alpha + 1)} = \mathbb{E}_\alpha \left(r(\psi(b) - \psi(a))^\alpha \right).
$$

It follows that the mapping $\mathcal{T}^n$ is a contraction. Hence, by Weissinger fixed-point theorem, $\mathcal{T}$ has a unique fixed point. That is (14.2) has a unique solution.

Now we apply the method of successive approximations to prove that the integral equation (14.3) has the following integral representation of solution

$$
\begin{aligned}
x(t) = &\sum_{k=0}^{n-1} a_k [\psi(t) - \psi(a)]^k \mathbb{E}_{\alpha,k+1} \left(-r(\psi(t) - \psi(a))^\alpha \right) \\
&+ \int_a^t \psi'(s)(\psi(t) - \psi(s))^{\alpha-1} \mathbb{E}_{\alpha,\alpha} \left(-r(\psi(t) - \psi(a))^{\alpha-1} \right) h(s) \, ds.
\end{aligned}
$$

For this, we set

$$
\begin{cases}
x_0(t) = \displaystyle\sum_{k=0}^{n-1} \frac{a_k}{k!} [\psi(t) - \psi(a)]^k \\
x_m(t) = x_0(t) - \frac{r}{\Gamma(\alpha)} \int_a^t \psi'(s)(\psi(t) - \psi(s))^{\alpha-1} x_{m-1}(s) \, ds \\
\qquad + \frac{1}{\Gamma(\alpha)} \int_a^t \psi'(s)(\psi(t) - \psi(s))^{\alpha-1} h(s) \, ds.
\end{cases}
\tag{14.5}
$$

It follows from Eq. (14.5) and Lemma 2.6 that

$$x_1(t) = x_0(t) - r\mathcal{I}_{a+}^{\alpha;\psi} x_0(t) + \mathcal{I}_{a+}^{\alpha;\psi} h(t)$$

$$= \sum_{k=0}^{n-1} \frac{a_k}{k!} [\psi(t) - \psi(a)]^k - r \sum_{k=0}^{n-1} \frac{a_k}{\Gamma(\alpha+k+1)} [\psi(t) - \psi(a)]^{\alpha+k} + \mathcal{I}_{a+}^{\alpha;\psi} h(t). \qquad (14.6)$$

Similarly, Eqs. (14.5), (14.6), and Lemmas 2.4, 2.6 yield

$$x_2(t) = x_0(t) - r\mathcal{I}_{a+}^{\alpha;\psi} x_1(t) + \mathcal{I}_{a+}^{\alpha;\psi} h(t)$$

$$= \sum_{k=0}^{n-1} \frac{a_k}{k!} [\psi(t) - \psi(a)]^k - r\mathcal{I}_{a+}^{\alpha;\psi} \left(\sum_{k=0}^{n-1} \frac{a_k}{k!} [\psi(t) - \psi(a)]^k \right.$$

$$\left. - r \sum_{k=0}^{n-1} \frac{a_k}{\Gamma(\alpha+k+1)} [\psi(t) - \psi(a)]^{\alpha+k} + \mathcal{I}_{a+}^{\alpha;\psi} h(t) \right) + \mathcal{I}_{a+}^{\alpha;\psi} h(t)$$

$$= \sum_{k=0}^{n-1} \frac{a_k}{k!} [\psi(t) - \psi(a)]^k - r \sum_{k=0}^{n-1} \frac{a_k}{\Gamma(\alpha+k+1)} [\psi(t) - \psi(a)]^{\alpha+k}$$

$$+ r^2 \sum_{k=0}^{n-1} \frac{a_k}{\Gamma(2\alpha+k+1)} [\psi(t) - \psi(a)]^{2\alpha+k} - r\mathcal{I}_{a+}^{2\alpha;\psi} h(t) + \mathcal{I}_{a+}^{\alpha;\psi} h(t)$$

$$= \sum_{l=0}^{2} \sum_{k=0}^{n-1} \frac{(-r)^l a_k}{\Gamma(l\alpha+k+1)} [\psi(t) - \psi(a)]^{l\alpha+k}$$

$$+ \int_a^t \psi'(s) \sum_{l=0}^{1} \frac{(-r)^{l-1}(\psi(t) - \psi(s))^{l\alpha+\alpha-1}}{\Gamma(l\alpha+\alpha)} h(s) ds.$$

Continuing this process, we derive the following relation

$$x_m(t) = \sum_{l=0}^{m} \sum_{k=0}^{n-1} \frac{(-r)^l a_k}{\Gamma(l\alpha+k+1)} [\psi(t) - \psi(a)]^{l\alpha+k}$$

$$+ \int_a^t \psi'(s) \sum_{l=0}^{m-1} \frac{(-r)^{l-1}(\psi(t) - \psi(s))^{l\alpha+\alpha-1}}{\Gamma(l\alpha+\alpha)} h(s) ds.$$

Taking the limit as $n \to \infty$, we obtain the following explicit solution $x(t)$ to the integral equation (14.3):

$$x(t) = \sum_{l=0}^{\infty} \sum_{k=0}^{n-1} \frac{(-r)^l a_k}{\Gamma(l\alpha+k+1)} [\psi(t) - \psi(a)]^{l\alpha+k}$$

$$+ \int_a^t \psi'(s) \sum_{l=0}^{\infty} \frac{(-r)^{l-1}(\psi(t) - \psi(s))^{l\alpha+\alpha-1}}{\Gamma(l\alpha+\alpha)} h(s) ds$$

$$= \sum_{k=0}^{n-1} a_k \, (\psi(t) - \psi(a))^k \sum_{l=0}^{\infty} \frac{(-r)^l}{\Gamma(l\alpha + k + 1)} [\psi(t) - \psi(a)]^{l\alpha}$$

$$+ \int_a^t \psi'(s)(\psi(t) - \psi(s))^{\alpha-1} \sum_{l=0}^{\infty} \frac{(-r)^{l-1}(\psi(t) - \psi(s))^{l\alpha}}{\Gamma(l\alpha + \alpha)} h(s) \mathrm{d}s.$$

Taking into account (2.11), we get

$$x(t) = \sum_{k=0}^{n-1} a_k \, [\psi(t) - \psi(a)]^k \, \mathbb{E}_{\alpha,k+1}\left(-r(\psi(t) - \psi(a))^\alpha\right)$$

$$+ \int_a^t \psi'(s)(\psi(t) - \psi(s))^{\alpha-1} \mathbb{E}_{\alpha,\alpha}\left(-r(\psi(t) - \psi(s))^\alpha\right) h(s) \mathrm{d}s.$$

Then the proof is completed. □

The following result will play a very important role in this work.

Lemma 14.3. *(Comparison result). Let $\alpha \in (0, 1]$ be fixed and $r \in \mathbb{R}$. If $\rho \in C(J, \mathbb{R})$ satisfies the following inequalities*

$$\begin{cases} {}^c D_{a^+}^{\alpha;\psi} \rho(t) \geq -r\rho(t), & t \in [a, b], \\ \rho(a) \geq 0, \end{cases} \tag{14.7}$$

then $\rho(t) \geq 0$ for all $t \in J$.

Proof. Using the integral representation (14.4) and the fact that $\mathbb{E}_{\alpha,1}(z) \geq 0$ and $\mathbb{E}_{\alpha,\alpha}(z) \geq 0$, for all $\alpha \in (0, 1]$, and $z \in \mathbb{R}$ (see, [319]), it suffices to take $h(t) = {}^c\mathcal{D}_{a^+}^{\alpha;\psi} \rho(t) + r\rho(t) \geq 0$ with initial conditions $\rho(a) = a^* \geq 0$. □

Next, we give the definition of lower and upper solutions of problem (14.1).

Definition 14.4. A function $x_0 \in C(J, \mathbb{R})$ is called a lower solution of problem (14.1), if it satisfies

$$\begin{cases} {}^c D_{a^+}^{\alpha;\psi} x_0(t) \leq f(t, x_0), & t \in (a, b], \\ x_0(a) \leq a^*. \end{cases} \tag{14.8}$$

Definition 14.5. A function $y_0 \in C(J, \mathbb{R})$ is called an upper solution of problem (14.1), if it satisfies

$$\begin{cases} {}^c D_{a^+}^{\alpha;\psi} y_0(t) \geq f(t, y_0), & t \in (a, b], \\ y_0(a) \geq a^*. \end{cases} \tag{14.9}$$

In this work we will apply the monotone iterative method to present a result on the existence and uniqueness of the solution of problem (14.1).

Theorem 14.6. *Assume that $f \in C(J \times \mathbb{R}, \mathbb{R})$, and*

(14.6.1) *There exist x_0, $y_0 \in C(J, \mathbb{R})$ such that x_0 and y_0 are lower and upper solutions of problem (14.1), respectively, with $x_0(t) \le y_0(t)$, $t \in J$.*

(14.6.2) *There exists a constant $r \in \mathbb{R}$ such that*

$$f(t, y) - f(t, x) \ge -r(y - x) \quad \text{for } x_0 \le x \le y \le y_0.$$

Then there exist monotone iterative sequences $\{x_n\}$ and $\{y_n\}$, which converge uniformly on the interval J to the extremal solutions of (14.1) in the sector $[x_0, y_0]$, where

$$[x_0, y_0] = \{z \in C(J, \mathbb{R}) : x_0(t) \le z(t) \le y_0(t), \quad t \in J\}.$$

Proof. First, for any $x_0(t)$, $y_0(t) \in C(J, \mathbb{R})$, we consider the following linear initial value problems of fractional order:

$$\begin{cases} {}^c D_{a+}^{\alpha;\psi} x_{n+1}(t) = f(t, x_n(t)) - r(x_{n+1}(t) - x_n(t)), & t \in J, \\ x_{n+1}(a) = a^*, \end{cases} \tag{14.10}$$

and

$$\begin{cases} {}^c D_{a+}^{\alpha;\psi} y_{n+1}(t) = f(t, y_n(t)) - r(y_{n+1}(t) - y_n(t)), & t \in J, \\ y_{n+1}(a) = a^*. \end{cases} \tag{14.11}$$

By Lemma 14.2, we know that (14.10) and (14.11) have unique solutions in $C(J, \mathbb{R})$ which are defined as follows:

$$\begin{aligned} x_{n+1}(t) = a^* \mathbb{E}_{\alpha,1}\left(-r(\psi(t) - \psi(a))^\alpha\right) + \int_a^t \psi'(s)(\psi(t) - \psi(s))^{\alpha-1} \\ \times \mathbb{E}_{\alpha,\alpha}\left(-r(\psi(t) - \psi(s))^\alpha\right)\left(f(s, x_n(s)) + rx_n(s)\right)ds, \ t \in J, \end{aligned} \tag{14.12}$$

$$\begin{aligned} y_{n+1}(t) = a^* \mathbb{E}_{\alpha,\alpha}\left(-r(\psi(t) - \psi(a))^\alpha\right) + \int_a^t \psi'(s)(\psi(t) - \psi(s))^{\alpha-1} \\ \times \mathbb{E}_{\alpha,\alpha}\left(-r(\psi(t) - \psi(s))^\alpha\right)\left(f(s, y_n(s)) + ry_n(s)\right)ds, \ t \in J. \end{aligned} \tag{14.13}$$

We will divide the proof in the following steps.

Step 1: We show that the sequences $x_n(t)$, $y_n(t)(n \ge 1)$ are lower and upper solutions of problem (14.1), respectively, and the following relation holds

$$x_0(t) \le x_1(t) \le \cdots \le x_n(t) \le \cdots \le y_n(t) \le \cdots \le y_1(t) \le y_0(t), \quad t \in J. \tag{14.14}$$

First, we prove that

$$x_0(t) \le x_1(t) \le y_1(t) \le y_0(t), \quad t \in J. \tag{14.15}$$

Set $\rho(t) = x_1(t) - x_0(t)$. From (14.10) and Definition 14.4, we obtain

$$
\begin{aligned}
{}^cD_{a^+}^{\alpha;\psi}\rho(t) &= {}^cD_{a^+}^{\alpha;\psi}x_1(t) - {}^cD_{a^+}^{\alpha;\psi}x_0(t) \\
&\geq f\big(t, x_0(t)\big) - r(x_1(t) - x_0(t)) - f\big(t, x_0(t)\big) \\
&= -r\rho(t).
\end{aligned}
$$

Again, since

$$
\rho(a) = x_1(a) - x_0(a) = a^* - x_0(a) \geq 0.
$$

By Lemma 14.3, $\rho(t) \geq 0$, for $t \in J$. That is, $x_0(t) \leq x_1(t)$. Similarly, we can show that $y_1(t) \leq y_0(t)$, $t \in J$.

Now, let $\rho(t) = y_1(t) - x_1(t)$. From (14.10), (14.11), and (14.6.2), we get

$$
\begin{aligned}
{}^cD_{a^+}^{\alpha;\psi}\rho(t) &= {}^cD_{a^+}^{\alpha;\psi}y_1(t) - {}^cD_{a^+}^{\alpha;\psi}x_1(t) \\
&= f\big(t, y_0(t)\big) - r\big(y_1(t) - y_0(t)\big) - f\big(t, x_0(t)\big) + r\big(x_1(t) - x_0(t)\big) \\
&= f\big(t, y_0(t)\big) - f\big(t, x_0(t)\big) - r\big(y_1(t) - y_0(t)\big) + r\big(x_1(t) - x_0(t)\big) \\
&\geq -r\big(y_0(t) - x_0(t)\big) - r\big(y_1(t) - y_0(t)\big) + r\big(x_1(t) - x_0(t)\big) \\
&= -r\rho(t).
\end{aligned}
$$

Since, $\rho(a) = x_1(a) - y_1(a) = a^* - a^* = 0$. By Lemma 14.3, we get $x_1(t) \leq y_1(t)$, $t \in J$.

Secondly, we show that $x_1(t)$, $y_1(t)$ are lower and upper solutions of problem (14.1), respectively. Since x_0 and y_0 are lower and upper solutions of problem (14.1), by (14.6.2), it follows that

$$
{}^cD_{a^+}^{\alpha;\psi}x_1(t) = f\big(t, x_0(t)\big) - r\big(x_1(t) - x_0(t)\big) \leq f\big(t, x_1(t)\big),
$$

also $x_1(a) = a^*$. Therefore $x_1(t)$ is a lower solution of problem (14.1). Similarly, it can be obtained that $y_1(t)$ is an upper solution of problem (14.1).

By the above arguments and mathematical induction, we can show that the sequences $x_n(t)$, $y_n(t)$, $(n \geq 1)$ are lower and upper solutions of problem (14.1), respectively, and the following relation holds

$$
x_0(t) \leq x_1(t) \leq \cdots \leq x_n(t) \leq \cdots \leq y_n(t) \leq \cdots \leq y_1(t) \leq y_0(t), \quad t \in J.
$$

Step 2: The sequences $\{x_n(t)\}$, $\{y_n(t)\}$ converge uniformly to their limit functions $x^*(t)$, $y^*(t)$, respectively.

Note that the sequence $\{x_n(t)\}$ is monotone nondecreasing and is bounded from above by $y_0(t)$. Also, since the sequence $\{y_n(t)\}$ is monotone nonincreasing and is bounded from below by $x_0(t)$, therefore the pointwise limits exist and these limits are denoted by x^* and y^*. Moreover, since $\{x_n(t)\}$, $\{y_n(t)\}$ are sequences of continuous functions defined on the compact set $[a, b]$, hence, by Dini theorem [354], the convergence is uniform. This is

$$
\lim_{n\to\infty} x_n(t) = x^*(t) \quad \text{and} \quad \lim_{n\to\infty} y_n(t) = y^*(t),
$$

uniformly on $t \in J$ and the limit functions x^*, y^* satisfy problem (14.1). Furthermore, x^* and y^* satisfy the relation

$$x_0 \le x_1 \le \cdots \le x_n \le x^* \le y^* \le \cdots \le y_n \le \cdots \le y_1 \le y_0.$$

Step 3: We prove that x^* and y^* are extremal solutions of problem (14.1) in $[x_0, y_0]$.

Let $z \in [x_0, y_0]$ be any solution of (14.1). We assume that the following relation holds for some $n \in \mathbb{N}$:

$$x_n(t) \le z(t) \le y_n(t), \quad t \in J. \tag{14.16}$$

Let $\rho(t) = z(t) - x_{n+1}(t)$. We have

$$
\begin{aligned}
{}^c D_{a+}^{\alpha;\psi} \rho(t) &= {}^c D_{a+}^{\alpha;\psi} z(t) - {}^c D_{a+}^{\alpha;\psi} x_{n+1}(t) \\
&= f(t, z(t)) - f(t, x_n(t)) + r(x_{n+1}(t) - x_n(t)) \\
&\ge -r(z(t) - x_n(t)) + r(x_{n+1}(t) - x_n(t)) \\
&= -r\rho(t).
\end{aligned}
$$

Furthermore, $\rho(a) = z(a) - x_{n+1}(a) = a^* - a^* = 0$. By Lemma 14.3, we obtain $\rho(t) \ge 0, t \in J$, which means

$$x_{n+1}(t) \le z(t), \ t \in J.$$

Using the same method, we can show that

$$z(t) \le y_{n+1}(t), \ t \in J.$$

Hence we have

$$x_{n+1}(t) \le z(t) \le y_{n+1}(t), \ t \in J.$$

Therefore, (14.16) holds on J for all $n \in \mathbb{N}$. Taking the limit as $n \to \infty$ on both sides of (14.16), we get

$$x^* \le z \le y^*.$$

Therefore x^*, y^* are the extremal solutions of (14.1) in $[x_0, y_0]$. This completes the proof. $\square$

Now, we shall prove the uniqueness of the solution of the system (14.1) by monotone iterative technique.

Theorem 14.7. *Suppose that* (14.6.1) *and* (14.6.2) *are satisfied. Furthermore, we impose that:*

(14.7.1) *There exists a constant $r^* \ge r$ such that*

$$f(t, z_2) - f(t, z_1) \le r^*(z_2 - z_1),$$

for every $x_0 \leq z_1 \leq z_2 \leq y_0$, $t \in J$. *Then problem* (14.1) *has a unique solution between* x_0 *and* y_0.

Proof. From Theorem 14.6, we know that $x^*(t)$ and $y^*(t)$ are the extremal solutions of the IVP (14.1) and $x^*(t) \leq y^*(t)$, $t \in J$. It is sufficient to prove $x^*(t) \geq y^*(t)$, $t \in J$. In fact, let $\rho(t) = x^*(t) - y^*(t)$, $t \in J$, in view of (14.7.1), we have

$$
\begin{aligned}
{}^cD_{a+}^{\alpha;\psi}\rho(t) &= {}^cD_{a+}^{\alpha;\psi}x^*(t) - {}^cD_{a+}^{\alpha;\psi}y^*(t) \\
&= f\big(t, x^*(t)\big) - f\big(t, y^*(t)\big) \\
&\geq r^*\big(x^*(t) - y^*(t)\big) \\
&= r^*\rho(t).
\end{aligned}
$$

Furthermore, $\rho(a) = x^*(a) - y^*(a) = a^* - a^* = 0$. From Lemma 14.3, it follows that $\rho(t) \geq 0$, $t \in J$. Hence we obtain

$$
x^*(t) \geq y^*(t), \quad t \in J.
$$

Therefore $x^* \equiv y^*$ is the unique solution of the Cauchy problem (14.1) in $[x_0, y_0]$. This ends the proof of Theorem 14.7.   □

As a direct consequence of the previous result, we arrive at the following one

Corollary 14.8. *Suppose that* (14.6.1) *is satisfied and that $f \in C(E, \mathbb{R})$, is differentiable with respect to x and $\partial f/\partial x \in C(E, \mathbb{R})$, with*

$$
E = \{(t, x) \in \mathbb{R}^2, \quad such\ that \quad x_0(t) \leq x \leq y_0(t)\}.
$$

Then problem (14.1) *has a unique solution between x_0 and y_0.*

Proof. The proof follows immediately from the fact that E is a compact set and, as a consequence, $\partial f/\partial x$ is bounded in E.   □

14.1.3 An example

In this section we give an example to illustrate the usefulness of our main result.

Example 14.9. Consider the following initial value problem:

$$
\begin{cases}
{}^cD_{0+}^{\frac{1}{2}}x(t) = 1 - x^2(t) + 2t, & t \in J := [0, 1], \\
x(0) = 1
\end{cases}
\tag{14.17}
$$

Note that this problem is a particular case of IVP (14.1), where

$$
\alpha = \frac{1}{2}, \ a = 0, \ b = 1, \ a^* = 1, \ \psi(t) = t,
$$

and $f: J \times \mathbb{R} \longrightarrow \mathbb{R}$ given by

$$f(t,x) = 1 - x^2 + 2t, \quad \text{for } t \in J, x \in \mathbb{R}.$$

Taking $x_0(t) \equiv 0$ and $y_0(t) = 1 + t$, it is not difficult to verify that x_0, y_0 are lower and upper solutions of (14.17), respectively, and $x_0 \le y_0$. So (14.6.1) of Theorem 14.6 holds
On the other hand, it is clear that the function f is continuous and satisfies

$$\left| \frac{f(t,x)}{\partial x}(t,x) \right| = |-2x| \le 4 \quad \text{for all } t \in [0,1] \text{ and } 0 \le x \le t+1.$$

Hence by Corollary 14.8 the initial value problem (14.17) has a unique solution u^* and there exist monotone iterative sequences $\{x_n\}$ and $\{y_n\}$ converging uniformly to u^*.

14.2 Sequential ψ-Caputo fractional differential equations with nonlinear boundary conditions

14.2.1 Introduction

In this section we investigate the existence of extremal solutions for the following boundary value problem of ψ-Caputo sequential fractional differential equations involving nonlinear boundary conditions:

$$\begin{cases} (^c D_{a^+}^{\alpha+1;\psi} + \lambda^c D_{a^+}^{\alpha;\psi})x(t) = f(t,x(t)), & t \in J := [a,b], \\ g(x(a), x(b)) = 0, & x'(a) = 0, \end{cases} \tag{14.18}$$

where $^c D_{a^+}^{\alpha;\psi}$ is the ψ-Caputo fractional derivative of order $\alpha \in (0,1]$, $f: [a,b] \times \mathbb{R} \longrightarrow \mathbb{R}$, $g: \mathbb{R} \times \mathbb{R} \longrightarrow \mathbb{R}$ are both continuous functions and λ is a positive real number.

14.2.2 Existence results

First, we give the definitions of lower and upper solutions of problem (14.18).

Definition 14.10. A function $x_0 \in C(J,\mathbb{R})$ is called a lower solution of problem (14.18) if it satisfies

$$\begin{cases} (^c D_{a^+}^{\alpha+1;\psi} + \lambda^c D_{a^+}^{\alpha;\psi})x_0(t) \le f(t,x_0(t)), & t \in (a,b], \\ g(x_0(a), x_0(b)) \le 0, & x_0'(a) = 0. \end{cases} \tag{14.19}$$

Definition 14.11. A function $y_0 \in C(J,\mathbb{R})$ is called an upper solution of problem (14.18) if it satisfies

$$\begin{cases} (^c D_{a^+}^{\alpha+1;\psi} + \lambda^c D_{a^+}^{\alpha;\psi})y_0(t) \ge f(t,y_0(t)), & t \in (a,b], \\ g(y_0(a), y_0(b)) \ge 0, & y_0'(a) = 0. \end{cases} \tag{14.20}$$

Lemma 14.12. *For any $h \in C(J, \mathbb{R})$, the unique solution of the following sequential fractional differential equation,*

$$\left({}^c D_{a+}^{\alpha+1;\psi} + \lambda {}^c D_{a+}^{\alpha;\psi}\right) x(t) = h(t), \quad t \in J = [a, b], \tag{14.21}$$

supplemented with the initial conditions

$$x(a) = x_a, \quad x'(a) = 0, \tag{14.22}$$

is given by:

$$x(t) = x_a + \int_a^t \psi'(s) e^{-\lambda(\psi(t)-\psi(s))} \left(\int_a^s \frac{\psi'(r)(\psi(s) - \psi(r))^{\alpha-1}}{\Gamma(\alpha)} h(r) dr\right) ds. \tag{14.23}$$

Proof. Applying the ψ–Riemann–Liouville fractional integral of order α to both sides of (14.21) and using Lemma 2.5, we get

$$x_\psi^{[1]}(t) + \lambda x(t) = \mathcal{I}_{a+}^{\alpha;\psi} h(t) + c_0, \quad c_0 \in \mathbb{R}.$$

Using the notation of $x_\psi^{[1]}$ given by Eq. (2.2), we obtain

$$x'(t) + \psi'(t)\lambda x(t) = \psi'(t)\left(\mathcal{I}_{a+}^{\alpha;\psi} h(t) + c_0\right), \tag{14.24}$$

by multiplying $e^{\lambda(\psi(t)-\psi(a))}$ to both sides of (14.24), we can write

$$\left(x(t)e^{\lambda(\psi(t)-\psi(a))}\right)' = \psi'(t)e^{\lambda(\psi(t)-\psi(a))}\mathcal{I}_{a+}^{\alpha;\psi} h(t) + c_0\psi'(t)e^{\lambda(\psi(t)-\psi(a))}.$$

Integrating from a to t, we have

$$x(t) = c_1 e^{-\lambda(\psi(t)-\psi(a))} + \frac{c_0}{\lambda} + \int_a^t \psi'(s)e^{-\lambda(\psi(t)-\psi(s))}\mathcal{I}_{a+}^{\alpha;\psi} h(s) ds, \tag{14.25}$$

where c_1 is an arbitrary constant. Differentiating (14.25), we obtain

$$\begin{aligned} x'(t) = {} & -\lambda c_1 \psi'(t) e^{-\lambda(\psi(t)-\psi(a))} + \psi'(t)\mathcal{I}_{a+}^{\alpha;\psi} h(t) \\ & - \lambda\psi'(t)\int_a^t \psi'(s)e^{-\lambda(\psi(t)-\psi(s))}\mathcal{I}_{a+}^{\alpha;\psi} h(s) ds. \end{aligned} \tag{14.26}$$

Using the initial conditions given by Eq. (14.22) together with Eqs. (14.25) and (14.26), we obtain

$$c_0 = \lambda x_a, \quad c_1 = 0.$$

Substituting the value of c_0, c_1 in (14.25), we get (14.23). The converse of the lemma follows by direct computation. This completes the proof. $\square$

Now consider the following linear fractional initial value problem

Lemma 14.13. *Let $0 < \alpha \leq 1$ and $p, q \in C(J, \mathbb{R})$. Then the following linear fractional initial value problem*

$$\begin{cases} ({}^c D_{a+}^{\alpha+1;\psi} + \lambda^c D_{a+}^{\alpha;\psi})x(t) - p(t)x(t) = q(t), & t \in J := [a, b], \\ x(a) = x_a, \quad x'(a) = 0, \end{cases} \tag{14.27}$$

has a unique solution $x \in C(J, \mathbb{R})$, provided that

$$\|p\| < \frac{\lambda \Gamma(\alpha + 1)}{(\psi(b) - \psi(a))^\alpha}. \tag{14.28}$$

Proof. It follows from Lemma 14.12 that problem (14.27) is equivalent to the following integral equation:

$$x(t) = x_a + \int_a^t \psi'(s)e^{-\lambda(\psi(t) - \psi(s))}$$
$$\times \left(\int_a^s \frac{\psi'(r)(\psi(s) - \psi(r))^{\alpha-1}}{\Gamma(\alpha)} (p(r)x(r) + q(r))dr \right) ds.$$

Define the operator $T: C(J, \mathbb{R}) \longrightarrow C(J, \mathbb{R})$ as follows

$$(Tx)(t) = x_a + \int_a^t \psi'(s)e^{-\lambda(\psi(t) - \psi(s))}$$
$$\times \left(\int_a^s \frac{\psi'(r)(\psi(s) - \psi(r))^{\alpha-1}}{\Gamma(\alpha)} (p(r)x(r) + q(r))dr \right) ds, \ t \in J.$$

Now, we have to show that the operator T has a unique fixed point. To do this, we will prove that T is a contraction map.

Let $x, y \in C(J, \mathbb{R})$ and $t \in [a, b]$, then we have

$$|(Tx)(t) - (Ty)(t)| \leq \int_a^t \psi'(s)e^{-\lambda(\psi(t) - \psi(s))}$$
$$\times \left(\int_a^s \frac{\psi'(r)(\psi(s) - \psi(r))^{\alpha-1}}{\Gamma(\alpha)} |p(r)||x(r) - y(r)|dr \right) ds$$
$$\leq \frac{(\psi(b) - \psi(a))^\alpha}{\Gamma(\alpha + 1)} \|p\| \|x - y\| \int_a^t \psi'(s)e^{-\lambda(\psi(t) - \psi(s))} ds$$
$$\leq \frac{(\psi(b) - \psi(a))^\alpha}{\lambda \Gamma(\alpha + 1)} \|p\| \|x - y\|.$$

By (14.28) it follows that the operator T is a contraction. Consequently, by Banach fixed-point theorem, the operator T has a unique fixed point. That is, problem (14.27) has a unique solution. This completes the proof. $\qquad \square$

The following result will play a very important role in this work.

Lemma 14.14 (Comparison result). *Assume that $p \in C(J, (0, \infty))$ and satisfies (14.28). If $\theta \in C(J, \mathbb{R})$ satisfies the following inequalities*

$$\begin{cases} (^c D_{a+}^{\alpha+1;\psi} + \lambda^c D_{a+}^{\alpha;\psi})\theta(t) \geq p(t)\theta(t), & t \in J := [a, b], \\ \theta(a) \geq 0, \quad \theta'(a) = 0, \end{cases} \tag{14.29}$$

then $\theta(t) \geq 0$ on $[a, b]$.

Proof. Let $\left(^c D_{a+}^{\alpha+1;\psi} + \lambda^c D_{a+}^{\alpha;\psi}\right)\theta(t) - p(t)\theta(t) = q(t)$, $\theta(a) = x_a$ and $\theta'(a) = 0$ we know that

$$q(t) \geq 0, \ x_a \geq 0.$$

Suppose that the inequality $\theta(t) \geq 0$, $t \in [a, b]$ is not true. It means that there exists at least a $t_0 \in [a, b]$ such that $\theta(t_0) < 0$. Without loss of generality, we assume $\theta(t_0) = \min_{t \in [a,b]} \theta(t)$. Then, by Lemma 14.13, we have

$$\begin{aligned} \theta(t) &= x_a + \int_a^t \psi'(s)e^{-\lambda(\psi(t)-\psi(s))} \\ &\quad \times \left(\int_a^s \frac{\psi'(r)(\psi(s) - \psi(r))^{\alpha-1}}{\Gamma(\alpha)}(p(r)\theta(r) + q(r))dr \right) ds \\ &\geq \theta(t_0) \int_a^t \psi'(s)e^{-\lambda(\psi(t)-\psi(s))} \\ &\quad \times \left(\int_a^s \frac{\psi'(r)(\psi(s) - \psi(r))^{\alpha-1}}{\Gamma(\alpha)}p(r)dr \right) ds. \end{aligned}$$

For $t = t_0$, we can get

$$\theta(t_0) \geq \theta(t_0) \int_a^{t_0} \psi'(s)e^{-\lambda(\psi(t_0)-\psi(s))} \left(\int_a^s \frac{\psi'(r)(\psi(s) - \psi(r))^{\alpha-1}}{\Gamma(\alpha)}p(r)dr \right) ds.$$

Therefore keeping in mind $\theta(t_0) < 0$, we have

$$1 \leq \int_a^{t_0} \psi'(s)e^{-\lambda(\psi(t_0)-\psi(s))} \left(\int_a^s \frac{\psi'(r)(\psi(s) - \psi(r))^{\alpha-1}}{\Gamma(\alpha)}p(r)dr \right) ds.$$

Hence

$$\|p\| \geq \frac{\lambda\Gamma(\alpha + 1)}{(\psi(b) - \psi(a))^\alpha},$$

which is in contradiction to (14.28). Hence $\theta(t) \geq 0$ for all $t \in [a, b]$. $\qquad \square$

We will apply the monotone iterative method to present a result on the existence of the solution of problem (14.18).

Theorem 14.15. *Let the function $f \in C(J \times \mathbb{R}, \mathbb{R})$. In addition assume that:*

(14.15.1) *There exist x_0, $y_0 \in C(J, \mathbb{R})$ such that x_0 and y_0 are lower and upper solutions of problem (14.18), respectively, with $x_0(t) \le y_0(t)$, $t \in J$.*

(14.15.2) *There exists $p \in C(J, \mathbb{R}_+)$ satisfies (14.28) such that*

$$f(t, z_2) - f(t, z_1) \ge p(t)(z_2 - z_1) \quad \text{for} \quad x_0 \le z_1 \le z_2 \le y_0.$$

(14.15.3) *There exist constants $c > 0$ and $d \ge 0$, such that for $x_0(a) \le u_1 \le u_2 \le y_0(a)$, $x_0(b) \le v_1 \le v_2 \le y_0(b)$,*

$$g(u_2, v_2) - g(u_1, v_1) \le c(u_2 - u_1) - d(v_2 - v_1).$$

Then there exist monotone iterative sequences $\{x_n\}$ and $\{y_n\}$, which converge uniformly on the interval J to the extremal solutions of (14.18) in the sector $[x_0, y_0]$, where

$$[x_0, y_0] = \{z \in C(J, \mathbb{R}) : x_0(t) \le z(t) \le y_0(t), \quad t \in J\}.$$

Proof. First, for any $x_0, y_0 \in C(J, \mathbb{R})$, we consider the following boundary value problems of fractional order:

$$\begin{cases} \left({}^{c}D_{a^+}^{\alpha+1;\psi} + \lambda^c D_{a^+}^{\alpha;\psi}\right) x_{n+1}(t) = f\left(t, x_n(t)\right) + p(t)\left(x_{n+1}(t) - x_n(t)\right), & t \in J, \\ x_{n+1}(a) = x_n(a) - \frac{1}{c}g\left(x_n(a), x_n(b)\right), & x'_{n+1}(a) = 0, \end{cases} \quad (14.30)$$

and

$$\begin{cases} \left({}^{c}D_{a^+}^{\alpha+1;\psi} + \lambda^c D_{a^+}^{\alpha;\psi}\right) y_{n+1}(t) = f\left(t, y_n(t)\right) + p(t)\left(y_{n+1}(t) - y_n(t)\right), & t \in J, \\ y_{n+1}(a) = y_n(a) - \frac{1}{c}g\left(y_n(a), y_n(b)\right), & y'_{n+1}(a) = 0 \end{cases} \quad (14.31)$$

By Lemma 14.13, we know that (14.30) and (14.31) have a unique solutions in $C(J, \mathbb{R})$. We will divide the proof in the following steps.

Step 1: We show that the sequences $x_n, y_n (n \ge 1)$ are lower and upper solutions of problem (14.18), respectively, and the following relation holds

$$x_0(t) \le x_1(t) \le \cdots \le x_n(t) \le \cdots \le y_n(t) \le \cdots \le y_1(t) \le y_0(t), \quad t \in J. \quad (14.32)$$

First, we prove that

$$x_0(t) \le x_1(t) \le y_1(t) \le y_0(t), \quad t \in J. \quad (14.33)$$

Set $\theta(t) = x_1(t) - x_0(t)$. From (14.30) and Definition 14.10, we obtain

$$\left({}^{c}D_{a^+}^{\alpha+1;\psi} + \lambda^c D_{a^+}^{\alpha;\psi}\right)\theta(t) = f\left(t, x_0(t)\right) - \left({}^{c}D_{a^+}^{\alpha+1;\psi} + \lambda^c D_{a^+}^{\alpha;\psi}\right)x_0(t) + p(t)\theta(t) \ge p(t)\theta(t).$$

Again, since $\theta'(a) = 0$ and

$$\theta(a) = -\frac{1}{c}g\left(x_0(a), x_0(b)\right) \ge 0.$$

By Lemma 14.14, $\theta(t) \geq 0$, for $t \in J$. That is, $x_0(t) \leq x_1(t)$. Similarly, we can show that $y_1(t) \leq y_0(t)$, $t \in J$.

Now, let $\theta(t) = y_1(t) - x_1(t)$. From (14.30), (14.31), and (14.15.2), we get

$$
\begin{aligned}
\left({}^c D_{a+}^{\alpha+1;\psi} + \lambda^c D_{a+}^{\alpha;\psi}\right)\theta(t) \\
= f\left(t, y_0(t)\right) - f\left(t, x_0(t)\right) + p(t)\left(y_1(t) - y_0(t)\right) - p(t)\left(x_1(t) - x_0(t)\right) \\
\geq p(t)\left(y_0(t) - x_0(t)\right) + p(t)\left(y_1(t) - y_0(t)\right) - p(t)\left(x_1(t) - x_0(t)\right) \\
= p(t)\theta(t).
\end{aligned}
$$

Since, $\theta'(a) = 0$ and

$$
\begin{aligned}
\theta(a) = \left(y_0(a) - x_0(a)\right) - \frac{1}{c}\left(g\left(y_0(a), y_0(b)\right) - g\left(x_0(a), x_0(b)\right)\right) \\
\geq \frac{d}{c}\left(y_0(b) - x_0(b)\right) \geq 0.
\end{aligned}
$$

By Lemma 14.14, we get $x_1(t) \leq y_1(t)$, $t \in J$.

Secondly, we show that $x_1(t)$, $y_1(t)$ are lower and upper solutions of problem (14.18), respectively. Since x_0 and y_0 are lower and upper solutions of problem (14.18), by (14.15.2), it follows that

$$
\left({}^c D_{a+}^{\alpha+1;\psi} + \lambda^c D_{a+}^{\alpha;\psi}\right)x_1(t) = f\left(t, x_0(t)\right) + p(t)\left(x_1(t) - x_0(t)\right) \leq f\left(t, x_1(t)\right),
$$

also $x_1'(a) = 0$ and

$$
0 = c\left(x_1(a) - x_0(a)\right) + g\left(x_0(a), x_0(b)\right) \geq g\left(x_1(a), x_1(b)\right) + d\left(x_1(b) - x_0(b)\right).
$$

Thus

$$
g(x_1(a), x_1(b)) \leq 0.
$$

Therefore $x_1(t)$ is a lower solution of problem (14.18). Similarly, it can be obtained that $y_1(t)$ is an upper solution of problem (14.18).

By the above arguments and mathematical induction, we can show that the sequences $x_n(t)$, $y_n(t)$, $(n \geq 1)$ are lower and upper solutions of problem (14.18), respectively, and the following relation holds

$$
x_0(t) \leq x_1(t) \leq \cdots \leq x_n(t) \leq \cdots \leq y_n(t) \leq \cdots \leq y_1(t) \leq y_0(t), \quad t \in J.
$$

Step 2: The sequences $\{x_n(t)\}$, $\{y_n(t)\}$ uniformly converge to their limit functions $x^*(t)$, $y^*(t)$, respectively.

Note that the sequence $\{x_n(t)\}$ is monotone nondecreasing and is bounded from above by $y_0(t)$. Also, since the sequence $\{y_n(t)\}$ is monotone nonincreasing and is bounded from below by $x_0(t)$, therefore the pointwise limits exist and these limits are denoted by x^*

and y^*. Moreover, since $\{x_n(t)\}$, $\{y_n(t)\}$ are sequences of continuous functions defined on the compact set $[a, b]$, hence by Dini theorem [354], the convergence is uniform. This is

$$\lim_{n\to\infty} x_n(t) = x^*(t) \quad \text{and} \quad \lim_{n\to\infty} y_n(t) = y^*(t),$$

uniformly on $t \in J$ and the limit functions x^*, y^* satisfy problem (14.18). Furthermore, x^* and y^* satisfy the relation

$$x_0 \le x_1 \le \cdots \le x_n \le x^* \le y^* \le \cdots \le y_n \le \cdots \le y_1 \le y_0.$$

Step 3: We prove that x^* and y^* are extremal solutions of problem (14.18) in $[x_0, y_0]$.

Let $z \in [x_0, y_0]$ be any solution of (14.18). We assume that the following relation holds for some $n \in \mathbb{N}$:

$$x_n(t) \le z(t) \le y_n(t), \quad t \in J. \tag{14.34}$$

Let $\theta(t) = z(t) - x_{n+1}(t)$. We have

$$
\begin{aligned}
\left({}^c D_{a+}^{\alpha+1;\psi} + \lambda^c D_{a+}^{\alpha;\psi}\right)\theta(t) &= f\left(t, z(t)\right) - f\left(t, x_n(t)\right) - p(t)\left(x_{n+1}(t) - x_n(t)\right) \\
&\ge p(t)\left(z(t) - x_n(t)\right) - p(t)\left(x_{n+1}(t) - x_n(t)\right) \\
&= p(t)\theta(t).
\end{aligned} \tag{14.35}
$$

Furthermore, $\theta'(a) = 0$ and

$$
\begin{aligned}
0 &= g\left(z(a), z(b)\right) - g\left(x_n(a), x_n(b)\right) + c\left(x_{n+1}(a) - x_n(a)\right) \\
&\ge c\left(z(a) - x_n(a)\right) - d\left(z(b) - x_n(b)\right) + c\left(x_{n+1}(a) - x_n(a)\right) \\
&= c\theta(a) - d\left(z(b) - x_n(b)\right).
\end{aligned}
$$

That is

$$\theta(a) \ge \frac{d}{c}\left(z(b) - x_n(b)\right) \ge 0.$$

By Lemma 14.14, we obtain $\theta(t) \ge 0$, $t \in J$, which means

$$x_{n+1}(t) \le z(t), \ t \in J.$$

Using the same method, we can show that

$$z(t) \le y_{n+1}(t), \ t \in J.$$

Hence we have

$$x_{n+1}(t) \le z(t) \le y_{n+1}(t), \ t \in J.$$

Therefore (14.34) holds on J for all $n \in \mathbb{N}$. Taking the limit as $n \to \infty$ on both sides of (14.34), we get

$$x^* \leq z \leq y^*.$$

Therefore x^*, y^* are the extremal solutions of (14.18) in $[x^*, y^*]$. This completes the proof.   □

14.2.3 An example

Example 14.16. Consider the following boundary value problem:

$$\begin{cases} \left({}^{c}D_{a^{+}}^{\frac{3}{2}} + \frac{2}{\sqrt{\pi}} {}^{c}D_{a^{+}}^{\frac{1}{2}}\right) x(t) = \sin(t)(x - 1) + e^{-t}, & t \in J := [0, 1], \\ x(0) = 1, \quad x'(0) = 0. \end{cases} \tag{14.36}$$

Note that, this problem is a particular case of BVP (14.18), where

$$\alpha = \frac{1}{2}, \quad \lambda = \frac{2}{\sqrt{\pi}}, \quad \psi(t) = t, \quad f(t, x) = \sin(t)(x - 1) + e^{-t}, \quad g(x, y) = x - 1.$$

Obviously, $f \in C([0, 1] \times \mathbb{R}, \mathbb{R})$, $g \in C(\mathbb{R} \times \mathbb{R}, \mathbb{R})$. On the other hand, taking $x_0(t) = 1$ and $y_0(t) = 1 + t\sqrt{t}$, it is not difficult to verify that x_0, y_0 are lower and upper solutions of (14.36), respectively, and $x_0 \leq y_0$. So condition (14.15.1) holds.

Moreover, for $x_0 \leq x \leq y \leq y_0$, we have

$$f(t, y) - f(t, x) \geq \sin t (y - x). \tag{14.37}$$

And if $x_0(a) \leq u_1 \leq u_2 \leq y_0(a)$, $x_0(b) \leq v_1 \leq v_2 \leq y_0(b)$, we have

$$g(u_2, v_2) - g(u_1, v_1) \leq (u_2 - u_1). \tag{14.38}$$

In view of (14.37) and (14.38), we can choose $p(t) = \sin t$, $c = 1$ and $d = 0$ in Theorem 14.15. At last, by a simple computation, we have

$$\frac{(\psi(b) - \psi(a))^{\alpha}}{\lambda \Gamma(\alpha + 1)} \|p\| < 1.$$

Hence, all conditions of Theorem 14.15 are satisfied and consequently the problem (14.36) has extremal solutions on $[x_0, y_0]$.

14.3 Hyperbolic fractional partial differential equation

14.3.1 Introduction

The objective of this section is to develop the existence and uniqueness of the extremal solutions for the following hyperbolic fractional partial differential equation:

$$\left({}^{c}D_{\theta}^{\alpha;\psi} x\right)(t, u) = f(t, u, x(t, u)), \quad (t, u) \in \tilde{I} := [a, b] \times [a, c], \tag{14.39}$$

under the initial conditions

$$\begin{cases} x(t,a) = \eta(t), & t \in [a,b], \\ x(a,u) = v(u), & u \in [a,c], \\ x(a,a) = \eta(t) = v(u), \end{cases} \tag{14.40}$$

where a, b and c are positive constants, $f : \tilde{I} \times \mathbb{R} \longrightarrow \mathbb{R}$ is a given continuous function and $\eta : [a,b] \longrightarrow \mathbb{R}$, $v : [a,c] \longrightarrow \mathbb{R}$ are given absolutely continuous functions. ${}^{c}D_{\theta}^{\alpha;\psi}$ is the ψ-Caputo fractional derivative of order $\alpha = (\alpha_1, \alpha_2) \in (0,1] \times (0,1]$ and $\theta = (a,a)$.

14.3.2 Existence results

In this section we prove an uniqueness result and approximation of solutions for our proposed problem (14.39)–(14.40).

Let us start by defining what we mean by a solution of problem (14.39)–(14.40).

Definition 14.17 ([407]). *A function $x \in C(\tilde{I}, \mathbb{R})$ is said to be a solution of (14.39)–(14.40) if x satisfies Eq. (14.39) and conditions (14.40) on $\tilde{I}$.*

We state the next lemma, which has an important role in this work.

Lemma 14.18 ([284]). *Suppose that $\mathbb{S}$ is a linear bounded operator defined on a Banach space $\mathbb{X}$, and assume that $\|\mathbb{S}\| < 1$. Then $(I - \mathbb{S})^{-1}$ is linear and bounded. Also*

$$(I - \mathbb{S})^{-1} = \sum_{n=0}^{\infty} \mathbb{S}^n,$$

the convergence of the series being in the operator norm and

$$\|(I - \mathbb{S})^{-1}\| \leq \frac{1}{1 - \|\mathbb{S}\|}.$$

To prove our main result, we need the following lemma:

Lemma 14.19 ([407]). *A function $x \in C(\tilde{I}, \mathbb{R})$ is a solution of the problem (14.39)–(14.40) if and only if x satisfies the following integral equation*

$$x(t,u) = \zeta(t,u) + \int_a^t \int_a^u \frac{\psi'(s)\psi'(t)(\psi(t) - \psi(s))^{\alpha_1-1}(\psi(u) - \psi(t))^{\alpha_2-1}}{\Gamma(\alpha_1)\Gamma(\alpha_2)} \tag{14.41}$$
$$\times f(s,t,x(s,t))\,dtds, \quad (t,u) \in \tilde{I} := [a,b] \times [a,c],$$

where

$$\zeta(t,u) = \eta(t) + v(u) - \eta(a).$$

The first existence result is based on the Banach fixed-point theorem combined with a Bielecki type norm.

Theorem 14.20. *Assume that the following hypotheses hold.*

(14.20.1) *The function $f : \tilde{I} \times \mathbb{R} \longrightarrow \mathbb{R}$ is continuous.*

(14.20.2) *There exists $p \in C(\tilde{I}, \mathbb{R}^+)$ such that*

$$|f(t, u, x) - f(t, u, y)| \le p(t, u)|x - y|, \quad (t, u) \in \tilde{I}, x, y \in \mathbb{R}.$$

Then there exists a unique solution of problem (14.39)–(14.40) *on $\tilde{I}$.*

Set

$$p^* = \sup_{(t,u) \in \tilde{I}} p(t, u).$$

Proof. In view of Lemma 14.19, we transform the integral representation (14.41) of the problem (14.39)–(14.40) into

$$x = \mathbb{T}x, \quad x \in C(\tilde{I}, \mathbb{R}),$$

where $\mathbb{T} : C(\tilde{I}, \mathbb{R}) \longrightarrow C(\tilde{I}, \mathbb{R})$ is defined by

$$
\begin{aligned}
\mathbb{T}x(t, u) = \zeta(t, u) + \int_a^t \int_a^u &\frac{\psi'(s)\psi'(t)(\psi(t) - \psi(s))^{\alpha_1 - 1}(\psi(u) - \psi(t))^{\alpha_2 - 1}}{\Gamma(\alpha_1)\Gamma(\alpha_2)} \\
&\times f(s, t, x(s, t)) \, dt \, ds.
\end{aligned}
\tag{14.42}
$$

Clearly the operator $\mathbb{T}$ is well defined. Now, we apply the Banach fixed-point theorem to prove that $\mathbb{T}$ has a unique fixed point. Indeed, it is enough to show that $\mathbb{T}$ is a contraction operator on $C(\tilde{I}, \mathbb{R})$ via the Bielecki norm. To this end, given $x, y \in C(\tilde{I}, \mathbb{R})$ and $(t, u) \in \tilde{I}$, using (14.20.2), we can get

$$
\begin{aligned}
\big|\mathbb{T}x(t, u)) &- \mathbb{T}y(t, u))\big| \\
\le \int_a^t \int_a^u &\frac{\psi'(s)\psi'(t)(\psi(t) - \psi(s))^{\alpha_1 - 1}(\psi(u) - \psi(t))^{\alpha_2 - 1}}{\Gamma(\alpha_1)\Gamma(\alpha_2)} \\
&\times \big|f(s, t, x(s, t)) - f(s, t, y(s, t))\big| \, dt \, ds \\
\le \int_a^t \int_a^u &\frac{\psi'(s)\psi'(t)(\psi(t) - \psi(s))^{\alpha_1 - 1}(\psi(u) - \psi(t))^{\alpha_2 - 1}}{\Gamma(\alpha_1)\Gamma(\alpha_2)} \\
&\times p(s, t)|x(s, t) - y(s, t)| \, dt \, ds \\
\le p^* \|x - y\|_{\mathfrak{B}} \int_a^t \int_a^u &\frac{\psi'(s)\psi'(t)(\psi(t) - \psi(s))^{\alpha_1 - 1}(\psi(u) - \psi(t))^{\alpha_2 - 1}}{\Gamma(\alpha_1)\Gamma(\alpha_2)} \\
&\times e^{\lambda(\psi(s) + \psi(t))} \, dt \, ds \\
\le \frac{e^{\lambda(\psi(t) + \psi(u))}}{\lambda^{\alpha_1 + \alpha_2}} &p^* \|x - y\|_{\mathfrak{B}}
\end{aligned}
$$

where

$$\|x\|_{\mathfrak{B}} := \sup_{(t,u)\in \tilde{I}} \frac{|x(t,u)|}{e^{\lambda(\psi(t)+\psi(u))}}, \quad \lambda > 0, \tag{14.43}$$

denotes the Bielecki-type norm on the Banach space $C(\tilde{I}, \mathbb{R})$ (for more properties on Bielecki type norm, see [400]).

Thus, we obtain

$$\|\mathbb{T}x - \mathbb{T}y\|_{\mathfrak{B}} \le \frac{p^*}{\lambda^{\alpha_1 + \alpha_2}} \|x - y\|_{\mathfrak{B}}.$$

Taking $\lambda > 0$ large enough such that

$$\frac{p^*}{\lambda^{\alpha_1 + \alpha_2}} < 1,$$

it follows that the mapping $\mathbb{T}$ is a contraction with respect to the Bielecki norm. Therefore by the Banach fixed-point theorem, $\mathbb{T}$ has a unique fixed point. That is (14.39)–(14.40) has a unique solution. This completes the proof. $\qquad\square$

An immediate consequence of Theorem 14.20 is the following result, which plays an important role in our further discussion.

Corollary 14.21. *Let* $\alpha = (\alpha_1, \alpha_2) \in (0, 1] \times (0, 1]$ *be fixed and let* $p, q \in C(\tilde{I}, \mathbb{R})$. *Then the following linear fractional initial value problem*

$$\left({}^c D_\theta^{\alpha;\psi} x\right)(t, u) - p(t, u)x(t, u) = q(t, u), \quad (t, u) \in \tilde{I} \tag{14.44}$$

under the initial conditions

$$\begin{cases} x(t, a) = \eta(t), & t \in [a, b], \\ x(a, u) = v(u), & u \in [a, c], \\ x(a, a) = \eta(a) = v(a), \end{cases} \tag{14.45}$$

has a unique solution given by

$$x(t, u) = \zeta(t, u) + \int_a^t \int_a^u \frac{\psi'(s)\psi'(t)(\psi(t) - \psi(s))^{\alpha_1 - 1}(\psi(u) - \psi(t))^{\alpha_2 - 1}}{\Gamma(\alpha_1)\Gamma(\alpha_2)} \tag{14.46}$$
$$\times [p(s, t)x(s, t) + q(s, t)] \, dtds.$$

The following lemma is fundamental to our results.

Lemma 14.22 (Comparison result). *Let* $\alpha = (\alpha_1, \alpha_2) \in (0, 1] \times (0, 1]$ *be fixed and* $p \in C(\tilde{I}, \mathbb{R}_+)$. *If* $\omega \in C(\tilde{I}, \mathbb{R})$ *satisfies the inequality,*

$$\omega(t, u) \le \int_a^t \int_a^u \frac{\psi'(s)\psi'(t)(\psi(t) - \psi(s))^{\alpha_1 - 1}(\psi(u) - \psi(t))^{\alpha_2 - 1}}{\Gamma(\alpha_1)\Gamma(\alpha_2)} p(s, t)\omega(s, t) dtds, \tag{14.47}$$

then $\omega(t, u) \le 0$ *for all* $(t, u) \in \tilde{I}$.

Proof. Consider the Banach space $C(\tilde{I}, \mathbb{R})$ equipped with a Bielecki norm $\| \cdot \|_{\mathfrak{B}}$ defined as in (14.43).

Define an operator $\mathbb{S} : C(\tilde{I}, \mathbb{R}) \longrightarrow C(\tilde{I}, \mathbb{R})$ by

$$\mathbb{S}\omega(t, u)$$
$$= r \int_a^t \int_a^u \frac{\psi'(s)\psi'(t)(\psi(t) - \psi(s))^{\alpha_1 - 1}(\psi(u) - \psi(t))^{\alpha_2 - 1}}{\Gamma(\alpha_1)\Gamma(\alpha_2)} p(s, t)\omega(s, t) dt ds. \qquad (14.48)$$

Note that $\mathbb{S}\omega(t, u) \leq 0$ for all $\omega \in C(\tilde{I}, \mathbb{R})$, if $\omega(t, u) \leq 0$ for all $(t, u) \in \tilde{I}$. Now

$$|\mathbb{S}\omega(t, u)|$$
$$\leq \int_a^t \int_a^u \frac{\psi'(s)\psi'(t)(\psi(t) - \psi(s))^{\alpha_1 - 1}(\psi(u) - \psi(t))^{\alpha_2 - 1}}{\Gamma(\alpha_1)\Gamma(\alpha_2)} p(s, t)|\omega(s, t)| dt ds$$
$$\leq p^* \|\omega\|_{\mathfrak{B}} \int_a^t \int_a^u \frac{\psi'(s)\psi'(t)(\psi(t) - \psi(s))^{\alpha_1 - 1}(\psi(u) - \psi(t))^{\alpha_2 - 1}}{\Gamma(\alpha_1)\Gamma(\alpha_2)}$$
$$\times e^{\lambda(\psi(s) + \psi(t))} dt ds$$
$$\leq \frac{e^{\lambda(\psi(t) + \psi(u))}}{\lambda^{\alpha_1 + \alpha_2}} p^* \|\omega\|_{\mathfrak{B}},$$

which implies that

$$\|\mathbb{S}\| \leq \frac{p^*}{\lambda^{\alpha_1 + \alpha_2}}.$$

Since we can choose $\lambda > 0$ sufficiently large such that

$$\frac{p^*}{\lambda^{\alpha_1 + \alpha_2}} < 1,$$

it follows that

$$\|\mathbb{S}\| < 1.$$

Hence, by Lemma 14.18, we conclude that $(I - \mathbb{S})^{-1}$ is a bounded linear operator satisfying

$$(I - \mathbb{S})^{-1} = \sum_{n=0}^{\infty} \mathbb{S}^n.$$

Now, if $x(t, u) \leq 0$, $(t, u) \in \tilde{I}$, then

$$(I - \mathbb{S})^{-1}x(t, u) = \sum_{n=0}^{\infty} \mathbb{S}^n x(t, u) \leq 0.$$

Moreover, Eq. (14.47) can be written as

$$(I - \mathbb{S})\omega(t, u) \leq 0.$$

Applying $(I - \mathbb{S})^{-1}$ on both sides of the above inequality, we obtain

$$\omega(t, u) \leq 0. \qquad \qquad \square$$

Now, we will apply the monotone iterative method to present a result on the existence and uniqueness of the extremal solution for the problem (14.39)–(14.40).

Definition 14.23. A function $x_0 \in C(\tilde{I}, \mathbb{R})$ is called a lower solution of the operator equation (14.42), if it satisfies

$$x_0(t, u) \leq \mathbb{T}x_0(t, u)$$
$$= \zeta(t, u) + \int_a^t \int_a^u \frac{\psi'(s)\psi'(t)(\psi(t) - \psi(s))^{\alpha_1 - 1}(\psi(u) - \psi(t))^{\alpha_2 - 1}}{\Gamma(\alpha_1)\Gamma(\alpha_2)}$$
$$\times f(s, t, x_0(s, t)) \, dt\, ds.$$

Otherwise, x_0 is said to be an upper solution if the inequality is reversed.

The following functional interval plays a fundamental role in our discussion

$$[x_0, y_0] := \{x \in C(\tilde{I}, \mathbb{R}) : x_0(t, u) \leq x(t, u) \leq y_0(t, u), \quad (t, u) \in \tilde{I}\}.$$

Theorem 14.24. *Assume that (14.20.1) and the following conditions hold:*

(14.24.1) *There exist x_0, $y_0 \in C(\tilde{I}, \mathbb{R})$ such that x_0 and y_0 are lower and upper solutions of the operator equation (14.42), respectively, with $x_0(t, u) \leq y_0(t, u)$, $(t, u) \in \tilde{I}$.*
(14.24.2) *There exists $p \in C(\tilde{I}, \mathbb{R}^+)$ such that*

$$f(t, u, y) - f(t, u, x) \geq p(t, u)(y - x), \quad (t, u) \in \tilde{I},$$

and $x_0 \leq x \leq y \leq y_0$.

Then there exist monotone iterative sequences $\{x_n\}, \{y_n\} \subset [x_0, y_0]$ such that $x_n \to x^*$, $y_n \to y^*$ as $n \to \infty$ uniformly in $[x_0, y_0]$, and x^*, y^* are a minimal and a maximal solution of the operator equation (14.42) in $[x_0, y_0]$, respectively, and the following relation holds

$$x_0(t, u) \leq x_1(t, u) \leq \cdots \leq x_n(t, u) \leq \cdots \leq y_n(t, u) \leq \cdots \leq y_1(t, u) \leq y_0(t, u). \qquad (14.49)$$

Proof. First, for any $x_0, y_0 \in C(\tilde{I}, \mathbb{R})$, we define two sequences $\{x_n\}$ and $\{y_n\}$ satisfying the following, respective, linear integral equations,

$$x_{n+1}(t, u) = \zeta(t, u) + \int_a^t \int_a^u \frac{\psi'(s)\psi'(t)(\psi(t) - \psi(s))^{\alpha_1 - 1}(\psi(u) - \psi(t))^{\alpha_2 - 1}}{\Gamma(\alpha_1)\Gamma(\alpha_2)} \qquad (14.50)$$
$$\times \big(f(s, t, x_n(s, t)) + p(s, t)(x_{n+1}(s, t) - x_n(s, t)) \big) dt\, ds,$$

and

$$y_{n+1}(t, \mathfrak{u}) = \zeta(t, \mathfrak{u}) + \int_a^t \int_a^\mathfrak{u} \frac{\psi'(s)\psi'(t)(\psi(t) - \psi(s))^{\alpha_1 - 1}(\psi(\mathfrak{u}) - \psi(t))^{\alpha_2 - 1}}{\Gamma(\alpha_1)\Gamma(\alpha_2)}$$
$$\times \left(f(s, t, y_n(s, t)) + p(s, t)(y_{n+1}(s, t) - y_n(s, t)) \right) dt ds. \tag{14.51}$$

By Corollary 14.21, we know that (14.50) and (14.51) have unique solutions in $C(\tilde{I}, \mathbb{R})$.

For clarity, we will divide the remainder of the proof into several steps.

Step 1: We show that the sequences $x_n(t, \mathfrak{u})$ and $y_n(t, \mathfrak{u}), n \geq 1$, are lower and upper solutions of the operator equation (14.42), respectively, and the relation (14.49) holds.
First, we prove that

$$x_0(t, \mathfrak{u}) \leq x_1(t, \mathfrak{u}) \leq y_1(t, \mathfrak{u}) \leq y_0(t, \mathfrak{u}), \quad (t, \mathfrak{u}) \in \tilde{I}. \tag{14.52}$$

Set $\omega(t, \mathfrak{u}) = x_0(t, \mathfrak{u}) - x_1(t, \mathfrak{u})$. From (14.50) and Definition 14.23, we obtain

$$\omega(t, \mathfrak{u}) \leq \int_a^t \int_a^\mathfrak{u} \frac{\psi'(s)\psi'(t)(\psi(t) - \psi(s))^{\alpha_1 - 1}(\psi(\mathfrak{u}) - \psi(t))^{\alpha_2 - 1}}{\Gamma(\alpha_1)\Gamma(\alpha_2)} p(s, t)\omega(s, t)ds.$$

By Lemma 14.22, $\omega(t, \mathfrak{u}) \leq 0$, for $(t, \mathfrak{u}) \in \tilde{I}$. That is, $x_0(t, \mathfrak{u}) \leq x_1(t, \mathfrak{u})$. Similarly, we can show that $y_1(t, \mathfrak{u}) \leq y_0(t, \mathfrak{u})$, $(t, \mathfrak{u}) \in \tilde{I}$.
Now, let $\omega(t, \mathfrak{u}) = x_1(t, \mathfrak{u}) - y_1(t, \mathfrak{u})$. From (14.50), (14.51), and (14.24.2), we get

$$\omega(t, \mathfrak{u})$$
$$= \int_a^t \int_a^\mathfrak{u} \frac{\psi'(s)\psi'(t)(\psi(t) - \psi(s))^{\alpha_1 - 1}(\psi(\mathfrak{u}) - \psi(t))^{\alpha_2 - 1}}{\Gamma(\alpha_1)\Gamma(\alpha_2)}$$
$$\times \left[(f(s, t, x_0(s, t)) - f(s, t, y_0(s, t)) + p(s, t)(y_0(s, t) - x_0(s, t)) \right] dt ds$$
$$+ \int_a^t \int_a^\mathfrak{u} \frac{\psi'(s)\psi'(t)(\psi(t) - \psi(s))^{\alpha_1 - 1}(\psi(\mathfrak{u}) - \psi(t))^{\alpha_2 - 1}}{\Gamma(\alpha_1)\Gamma(\alpha_2)}$$
$$\times p(s, t)(x_1(s, t) - y_1(s, t)) dt ds$$
$$\leq \int_a^t \int_a^\mathfrak{u} \frac{\psi'(s)\psi'(t)(\psi(t) - \psi(s))^{\alpha_1 - 1}(\psi(\mathfrak{u}) - \psi(t))^{\alpha_2 - 1}}{\Gamma(\alpha_1)\Gamma(\alpha_2)} p(s, t)\omega(s, t) dt ds.$$

By Lemma 14.22, we get $x_1(t, \mathfrak{u}) \leq y_1(t, \mathfrak{u})$, $(t, \mathfrak{u}) \in \tilde{I}$.
Second, we show that $x_1(t, \mathfrak{u})$ and $y_1(t, \mathfrak{u})$ are lower and upper solutions of the operator equation (14.42), respectively. Since x_0 and y_0 are lower and upper solutions of the operator

equation (14.42), by (14.24.2), it follows that

$$
\begin{aligned}
x_1(t, u) &= \zeta(t, u) + \int_a^t \int_a^u \frac{\psi'(s)\psi'(t)(\psi(t) - \psi(s))^{\alpha_1-1}(\psi(u) - \psi(t))^{\alpha_2-1}}{\Gamma(\alpha_1)\Gamma(\alpha_2)} \\
&\quad \times \big(f(s, t, x_0(s, t)) + p(s, t)(x_1(s, t) - x_0(s, t))\big) dtds \\
&\leq \zeta(t, u) + \int_a^t \int_a^u \frac{\psi'(s)\psi'(t)(\psi(t) - \psi(s))^{\alpha_1-1}(\psi(u) - \psi(t))^{\alpha_2-1}}{\Gamma(\alpha_1)\Gamma(\alpha_2)} \\
&\quad \times f(s, t, x_1(s, t)) dtds.
\end{aligned}
$$

Therefore $x_1(t, u)$ is a lower solution of the operator equation (14.42). Similarly, it can be obtained that $y_1(t, u)$ is an upper solution of the operator equation (14.42).

By the above arguments and mathematical induction, we can show that the sequences $x_n(t, u), y_n(t, u), (n \geq 1)$ are lower and upper solutions of the operator equation (14.42), respectively, and the relation (14.49) holds.

Step 2: Our aim in this step is to prove that sequences $\{x_n\}$ and $\{y_n\}$ converge uniformly to their limit functions x^*, y^*, respectively. It is enough to show that $\{x_n\}$ and $\{y_n\}$ are uniformly bounded and equicontinuous on $\tilde{I}$.

First, observe that the uniform boundedness of sequences $\{x_n\}$ and $\{y_n\}$ follows from (14.49).

Next, we need to prove that the sequences $\{x_n\}$ and $\{y_n\}$ are equicontinuous on $\tilde{I}$. To this end, let $(t_1, u_1), (t_2, u_2) \in \tilde{I}$, with $t_1 < t_2$ and $u_1 < u_2$. Then, from (14.50), we can get

$$
\begin{aligned}
|x_{n+1}(t_2, u_2) &- x_{n+1}(t_1, u_1)| \leq |\zeta(t_2, u_2) - \zeta(t_1, u_1)| \\
&+ \frac{1}{\Gamma(\alpha_1)\Gamma(\alpha_2)} \int_a^{t_1} \int_a^{u_1} \psi'(s)\psi'(t)\big[(\psi(t_1) - \psi(s))^{\alpha_1-1}(\psi(u_1) - \psi(t))^{\alpha_2-1} \\
&- (\psi(t_2) - \psi(s))^{\alpha_1-1}(\psi(u_2) - \psi(t))^{\alpha_2-1}\big] \\
&\times \big|f(s, t, x_n(s, t)) + p(s, t)(x_{n+1}(s, t) - x_n(s, t))\big| dtds \\
&+ \frac{1}{\Gamma(\alpha_1)\Gamma(\alpha_2)} \int_{t_1}^{t_2} \int_{u_1}^{u_2} \psi'(s)\psi'(t)(\psi(t_2) - \psi(s))^{\alpha_1-1}(\psi(u_2) - \psi(t))^{\alpha_2-1} \\
&\times \big|f(s, t, x_n(s, t)) + p(s, t)(x_{n+1}(s, t) - x_n(s, t))\big| dtds \\
&+ \frac{1}{\Gamma(\alpha_1)\Gamma(\alpha_2)} \int_{t_1}^{t_2} \int_a^{u_1} \psi'(s)\psi'(t)(\psi(t_2) - \psi(s))^{\alpha_1-1}(\psi(u_2) - \psi(t))^{\alpha_2-1} \\
&\times \big|f(s, t, x_n(s, t)) + p(s, t)(x_{n+1}(s, t) - x_n(s, t))\big| dtds \\
&+ \frac{1}{\Gamma(\alpha_1)\Gamma(\alpha_2)} \int_a^{t_1} \int_{u_1}^{u_2} \psi'(s)\psi'(t)(\psi(t_2) - \psi(s))^{\alpha_1-1}(\psi(u_2) - \psi(t))^{\alpha_2-1} \\
&\times \big|f(s, t, x_n(s, t)) + p(s, t)(x_{n+1}(s, t)x_n(s, t))\big| dtds.
\end{aligned}
$$

Since $\{x_n\}$ is uniformly bounded and f is continuous on $\tilde{I}$, there exists M independent of n such that

$$|f(s, t, x_n(t, u)) + p(t, u)(x_{n+1}(t, u) - x_{n+1}(t, u)))| \le |f(s, t, x_{n+1}(t, u))| \le M.$$

Using the above estimation, we obtain

$$|x_{n+1}(t_2, u_2) - x_{n+1}(t_1, u_1)| \le |\zeta(t_2, u_2) - \zeta(t_1, u_1)|$$

$$+ \frac{M}{\Gamma(\alpha_1)\Gamma(\alpha_2)} \int_a^{t_1} \int_a^{u_1} \psi'(s)\psi'(t) \big[(\psi(t_1) - \psi(s))^{\alpha_1 - 1} (\psi(u_1) - \psi(t))^{\alpha_2 - 1}$$

$$- (\psi(t_2) - \psi(s))^{\alpha_1 - 1} (\psi(u_2) - \psi(t))^{\alpha_2 - 1} \big] dt ds$$

$$+ \frac{M}{\Gamma(\alpha_1)\Gamma(\alpha_2)} \int_{t_1}^{t_2} \int_a^{u_2} \psi'(s)\psi'(t)(\psi(t_2) - \psi(s))^{\alpha_1 - 1} (\psi(u_2) - \psi(t))^{\alpha_2 - 1} dt ds$$

$$+ \frac{M}{\Gamma(\alpha_1)\Gamma(\alpha_2)} \int_a^{t_1} \int_{u_1}^{u_2} \psi'(s)\psi'(t)(\psi(t_2) - \psi(s))^{\alpha_1 - 1} (\psi(u_2) - \psi(t))^{\alpha_2 - 1} dt ds$$

$$\le |\zeta(t_2, u_2) - \zeta(t_1, u_1)|$$

$$+ \frac{2M(\psi(u_2) - \psi(u_1))^{\alpha_2} \big[(\psi(t_2) - \psi(a))^{\alpha_1} - (\psi(t_2) - \psi(t_1))^{\alpha_1} \big]}{\Gamma(\alpha_1 + 1)\Gamma(\alpha_2 + 1)}$$

$$+ \frac{M \big[(\psi(u_2) - \psi(a))^{\alpha_2} (\psi(t_2) - \psi(t_1))^{\alpha_1} - (\psi(u_1) - \psi(a))^{\alpha_2} (\psi(t_1) - \psi(a))^{\alpha_1} \big]}{\Gamma(\alpha_1 + 1)\Gamma(\alpha_2 + 1)}.$$

As $t_1 \to t_2$ and $u_1 \to u_2$, the right-hand side of the above inequality tends to zero. So $\{x_n\}$ is equicontinuous on $\tilde{I}$. Similarly, we can prove $\{y_n\}$ is equicontinuous. Hence, by the Ascoli–Arzelá theorem, $\{x_n\}$ and $\{y_n\}$ have convergent subsequences. Combining this with the monotonicity of (14.49) it follows that $\{x_n\}$ and $\{y_n\}$ are uniformly convergent. That is,

$$\lim_{n \to \infty} x_n(t, u) = x^*(t, u) \quad \text{and} \quad \lim_{n \to \infty} y_n(t, u) = y^*(t, u), \quad (t, u) \in \tilde{I},$$

and the limit functions x^* and y^* satisfy the operator equation (14.42)

Step 3: We prove that x^* and y^* are minimal and maximal solutions of the operator equation (14.42) in $[x_0, y_0]$.

Let $z \in [x_0, y_0]$ be any solution of (14.42). We assume that the following relation holds for some $n \in \mathbb{N}$:

$$x_n(t, u) \le z(t, u) \le y_n(t, u), \quad (t, u) \in \tilde{I}. \tag{14.53}$$

Let $\omega(t, u) = x_{n+1}(t, u) - z(t, u)$. We have

$$\omega(t, u) = \int_a^t \int_a^u \frac{\psi'(s)\psi'(t)(\psi(t) - \psi(s))^{\alpha_1 - 1} (\psi(u) - \psi(t))^{\alpha_2 - 1}}{\Gamma(\alpha_1)\Gamma(\alpha_2)}$$

$$\times \big[f(s, t, x_n(s, t)) - f(s, t, z(s, t)) + p(s, t)(z(s, t) - x_n(s, t)) \big] dt ds$$

$$+ \int_a^t \int_a^u \frac{\psi'(s)\psi'(t)(\psi(t) - \psi(s))^{\alpha_1 - 1}(\psi(u) - \psi(t))^{\alpha_2 - 1}}{\Gamma(\alpha_1)\Gamma(\alpha_2)}$$

$$\times\, p(s,t)\big(x_{n+1}(s,t) - z(s,t)\big)dtds$$

$$\leq \int_a^t \int_a^u \frac{\psi'(s)\psi'(t)(\psi(t) - \psi(s))^{\alpha_1 - 1}(\psi(u) - \psi(t))^{\alpha_2 - 1}}{\Gamma(\alpha_1)\Gamma(\alpha_2)} p(s,t)\omega(s,t)dtds.$$

By Lemma 14.22, we obtain $\omega(t, \mathrm{u}) \leq 0$, $(t, \mathrm{u}) \in \tilde{I}$, which means

$$x_{n+1}(t, \mathrm{u}) \leq z(t, \mathrm{u}), \quad (t, \mathrm{u}) \in \tilde{I}.$$

Using the same method, we can show that

$$z(t, \mathrm{u}) \leq y_{n+1}(t, \mathrm{u}), \quad (t, \mathrm{u}) \in \tilde{I}.$$

Hence we have

$$x_{n+1}(t, \mathrm{u}) \leq z(t, \mathrm{u}) \leq y_{n+1}(t, \mathrm{u}), \quad (t, \mathrm{u}) \in \tilde{I}.$$

Therefore (14.53) holds on $\tilde{I}$ for all $n \in \mathbb{N}$. Taking the limit as $n \to \infty$ on both sides of (14.53), we get

$$x^* \leq z \leq y^*,$$

which shows that x^* and y^* are minimal and maximal solutions of (14.42) in $[x_0, y_0]$, respectively. Hence x^* and y^* are minimal and maximal solutions of the problem (14.39)–(14.40) in $[x_0, y_0]$, respectively. This completes the proof. □

Finally, we shall prove the uniqueness of the solution of the problem (14.39)–(14.40) by the monotone iterative technique.

Theorem 14.25. *Assume that all assumptions of Theorem 14.24 hold. In addition, we assume that*

(14.25.1) *There exists $\tilde{p} \in C(\tilde{I}, \mathbb{R})$ such that*

$$f(t, \mathrm{u}, y) - f(t, \mathrm{u}, x) \leq \tilde{p}(t, \mathrm{u})(y - x), \quad (t, \mathrm{u}) \in \tilde{I},$$

and $x_0 \leq x \leq y \leq y_0$.

Then the operator equation (14.42) has a unique solution between x_0 and y_0, which can be obtained by a monotone iterative procedure starting from x_0 or y_0.

Proof. From Theorem 14.24, we know that x^* and y^* are the extremal solutions of (14.42) and $x^*(t, \mathrm{u}) \leq y^*(t, \mathrm{u})$, $(t, \mathrm{u}) \in \tilde{I}$. It is sufficient to prove $x^*(t, \mathrm{u}) \geq y^*(t, \mathrm{u})$, $(t, \mathrm{u}) \in \tilde{I}$. In fact, let $\omega(t, u) = y^*(t, u) - x^*(t, u)$, $(t, u) \in \tilde{I}$. In view of (14.25.1), we have

$$
\begin{aligned}
\omega(t, u) \\
= \int_a^t \int_a^u & \frac{\psi'(s)\psi'(t)(\psi(t) - \psi(s))^{\alpha_1-1}(\psi(u) - \psi(t))^{\alpha_2-1}}{\Gamma(\alpha_1)\Gamma(\alpha_2)} \\
& \times \left(f(s, t, y^*(s, t)) - f(s, t, x^*(s, t))\right)dtds \\
\leq \int_a^t \int_a^u & \frac{\psi'(s)\psi'(t)(\psi(t) - \psi(s))^{\alpha_1-1}(\psi(u) - \psi(t))^{\alpha_2-1}}{\Gamma(\alpha_1)\Gamma(\alpha_2)} \tilde{p}(s, t)\omega(s, t)dtds.
\end{aligned}
$$

From Lemma 14.22, it follows that $\omega(t, u) \leq 0$, $(t, u) \in \tilde{I}$. Hence we obtain

$$
x^*(t, u) \geq y^*(t, u), \quad (t, u) \in \tilde{I}.
$$

Therefore $x^* \equiv y^*$ is the unique solution of (14.42) in $[x_0, y_0]$. That is, the Cauchy problem (14.39)–(14.40) has a unique solution in $[x_0, y_0]$. This ends the proof of Theorem 14.25.   □

14.3.3 Some examples

We conclude this section with two examples to illustrate our main result.

Example 14.26. Let us consider problem (14.39) with specific data:

$$
\alpha = (\alpha_1, \alpha_2) = (0.5, 0.5), \quad a = 0, \quad b = c = 1. \tag{14.54}
$$

To illustrate Theorem 14.20, we take $\psi(t) = \sigma(t)$, where $\sigma(t)$ is the Sigmoid function, which can be expressed as in the following form

$$
\sigma(t) = \frac{1}{1 + e^{-t}}, \tag{14.55}
$$

and a convenience of the Sigmoid function is its derivative

$$
\sigma'(t) = \sigma(t)(1 - \sigma(t)).
$$

Taking also

$$
\begin{aligned}
f(t, u, x(t, u)) &= \frac{t + u}{2}(1 + \sin(x(t, u))), \quad (t, u) \in [0, 1] \times [0, 1], \\
\eta(t) &= te^t, \quad t \in [0, 1], \\
v(u) &= u^2, \quad u \in [0, 1],
\end{aligned} \tag{14.56}
$$

in (14.39). Clearly, the function f is continuous. Moreover, for any $x, y \in \mathbb{R}$ and $(t, u) \in [0, 1] \times [0, 1]$, we have

$$
|f(t, u, x) - f(t, u, y)| \leq \frac{t + u}{2}|x - y|.
$$

Thus hypothesis (14.20.2) is satisfied with $p(t, u) = \frac{t+u}{2}$. Evidently, $p^* = \sup_{(t,u)\in\bar{I}} p(t, u) = 1$. Moreover, if we choose $\lambda > 1$, it follows that the mapping $\mathbb{T}$ defined by (14.41) is a contraction. An application of Theorem 14.20 shows that the problem (14.39)–(14.40) with the data (14.54), (14.55), and (14.56) has a unique solution in $C([0, 1] \times [0, 1], \mathbb{R})$.

Example 14.27. Now, we apply Theorem 14.24 to the partial hyperbolic differential equation of the form,

$$
\begin{cases}
({}^cD_\theta^{\alpha;\psi}x)(t, u) = \frac{e^{-(t+u)}}{4}(x^2(t, u) - 0.5), & (t, u) \in \tilde{I} := [0, 1] \times [0, 1], \\
x(t, 0) = 1, & t \in [0, 1], \\
x(0, u) = 1, & u \in [0, 1],
\end{cases}
\tag{14.57}
$$

where

$$
\alpha = (\alpha_1, \alpha_2) = (0.5, 0.5), \quad \psi(t) = t, \quad f(t, u, x(t, u)) = \frac{e^{-(t+u)}}{4}(x^2 - 0.5)
$$

and $(t, u) \in \tilde{I}$. Obviously, $f \in C(\tilde{I} \times \mathbb{R}, \mathbb{R})$. On the one hand, taking $x_0(t, u) = 1$ and $y_0(t, u) = 1 + \sqrt{tu}$, it is not difficult to verify that x_0 and y_0 are lower and upper solutions of (14.57), respectively, and $x_0 \le y_0$. So condition (14.24.1) holds. Moreover, for $x_0 \le x \le y \le y_0$, we have

$$
f(t, u, y) - f(t, u, x) \ge p(t, u)(y - x), \quad (t, u) \in \tilde{I}.
\tag{14.58}
$$

In view of (14.58), we can choose $p(t, u) = \frac{e^{-(t+u)}}{2}$ in Theorem 14.24. Hence, all conditions of Theorem 14.24 are satisfied and consequently the problem (14.57) has extremal solutions on $[x_0, y_0]$. Furthermore, we have the following iterative sequences

$$
x_{n+1}(t, u) = 1 + \int_0^t \int_0^u \frac{e^{-(s+t)}}{4\pi\sqrt{(t-s)(u-t)}}(x_n^2(s, t) - 0.5 + 2(x_{n+1}(s, t) - x_n(s, t)))dtds,
$$

and

$$
y_{n+1}(t, u) = 1 + \int_0^t \int_0^u \frac{e^{-(s+t)}}{4\pi\sqrt{(t-s)(u-t)}}(y_n^2(s, t) - 0.5 + 2(y_{n+1}(s, t) - y_n(s, t)))dtds.
$$

On the other hand, for $x_0 \le x \le y \le y_0$, we have

$$
f(t, u, y) - f(t, u, x) \le e^{-(t+u)}(y - x), \quad (t, u) \in \tilde{I}.
\tag{14.59}
$$

In view of (14.59), we can choose $\tilde{p}(t, u) = e^{-(t+u)}$ in Theorem 14.25. Therefore all conditions of Theorem 14.25 are satisfied and consequently the problem (14.57) has unique extremal solutions.

14.4 Notes and remarks

Using the papers [149,151,220], we substantiated all of the results in this chapter. For further knowledge on the topic, see the monographs of Kilbas et al. [281], Miller and Ross [312], Podlubny [332], and Zhou [438], as well as to the papers by Agarwal et al. [88], Benchohra et al. [163,166,174], and the references therein.

References

[1] S. Abbas, R.P. Agarwal, M. Benchohra, N. Benkhettou, Hilfer-Hadamard fractional differential equations and inclusions under weak topologies, Prog. Fract. Differ. Appl. 4 (4) (2018) 247–261.

[2] S. Abbas, E. Alaidarous, W. Albarakati, M. Benchohra, Upper and lower solutions method for partial Hadamard fractional integral equations and inclusions, Discuss. Math., Differ. Incl. Control Optim. 35 (2015) 105–122.

[3] S. Abbas, E. Alaidarous, M. Benchohra, J.J. Nieto, Existence and stability of solutions for Hadamard-Stieltjes fractional integral equations, Discrete Dyn. Nat. Soc. 2015 (2015) 317094.

[4] S. Abbas, N. Al Arifi, M. Benchohra, J.R. Graef, Random coupled systems of implicit Caputo-Hadamard fractional differential equations with multi-point boundary conditions in generalized Banach spaces, Dyn. Syst. Appl. 28 (2) (2019) 229–350.

[5] S. Abbas, N. Al Arifi, M. Benchohra, J. Henderson, Coupled Hilfer and Hadamard random fractional differential systems with finite delay in generalized Banach spaces, Differ. Equ. Appl. 12 (4) (2020) 337–353.

[6] S. Abbas, N. Al Arifi, M. Benchohra, G.M. N'Guérékata, Random coupled Caputo–Hadamard fractional differential systems with four-point boundary conditions in generalized Banach spaces, Ann. Commun. Math. 2 (1) (2019) 1–15.

[7] S. Abbas, N. Al Arifi, M. Benchohra, Y. Zhou, Random coupled Hilfer and Hadamard fractional differential systems in generalized Banach spaces, Mathematics 7 (285) (2019) 1–15.

[8] S. Abbas, W. Albarakati, M. Benchohra, Existence and attractivity results for Volterra type nonlinear multi-delay Hadamard-Stieltjes fractional integral equations, Panam. Math. J. 16 (1) (2016) 1–17.

[9] S. Abbas, W.A. Albarakati, M. Benchohra, M.A. Darwish, E.M. Hilal, New existence and stability results for partial fractional differential inclusions with multiple delay, Ann. Pol. Math. 114 (2015) 81–100.

[10] S. Abbas, W.A. Albarakati, M. Benchohra, E.M. Hilal, Global existence and stability results for partial fractional random differential equations, J. Appl. Anal. 21 (2) (2015) 79–87.

[11] S. Abbas, W.A. Albarakati, M. Benchohra, J. Henderson, Existence and Ulam stabilities for Hadamard fractional integral equations with random effects, Electron. J. Differ. Equ. 2016 (25) (2016) 1–12.

[12] S. Abbas, W. Albarakati, M. Benchohra, G.M. N'Guérékata, Existence and Ulam stabilities for Hadamard fractional integral equations in Fréchet spaces, J. Fract. Calc. Appl. 7 (2) (2016) 1–12.

[13] S. Abbas, W. Albarakati, M. Benchohra, J.J. Nieto, Existence and global stability results for Volterra type fractional Hadamard partial integral equations, Commun. Math. Anal. 21 (1) (2018) 42–53.

[14] S. Abbas, W. Albarakati, M. Benchohra, A. Petruşel, Existence and Ulam stability results for Hadamard partial fractional integral inclusions via Picard operators, Stud. Univ. Babeş–Bolyai, Math. 61 (4) (2016) 409–420.

[15] S. Abbas, W.A. Albarakati, M. Benchohra, S. Sivasundaram, Dynamics and stability of Fredholm type fractional order Hadamard integral equations, J. Nonlinear Stud. 22 (4) (2015) 673–686.

[16] S. Abbas, W.A. Albarakati, M. Benchohra, S. Sivasundaram, On the solutions of Pettis partial Hadamard-Stieltjes fractional integral equations, Nonlinear Stud. 23 (2) (2016) 333–344.

[17] S. Abbas, W. Albarakati, M. Benchohra, J.J. Trujillo, Ulam stabilities for partial Hadamard fractional integral equations, Arab. J. Math. 5 (1) (2016) 1–7.

[18] S. Abbas, W. Albarakati, M. Benchohra, Y. Zhou, Weak solutions for partial Pettis Hadamard fractional integral equations with random effects, J. Integral Equ. Appl. 29 (4) (2017) 473–491.

[19] S. Abbas, M. Benchohra, Advanced Functional Evolution Equations and Inclusions, Developments in Mathematics, Springer, Cham, 2015.

[20] S. Abbas, M. Benchohra, Some stability concepts for Darboux problem for partial fractional differential equations on unbounded domain, Fixed Point Theory 16 (1) (2015) 3–14.

[21] S. Abbas, M. Benchohra, Uniqueness and Ulam stabilities results for partial fractional differential equations with not instantaneous impulses, Appl. Math. Comput. 257 (2015) 190–198.

[22] S. Abbas, M. Benchohra, Nonlinear fractional order Riemann-Liouville Volterra-Stieltjes partial integral equations on unbounded domains, Commun. Math. Anal. 14 (1) (2013) 104–117.

[23] S. Abbas, M. Benchohra, Existence and stability of nonlinear fractional order Riemann-Liouville Volterra-Stieltjes multi-delay integral equations, J. Integral Equ. Appl. 25 (2) (2013) 143–158.

[24] S. Abbas, M. Benchohra, Global stability results for nonlinear partial fractional order Riemann-Liouville Volterra-Stieltjes functional integral equations, Math. Sci. Res. J. 16 (4) (2012) 82–92.

[25] S. Abbas, M. Benchohra, Ulam-Hyers stability for the Darboux problem for partial fractional differential and integro-differential equations via Picard operators, Results Math. 65 (1–2) (2014) 67–79.

[26] S. Abbas, M. Benchohra, Ulam stabilities for the Darboux problem for partial fractional differential inclusions, Demonstr. Math. XLVII (4) (2014) 826–838.

[27] S. Abbas, M. Benchohra, A Filippov's Theorem and Topological Structure of Solution Sets for Fractional q-Difference Inclusions, (Submitted).

[28] S. Abbas, M. Benchohra, Existence theory for implicit fractional q-difference equations in Banach spaces, (Submitted).

[29] S. Abbas, M. Benchohra, Existence and attractivity for fractional q-difference equations, (Submitted).

[30] S. Abbas, M. Benchohra, Existence theory for fractional q-difference equations in Fréchet spaces, (Submitted).

[31] S. Abbas, B. Ahmad, M. Benchohra, S. Ntouyas, Weak solutions for Caputo Pettis fractional q-difference inclusions, Fract. Differ. Calc. 10 (1) (2020) 141–152.

[32] S. Abbas, M. Benchohra, Uniqueness and Ulam stabilities results for partial fractional differential equations with not instantaneous impulses, Appl. Math. Comput. 257 (2015) 190–198.

[33] S. Abbas, M. Benchohra, Fractional order partial hyperbolic differential equations involving Caputo's derivative, Stud. Univ. Babeş–Bolyai, Math. 57 (4) (2012) 469–479.

[34] S. Abbas, M. Benchohra, On the generalized Ulam-Hyers-Rassias stability for Darboux problem for partial fractional implicit differential equations, Appl. Math. E-Notes 14 (2014) 20–28.

[35] S. Abbas, M. Benchohra, Coupled Caputo–Hadamard fractional differential systems with multipoint boundary conditions in generalized Banach spaces, (Submitted).

[36] S. Abbas, M. Benchohra, A. Alsaedi, Y. Zhou, Weak solutions for partial random Hadamard fractional integral equations with multiple delay, Discrete Dyn. Nat. Soc. 2017 (2017) 8607946.

[37] S. Abbas, M. Benchohra, A. Alsaedi, Y. Zhou, Weak solutions for a coupled system of Pettis-Hadamard fractional differential equations, Adv. Differ. Equ. 2017 (2017) 332, https://doi.org/10.1186/s13662-017-1391-z.

[38] S. Abbas, M. Benchohra, F. Berhoun, J. Henderson, Caputo-Hadamard fractional differential Cauchy problem in Fréchet spaces, Rev. R. Acad. Cienc. Exactas Fís. Nat., Ser. A Mat. RACSAM 113 (3) (2019) 2335–2344.

[39] S. Abbas, M. Benchohra, F. Berhoun, J.J. Nieto, Weak solutions for impulsive implicit Hadamard fractional differential equations, Adv. Dyn. Syst. Appl. 13 (1) (2018) 1–18.

[40] S. Abbas, M. Benchohra, A. Cabada, Implicit Caputo fractional q-difference equations with non instantaneous impulses, (Submitted).

[41] S. Abbas, M. Benchohra, M.A. Darwish, New stability results for partial fractional differential inclusions with not instantaneous impulses, Fract. Calc. Appl. Anal. 18 (1) (2015) 172–191.

[42] S. Abbas, M. Benchohra, M.A. Darwish, Fractional differential inclusions of Hilfer and Hadamard types in Banach spaces, Discuss. Math., Differ. Incl. Control Optim. 37 (2) (2017) 187–204.

[43] S. Abbas, M. Benchohra, M.A. Darwish, Asymptotic stability for implicit differential equations involving Hilfer fractional derivative, Panam. Math. J. 27 (3) (2017) 40–52.

[44] S. Abbas, M. Benchohra, T. Diagana, Existence and attractivity results for some fractional order partial integro-differential equations with delay, Afr. Diaspora J. Math. 15 (2) (2013) 87–100.

[45] S. Abbas, M. Benchohra, J.R. Graef, Weak solutions to implicit differential equations involving the Hilfer fractional derivative, Nonlinear Dyn. Syst. Theory 18 (1) (2018) 1–11.

[46] S. Abbas, M. Benchohra, J. Graef, Oscillation and nonoscillation for the Caputo fractional q-difference equations and inclusions, J. Math. Sci. 258 (5) (2021) 1–17.

[47] S. Abbas, M. Benchohra, J. Graef, Upper and lower solutions for fractional q-difference inclusions, Nonlinear Dyn. Syst. Theory 21 (1) (2021) 1–12.

[48] S. Abbas, M. Benchohra, J.R. Graef, J. Henderson, Implicit Fractional Differential and Integral Equations: Existence and Stability, De Gruyter, Berlin, 2018.

[49] S. Abbas, M. Benchohra, J. Graef, N. Laledj, Uniqueness and Ulam stability for implicit fractional q-difference equations via Picard operators theory, Int. J. Dyn. Syst. Differ. Equ. 13 (2023) 58–75.

[50] S. Abbas, M. Benchohra, J.R. Graef, J.E. Lazreg, Implicit Hadamard fractional differential equations with impulses under weak topologies, Dyn. Contin. Discrete Impuls. Syst. Ser. A: Math. Anal. 26 (2019) 89–112.

[51] S. Abbas, M. Benchohra, S. Hamani, J. Henderson, Upper and lower solutions method for Caputo-Hadamard fractional differential inclusions, Math. Morav. 23 (1) (2019) 107–118.

[52] S. Abbas, M. Benchohra, N. Hamidi, Hilfer and Hilfer-Hadamard fractional differential equations with random effects, Libertas Math. 37 (1) (2018) 45–64.

[53] S. Abbas, M. Benchohra, N. Hamidi, J. Henderson, Caputo-Hadamard fractional differential equations in Banach spaces, Fract. Calc. Appl. Anal. 21 (4) (2018) 1027–1045.

[54] S. Abbas, M. Benchohra, J. Henderson, Asymptotic attractive nonlinear fractional order Riemann-Liouville integral equations in Banach algebras, Nonlinear Stud. 20 (1) (2013) 1–10.

[55] S. Abbas, M. Benchohra, J. Henderson, Weak solutions for implicit fractional differential equations of Hadamard type, Adv. Dyn. Syst. Appl. 11 (1) (2016) 1–13.

[56] S. Abbas, M. Benchohra, J. Henderson, Weak solutions for implicit Hilfer fractional differential equations with not instantaneous impulses, Commun. Math. Anal. 20 (2) (2017) 48–61.

[57] S. Abbas, M. Benchohra, J. Henderson, Existence and oscillation for coupled fractional q-difference systems, J. Fract. Calc. Appl. 12 (1) (2021) 143–155.

[58] S. Abbas, M. Benchohra, J. Henderson, J.E. Lazreg, Measure of noncompactness and impulsive Hadamard fractional implicit differential equations in Banach spaces, Math. Eng. Sci. Aerosp. 8 (3) (2017) 1–19.

[59] S. Abbas, M. Benchohra, J. Henderson, J.E. Lazreg, Weak solution for a coupled system of partial Pettis Hadamard fractional integral equations, Adv. Theory Nonlinear Anal. Appl. 1 (2) (2017) 136–146.

[60] S. Abbas, M. Benchohra, N. Laledj, Y. Zhou, Existence and Ulam stability for implicit fractional q-difference equations, Adv. Differ. Equ. 2019 (2019) 480.

[61] S. Abbas, M. Benchohra, N. Laledj, Y. Zhou, Fractional q-difference equations on the half line, Arch. Math. 56 (2020) 207–223.

[62] S. Abbas, M. Benchohra, J.E. Lazreg, A. Alsaedi, Y. Zhou, Existence and Ulam stability for fractional differential equations of Hilfer-Hadamard type, Adv. Differ. Equ. 2017 (2017) 180, https://doi.org/10.1186/s13662-017-1231-1.

[63] S. Abbas, M. Benchohra, J.E. Lazreg, G.M. N'Guérékata, Weak solution for implicit Pettis-Hadamard fractional differential equations with retarded and advanced arguments, Nonlinear Stud. 24 (2017) 355–365.

[64] S. Abbas, M. Benchohra, J.E. Lazreg, G.M. N'Guérékata, Hilfer and Hadamard functional random fractional differential inclusions, CUBO 19 (1) (2017) 17–38.

[65] S. Abbas, M. Benchohra, J.E. Lazreg, J.J. Nieto, On a coupled system of Hilfer and Hilfer-Hadamard fractional differential equations in Banach spaces, J. Nonlinear Funct. Anal. 2018 (2018) 12.

[66] S. Abbas, M. Benchohra, J.E. Lazreg, J.J. Nieto, Y. Zhou, Fractional Differential Equations and Inclusions: Classical and Advanced Topics, Series on Analysis, Applications and Computation, vol. 10, World Scientific, 2023.

[67] S. Abbas, M. Benchohra, J.E. Lazreg, Y. Zhou, A survey on Hadamard and Hilfer fractional differential equations: analysis and stability, Chaos Solitons Fractals 102 (2017) 47–71.

[68] S. Abbas, M. Benchohra, J.E. Lazreg, Y. Zhou, Caputo fractional q-difference equations in Banach spaces, (Submitted).

[69] S. Abbas, M. Benchohra, G.M. N'Guérékata, Advanced Fractional Differential and Integral Equations, Nova Science Publishers, New York, 2015.

[70] S. Abbas, M. Benchohra, G.M. N'Guérékata, Topics in Fractional Differential Equations, Developments in Mathematics, vol. 27, Springer, New York, 2012.

[71] S. Abbas, M. Benchohra, G.M. N'Guérékata, Impulsive implicit Caputo fractional q-difference equations in finite and infinite dimensional Banach spaces, in: Gaston M. N'Guérékata, Bourama Toni (Eds.), Studies in Evolution Equations and Related Topics, in: STEAM-H: Science, Technology, Engineering, Agriculture, Mathematics and Health, Springer, Cham, 2021, pp. 187–209.

[72] S. Abbas, M. Benchohra, G.M. N'Guérékata, Instantaneous and noninstantaneous impulsive integro-differential equations in Banach spaces, Abstr. Appl. Anal. (2020) 2690125.

[73] S. Abbas, M. Benchohra, A. Petruşel, Ulam stabilities for the Darboux problem for partial fractional differential inclusions via Picard operators, Electron. J. Qual. Theory Differ. Equ. 2014 (51) (2014) 1–13.

[74] S. Abbas, M. Benchohra, A. Petrusel, Coupled Hilfer and Hadamard fractional differential systems in generalized Banach spaces, Fixed Point Theory 23 (1) (2022) 21–34.

[75] S. Abbas, M. Benchohra, B. Samet, Y. Zhou, Coupled implicit Caputo fractional q-difference systems, Adv. Differ. Equ. 2019 (2019) 527.

[76] S. Abbas, M. Benchohra, S. Sivasundaram, Coupled Pettis Hadamard fractional differential systems with retarded and advanced arguments, J. Math. Stat. 14 (1) (2018) 56–63.

[77] S. Abbas, M. Benchohra, J.J. Trujillo, Upper and lower solutions method for partial fractional differential inclusions with not instantaneous impulses, Prog. Fract. Differ. Appl. 1 (1) (2015) 11–22.

[78] S. Abbas, M. Benchohra, A.N. Vityuk, On fractional order derivatives and Darboux problem for implicit differential equations, Fract. Calc. Appl. Anal. 15 (2012) 168–182.

[79] S. Abbas, M. Benchohra, Y. Zhou, Hilfer and Hadamard random fractional differential equations in Fréchet spaces, (Submitted).

[80] S. Abbas, M. Benchohra, Y. Zhou, Coupled Hilfer fractional differential systems with random effects, Adv. Differ. Equ. 2018 (2018) 369.

[81] T. Abdeljawad, J. Alzabut, On Riemann-Liouville fractional q-difference equations and their application to retarded logistic type model, Math. Methods Appl. Sci. 41 (18) (2018) 8953–8962.

[82] T. Abdeljawad, J. Alzabut, On Riemann-Liouville fractional q-difference equations and their application to retarded logistic type model, Math. Methods Appl. Sci. 2018 (2018) 110.

[83] T. Abdeljawad, J. Alzabut, D. Baleanu, A generalized q-fractional Gronwall inequality and its applications to nonlinear delay q-fractional difference systems, J. Inequal. Appl. 2016 (2019) 240.

[84] M.S. Abdo, S.K. Panchal, A.M. Saeed, Fractional boundary value problem with ψ-Caputo fractional derivative, Proc. Indian Acad. Sci. Math. Sci. 129 (2019) 65.

[85] R. Agarwal, Certain fractional q-integrals and q-derivatives, Proc. Camb. Philos. Soc. 66 (1969) 365–370.

[86] R.P. Agarwal, H. Al-Hutami, B. Ahmad, A Langevin-type q-variant system of nonlinear fractional integro-difference equations with nonlocal boundary conditions, Fractal Fract. 6 (1) (2022) 45.

[87] R.P. Agarwal, B. Ahmad, Existence theory for anti-periodic boundary value problems of fractional differential equations and inclusions, Comput. Math. Appl. 62 (2011) 1200–1214.

[88] R.P. Agarwal, M. Benchohra, S. Hamani, A survey on existence results for boundary value problems of nonlinear fractional differential equations and inclusions, Acta Appl. Math. 109 (2010) 973–1033.

[89] R.P. Agarwal, M. Benchohra, D. Seba, On the application of measure of noncompactness to the existence of solutions for fractional differential equations, Results Math. 55 (2009) 221–230.

[90] R.P. Agarwal, M. Benchohra, B.A. Slimani, Existence results for differential equations with fractional order and impulses, Mem. Differ. Equ. Math. Phys. 44 (1) (2008) 1–21.

[91] R.P. Agarwal, M. Bohner, W.T. Li, Nonoscillation and Oscillation: Theory for Functional Differential Equations, Pure and Applied Mathematics, 2004.

[92] R.P. Agarwal, M. Meehan, D. O'Regan, Fixed Point Theory and Applications, Cambridge University Press, Cambridge, 2001.

[93] A. Aghajani, J. Banaś, N. Sebzali, Some generalizations of Darbo fixed point theorem and applications, Bull. Belg. Math. Soc. Simon Stevin 20 (2013) 345–358.

[94] A. Aghajani, M. Mursaleen, A. Shole Haghighi, Fixed point theorems for Meir–Keeler condensing operators via measure of noncompactness, Acta Math. Sci. Ser. B Engl. Ed. 35 (2015) 552–566.

[95] A. Aghajani, E. Pourhadi, J.J. Trujillo, Application of measure of noncompactness to a Cauchy problem for fractional differential equations in Banach spaces, Fract. Calc. Appl. Anal. 16 (2013) 962–977.

[96] B. Ahmad, Boundary value problem for nonlinear third order q-difference equations, Electron. J. Differ. Equ. 2011 (94) (2011) 1–7.

[97] B. Ahmad, A. Alsaedi, S.K. Ntouyas, J. Tariboon, Hadamard-Type Fractional Differential Equations, Inclusions and Inequalities, Springer, Cham, 2017.

[98] B. Ahmad, N. Alghamdi, A. Alsaedi, S.K. Ntouyas, Multi-term fractional differential equations with nonlocal boundary conditions, Open Math. 16 (2018) 1519–1536.

[99] B. Ahmad, J.R. Graef, Coupled systems of nonlinear fractional differential equations with nonlocal boundary conditions, Panam. Math. J. 19 (2009) 29–39.

[100] B. Ahmad, J.J. Nieto, Boundary value problems for a class of sequential integrodifferential equations of fractional order, J. Funct. Spaces Appl. 2013 (2013) 149659.

[101] B. Ahmad, J.J. Nieto, Existence of solutions for nonlocal boundary value problems of higher-order nonlinear fractional differential equations, Abstr. Appl. Anal. 2009 (2009) 494720.

[102] B. Ahmad, J.J. Nieto, Existence of solutions for impulsive anti-periodic boundary value problems of fractional order, Taiwan. J. Math. 15 (3) (2011) 981–993.

[103] B. Ahmad, S.K. Ntouyas, L.K. Purnaras, Existence results for nonlocal boundary value problems of nonlinear fractional q-difference equations, Adv. Differ. Equ. 2012 (2012) 140.

[104] B. Ahmad, J.J. Nieto, Anti-periodic fractional boundary value problems with nonlinear term depending on lower order derivative, Fract. Calc. Appl. Anal. 15 (2012) 451–462.

[105] B. Ahmad, J.J. Nieto, Anti-periodic fractional boundary value problems, Comput. Math. Appl. 62 (3) (2011) 1150–1156.

[106] B. Ahmad, J.J. Nieto, A. Alsaedi, N. Mohamad, On a new class of anti-periodic fractional boundary value problems, Abstr. Appl. Anal. 2013 (2013), 7 p.

[107] B. Ahmad, S.K. Ntouyas, A. Alsaedi, R.P. Agarwal, A study of nonlocal integro-multi-point boundary value problems of sequential fractional integro-differential inclusions, Dyn. Contin. Discrete Impuls. Syst. Ser. A: Math. Anal. 25 (2) (2018) 125–140.

[108] B. Ahmad, S. Ntouyas, J. Tariboon, Quantum Calculus. New Concepts, Impulsive IVPs and BVPs, Inequalities, Trends in Abstract and Applied Analysis, World Scientific Publishing Co. Pte. Ltd., Hackensack, NJ, 2016.

[109] B. Ahmad, S.K. Ntouyas, Y. Zhou, A. Alsaedi, A study of fractional differential equations and inclusions with nonlocal Erdélyi-Kober type integral boundary conditions, Bull. Iranian Math. Soc. 44 (5) (2018) 1315–1328.

[110] B. Ahmad, S. Sivasundaram, Existence results for nonlinear impulsive hybrid boundary value problems involving fractional differential equations, Nonlinear Anal. Hybrid Syst. 3 (3) (2009) 251–258.

[111] B. Ahmad, S. Sivasundaram, Theory of fractional differential equations with three-point boundary conditions, Commun. Appl. Anal. 12 (2008) 479–484.

[112] M. Ahmad, A. Zada, J. Alzabut, Hyers-Ulam stability of a coupled system of fractional differential equations of Hilfer-Hadamard type, Demonstr. Math. 52 (1) (2019) 283–295.

[113] K.K. Akhmerov, M.I. Kamenskii, A.S. Potapov, A.E. Rodkina, B.N. Sadovskii, Measures of Noncompactness and Condensing Operators, Birkhäuser Verlag, Basel, Boston, Berlin, 1992.

[114] W. Albarakati, M. Benchohra, J.E. Lazreg, J.J. Nieto, Anti-periodic boundary value problem for nonlinear implicit fractional differential equations with impulses, An. Univ. Oradea, Fasc. Mat. XXV (1) (2018) 13–24.

[115] A. Ali, K. Shah, F. Jarad, V. Gupta, T. Abdeljawad, Existence and stability analysis to a coupled system of implicit type impulsive boundary value problems of fractional-order differential equations, Adv. Differ. Equ. 2019 (2019) 101.

[116] S. Aljoudi, B. Ahmad, J.J. Nieto, A. Alsaedi, A coupled system of Hadamard type sequential fractional differential equations with coupled strip conditions, Chaos Solitons Fractals 91 (2016) 39–46.

[117] S. Aljoudi, B. Ahmad, J.J. Nieto, A. Alsaedi, On coupled Hadamard type sequential fractional differential equations with variable coefficients and nonlocal integral boundary conditions, Filomat 31 (2017) 6041–6049.

[118] G. Allaire, S.M. Kaber, Numerical Linear Algebra, Texts in Applied Mathematics, Springer, New York, 2008.

[119] R. Almeida, A Caputo fractional derivative of a function with respect to another function, Commun. Nonlinear Sci. 44 (2017) 460–481.

[120] R. Almeida, Functional differential equations involving the ψ-Caputo fractional derivative, Fractal Fract. 4 (2) (2020).

[121] R. Almeida, Fractional differential equations with mixed boundary conditions, Bull. Malays. Math. Sci. Soc. 42 (2019) 1687–1697.

[122] R. Almeida, A.B. Malinowska, M.T.T. Monteiro, Fractional differential equations with a Caputo derivative with respect to a kernel function and their applications, Math. Methods Appl. Sci. 41 (2018) 336–352.

[123] R. Almeida, A.B. Malinowska, T. Odzijewicz, Optimal leader-follower control for the fractional opinion formation model, J. Optim. Theory Appl. 182 (2019) 1171–1185.

[124] R. Almeida, M. Jleli, B. Samet, A numerical study of fractional relaxation-oscillation equations involving ψ-Caputo fractional derivative, Rev. R. Acad. Cienc. Exactas Fís. Nat., Ser. A Mat. RACSAM 113 (2019) 1873–1891.

[125] S. Almezel, Q.H. Ansari, M.A. Khamsi, Topics in Fixed Point Theory, Springer-Verlag, New York, 2014.

[126] B. Alqahtani, S. Abbas, M. Benchohra, S.S. Alzaid, Fractional q-difference inclusions in Banach spaces, Mathematics 8 (91) (2020) 1–12.

[127] M. Al-Refai, M. Ali Hajji, Monotone iterative sequences for nonlinear boundary value problems of fractional order, Nonlinear Anal. 74 (11) (2011) 3531–3539.

[128] A. Alsaedi, H. Al-Hutami, B. Ahmad, R.P. Agarwal, Existence results for a coupled system of nonlinear fractional q-integro-difference equations with q-integral coupled boundary conditions, Fractals 30 (1) (2022) 2240042.

[129] W.A. Al-Salam, q-Analogues of Cauchy's formula, Proc. Am. Math. Soc. 17 (1824) 1952–1953.

[130] W.A. Al-Salam, Some fractional q-integrals and q-derivatives, Proc. Edinb. Math. Soc. 15 (1969) 13540.

[131] W.A. Al-Salam, A. Verma, A fractional Leibniz q-formula, Pac. J. Math. 60 (1975) 19.

[132] C. Alsina, R. Ger, On some inequalities and stability results related to the exponential function, J. Inequal. Appl. 2 (1998) 373–380.

[133] J.C. Alvàrez, Measure of noncompactness and fixed points of nonexpansive condensing mappings in locally convex spaces, Rev. Real. Acad. Cienc. Exact. Fis. Natur. Madrid 79 (1985) 53–66.

[134] R. Ameena, F. Jaradb, T. Abdeljawad, Ulam stability for delay fractional differential equations with a generalized Caputo derivative, Filomat 32 (15) (2018) 5265–5274.

[135] G.A. Anastassiou, I.K. Argyros, Functional Numerical Methods: Applications to Abstract Fractional Calculus, Springer, 2018.

[136] J. Andres, L. Górniewicz, Topological Fixed Point Principles for Boundary Value Problems, Kluwer Academic Publishers, Dordrecht, 2003.

[137] J.P. Aubin, A. Cellina, Differential Inclusions, Springer-Verlag, Berlin-Heidelberg, New York, 1984.

[138] A. Atangana, Fractional Operators with Constant and Variable Order with Application to Geo-Hydrology, Academic Press, London, 2018.

[139] J.-P. Aubin, H. Frankowska, Set-Valued Analysis, Birkhauser, Basel, 1990.

[140] C. Avramescu, Some remarks on a fixed point theorem of Krasnoselskii, Electron. J. Qual. Theory Differ. Equ. 5 (2003) 1–15.

[141] H. Aydi, M. Jleli, B. Samet, On positive solutions for a fractional thermostat model with a convex–concave source term via ψ–Caputo fractional derivative, Mediterr. J. Math. 17 (2020) 16, https://doi.org/10.1007/s00009-019-1450-7.

[142] J.M. Ayerbee Toledano, T. Dominguez Benavides, G. Lopez Acedo, Measures of Noncompactness in Metric Fixed Point Theory, Operator Theory, Advances and Applications, vol. 99, Birkhäuser, Basel, Boston, Berlin, 1997.

[143] D. Baleanu, M.M. Arjunan, M. Nagaraj, S. Suganya, Approximate controllability of second-order nonlocal impulsive functional integro-differential systems in Banach spaces, Bull. Korean Math. Soc. 55 (4) (2018) 1065–1092.

[144] D.D. Bainov, S.G. Hristova, Integral inequalities of Gronwall type for piecewise continuous functions, J. Appl. Math. Stoch. Anal. 10 (1997) 89–94.

[145] L. Bai, J.J. Nieto, Variational approach to differential equations with not instantaneous impulses, Appl. Math. Lett. 73 (2017) 44–48.

[146] L. Bai, J.J. Nieto, X. Wang, Variational approach to non-instantaneous impulsive nonlinear differential equations, J. Nonlinear Sci. Appl. 10 (2017) 2440–2448.

[147] Z. Baitiche, C. Derbazi, M. Benchohra, A. Cabada, The application of Meir-Keeler condensing operators to a new class of fractional differential equations involving Ψ-Caputo fractional derivative, J. Nonlinear Var. Anal. 5 (2021) 561–572.

[148] Z. Baitiche, C. Derbazi, M. Benchohra, A. Cabada, Application of Meir–Keeler condensing operators to a new class of fractional differential equations involving ψ Caputo fractional derivative, (Submitted).

[149] Z. Baitiche, C. Derbazi, M. Benchohra, J. Henderson, Monotone iterative technique for a hyperbolic fractional partial differential equation involving the ψ–Caputo derivative with initial conditions, Commun. Appl. Nonlinear Anal. 28 (2021) 11–26.

[150] Z. Baitiche, C. Derbazi, M. Benchohra, J. Henderson, Qualitative analysis of ψ–Caputo fractional partial differential equations in Banach spaces, (Submitted).

[151] Z. Baitiche, C. Derbazi, M. Benchohra, J.J. Nieto, Monotone iterative technique for a sequential ψ-Caputo fractional differential equations with nonlinear boundary conditions, (Submitted).

[152] Z. Baitiche, C. Derbazi, M. Benchohra, Y. Zhou, A new class of coupled system of nonlinear hyperbolic partial fractional differential equations in generalized Banach spaces involving the ψ–Caputo fractional derivative, (Submitted).

[153] J. Banaš, K. Goebel, Measures of Noncompactness in Banach Spaces, Marcel Dekker, New York, 1980.

[154] J. Banaš, L. Olszowy, Measures of noncompactness related to monotonicity, Comment. Math. (Prace Mat.) 41 (2001) 13–23.

[155] T.D. Benavides, An existence theorem for implicit differential equations in a Banach space, Ann. Mat. Pura Appl. 4 (1978) 119–130.

[156] M. Benchohra, F. Berhoun, G.M. N'Guérékata, Bounded solutions for fractional order differential equations on the half-line, Bull. Math. Anal. Appl. 146 (4) (2012) 62–71.

[157] M. Benchohra, F. Bouazzaoui, E. Karapinar, A. Salim, Controllability of second order functional random differential equations with delay, Mathematics 10 (2022) 1–16, https://doi.org/10.3390/math10071120.

[158] M. Benchohra, S. Bouriah, Existence and stability results for nonlinear implicit fractional differential equations with impulses, Mem. Differ. Equ. Math. Phys. 69 (2016) 15–31.

[159] M. Benchohra, S. Bouriah, M. Darwish, Nonlinear boundary value problem for implicit differential equations of fractional order in Banach spaces, Fixed Point Theory 18 (2) (2017) 457–470.

[160] M. Benchohra, S. Bouriah, J. Henderson, Existence and stability results for nonlinear implicit neutral fractional differential equations with finite delay and impulses, Commun. Appl. Nonlinear Anal. 22 (1) (2015) 46–67.

[161] M. Benchohra, S. Bouriah, J.E. Lazreg, J.J. Nieto, Nonlinear implicit Hadamard fractional differential equations with delay in Banach spaces, Acta Univ. Palack. Olomuc. Fac. Rerum Natur. Math. 55 (2016) 15–26.

[162] M. Benchohra, S. Bouriah, J.J. Nieto, Existence of periodic solutions for nonlinear implicit Hadamard's fractional differential equations, Rev. R. Acad. Cienc. Exactas Fís. Nat., Ser. A Mat. 112 (2018) 25–35.

[163] M. Benchohra, J.R. Graef, S. Hamani, Existence results for boundary value problems with nonlinear fractional differential equations, Appl. Anal. 87 (7) (2008) 851–863.

[164] M. Benchohra, J. Graef, F-Z. Mostefai, Weak solutions for boundary-value problems with nonlinear fractional differential inclusions, Nonlinear Dyn. Syst. Theory 11 (3) (2011) 227–237.

[165] M. Benchohra, S. Hamani, S.K. Ntouyas, Boundary value problems for differential equations with fractional order, Surv. Math. Appl. 3 (2008) 1–12.

[166] M. Benchohra, S. Hamani, S.K. Ntouyas, Boundary value problems for differential equations with fractional order and nonlocal conditions, Nonlinear Anal. 71 (2009) 2391–2396.

[167] M. Benchohra, S. Hamani, Y. Zhou, Oscillation and nonoscillation for Caputo-Hadamard impulsive fractional differential inclusions, Adv. Differ. Equ. 2019 (2019) 74.

[168] M. Benchohra, J. Henderson, F-Z. Mostefai, Weak solutions for hyperbolic partial fractional differential inclusions in Banach spaces, Comput. Math. Appl. 64 (2012) 3101–3107.

[169] M. Benchohra, J. Henderson, S.K. Ntouyas, A. Ouahab, Existence results for functional differential equations of fractional order, J. Math. Anal. Appl. 338 (2008) 1340–1350.

[170] M. Benchohra, J. Henderson, D. Seba, Measure of noncompactness and fractional differential equations in Banach spaces, Commun. Appl. Anal. 12 (4) (2008) 419–428.

[171] M. Benchohra, J.E. Lazreg, Nonlinear fractional implicit differential equations, Commun. Appl. Anal. 17 (2013) 471–482.

[172] M. Benchohra, J.E. Lazreg, Existence and uniqueness results for nonlinear implicit fractional differential equations with boundary conditions, Rom. J. Math. Comput. Sci. 4 (1) (2014) 60–72.

[173] M. Benchohra, J.E. Lazreg, Existence results for nonlinear implicit fractional differential equations, Surv. Math. Appl. 9 (2014) 79–92.

[174] M. Benchohra, J.E. Lazreg, Existence results for nonlinear implicit fractional differential equations, Surv. Math. Appl. 9 (2014) 79–92.

[175] M. Benchohra, J.E. Lazreg, Existence results for nonlinear implicit fractional differential equations with impulses, Commun. Appl. Anal. 19 (2015) 413–426.

[176] M. Benchohra, J.E. Lazreg, On stability for nonlinear implicit fractional differential equations, Matematiche (Catania) 70 (2) (2015) 49–61.

[177] M. Benchohra, J.E. Lazreg, Existence and Ulam stability for nonlinear implicit fractional differential equations with Hadamard derivatives, Stud. Univ. Babeş–Bolyai, Math. 62 (1) (2017) 27–38.

[178] M. Benchohra, J.E. Lazreg, G.M. N'Guérékata, Nonlinear implicit Hadamard's fractional differential equations on Banach space with retarded and advanced arguments, Int. J. Evol. Equ. 10 (2018) 283–295.

[179] M. Benchohra, S.K. Ntouyas, The lower and upper method for first order differential inclusions with nonlinear boundary condition, JIPAM. J. Inequal. Pure Appl. Math. 3 (1) (2002) 14.

[180] M. Benchohra, M.S. Souid, Integrable solutions for implicit fractional order differential equations, Transylv. J. Math. Mech. 6 (2) (2014) 101–107.

[181] M. Benchohra, M.S. Souid, L^1-solutions for implicit fractional order differential equations with nonlocal condition, Filomat 30 (6) (2016) 1485–1492.

[182] N. Benkhettou, K. Aissani, A. Salim, M. Benchohra, C. Tunc, Controllability of fractional integro-differential equations with infinite delay and non-instantaneous impulses, Appl. Anal. Optim. 6 (2022) 79–94.

[183] N. Benkhettou, A. Salim, K. Aissani, M. Benchohra, E. Karapınar, Non-instantaneous impulsive fractional integro-differential equations with state-dependent delay, Sahand Commun. Math. Anal. 19 (2022) 93–109, https://doi.org/10.22130/scma.2022.542200.1014.

[184] N. Benkhettou, A. Salim, J.E. Lazreg, S. Abbas, M. Benchohra, Lakshmikantham monotone iterative principle for hybrid Atangana-Baleanu-Caputo fractional differential equations, An. Univ. Vest. Timiş., Ser. Mat.-Inform. 59 (1) (2023) 79–91, https://doi.org/10.2478/awutm-2023-0007.

[185] A. Bensalem, A. Salim, M. Benchohra, Ulam-Hyers-Rassias stability of neutral functional integrodifferential evolution equations with non-instantaneous impulses on an unbounded interval, Qual. Theory Dyn. Syst. 22 (2023), 29 pages, https://doi.org/10.1007/s12346-023-00787-y.

[186] A. Bensalem, A. Salim, M. Benchohra, M. Fečkan, Approximate controllability of neutral functional integro-differential equations with state-dependent delay and non-instantaneous impulses, Mathematics 11 (2023) 1–17, https://doi.org/10.3390/math11071667.

[187] A. Bensalem, A. Salim, M. Benchohra, G. N'Guérékata, Functional integro-differential equations with state-dependent delay and non-instantaneous impulsions: existence and qualitative results, Fractal Fract. 6 (2022) 1–27, https://doi.org/10.3390/fractalfract6100615.

[188] H.F. Bohnenblust, S. Karlin, On a Theorem of Ville. Contribution to the Theory of Games, Annals of Mathematics Studies, vol. 24, Princeton University Press, Princeton. N. G., 1950, pp. 155–160.

[189] D. Bothe, Multivalued perturbation of m-accretive differential inclusions, Isr. J. Math. 108 (1998) 109–138.

[190] S. Bouriah, A. Salim, M. Benchohra, On nonlinear implicit neutral generalized Hilfer fractional differential equations with terminal conditions and delay, Topol. Algebra Appl. 10 (2022) 77–93, https://doi.org/10.1515/taa-2022-0115.

[191] Z. Bouteffal, A. Salim, S. Litimein, M. Benchohra, Uniqueness results for fractional integro-differential equations with state-dependent nonlocal conditions in Fréchet spaces, An. Univ. Vest. Timiş., Ser. Mat.-Inform. 59 (1) (2023) 35–44, https://doi.org/10.2478/awutm-2023-0004.

[192] F. Browder, On the convergence of successive approximations for nonlinear functional equations, Indag. Math. 30 (1968) 27–35.

[193] J. Brzdek, D. Popa, I. Rasa, B. Xu, Ulam Stability of Operators, Mathematical Analysis and Its Applications, Academic Press, London, 2018.

[194] D. Bugajewski, S. Szufla, Kneser's theorem for weak solutions of the Darboux problem in a Banach space, Nonlinear Anal. 20 (2) (1993) 169–173.

[195] T.A. Burton, C. Kirk, A fixed point theorem of Krasnoselskii-Schaefer type, Math. Nachr. 189 (1989) 23–31.

[196] P.L. Butzer, A.A. Kilbas, J.J. Trujillo, Fractional calculus in the Mellin setting and Hadamard-type fractional integrals, J. Math. Anal. Appl. 269 (2002) 1–27.

[197] P.L. Butzer, A.A. Kilbas, J.J. Trujillo, Mellin transform analysis and integration by parts for Hadamard-type fractional integrals, J. Math. Anal. Appl. 270 (2002) 1–15.

[198] L. Byszewski, Theorems about existence and uniqueness of solutions of a semilinear evolution nonlocal Cauchy problem, J. Math. Anal. Appl. 162 (1991) 494–505.

[199] J. Caballero, J. Harjani, K. Sadarangani, A fixed point theorem for operators of Meir-Keeler type via the degree of nondensifiability and its application in dynamic programming, J. Fixed Point Theory Appl. 22 (2020) 1–14.

[200] R.D. Carmichael, The general theory of linear q-difference equations, Am. J. Math. 34 (1912) 147–168.

[201] C. Castaing, M. Valadier, Convex Analysis and Measurable Multifunctions, Lecture Notes in Mathematics, vol. 580, Springer-Verlag, Berlin-Heidelberg-New York, 1977.

[202] A. Cernea, Arcwise connectedness of the solution set of a nonclosed nonconvex integral inclusion, Miskolc Math. Notes 9 (1) (2008) 33–39.

[203] A. Cernea, A Filippov-type existence theorem for some nonlinear q-difference inclusions, in: Differential and Difference Equations with Applications, in: Springer Proc. Math. Stat., vol. 164, Springer, Cham, 2016, pp. 71–77.

[204] A. Cernea, Filippov lemma for a class of Hadamard-type fractional differential inclusions, Fract. Calc. Appl. Anal. 18 (1) (2015) 163–171.

[205] A. Cernea, On the mild solutions of a class of second order integro-differential inclusions, J. Nonlinear Var. Anal. 3 (3) (2019) 247–256.

[206] S. Chakraverty, B. Tapaswini, D. Behera, Fuzzy Arbitrary Order System. Fuzzy Fractional Differential Equations and Applications, John Wiley & Sons, Inc., Hoboken, NJ, 2016.

[207] Y. Chang, A. Anguraj, P. Karthikeyan, Existence results for initial value problems with integral condition for impulsive fractional differential equations, J. Fract. Calc. Appl. 2 (7) (2012) 1–10.

[208] C. Chen, M. Bohner, B. Jia, Method of upper and lower solutions for nonlinear Caputo fractional difference equations and its applications, Fract. Calc. Appl. Anal. 22 (5) (2019) 1307–1320.

[209] Y. Cherruault, A. Guillez, Une méthode pour la recherche du minimum global d'une fonctionnelle, C. R. Acad. Sci., Sér. 1 Math. 296 (1983) 175–178.

[210] Y. Cherruault, G. Mora, Optimisation globale: théorie des courbes [alpha]-denses, Economica, Paris, 2005.

[211] M. Chohri, S. Bouriah, A. Salim, M. Benchohra, On nonlinear periodic problems with Caputo's exponential fractional derivative, Adv. Theory Nonlinear Anal. Appl. 7 (2023) 103–120, https://doi.org/10.31197/atnaa.1130743.

[212] V. Colao, L. Muglia, H.-K. Xu, Existence of solutions for a second-order differential equation with non-instantaneous impulses and delay, Ann. Mat. Pura Appl. 195 (2016) 697–716.

[213] N.D. Cong, H.T. Tuan, Existence, uniqueness, and exponential boundedness of global solutions to delay fractional differential equations, Mediterr. J. Math. 14 (5) (2017) 193.

[214] C. Corduneanu, Integral Equations and Applications, Cambridge University Press, 1991.

[215] H. Covitz, S.B. Nadler Jr., Multivalued contraction mappings in generalized metric spaces, Isr. J. Math. 8 (1970) 5–11.

[216] A. Das, B. Hazarika, M. Mursaleen, Application of measure of noncompactness for solvability of the infinite system of integral equations in two variables in ℓ_p $(1 < p < \infty)$, Rev. R. Acad. Cienc. Exactas Fís. Nat., Ser. A Mat. RACSAM 113 (2019) 31–40.

[217] F.S. De Blasi, On the property of the unit sphere in a Banach space, Bull. Math. Soc. Sci. Math. R.S. Roum. 21 (1977) 259–262.

[218] K. Deimling, Multivalued Differential Equations, Walter De Gruyter, Berlin-New York, 1992.

[219] C. Derbazi, Z. Baitiche, M. Benchohra, Coupled system of ψ–Caputo fractional differential equations without and with delay in generalized Banach spaces, (Submitted).

[220] C. Derbazi, Z. Baitiche, M. Benchohra, A. Cabada, Initial value problem for nonlinear fractional differential equations with ψ–Caputo derivative via monotone iterative technique, (Submitted).

[221] C. Derbazi, Z. Baitiche, M. Benchohra, Y. Zhou, Boundary value problem for ψ-Caputo fractional differential equations in Banach spaces via densifiability techniques, (Submitted).

[222] C. Derbazi, H. Hammouche, A. Salim, M. Benchohra, Measure of noncompactness and fractional hybrid differential equations with hybrid conditions, Differ. Equ. Appl. 14 (2022) 145–161, https://doi.org/10.7153/dea-2022-14-09.

[223] C. Derbazi, H. Hammouche, A. Salim, M. Benchohra, Weak solutions for fractional Langevin equations involving two fractional orders in Banach spaces, Afr. Mat. 34 (2023), 10 pages, https://doi.org/10.1007/s13370-022-01035-3.

[224] B.C. Dhage, Multi-valued condensing random operators and functional random integral inclusions, Opusc. Math. 31 (1) (2011) 27–48.

[225] S. Djebali, L. Górniewicz, A. Ouahab, Existence and Structure of Solution Sets for Impulsive Differential Inclusions: A Survey, Lecture Notes in Nonlinear Analysis, vol. 13, Julius Schauder University Centre for Nonlinear Studies - Nicolaus Copernicus University, 2012.

[226] J.B. Diaz, B. Margolis, A fixed point theorem of the alternative for contractions on generalized complete metric space, Bull. Am. Math. Soc. 74 (1968) 305–309.

[227] M. Dieye, M.A. Diop, K. Ezzinbi, H. Hmoyed, On the existence of mild solutions for nonlocal impulsive integro-differential equations in Banach spaces, Matematiche LXXIV (1) (2019) 13–34.

[228] K. Diethelm, N.J. Ford, Analysis of fractional differential equations, J. Math. Anal. Appl. 265 (2) (2002) 229–248.

[229] S. Dudek, Fixed point theorems in Fréchet algebras and Fréchet spaces and applications to nonlinear integral equations, Appl. Anal. Discrete Math. 11 (2017) 340–357.

[230] S. Dudek, L. Olszowy, Continuous dependence of the solutions of nonlinear integral quadratic Volterra equation on the parameter, J. Funct. Spaces 2015 (2015) 471235.

[231] N. Eghbali, V. Kalvandi, J.M. Rassias, A fixed point approach to the Mittag-Leffler-Hyers-Ulam stability of a fractional integral equation, Open Math. 14 (1) (2016) 237–246.

[232] M. El-Shahed, H.A. Hassan, Positive solutions of q-difference equation, Proc. Am. Math. Soc. 138 (2010) 1733–1738.

[233] T. Ernst, A Comprehensive Treatment of q-Calculus, Birkhäuser, Basel, 2012.

[234] S. Etemad, S.K. Ntouyas, B. Ahmad, Existence theory for a fractional q-integro-difference equation with q-integral boundary conditions of different orders, Mathematics 7 (659) (2019) 1–15.

[235] H. Fazli, H. Sun, S. Aghchi, Existence of extremal solutions of fractional Langevin equation involving nonlinear boundary conditions, Int. J. Comput. Math. 98 (2021) 1–10, https://doi.org/10.1080/00207160.2020.1720662.

[236] H. Frankowska, A priori estimates for operational differential inclusions, J. Differ. Equ. 84 (1990) 100–128.

[237] M. Frigon, A. Granas, Théorèmes dexistence pour des inclusions différentielles sans convexité, C. R. Acad. Sci. Paris, Ser. I 310 (1990) 819–822.

[238] A. Fryszkowski, Fixed Point Theory for Decomposable Sets, Topological Fixed Point Theory and Its Applications, vol. 2, Kluwer Academic Publishers, Dordrecht, 2004.

[239] K.M. Furati, M.D. Kassim, Non-existence of global solutions for a differential equation involving Hilfer fractional derivative, Electron. J. Differ. Equ. 2013 (235) (2013) 1–10.

[240] K.M. Furati, M.D. Kassim, N.e-. Tatar, Existence and uniqueness for a problem involving Hilfer fractional derivative, Comput. Math. Appl. 64 (2012) 1616–1626.

[241] Z. Gao, A computing method on stability intervals of time-delay for fractional-order retarded systems with commensurate time-delays, Automatica 50 (2014) 1611–1616.

[242] G.R. Gautam, J. Dabas, Mild solution for fractional functional integro-differential equation with not instantaneous impulse, Malaya J. Mat. 2 (2014) 428–437.

[243] G. García, Solvability of an initial value problem with fractional order differential equations in Banach space by ϵ-dense curves, Fract. Calc. Appl. Anal. 20 (2017) 646–661.

[244] G. García, G. Mora, A fixed point result in Banach algebras based on the degree of nondensifiability and applications to quadratic integral equations, J. Math. Anal. Appl. 472 (2019) 1220–1235.

[245] G. García, Existence of solutions for infinite systems of differential equations by densifiability techniques, Filomat 32 (2018) 3419–3428.

[246] G. García, A quantitative version of the Arzelà-Ascoli theorem based on the degree of nondensifiability and applications, Appl. Gen. Topol. 20 (2019) 265–279.

[247] G. García, G. Mora, The degree of convex nondensifiability in Banach spaces, J. Convex Anal. 22 (2015) 871–888.

[248] S.G. Georgiev, Fractional Dynamic Calculus and Fractional Dynamic Equations on Time Scales, Springer Nature, 2018.

[249] R. Gorenflo, A.A. Kilbas, F. Mainardi, S.V. Rogosin, Mittag-Leffler Functions, Related Topics and Applications, Springer Monographs in Mathematics, Springer, Heidelberg, 2014.

[250] L. Górniewicz, Topological Fixed Point Theory of Multivalued Mappings, Mathematics and Its Applications, vol. 495, Kluwer Academic Publishers, Dordrecht, 1999.

[251] J.R. Graef, J. Henderson, A. Ouahab, Impulsive Differential Inclusions. A Fixed Point Approach, De Gruyter, Berlin/Boston, 2013.

[252] J.R. Graef, J. Henderson, A. Ouahab, Some Krasnosel'skii type random fixed point theorems, J. Nonlinear Funct. Anal. 2017 (2017) 46.

[253] A. Granas, J. Dugundji, Fixed Point Theory, Springer-Verlag, New York, 2003.

[254] R.C. Grimmer, Resolvent operators for integral equations in a Banach space, Trans. Am. Math. Soc. 273 (1982) 333–349.

[255] B. Guo, X. Pu, F. Huang, Fractional Partial Differential Equations and Their Numerical Solutions, World Scientific Publishing Co. Pte. Ltd., Hackensack, 2015.

[256] D.J. Guo, V. Lakshmikantham, X. Liu, Nonlinear Integral Equations in Abstract Spaces, Kluwer Academic Publishers, Dordrecht, 1996.

[257] Z. Guo, M. Liu, Existence and uniqueness of solutions for fractional order integrodifferential equations with nonlocal initial conditions, Panam. Math. J. 21 (2011) 51–61.

[258] J. Hadamard, Essai sur l'étude des fonctions données par leur développment de Taylor, J. Pure Appl. Math. 4 (8) (1892) 101–186.

[259] S. Harikrishnan, P. Prakash, J.J. Nieto, Forced oscillation of solutions of a nonlinear fractional partial differential equation, Appl. Math. Comput. 254 (2015) 14–19.

[260] S. Harikrishnan, K. Shah, K. Kanagarajan, Existence theory of fractional coupled differential equations via Ψ-Hilfer fractional derivative, Random Oper. Stoch. Equ. 27 (4) (2019) 207–212.

[261] A. Harrat, J.J. Nieto, A. Debbouche, Solvability and optimal controls of impulsive Hilfer fractional delay evolution inclusions with Clarke subdifferential, J. Comput. Appl. Math. 344 (2018) 725–737.

[262] J. Henderson, R. Luca, Boundary Value Problems for Systems of Differential, Difference and Fractional Equations - Positive Solutions, Elsevier, 2016.

[263] J. Henderson, A. Ouahab, A Filippov's theorem, some existence results and the compactness of solution sets of impulsive fractional order differential inclusions, Mediterr. J. Math. 9 (3) (2012) 453–485.

[264] A. Heris, A. Salim, M. Benchohra, E. Karapinar, Fractional partial random differential equations with infinite delay, Results Phys. 37 (2022) 105557, https://doi.org/10.1016/j.rinp.2022.105557.

[265] E. Hernández, D. O'Regan, On a new class of abstract impulsive differential equations, Proc. Am. Math. Soc. 141 (2013) 1641–1649.

[266] R. Hilfer, Applications of Fractional Calculus in Physics, World Scientific, Singapore, 2000.

[267] R. Hilfer, Application of Fractional Calculus in Physics, World Scientific, New Jersey, 2001.

[268] Sh. Hu, N. Papageorgiou, Handbook of Multivalued Analysis, Volume I: Theory, Kluwer Academic Publishers, Dordrecht, 1997.

[269] D.H. Hyers, On the stability of the linear functional equation, Proc. Natl. Acad. Sci. USA 27 (1941) 222–224.

[270] S. Itoh, Random fixed point theorems with applications to random differential equations in Banach spaces, J. Math. Anal. Appl. 67 (1979) 261–273.

[271] F. Jarad, T. Abdeljawad, D. Baleanu, Stability of q-fractional non-autonomous systems, Nonlinear Anal., Real World Appl. 14 (1) (2013) 780–784.

[272] S.-M. Jung, Hyers-Ulam-Rassias Stability of Functional Equations in Mathematical Analysis, Hadronic Press, Palm Harbor, 2001.

[273] S.-M. Jung, Hyers-Ulam-Rassias Stability of Functional Equations in Nonlinear Analysis, Springer, New York, 2011.

[274] V. Kac, P. Cheung, Quantum Calculus, Springer, New York, 2002.

[275] P. Kalamani, D. Baleanu, M. Mallika Arjunan, Local existence for an impulsive fractional neutral integro-differential system with Riemann-Liouville fractional derivatives in a Banach space, Adv. Differ. Equ. 2018 (2018) 416.

[276] V. Kalvandi, N. Eghbali, J.M. Rassias, Mittag-Leffler-Hyers-Ulam stability of fractional differential equations of second order, J. Math. Ext. 13 (1) (2019) 1–15.

[277] R. Kamocki, C. Obczński, On fractional Cauchy-type problems containing Hilfer's derivative, Electron. J. Qual. Theory Differ. Equ. 2016 (50) (2016) 1–12.

[278] A. Khan, H. Khan, J.F. Gómez-Aguilar, T. Abdeljawad, Existence and Hyers-Ulam stability for a nonlinear singular fractional differential equation with Mittag-Leffler kernel, Chaos Solitons Fractals 127 (2019) 422–427.

[279] H. Khan, T. Abdeljawad, M. Aslam, R.A. Khan, A. Khan, Existence of positive solution and Hyers-Ulam stability for a nonlinear singular-delay-fractional differential equation, Adv. Differ. Equ. 2019 (2019) 104.

[280] A.A. Kilbas, Hadamard-type fractional calculus, J. Korean Math. Soc. 38 (6) (2001) 1191–1204.

[281] A.A. Kilbas, H.M. Srivastava, Juan J. Trujillo, Theory and Applications of Fractional Differential Equations, North-Holland Mathematics Studies, vol. 204, Elsevier Science B.V., Amsterdam, 2006.

[282] M. Kisielewicz, Differential Inclusions and Optimal Control, Kluwer, Dordrecht, The Netherlands, 1991.

[283] W.A. Kirk, B. Sims, Handbook of Metric Fixed Point Theory, Springer-Science + Business Media, B.V., Dordrecht, 2001.

[284] E. Kreyszig, Introductory Functional Analysis with Applications, Wiley, 1978.

[285] S. Krim, A. Salim, S. Abbas, M. Benchohra, On implicit impulsive conformable fractional differential equations with infinite delay in b-metric spaces, Rend. Circ. Mat. Palermo 2 (2022) 1–14, https://doi.org/10.1007/s12215-022-00818-8.

[286] S. Krim, A. Salim, S. Abbas, M. Benchohra, Functional k-generalized ψ-Hilfer fractional differential equations in b-metric spaces, Panam. Math. J. 2 (2023), 10 pages, https://doi.org/10.28919/cpr-pajm/2-5.

[287] S. Krim, A. Salim, M. Benchohra, On implicit Caputo tempered fractional boundary value problems with delay, Lett. Nonlinear Anal. Appl. 1 (2023) 12–29.

[288] K.D. Kucche, A.D. Mali, J.V.C. Sousa, On the nonlinear Ψ-Hilfer fractional differential equations, Comput. Appl. Math. 38 (2) (2019) 73.

[289] V. Lakshmikantham, D.D. Bainov, P.S. Simeonov, Theory of Impulsive Differential Equations, World Scientific, Singapore, 1989.

[290] V. Lakshmikantham, J. Vasundhara Devi, Theory of fractional differential equations in a Banach space, Eur. J. Pure Appl. Math. 1 (2008) 38–45.

[291] V. Lakshmikantham, A.S. Vatsala, Basic theory of fractional differential equations, Nonlinear Anal. 69 (2008) 2677–2682.

[292] V. Lakshmikantham, A.S. Vatsala, General uniqueness and monotone iterative technique for fractional differential equations, Appl. Math. Lett. 21 (2008) 828–834.

[293] N. Laledj, A. Salim, J.E. Lazreg, S. Abbas, B. Ahmad, M. Benchohra, On implicit fractional q-difference equations: analysis and stability, Math. Methods Appl. Sci. 45 (17) (2022) 10775–10797, https://doi.org/10.1002/mma.8417.

[294] A. Lasota, Z. Opial, An application of the Kakutani-Ky Fan theorem in the theory of ordinary differential equations, Bull. Acad. Pol. Sci., Sér. Sci. Math. Astron. Phys. 13 (1965) 781–786.

[295] J.E. Lazreg, M. Benchohra, A. Salim, Existence and Ulam stability of k-generalized ψ-Hilfer fractional problem, J. Innov. Appl. Math. Comput. Sci. 2 (2022) 01.

[296] Y. Li, Y.Q. Chen, I. Podlubny, Stability of fractional-order nonlinear dynamic systems: Lyapunov direct method and generalized Mittag-Leffler stability, Comput. Math. Appl. 59 (2010) 1810–1821.

[297] Y. Li, Y.Q. Chen, I. Podlubny, Mittag-Leffler stability of fractional order nonlinear dynamic systems, Automatica 45 (2009) 1965–1969.

[298] J. Liang, Z. Liu, X. Wang, Solvability for a couple system of nonlinear fractional differential equations in a Banach space, Fract. Calc. Appl. Anal. 16 (1) (2013) 51–63.

[299] X. Lin, Z. Zhao, Iterative technique for a third-order differential equation with three-point nonlinear boundary value conditions, Electron. J. Qual. Theory Differ. Equ. (2016) 12.

[300] L. Liu, F. Guo, C. Wu, Y. Wu, Existence theorems of global solutions for nonlinear Volterra type integral equations in Banach spaces, J. Math. Anal. Appl. 309 (2005) 638–649.

[301] L. Liu, C. Wu, F. Guo, Existence theorems of global solutions of initial value problems for nonlinear integrodifferential equations of mixed type in Banach spaces and applications, Comput. Math. Appl. 47 (2004) 13–22.

[302] K. Liu, J. Wang, D. O'Regan, Ulam-Hyers-Mittag-Leffler stability for ψ-Hilfer fractional-order delay differential equations, Adv. Differ. Equ. (2019) 50.

[303] N.I. Mahmudov, A. Al-Khateeb, Existence and Ulam-Hyers stability of coupled sequential fractional differential equations with integral boundary conditions, J. Inequal. Appl. (2019) 165.

[304] F. Mainardi, Fractional Calculus and Waves in Linear Viscoelasticity, Imperial College Press, London, 2010.

[305] M. Martelli, A Rothe's type theorem for noncompact acyclic-valued map, Boll. Unione Math. Ital. 11 (1975) 70–76.

[306] M.M. Matar, Solution of sequential Hadamard fractional differential equations by variation of parameter technique, Abstr. Appl. Anal. 2018 (2018) 9605353.

[307] J. Matkowski, Integrable solutions of functional equations, Diss. Math. 127 (1975) 1–68.

[308] M. Mckibben, Discovering Evolution Equations with Applications: Volume 1 Deterministic Models, Chapman and Hall/CRC Appl. Math. Nonlinear Sci. Ser., 2011.

[309] M. Meghnafi, M. Benchohra, K. Aissani, Impulsive fractional evolution equations with state-dependent delay, Nonlinear Stud. 22 (4) (2015) 659–671.

[310] A. Meir, E. Keeler, A theorem on contraction mappings, J. Math. Anal. Appl. 28 (1969) 326–329.

[311] V.D. Milman, A.A. Myshkis, On the stability of motion in the presence of impulses, Sib. Math. J. 1 (1960) 233–237 (in Russian).

[312] K.S. Miller, B. Ross, An Introduction to the Fractional Calculus and Fractional Differential Equations, A Wiley-Interscience Publication, John Wiley & Sons, Inc., New York, 1993.

[313] H. Mönch, Boundary value problems for nonlinear ordinary differential equations of second order in Banach spaces, Nonlinear Anal. 4 (1980) 985–999.

[314] G. Mora, Y. Cherruault, Characterization and generation of ϵ-dense curves, Comput. Math. Appl. 33 (1997) 83–91.

[315] G. Mora, J.A. Mira, Alpha-dense curves in infinite dimensional spaces, Int. J. Pure Appl. Math. Sci. 5 (2003) 437–449.

[316] G. Mora, D.A. Redtwitz, Densifiable metric spaces, Rev. R. Acad. Cienc. Exactas Fís. Nat., Ser. A Mat. 105 (2011) 71–83.

[317] M. Mursaleen, S.M.H. Rizvi, Solvability of infinite systems of second order differential equations in c_0 and ℓ_1 by Meir-Keeler condensing operators, Proc. Am. Math. Soc. 144 (2016) 4279–4289.

[318] J.J. Nieto, An abstract monotone iterative technique, Nonlinear Anal. 28 (1997) 1923–1933.

[319] J.J. Nieto, Maximum principles for fractional differential equations derived from Mittag-Leffler functions, Appl. Math. Lett. 23 (10) (2010) 1248–1251.

[320] A. Nowak, Applications of random fixed point theorem in the theory of generalized random differential equations, Bull. Pol. Acad. Sci. 34 (1986) 487–494.

[321] K.B. Oldham, J. Spanier, The Fractional Calculus: Theory and Application of Differentiation and Integration to Arbitrary Order, Academic Press, New York, London, 1974.

[322] L. Olszowy, Existence of mild solutions for the semilinear nonlocal problem in Banach spaces, Nonlinear Anal. 81 (2013) 211–223.

[323] D. O'Regan, Fixed point theory for weakly sequentially continuous mapping, Math. Comput. Model. 27 (5) (1998) 1–14.

[324] D. O'Regan, Weak solutions of ordinary differential equations in Banach spaces, Appl. Math. Lett. 12 (1999) 101–105.

[325] D. O'Regan, R. Precup, Fixed point theorems for set-valued maps and existence principles for integral inclusions, J. Math. Anal. Appl. 245 (2000) 594–612.

[326] A. Ouahab, Filippov's theorem for impulsive differential inclusions with fractional order, Electron. J. Qual. Theory Differ. Equ. Spec. Ed. I (23) (2009) 1–23.

[327] A. Pazy, Semigroups of Linear Operators and Applications to Partial Differential Equations, Springer-Verlag, New York, 1983.

[328] A. Petruşel, Multivalued weakly Picard operators and applications, Sci. Math. Jpn. 59 (2004) 167–202.

[329] T.P. Petru, A. Petruşel, J.-C. Yao, Ulam-Hyers stability for operatorial equations and inclusions via nonself operators, Taiwan. J. Math. 15 (2011) 2169–2193.

[330] A.I. Perov, On the Cauchy problem for a system of ordinary differential equations, Pviblizhen. Met. Reshen. Differ. Uvavn. Vyp. 2 (1964) 115–134.

[331] I.-R. Petre, A. Petruşel, Krasnoselskii's theorem in generalized Banach spaces and applications, Electron. J. Qual. Theory Differ. Equ. 85 (2012) 1–20.

[332] I. Podlubny, Fractional Differential Equations, Mathematics in Science and Engineering, vol. 198, Academic Press, Inc., San Diego, CA, 1999.

[333] S. Pooseh, R. Almeida, D. Torres, Expansion formulas in terms of integer-order derivatives for the Hadamard fractional integral and derivative, Numer. Funct. Anal. Optim. 33 (3) (2012) 301–319.

[334] R. Precup, Methods in Nonlinear Integral Equations, Kluwer Academic Publishers, Dordrecht, 2002.

[335] R. Precup, The role of matrices that are convergent to zero in the study of semilinear operator systems, Math. Comput. Model. 49 (3–4) (2009) 703–708.

[336] R. Precup, A. Viorel, Existence results for systems of nonlinear evolution equations, Int. J. Pure Appl. Math. 47 (2) (2008) 199–206.

[337] M.D. Qassim, K.M. Furati, N.-e. Tatar, On a differential equation involving Hilfer-Hadamard fractional derivative, Abstr. Appl. Anal. 2012 (2012) 391062.

[338] M.D. Qassim, N.-e. Tatar, Well-posedness and stability for a differential problem with Hilfer-Hadamard fractional derivative, Abstr. Appl. Anal. 2013 (2013) 605029.

[339] W. Rahou, A. Salim, J.E. Lazreg, M. Benchohra, Existence and stability results for impulsive implicit fractional differential equations with delay and Riesz–Caputo derivative, Mediterr. J. Math. 20 (2023) 143, https://doi.org/10.1007/s00009-023-02356-8.

[340] W. Rahou, A. Salim, J.E. Lazreg, M. Benchohra, On fractional differential equations with Riesz-Caputo derivative and non-instantaneous impulses, Sahand Commun. Math. Anal. 20 (3) (2023) 109–132, https://doi.org/10.22130/scma.2023.563452.1186.

[341] P.M. Rajkovic, S.D. Marinkovic, M.S. Stankovic, Fractional integrals and derivatives in q-calculus, Appl. Anal. Discrete Math. 1 (2007) 311–323.

[342] P.M. Rajkovic, S.D. Marinkovic, M.S. Stankovic, On q-analogues of Caputo derivative and Mittag-Leffler function, Fract. Calc. Appl. Anal. 10 (2007) 359–373.

[343] Th.M. Rassias, On the stability of the linear mapping in Banach spaces, Proc. Am. Math. Soc. 72 (1978) 297–300.

[344] S.S. Ray, A.K. Gupta, Wavelet Methods for Solving Partial Differential Equations and Fractional Differential Equations, CRC Press, Taylor & Francis Group, 2018.

[345] I.A. Rus, Ulam stability of ordinary differential equations, Stud. Univ. Babeş–Bolyai, Math. LIV (4) (2009) 125–133.

[346] I.A. Rus, Remarks on Ulam stability of the operatorial equations, Fixed Point Theory 10 (2009) 305–320.

[347] I.A. Rus, Fixed points, upper and lower fixed points: abstract Gronwall lemmas, Carpath. J. Math. 20 (2004) 125–134.

[348] I.A. Rus, Picard operators and applications, Sci. Math. Jpn. 58 (2003) 191–219.

[349] I.A. Rus, Generalized Contractions and Applications, Cluj University Press, Cluj-Napoca, 2001.

[350] Th.M. Rassias, On the stability of linear mappings in Banach spaces, Proc. Am. Math. Soc. 72 (1978) 297–300.

[351] J. Ren, C. Zhai, Characteristic of unique positive solution for a fractional q-difference equation with multistrip boundary conditions, Math. Commun. 24 (2) (2019) 181–192.

[352] J. Ren, C. Zhai, A fractional q-difference equation with integral boundary conditions and comparison theorem, Int. J. Nonlinear Sci. Numer. Simul. 18 (7–8) (2017) 575–583.

[353] B. Ross, Fractional calculus and its applications, in: Proceedings of the International Conference, New Haven, Springer-Verlag, New York, 1974.

[354] H.L. Royden, Real Analysis, 3rd edition, Macmillan Publishing Company, New York, NY, USA, 1988.

[355] I. Rus, A. Petrusel, G. Petrusel, Fixed Point Theory, Cluj University Press, Cluj, 2008.

[356] R. Saadati, E. Pourhadi, M. Mursaleen, Solvability of infinite systems of third-order differential equations in c_0 by Meir-Keeler condensing operators, J. Fixed Point Theory Appl. 21 (2019), 16 pp.

[357] J. Sabatier, O.P. Agrawal, J.A.T. Machado, Advances in Fractional Calculus-Theoretical Developments and Applications in Physics and Engineering, Springer, Dordrecht, 2007.

[358] A. Sajjad, A. Thabet, S. Kamal, J. Fahd, A. Muhammad, Computation of iterative solutions along with stability analysis to a coupled system of fractional order differential equations, Adv. Differ. Equ. 2019 (2019) 215.

[359] A. Salim, S. Abbas, M. Benchohra, E. Karapinar, A Filippov's theorem and topological structure of solution sets for fractional q-difference inclusions, Dyn. Syst. Appl. 31 (2022) 17–34, https://doi.org/10.46719/dsa202231.01.02.

[360] A. Salim, S. Abbas, M. Benchohra, E. Karapinar, Global stability results for Volterra-Hadamard random partial fractional integral equations, Rend. Circ. Mat. Palermo 2 (2022) 1–13, https://doi.org/10.1007/s12215-022-00770-7.

[361] A. Salim, S. Abbas, M. Benchohra, J.E. Lazreg, Caputo fractional q-difference equations in Banach spaces, J. Innov. Appl. Math. Comput. Sci. 3 (1) (2023) 1–14, https://doi.org/10.58205/jiamcs.v3i1.67.

[362] A. Salim, B. Ahmad, M. Benchohra, J.E. Lazreg, Boundary value problem for hybrid generalized Hilfer fractional differential equations, Differ. Equ. Appl. 14 (2022) 379–391, https://doi.org/10.7153/dea-2022-14-27.

[363] A. Salim, J. Alzabut, W. Sudsutad, C. Thaiprayoon, On impulsive implicit ψ-Caputo hybrid fractional differential equations with retardation and anticipation, Mathematics 10 (2022), 20 pages, https://doi.org/10.3390/math10244821.

[364] A. Salim, M. Benchohra, Existence and uniqueness results for generalized Caputo iterative fractional boundary value problems, Fract. Differ. Calc. 12 (2022) 197–208, https://doi.org/10.7153/fdc-2022-12-12.

[365] A. Salim, M. Benchohra, J.R. Graef, J.E. Lazreg, Boundary value problem for fractional generalized Hilfer-type fractional derivative with non-instantaneous impulses, Fractal Fract. 5 (2021) 1–21, https://doi.org/10.3390/fractalfract5010001.

[366] A. Salim, M. Benchohra, J.R. Graef, J.E. Lazreg, Initial value problem for hybrid ψ-Hilfer fractional implicit differential equations, J. Fixed Point Theory Appl. 24 (2022), 14 pp., https://doi.org/10.1007/s11784-021-00920-x.

[367] A. Salim, M. Benchohra, E. Karapinar, J.E. Lazreg, Existence and Ulam stability for impulsive generalized Hilfer-type fractional differential equations, Adv. Differ. Equ. 2020 (2020), 21 pp., https://doi.org/10.1186/s13662-020-03063-4.

[368] A. Salim, M. Benchohra, J.E. Lazreg, Nonlocal k-generalized ψ-Hilfer impulsive initial value problem with retarded and advanced arguments, Appl. Anal. Optim. 6 (2022) 21–47.

[369] A. Salim, M. Benchohra, J.E. Lazreg, On implicit k-generalized ψ-Hilfer fractional differential coupled systems with periodic conditions, Qual. Theory Dyn. Syst. 22 (2023), 46 pages, https://doi.org/10.1007/s12346-023-00776-1.

[370] A. Salim, M. Benchohra, J.E. Lazreg, J. Henderson, Nonlinear implicit generalized Hilfer-type fractional differential equations with non-instantaneous impulses in Banach spaces, Adv. Theory Nonlinear Anal. Appl. 4 (2020) 332–348, https://doi.org/10.31197/atnaa.825294.

[371] A. Salim, M. Benchohra, J.E. Lazreg, J. Henderson, On k-generalized ψ-Hilfer boundary value problems with retardation and anticipation, Adv. Theory Nonlinear Anal. Appl. 6 (2022) 173–190, https://doi.org/10.31197/atnaa.973992.

[372] A. Salim, M. Benchohra, J.E. Lazreg, E. Karapinar, On k-generalized ψ-Hilfer impulsive boundary value problem with retarded and advanced arguments, J. Math. Ext. 15 (2021) 1–39, https://doi.org/10.30495/JME.SI.2021.2187.

[373] A. Salim, M. Benchohra, J.E. Lazreg, G. N'Guérékata, Boundary value problem for nonlinear implicit generalized Hilfer-type fractional differential equations with impulses, Abstr. Appl. Anal. 2021 (2021) 5592010, https://doi.org/10.1155/2021/5592010.

[374] A. Salim, M. Benchohra, J.E. Lazreg, G. N'Guérékata, Existence and k-Mittag-Leffler-Ulam-Hyers stability results of k-generalized ψ-Hilfer boundary value problem, Nonlinear Stud. 29 (2022) 359–379.

[375] A. Salim, M. Benchohra, J.E. Lazreg, J.J. Nieto, Y. Zhou, Nonlocal initial value problem for hybrid generalized Hilfer-type fractional implicit differential equations, Nonauton. Dyn. Syst. 8 (2021) 87–100, https://doi.org/10.1515/msds-2020-0127.

[376] A. Salim, M. Benchohra, J.E. Lazreg, Y. Zhou, On k-generalized ψ-Hilfer impulsive boundary value problem with retarded and advanced arguments in Banach spaces, J. Nonlinear Evol. Equ. Appl. 2022 (2023) 105–126.

[377] A. Salim, M. Boumaaza, M. Benchohra, Random solutions for mixed fractional differential equations with retarded and advanced arguments, J. Nonlinear Convex Anal. 23 (2022) 1361–1375.

[378] A. Salim, S. Bouriah, M. Benchohra, J.E. Lazreg, E. Karapinar, A study on k-generalized ψ-Hilfer fractional differential equations with periodic integral conditions, Math. Methods Appl. Sci. (2023) 1–18, https://doi.org/10.1002/mma.9056.

[379] A. Salim, S. Krim, S. Abbas, M. Benchohra, On deformable implicit fractional differential equations in b-metric spaces, J. Math. Ext. 17 (2023) 1–17, https://doi.org/10.30495/JME.2023.2468.

[380] A. Salim, S. Krim, J.E. Lazreg, M. Benchohra, On Caputo tempered implicit fractional differential equations in b-metric spaces, Analysis 43 (2) (2023) 129–139, https://doi.org/10.1515/anly-2022-1114.

[381] A. Salim, J.E. Lazreg, S. Abbas, M. Benchohra, Y. Zhou, Initial value problems for hybrid generalized Hilfer fractional differential equations, DNC 12 (2) (2023) 287–298, https://doi.org/10.5890/DNC.2023.06.005.

[382] A. Salim, J.E. Lazreg, B. Ahmad, M. Benchohra, J.J. Nieto, A study on k-generalized ψ-Hilfer derivative operator, Vietnam J. Math. (2022), https://doi.org/10.1007/s10013-022-00561-8.

[383] A. Salim, F. Mesri, M. Benchohra, C. Tunç, Controllability of second order semilinear random differential equations in Fréchet spaces, Mediterr. J. Math. 20 (84) (2023) 1–12, https://doi.org/10.1007/s00009-023-02299-0.

[384] B. Samet, H. Aydi, Lyapunov-type inequalities for an anti-periodic fractional boundary value problem involving ψ-Caputo fractional derivative, J. Inequal. Appl. (2018) 286.

[385] S.G. Samko, A.A. Kilbas, O.I. Marichev, Fractional Integrals and Derivatives. Theory and Applications, Gordon and Breach, Yverdon, 1993.

[386] A.M. Samoilenko, N.A. Perestyuk, Impulsive Differential Equations, World Scientific, Singapore, 1995.

[387] P.U. Shikhare, K.D. Kucche, Existence, uniqueness and Ulam stabilities for nonlinear hyperbolic partial integrodifferential equations, Int. J. Appl. Comput. Math. 5 (6) (2019) 156, Paper No. 156, 21 pp.

[388] M.L. Sinacer, J.J. Nieto, A. Ouahab, Random fixed point theorems in generalized Banach spaces and applications, Random Oper. Stoch. Equ. 24 (2016) 93–112.

[389] H.M. Srivastava, A. Das, B. Hazarika, S.A. Mohiuddine, Existence of solutions of infinite systems of differential equations of general order with boundary conditions in the spaces c_0 and ℓ_1 via the measure of noncompactness, Math. Methods Appl. Sci. 41 (2018) 3558–3569.

[390] A. Suechoei, P. Sa Ngiamsunthorn, Existence uniqueness and stability of mild solutions for semilinear ψ-Caputo fractional evolution equations, Adv. Differ. Equ. (2020) 114.

[391] J. Sun, X. Zhang, The fixed point theorem of convex-power condensing operator and applications to abstract semilinear evolution equations, Acta Math. Sinica (Chin. Ser.) 48 (2005) 439–446.

[392] V.E. Tarasov, Fractional Dynamics: Application of Fractional Calculus to Dynamics of Particles, Fields and Media, Springer/Higher Education Press, Heidelberg/Beijing, 2010.

[393] K. Tas, D. Baleanu, J.A. Tenreiro Machado, Mathematical Methods in Engineering: Applications in Dynamics of Complex Systems, Springer Nature Switzerland AG, 2019.

[394] J.A. Tenreiro Machado, V. Kiryakova, The chronicles of fractional calculus, Fract. Calc. Appl. Anal. 20 (2017) 307–336.

[395] Ž. Tomovski, R. Hilfer, H.M. Srivastava, Fractional and operational calculus with generalized fractional derivative operators and Mittag-Leffler type functions, Integral Transforms Spec. Funct. 21 (11) (2010) 797–814.

[396] S.M. Ulam, A Collection of Mathematical Problems, Interscience Publishers, New York, 1968.

[397] C. Urs, Coupled fixed point theorems and applications to periodic boundary value problems, Miskolc Math. Notes 14 (1) (2013) 323–333.

[398] J. Vanterler da C. Sousa, M.N.N. dos Santos, L.A. Magna, E. Capelas de Oliveira, Validation of a fractional model for erythrocyte sedimentation rate, Comput. Appl. Math. 37 (2018) 6903–6919.

[399] J. Vanterler da Costa Sousa, Existence results and continuity dependence of solutions for fractional equations, Differ. Equ. Appl. 12 (4) (2020) 377–396.

[400] J. Vanterler da C. Sousa, E. Capelas de Oliveira, Existence, uniqueness, estimation and continuous dependence of the solutions of a nonlinear integral and an integrodifferential equations of fractional order, arXiv:1806.01441, 2018.

[401] J. Vanterler da C. Sousa, E. Capelas de Oliveira, On the ψ-Hilfer fractional derivative, Commun. Nonlinear Sci. Numer. Simul. 60 (2018) 72–91.

[402] J. Vanterler da Costa Sousa, E. Capelas de Oliveira, A Gronwall inequality and the Cauchy-type problem by means of ψ-Hilfer operator, Differ. Equ. Appl. 11 (1) (2019) 87–106.

[403] J. Vanterler da Costa Sousa, E.C. de Oliveira, On the stability of a hyperbolic fractional partial differential equation, Differ. Equ. Dyn. Syst. 31 (2023) 31–52, https://doi.org/10.1007/s12591-019-00499-3.

[404] R.S. Varga, Matrix Iterative Analysis, second revised and expanded edition, Springer Series in Computational Mathematics, vol. 27, Springer-Verlag, Berlin, 2000.

[405] A. Viorel, Contributions to the Study of Nonlinear Evolution Equations, Ph.D. Thesis, Babes-Bolyai University Cluj-Napoca, Department of Mathematics, 2011.

[406] A.N. Vityuk, A.V. Golushkov, Existence of solutions of systems of partial differential equations of fractional order, Nonlinear Oscil. 7 (3) (2004) 318–325.

[407] D. Vivek, E.M. Elsayed, K. Kanagarajan, Theory and analysis of partial differential equations with a ψ-Caputo fractional derivative, Rocky Mt. J. Math. 49 (4) (2019) 1355–1370.

[408] V. Vyawahare, P.S.V. Nataraj, Fractional-Order Modeling of Nuclear Reactor: From Subdiffusive Neutron Transport to Control-Oriented Models. A Systematic Approach, Springer, Singapore, 2018.

[409] J. Wang, M. Fečkan, Y. Zhou, Ulam's type stability of impulsive ordinary differential equations, J. Math. Anal. Appl. 395 (2012) 258–264.

[410] J. Wang, X. Li, Periodic BVP for integer/fractional order nonlinear differential equations with non-instantaneous impulses, J. Appl. Math. Comput. 46 (2014) 321–334.

[411] J. Wang, A.G. Ibrahim, D. O'Regan, Global attracting solutions to Hilfer fractional differential inclusions of Sobolev type with noninstantaneous impulses and nonlocal conditions, Nonlinear Anal. Model. Control 24 (5) (2019) 775–803.

[412] J. Wang, Y. Zhang, Existence and stability of solutions to nonlinear impulsive differential equations in β-normed spaces, Electron. J. Differ. Equ. 2014 (83) (2014) 1–10.

[413] J. Wang, L. Lv, Y. Zhou, Ulam stability and data dependence for fractional differential equations with Caputo derivative, Electron. J. Qual. Theory Differ. Equ. 2011 (63) (2011) 1–10.

[414] J. Wang, L. Lv, Y. Zhou, Boundary value problems for fractional differential equations involving Caputo derivative in Banach spaces, J. Appl. Math. Comput. 38 (1–2) (2012) 209–224.

[415] G. Wang, X. Ren, L. Zhang, B. Ahmad, Explicit iteration and unique positive solution for a Caputo-Hadamard fractional turbulent flow model, IEEE Access 7 (2019) 109833–109839.

[416] J. Wang, Y. Zhou, Mittag-Leffler-Ulam stabilities of fractional evolution equations, Appl. Math. Lett. 25 (4) (2012) 723–728.

[417] J. Wang, X. Li, E_α-Ulam type stability of fractional order ordinary differential equations, J. Appl. Math. Comput. 45 (1–2) (2014) 449–459.

[418] J. Wang, Y. Zhang, Ulam-Hyers-Mittag-Leffler stability of fractional-order delay differential equations, Optimization 63 (8) (2014) 1181–1190.

[419] J. West, Nature's Patterns and the Fractional Calculus, De Gruyter, 2017.

[420] G.-C. Wu, D. Baleanu, Stability analysis of impulsive fractional difference equations, Fract. Calc. Appl. Anal. 21 (2) (2018) 354–375.

[421] X. Xue, Existence of semilinear differential equations with nonlocal initial conditions, Acta Math. Sin. 23 (2007) 983–988.

[422] W. Yang, Monotone iterative technique for a coupled system of nonlinear Hadamard fractional differential equations, J. Appl. Math. Comput. 59 (1–2) (2019) 585–596.

[423] X.-J. Yang, General Fractional Derivatives: Theory, Methods and Application, Chapman Hall-CRC, 2019.

[424] X.-J. Yang, D. Baleanu, H.M. Srivastava, Local Fractional Integral Transforms and Their Applications, Academic Press, Elsevier, 2016.

[425] X.-J. Yang, F. Gao, Y. Ju, General Fractional Derivatives with Applications in Viscoelasticity, Acad. Press, Elsevier, 2020.

[426] M. Yang, Q. Wang, Existence of mild solutions for a class of Hilfer fractional evolution equations with nonlocal conditions, Fract. Calc. Appl. Anal. 20 (2017) 679–705.

[427] D. Yang, J. Wang, D. O'Regan, A class of nonlinear non-instantaneous impulsive differential equations involving parameters and fractional order, Appl. Math. Comput. 321 (2018) 654–671.

[428] H. Ye, J. Gao, Y. Ding, A generalized Gronwall inequality and its application to a fractional differential equation, J. Math. Anal. Appl. 328 (2007) 1075–1081.

[429] K. Yosida, Functional Analysis, 6th ed., Springer-Verlag, Berlin, 1980.

[430] W. Yukunthorn, B. Ahmad, S.K. Ntouyas, J. Tariboon, On Caputo-Hadamard type fractional impulsive hybrid systems with nonlinear fractional integral conditions, Nonlinear Anal. Hybrid Syst. 19 (2016) 77–92.

[431] A. Zada, M. Yar, T. Li, Existence and stability analysis of nonlinear sequential coupled system of Caputo fractional differential equations with integral boundary conditions, Ann. Univ. Paedagog. Crac. Stud. Math. 17 (2018) 103–125.

[432] S. Zhang, Existence of solutions for a boundary value problem of fractional order, Acta Math. Sci. 26 (2006) 220–228.

[433] X. Zhang, Y. Li, P. Chen, Existence of extremal mild solutions for the initial value problem of evolution equations with non-instantaneous impulses, J. Fixed Point Theory Appl. 19 (2017) 3013–3027.

[434] S. Zhang, J. Sun, Existence of mild solutions for the impulsive semilinear nonlocal problem with random effects, Adv. Differ. Equ. 19 (2014) 1–11.

[435] T. Zhang, Y. Tang, A difference method for solving the q-fractional differential equations, Appl. Math. Lett. 98 (2019) 292–299.

[436] L. Zhang, G. Wang, Existence of solutions for nonlinear fractional differential equations with impulses and anti-periodic boundary conditions, Electron. J. Qual. Theory Differ. Equ. 2011 (7) (2011) 1–11.

[437] Y. Zhang, J. Wang, Nonlocal Cauchy problems for a class of implicit impulsive fractional relaxation differential systems, J. Appl. Math. Comput. 52 (1–2) (2016) 323–343.

[438] Y. Zhou, Fractional Evolution Equations and Inclusions: Analysis and Control, Elsevier/Academic Press, London, 2016.

[439] Y. Zhou, Basic Theory of Fractional Differential Equations, World Scientific, Singapore, 2014.

[440] Y. Zhou, J.-R. Wang, L. Zhang, Basic Theory of Fractional Differential Equations, second edition, World Scientific Publishing Co. Pte. Ltd., Hackensack, NJ, 2017.

[441] M. Zuo, X. Hao, Existence results for impulsive fractional q-difference equation with antiperiodic boundary conditions, J. Funct. Spaces 2018 (2018) 3798342.

Index

Printed in the United States
by Baker & Taylor Publisher Services

Printed in the United States
by Baker & Taylor Publisher Services